Terahertz Phonons and Nanomechanical Instabilities

Alberto Carpinteri

Terahertz Phonons and Nanomechanical Instabilities

Unlocking LENR in Condensed Matter with Insights into Geo, Astro, Electro, and Biochemical Systems

Alberto Carpinteri
College of Engineering
Shantou University
Shantou, Guangdong, China

ISBN 978-3-032-14691-5 ISBN 978-3-032-14692-2 (eBook)
https://doi.org/10.1007/978-3-032-14692-2

This work was supported by Politecnico di Torino.

This book was written by Prof. Alberto Carpinteri, based on the results obtained during the development of the European project 'Clean Energy from Hydrogen-Metal Systems (CleanHME)' within the framework of the EU research and innovation programme Horizon 2020 and under Grant Agreement no. 951974

This Springer imprint is published by the registered company Springer Nature Switzerland AG
The registered company address is: Gewerbestrasse 11, 6330 Cham, Switzerland

If disposing of this product, please recycle the paper.

Preface

The present book titled *Terahertz Phonons and Nanomechanical Instabilities: Unlocking LENR in Condensed Matter with Insights into Geo, Astro, Electro, and Biochemical Systems* is composed of 25 chapters. The contents of 14 chapters over the total 25 are corresponding to those of my previous coedited Springer book titled *Acoustic, Electromagnetic, Neutron Emissions from Fracture and Earthquakes* (2015), which was composed of a total of 17 chapters. The present version of the 14 corresponding chapters is updated and improved, as well as the text has been totally revised, reedited, and often partially rewritten. The fundamental reasons to propose this extension are in my further research activity on the same topic over the last ten years. This activity was rich of new important results and of several experimental and theoretical confirmations of my earlier results. The latter are reported in the remaining 11 totally new Chapters 7, 9, 12–14, 16, 20, 22–25.

TeraHertz phonons and/or plasmons are produced in solids and fluids by mechanical instabilities at the nanoscale (fracture and cavitation). Their frequency is close to the resonance frequency of atomic lattices (*Debye frequency*) and their energy is close to that of thermal neutrons. A series of fracture experiments on natural rocks and the systematic monitoring of seismic events have revealed that TeraHertz phonons and/or plasmons are able to induce fission reactions in medium-weight chemical elements (in particular, iron and calcium), with neutron and/or alpha particle emissions (without gamma radiation and radioactive wastes).

The same phenomenon appears to occur in several different situations and to explain puzzles regarding the history of our planet, like the primordial carbon "pollution" (and the correlated iron depletion with the well-known basaltic-to-sialic transformation in the Earth's Continental Crust) or the ocean formation (and the correlated calcium depletion), as well as scientific mysteries, like the so-called cold fusion or the correct radiocarbon dating of organic materials. In general, phono-fission nuclear reactions have a fundamental role in the chemical evolution of our planet and the planets of Solar System, through tectonics and seismicity (rocky planets), as well as atmospheric storms (gaseous planets and the Sun itself).

Three different forms of emitted energy might be used as earthquake precursors. At the macro-scale, acoustic emission (AE) prevails, as well as electromagnetic emission (EME) at the meso-scale, and neutron emission (NE) at the micro- and nanoscale. The three fracto-emissions tend to anticipate the incoming seismic event with an evident and chronologically ordered time shifting: small forming cracks, high frequencies, and neutron emission at least one week before, longer extended cracks, lower frequencies, as well as electromagnetic and acoustic waves later and temporally closer to the seismic event. The experimental observations reveal a high correlation between the three fracto-emission peaks and the major earthquakes occurring in the areas closest to the seismic station.

Regarding "cold fusion", despite the great amount of experimental results, the comprehension of this phenomenon still remains unsatisfactory. On the other hand, as reported by several authors, one of the common features is the appearance of microcracks on the electrode external surfaces after the electrolysis experiments. A mechanical explanation is proposed as a consequence of hydrogen embrittlement in the electrodes during the electrolysis. The preliminary experimental activity was conducted using a Ni-Fe anode and a Co-Cr cathode immersed in a potassium carbonate water solution. Emissions of neutrons and alpha particles were measured during the experiments as well as evident chemical composition changes in the electrodes, revealing the effects of fission reactions in the host lattices. The resultant helium atoms (or alpha particles) are not usually due to the nuclear fusion of couples of hydrogen or deuterium atoms, rather they simply represent some of the fission fragments.

In order to confirm the preliminary results, further electrolytic tests were conducted using palladium and nickel electrodes. As for the earlier experiments, relevant compositional changes and the appearance of lighter elements previously absent were observed. The most relevant process emerging from the experiments is the primary fission of palladium (decrement approximately equal to 30%) into iron and calcium. Then, secondary fissions appear, in turn producing oxygen atoms, alpha particles, and neutrons. The chemical composition changes are fully confirmed by four repetitions of the same identical experiment.

An extensive evaluation of the heat generation is carried out showing a positive energy balance in correspondence to the major neutron emission peaks. Analogous experimental results are also obtained from hydrodynamic cavitation experiments on iron salt water solutions. In that case, the triggering mechanical instability is represented by the implosion of micro- and nanobubbles.

Very important implications to and applications in earthquake precursors, early-stage fatigue diagnostics, geochemical evolution, climate change, and energy production are likely to develop in the next future. Scientists engaged in geochemistry, mineralogy, geology, astrochemistry, planetology, climatology, seismology, geophysics, condensed matter physics, and biology could receive a great benefit from the innovative and holistic vision of this book.

I would like to gratefully thank all the contributors to the different chapters of the volume, for their competent efforts in sharing with me the difficulties of this cutting-edge research project. The volume results to be particularly multi- and interdisciplinary. Only with the valid help of so many and scientifically diverse scientists was it possible for me to rebuild a so complex, complete, and consistent scenario.

Such tireless work began in 2008, 17 years ago, and has developed over the years with the fruitful collaboration of 18 contributors (listed in the acknowledgements section), and with the important support of additional experts, who are usually acknowledged at the end of the single chapters. The disciplines involved and the expertise of my coworkers are the most different: from theoretical and applied mechanics to condensed matter physics, from chemistry to mineralogy, from radiation measurements to thermodynamics, from acoustics to electromagnetism, from seismology to geophysics. My personal expertise in Solid Mechanics and in Fracture Mechanics, as well as my academic titles in mathematics and in nuclear engineering, allowed me to discuss and exchange ideas with all of them rather easily.

After seven years without any specific financial support (2008–2015), when our research work was only curiosity-driven, I and my research group received two major dedicated grants.

MetalWork S.p.A., a private industrial company with a technical interest and expertise in fluid flow cavitation, supported us for four years (2015–2019) in the project "Hydrodynamic Cavitation and Correlated Energy Aspects". The follow-up from this research activity is relevant and disseminated over the entire book. I feel deeply indebted to the President Erminio Bonatti and to the Manager Fausto Rodella for their enthusiastic and continued support.

A later and decisive financial support to accomplish my studies and write the present book came from the European Union, through the "Horizon 2020" Programme for Research and Innovation (2020–2024). The general title of this four-year project was "Clean Energy from Hydrogen-Metal Systems (CleanHME)" (Grant 951974). This book is subdivided into seven parts and each part into different chapters. Over the total of 25 chapters in the book, the contents of the following ten chapters were developed during the project and financially supported by it: 7, 9, 12–14, 16, 22–25. The Open Access Service Agreement related to the book was established between Politecnico di Torino and Springer Nature on the basis of the above mentioned grant. As for the previous grant, the emphasis of the project title was on the energy aspects, whereas the major outcomes and most groundbreaking results are perhaps in other correlated directions. It appears to be a typical case of *Serendipity*. I would like to thank the European Coordinator of the project, Konrad Czerski, as well as the Colleagues Jean-Paul Biberian and Andras Kovacs, who have always shown a pro-active interest in the development of my contribution.

A sincere thought of gratitude is also addressed to the several colleagues who considered, encouraged, and, in some cases, inspired my research work. In particular, I would like to recall: Fabio Cardone, Francesco Celani, Yogendra Srivastava, Allan Widom, Peter Hagelstein, and Yoshiaki Arata, who are among the major experts in low-energy nuclear reactions (LENR); Giuseppe Careglio, Piero Pizzi, and Stefano Re Fiorentin, who are members of the "Club of Technical Managers (CDT)" within the Industrial Union in Turin; Maurizio Maggiore, the European Union Officer who firstly proposed the call for our rather controversial project; Claudio Pace and Bill Collis, who nominated me for the *Giuliano Preparata Medal*, which is awarded by the International Society for Condensed Matter Nuclear Science and I could win at Assisi in 2022.

A special thought is then for a very nice person and outstanding scientist, who passed away some years ago; nevertheless he is still alive in this book through his effective lattice model for the atomic nucleus that we have updated and extensively applied: Norman Cook. He was Coauthor of a joint theoretical chapter in my edited book *Acoustic, Electromagnetic, Neutron Emissions from Fracture and Earthquakes*, Springer 2015.

I cannot forget to mention that major Scientific Institutions in Italy, through the action of their representatives, were interested in our research work, sometimes giving it relevant contributions: first of all, Politecnico di Torino, with my Department of Structural, Geotechnical, and Building Engineering, the Department of Environment, Land, and Infrastructure Engineering (Riccardo Sandrone), the Department of Applied Science and Technology (Monica Ferraris), and the Department of Mechanical and Aerospace Engineering (Francesca M. Curà). In addition, I wish to gratefully acknowledge the National Institute of Nuclear Physics, INFN (Alba Zanini), the National Agency for Atomic Energy, ENEA (Massimo Sepielli), the National Research Council, CNR (Fabio Cardone), the Italian Institute of Technology, IIT (Candido Pirri), and the National Research Institute of Metrology, INRIM (Riccardo Malvano).

On different occasions, the echoes of my scientific results were reflected by very popular national and international media. The following TV channels and productions, as well as their very professional and well-known journalists, focused the attention onto my preliminary results. I wish to gratefully acknowledge: Silvia Rosa Brusin (RAI3, Tg Leonardo, 2011); Sveva Sagramola (RAI3, Geo and Geo, 2014); Sante Altizio (Director of the documentary "La Passione e la Ragione", Luna Film Production, 50 min, 2015); NBC News (National Broadcasting Company, USA, 2015).

Shantou, China
September 2025

Alberto Carpinteri

Acknowledgements The author would like to gratefully thank all the **contributors** to the different chapters of the volume, for their competent efforts in sharing with him the difficulties of this cutting-edge research project. The volume results to be particularly multi- and inter-disciplinary. Only with the valid help of so many and scientifically diverse scientists was it possible to rebuild a so complex, complete, and consistent scenario.

The list of the contributors is ordered according to the number of chapters contributed by the single person:

Oscar Borla (Chapters 2–8, 10–13, 15–17, 22)

Amedeo Manuello (Chapters 3 and 4, 10–13, 18, 19, 21)

Giuseppe Lacidogna (Chapters 2–6, 8, 15, 17)

Domenico Scaramozzino (Chapters 23–25)

Gianni Niccolini (Chapters 2 and 20)

Salvatore Guastella (Chapters 3 and 11)

Francesco Montagnoli (Chapters 9 and 14)

Alessandro Goi (Chapters 10 and 11)

Diego Veneziano (Chapters 10 and 11)

Luca Negri (Chapters 19 and 21)

Riccardo Sandrone (Chapter 3)

Francesca M. Curà (Chapter 4)

Raffaella Sesana (Chapter 4)

Federico Accornero (Chapter 7)

Stefano Invernizzi (Chapter 8)

Stefano Roggeri (Chapter 14)

Umberto Lucia (Chapter 22)

Massimo Zucchetti (Chapter 22)

MetalWork S.p.A., a private industrial company with a technical interest and expertise in fluid flow cavitation, supported the author for four years (2015–2019) in the project "Hydrodynamic Cavitation and Correlated Energy Aspects" (460,000 Euro).

The follow-up from this research activity is relevant and disseminated over the entire book. The author feels deeply indebted to the President **Erminio Bonatti** and to the Manager **Fausto Rodella** for their enthusiastic and continued support.

A decisive financial support to accomplish the author's studies and to write the present book came from the **European Union**, through the **"Horizon 2020" Programme** for Research and Innovation (**Grant 951974**, 2020–2024). The general title of this four-year project was "Clean Energy from Hydrogen-Metal Systems (CleanHME)" (260,000 Euro).

The book is subdivided into seven parts and each part into different chapters. Over the total of 25 chapters in the book, the contents of the following ten chapters were developed during the project and financially supported by it: 7, 9, 12–14, 16, 22–25.

The **Open Access** Service Agreement related to the book was established between **Politecnico di Torino (Department of Structural, Geotechnical, and Building Engineering)** and Springer Nature on the basis of the above mentioned grant.

Competing Interests The author has no competing interests to declare that are relevant to the content of this manuscript.

Contents

About the Author

Alberto Carpinteri received his Doctoral Degrees in Nuclear Engineering and in Mathematics, both cum Laude, from the University of Bologna in Italy. Currently he is Chang Jiang (Blue River) Chair Professor at the Engineering College of Shantou University in China, after retiring from Politecnico di Torino in Italy and spending there the major part of his career as Chair Professor of the Mechanics of Solids and Structures (for 37 years). He has held several positions of service and responsibility, within Politecnico di Torino, nationally, and internationally. He was President of the National Research Institute of Metrology (INRIM) in Italy, Head of the Engineering Division in the European Academy of Sciences (sited in Brussels), President of the major International Associations on Fracture Mechanics (ICF, ESIS, IA-FraMCoS), and Member of the Executive Boards of IUTAM and SEM. He took also visiting positions at the: University of São Paulo in Brazil, Tsinghua University, Tongji University, and Architectural Institute of Nanjing in China, Lehigh University in USA, Universities and Research Institutions across South Africa (Cape Town, Johannesburg, Pretoria, Bloemfontein). He was Editor-in-Chief of the International Journal "Meccanica" (Springer Nature) and Member of the Editorial Boards of eleven Refereed International Journals.

Professor Carpinteri is Honorary Fellow of the International Congress on Fracture (ICF), Founding Fellow of the Indian Structural Integrity Society (InSIS), and Fellow of the European Structural Integrity Society (ESIS) and of the International Association of Fracture Mechanics for Concrete and Concrete Structures (IA-FraMCoS). He is Life Fellow of the European Academy of Sciences and of the Academia Europea, Fellow of the European Academy of Sciences and Arts, of the International Academy of Engineering, and Foreign Member of the Academy of Athens. He is Life Fellow of the American Society of Civil Engineers (ASCE), becoming Member of it four decades ago in 1985. He received numerous international awards and recognitions: Giuliano Preparata Medal from the International Society for Condensed Matter Nuclear Science (ISCMNS), Robert l'Hermite Medal (RILEM), Griffith Medal (ESIS), Swedlow Memorial Lecture Award (ASTM), Inaugural Paul Paris Gold Medal (ICF), Frocht Award (SEM), George Irwin Medal (ASTM), and Lifetime Achievement Medal (ICDM). In addition, he received the following honors:

Doctor Honoris Causa in Engineering from the Russian Academy of Sciences, Honorary Professor from Tianjin University and Shenyang Northeastern University, Guest Professor from Harbin Institute of Technology, and Honorary Editor of the International Journal "Smart Construction and Sustainable Cities" (Springer Nature).

Professor Carpinteri organized and chaired several international conferences, among which the following one should be in particular recalled: 11th International Conference on Fracture (ICF11), Turin-Italy, 2005 (Record in the ICF history: 1041 official participants).

His scientific achievements can be summarized by the following numbers:

H-Index (Google Scholar) = 98; Total Citations: over 39,000

H-Index (Scopus) = 69; Total Citations: over 17,000

Ranking order position as a Top Scientist according to Research.com (Microsoft Academic) in the area "Engineering and Technology": n. 3 in Italy, n. 36 in China and Hong Kong, n. 240 in the World. He is Author of over 1000 publications, of which around 500 appear as articles in Refereed International Journals, and 58 are authored or edited volumes. He is also Author of five single-authored books published by major International Publishers. His most cited work (1283 Citations according to Google Scholar) is the volume "Fractals and Fractional Calculus in Continuum Mechanics", coedited with Francesco Mainardi and published by Springer in 1996.

Several specific topics have been considered by Prof. Carpinteri, always giving them an original and personal contribution. In some cases, such a contribution resulted to be also innovative, anticipating even by years the trends in cutting-edge international research. Among these peculiar topics, it is significant to recall the following.

(1) *Static-kinematic duality* and its crucial role in Computational Mechanics.
(2) Application of Dimensional Analysis (Buckingham's Theorem) and definition of the dimensionless *Brittleness Number* in the scaling competition between plastic collapse and brittle fracture, which are failure mechanisms governed by generalized forces with different physical dimensions.
(3) Interpretation of the phenomena of mechanical instability, such as brittle crack propagation (*Cohesive Crack Model*), frictional stick-slip, and buckling in thin cylindrical and spherical shells, in the general context of Catastrophe Theory.
(4) Solution to the problem of propagation stability for cracks bridged by reinforcements and/or fibers on the basis of rigorous conditions of static equilibrium and kinematic compatibility on the fracturing beam cross-section (*Bridged Crack Model*).
(5) *Multi-fractal Scaling Laws* (MFSL) for tensile strength (lacunar fractals) and fracture energy (invasive fractals) of concrete, rocks, and ceramic materials.
(6) Application of *Fractional Calculus* to field and boundary equations for elastic bodies deformable only over fractal subsets.
(7) Scaling and fractality of Fatigue Limit (Woehler's Curve) and Fatigue Threshold (Paris Law). Solution to the *short crack problem* in the framework of Fractal Geometry.

(8) Mechanics of *complex materials*: hierarchical, nanostructured, layered, and functionally graded.
(9) Dynamics and stability of elastic structures: From mega-structures (long-span bridges and high-rise buildings) to *nanostructures* (proteins and macromolecular structures).
(10) Acoustic, electromagnetic, and subatomic particle emissions from fracture and earthquakes. Fracto-emissions as *seismic precursors.*

Professor Carpinteri was invited at several international conferences, institutions, and advanced courses to present his most relevant scientific results (in 36 different countries). Over 30 of these presentations are opening, closing, or honorary lectures.

Professor Carpinteri was Supervisor of 35 Ph.D. Candidates. Presently, most of them are taking University Faculty positions in Italy or abroad: Eight full professors, seven associate professors, two assistant professors, and six post-doctoral fellows. He was also Supervisor of 121 Master Candidates. His educational activities include also that of European Coordinator of the "Innovative Learning and Training on Fracture (ILTOF)" project, in the framework of the European Union Leonardo da Vinci Programme for Education and Culture. He has been officially affiliated to three different universities (in Bologna, Torino, Shantou), as well as to the National School of the Italian Army. During the last 20 years of his career, he has taught two very advanced M.Sc. courses that were important sources of inspiration for his research work: "Static and Dynamic Instabilities of Structures", and "Fracture and Plasticity".

September 2025

Part I
Introduction

Chapter 1
TeraHertz Phono-Fission Nuclear Reactions: Emergence from Different Nanomechanics Instabilities

Abstract TeraHertz phonons and/or plasmons are produced in condensed matter by mechanical instabilities at the nanoscale (fracture, turbulence, cavitation, dynamic buckling). They present a frequency that is close to the resonance frequency of atomic lattices and an energy that is close to that of thermal neutrons. A series of fracture experiments on natural rocks as well as the systematic monitoring of seismic events have recently demonstrated that the TeraHertz phonons and/or plasmons are able to induce fission reactions in medium-weight atoms with neutron and/or alpha particle emissions. The same phenomenon appears to occur in several different situations and to explain puzzles regarding the history of our planet, like the ocean formation or the primordial carbon "pollution", as well as scientific mysteries, like the so-called cold nuclear fusion or the correct radiocarbon dating of organic materials. Very important applications to earthquake precursors, early-stage fatigue diagnostics, climate change, energy production, and cell biology are likely in the future based on such fundamental discoveries.

Keywords Fracture · Turbulence · Cavitation · Dynamic buckling · TeraHertz phonons and/or plasmons · Ultrasonic pressure waves · Phono-fission reactions · Neutron emissions · Alpha particle emissions · Electromagnetic emissions · Acoustic emissions · Fracto-emissions as seismic precursors · Chemical composition changes · Great oxidation event (GOE) · Ocean formation · Calcium depletion · Primordial carbon "pollution" · Iron depletion · Basaltic-to-sialic transformation · Solar system chemical evolution · Nickel depletion on mars · Iron increment on mars · Great red spot of jupiter · Lithium depletion on the solar corona · Cold nuclear fusion · Turin shroud · Cell mechanotransduction · Protein folding and unfolding · Phono-fission of potassium · Phono-fusion of sodium and oxygen

From the Distinguished Lecture in Solid Mechanics presented by Professor Alberto Carpinteri at the California Institute of Technology on May 30, 2014

A. Carpinteri, *Terahertz Phonons and Nanomechanical Instabilities*,
https://doi.org/10.1007/978-3-032-14692-2_1

1.1 Fracture and Acoustic Emission: From Hertz to TeraHertz Pressure Wave (or Phonon) Frequencies

When you cut a stretched rubber band, it remains subject to rapid fluctuations for a few moments. The same phenomenon occurs in any solid body when it breaks in a brittle way, even if only partially. In the case of the formation or propagation of microcracks, such dynamic phenomenon appears under the form of longitudinal waves of expansion/contraction (tension/compression), in addition to transverse or shear waves. These are generally said pressure waves, or phonons when their particle nature is emphasized, and travel at a speed that is characteristic of the medium, and, for most of the solids and fluids, presents an order of magnitude of 10^3 m/second. On the other hand, the wavelength of pressure waves emitted by forming or propagating cracks appears to be of the same order of magnitude of crack size or crack advancement length. The wavelength cannot therefore exceed the maximum size of the body in which the crack is contained and may vary from the nanometer scale (10^{-9} m), for defects in crystal lattices such as vacancies and dislocations, up to the kilometer-scale, in the case of Earth's Crust faults. Applying the well-known relationship: frequency = speed/wavelength, one obtains the two extreme cases corresponding to the frequency of pressure waves: 10^{12} oscillations/second (TeraHertz), in the case of the formation of nanocracks, as well as one oscillation/second (Hertz), in the case of large-scale tectonic dynamics (Fig. 1.1).

In fact in solids, whatever their size, the cracks that are formed or propagate are of different lengths, sometimes belonging to different orders of magnitude. In particular, in nanoscopic bodies the only frequency present should be the TeraHertz, since higher frequencies would imply defects at the atomic or subatomic scales. On

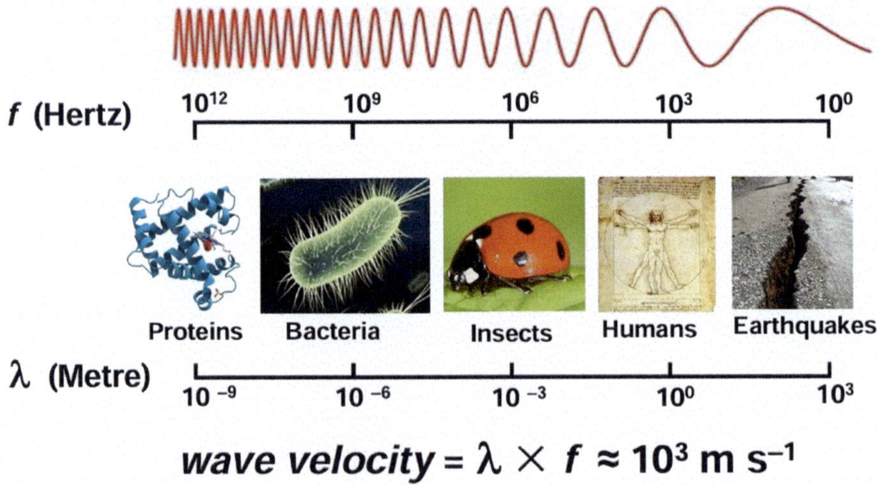

Fig. 1.1 Correlation between wavelength scale (scale of the forming crack) and frequency scale of the emitted pressure wave, by assuming a constant velocity of pressure wave

the contrary, in the Earth's Crust and during an earthquake, cracking is a multi-scale phenomenon as well as the frequencies of pressure waves are spread over a broad spectrum (Fig. 1.1). Moreover, while at the early stages of the seismic event, mainly small cracks will be present and active and therefore high frequencies, so at the end large cracks and low frequencies will prevail, the latter also typically in the audible field.

1.2 Seismic Precursors: Acoustic, Electromagnetic, Neutron Emissions

Further considering the very important case of earthquakes, you can complete the picture by stating that, as fracture at the nanoscale (10^{-9} m) emits phonons at the frequency scale of TeraHertz (10^{12} Hertz), so fracture at the microscale (10^{-6} m) emits phonons at the frequency scale of GigaHertz (10^{9} Hertz), at the meso-scale of millimeter emits phonons at the frequency scale of MegaHertz (10^{6} Hertz), at the macro-scale of meter emits phonons at the frequency scale of kiloHertz (10^{3} Hertz), and eventually faults at the kilometer-scale emit phonons at the frequency scale of the simple Hertz, which is the typical and most likely frequency of seismic oscillations (Fig. 1.1) [1].

The animals with sensitive hearing in the ultrasonic field (frequency > 20 kilo-Hertz) "feel" the earthquake up to one day in advance, when the active cracks are still below the meter scale. Ultrasounds are in fact a well-known seismic precursor [2, 3]. With frequencies between Mega- and GigaHertz, and therefore cracks between the micron and the millimeter scale, phonons can generate electromagnetic waves of the same frequency, which turn out to be even a more powerful seismic precursor (up to a few days before) [4, 5]. It is to be considered that there are not any instruments today able to directly measure such high mechanical frequencies. The secondary phenomenon of electromagnetic emissions is, on the other hand, able to reveal such vibrations.

When phonons show frequencies between Giga- and TeraHertz and then with cracks below the micron scale, we are witnessing a phenomenon partially unexpected: phonons resonate with the crystal lattices and, through a complex cascade of events (acceleration of electrons, bremsstrahlung gamma radiation, photo-fission, etc.), may produce nuclear fission reactions [6–14]. It can be shown experimentally how such fission reactions emit neutrons [15–17] like in the well-known case of uranium-235 but without gamma radiation and radioactive wastes.

Note that the Debye frequency, i.e., the fundamental frequency of free vibration in crystal lattices, is around the TeraHertz, and this is not a coincidence, since it is simply due to the fact that the inter-atomic distance is just around the nanometer, as indeed the minimum size of the lattice defects. As the chain reactions are sustained by thermal neutrons in a nuclear power plant, so the phono-fission reactions are triggered by phonons that have a frequency close to the resonance frequency of the crystal lattice and an energy close to that of thermal neutrons. Neutrons therefore appear to be as the most powerful earthquake precursor (up to three weeks before) [18–23].

1.3 Chemical Evolution of Our Planet and Its Replication at the Fracture Mechanics Laboratory Scale

The phono-fission reactions then appear to be induced by pressure waves at very high frequencies (TeraHertz). They are often accompanied and revealed by the emission of neutrons and/or alpha particles. However, gamma rays and radioactive wastes appear to be absent in the experiments. Ultrasonic pressure waves may in turn be produced by the most common mechanical instabilities in condensed matter, such as fracture in solids and turbulence (cavitation) in fluids. Both are hierarchical, multi-fractal, and dissipative phenomena, where cracks and vortexes (bubbles), respectively, are present at the different scales.

After the early experiments conducted at the National Research Council of Italy (CNR) [24, 25], soliciting with ultrasounds aqueous solutions of iron salts, the research group at Politecnico di Torino has conducted fracture experiments on solid samples, using iron-rich natural rocks like granite [26–36], basalt and magnetite [37, 38], and later marble [39], mortar [40], steel [41], gypsum and quartz (see Chap. 7). Different types of detectors have demonstrated the presence of significant neutron emissions, in some cases by different orders of magnitude higher than the usual environmental background (up to 10 times from granitic rocks, up to 100 times from basalt, up to 1000 times from magnetite).

The neutron flux was found to depend, besides on the iron content, on the size of the specimen through the well-known brittleness size-scale effect [42–45]: larger specimen sizes imply a higher brittleness, i.e., a more relevant strain energy release, and therefore more neutrons.

These studies have also been able to give an answer to some puzzles related to the history of our planet. It has been shown how the phono-fission reactions that have occurred between 3.8 and 2.5 billion years ago, during the periods of formation and most intense activity of tectonic plates, have resulted in the splitting of atoms of certain elements, which were so transformed into other lighter ones. Since the product-elements, i.e., the fragments of the fissions, appear to be stable isotopes, all the excess neutrons are therefore emitted. Several of the most abundant chemical elements have been involved in similar transformations, like a part of magnesium that transformed into carbon, forming the dense atmospheres of carbon dioxide (CO_2) and methane (CH_4) of the primordial terrestrial eras [46, 47]. In a similar way, calcium depletion contributed to the formation of oceans as a result of fracture phenomena in limestone rocks.

Considering the entire life of our planet and all the most abundant chemical elements [48–50], it can be seen how ferrous elements have dramatically decreased in the Earth's Crust (–12%), as well as at the same time aluminum and silicon have increased (+8.8%). An increment in magnesium (+3.2%), which then transformed into carbon, has been assumed as the origin of carbon-rich primordial atmospheres.

Similarly, alkaline-earth elements have strongly decreased (–8.7%), whereas alkaline elements (+5.4%) and oxygen (+3.3%) have increased. The appearance of a 3.3% oxygen represents the well-known Great Oxidation Event, a phenomenon that led to the formation of oceans and the origin of life on our planet.

These transformations, that have lasted for billion years in the Earth's Crust and atmosphere, have been reproduced in the laboratory during a fraction of a second by crushing different rock samples. We were able to confirm, through advanced micro-chemical analyses, the most relevant compositional variations described above at the geological and planetary scales: the transformation of iron into aluminum, or into magnesium and silicon (in iron-rich natural rocks [37]), as well as the transformation of calcium and magnesium into other lighter elements including carbon (in the samples of marble [39]). Such variations are shown to be not modest at all. The iron decrement in magnetite was found to be of 27.9%, compared to an overall increment of 27.7% in lighter elements. So in marble, carbon has increased by 13%, compared to an exactly equivalent overall decrement in heavier elements.

Since the natural carbon production in the primordial eras, although at a much slower rate, is going on even today, due to the seismic activity, the monitoring of carbon dioxide in relation to major earthquakes can be considered as a potential earthquake precursor [47], in addition to acoustic, electromagnetic, and neutron emissions.

1.4 Chemical Evolution in the Planets of Solar System

Even in the case of the other planets of Solar System, we are witnessing a series of evident experimental data that can be interpreted only in light of phono- fission reactions [51]. In particular, the data coming from different surveys on the crust of planet Mars, made available by the NASA space missions over the past 30 years, suggest that the increment in certain elements (iron, chlorine, and argon) and the concomitant decrement in others (nickel and potassium), together with the emission of neutrons from the major fault lines in the planet, should all be considered as phenomena directly and highly correlated. These data provide a clear confirmation that seismic activity has contributed to the chemical evolution of the Red Planet. Similar experimental evidence is concerning Mercury, Jupiter (with its relevant emission of neutrons from the Great Red Spot), and the Sun itself. The phono-fission reactions are triggered by earthquakes in rocky planets and by storms in gaseous planets. In the Sun, for example, the drastic decrement in lithium appears to be due to fission of the same lithium into helium and hydrogen.

1.5 A Plausible Explanation to the So-Called Cold Nuclear Fusion

Several evident data have been observed during the last 35 years of anomalous nuclear reactions in electrolytic experiments. The purpose was that of providing an explanation to the phenomena related to the so-called cold nuclear fusion and of evaluating the possibility of a heat generation from electrolytic cells [52–54]. Despite the large amount of positive experimental results, the understanding of these phenomena is still unsatisfactory. On the other hand, as reported in most of the articles on cold nuclear fusion, the appearance of microcracks on the surface of the electrodes used in the experiments is one of the most common observations. It is therefore possible to give an explanation of a mechanical nature, which takes into account the hydrogen embrittlement of the metallic electrodes.

In our earlier experiments [55, 56], electrolytic phenomena were produced by means of an anode of a nickel–iron alloy and a cathode of a cobalt-chromium alloy, immersed in an aqueous solution of potassium carbonate. During these experiments, emissions of neutrons were revealed. Furthermore, the chemical composition of the electrode surfaces was analyzed before and after the experiments, allowing to identify phono-fissions occurring in the electrodes. The primary process appears to be a symmetric fission of the atom of nickel into two atoms of silicon, or two atoms of magnesium. In the latter case, additional fragments were found to be constituted of alpha particles.

In our later experiments [57, 58], where a palladium electrode was used, the primary process appears to be the non-symmetric fission of palladium into iron and calcium, whereas the secondary processes appear to be the further fissions of both such products into oxygen atoms and alpha particles.

1.6 A Catastrophic Earthquake Behind the Mystery of Turin Shroud

A neutron radiation, produced by the historical earthquake of AD 33, may have caused the erroneous radiocarbon dating of the Shroud of Turin in 1988 [59, 60]. Neutron radiation could have also caused the image formation of a crucified man on the linen cloth. The Shroud has attracted a large interest since Secondo Pia took the first photograph in 1898. According to carbon-14 dating, the cloth would approximately be only 750 years old. From 1988 onwards, several researchers have instead argued that the Shroud would be much older and that the process of dating would have been wrong because of neutron radiation, so as to form new isotopes of carbon from nitrogen atoms previously present in the linen cloth. However, so far, any plausible physical reason has not yet been identified that can justify the origin of this neutron radiation [61]. The mechanical and chemical experiments described in the previous sections allow, on the other hand, to conjecture that high-frequency pressure waves, generated in the Earth's Crust by the historical earthquake of AD 33, which took place in the old Jerusalem area with a magnitude between 8 and 9 on the Richter scale, may have produced a neutron radiography on linen fibers and seemingly rejuvenated the same fabric. Let us consider that, although the calculated integral flux of 10^{13} neutrons per square centimeter is 10 times greater than the cancer therapy dose, nevertheless it is 100 times smaller than the lethal dose.

1.7 Future Applications also to Biology?

Regarding the living cells, the phono-fission reactions could explain the mechanism that governs the so-called sodium–potassium pump (ATP-ase protein) and, more in general, the metabolic processes. In the case of the ionic pump, the ions of sodium and potassium would be subject to a continuous, recursive, and mutual transformation, losing and gaining an oxygen atom at each passage through the cell membrane. As cells are microscopic objects, so proteins are nanoscopic and the typical mechanisms of *folding* and *unfolding*, which make the passage of ions through the cell membrane possible, are accompanied by vibrational phenomena of resonance at the frequency of TeraHertz [62, 63] (mechanotransduction [64]). More precisely, the folding and unfolding changes of configuration in the proteins should be interpreted as a dynamic nanobuckling (with possible snap-through) in complex-shaped thin shells. As in the case of acoustic emission from cracks, vortexes, or bubbles at the nanoscale, also the resonance frequency of nanostructures is evaluated in the TeraHertz range [65–67].

Analogous reasons could explain also the "digestion" of radioactive isotopes, intended as their natural transformation into stable isotopes of chemical elements that are essential for the life of microbial cultures [68].

References

1. Ashcroft NW, Mermin DN (2013) Solid state physics. Cengage learning, Delhi
2. Lockner DA et al (1991) Quasi-static fault growth and shear fracture energy in granite. Nature 350:39–42
3. Carpinteri A, Lacidogna G, Niccolini G (2007) Acoustic emission monitoring of medieval towers considered as sensitive earthquake receptors. Nat Hazard 7:251–261
4. Rabinovitch A, Frid V, Bahat D (2007) Surface oscillations. A possible source of fracture induced electromagnetic oscillations. Tectonophysics 431:15–21
5. Carpinteri A et al (2012) Mechanical and electromagnetic emissions related to stress-induced cracks. Exp Tech 36:53–64
6. Bridgman PW (1927) The breakdown of atoms at high pressures. Phys Rev 29:188–191
7. Batzel RE, Seaborg GT (1951) Fission of medium weight elements. Phys Rev 82:607–615
8. Fulmer CB et al (1967) Evidence for photofission of iron. Phys Rev Lett 19:522–523
9. Widom A, Swain J, Srivastava YN (2013) Neutron production from the fracture of piezoelectric rocks. J Phys G: Nucl Part Phys 40(15006):1–8
10. Widom A, Swain J, Srivastava YN (2014) Photo-disintegration of the iron nucleus in fractured magnetite rocks with magnetostriction. Meccanica 50:1205–1216

11. Hagelstein PL, Letts D, Cravens D (2010) Terahertz difference frequency response of PdD in two-laser experiments. J Cond Mat Nucl Sci 3:59–76
12. Hagelstein PL, Chaudhary IU (2014) Anomalies in fracture experiments and energy exchange between vibrations and nuclei. Meccanica 50:1189–1203
13. Cook ND (2010) Models of the atomic nucleus, 2nd ed, Springer, Dordrecht
14. Cook ND, Manuello A, Veneziano D, Carpinteri A (2015) Piezonuclear fission reactions simulated by the lattice model of the atomic nucleus. Acoustic, electromagnetic, neutron emissions from fracture and earthquakes, Chapter 15, Springer
15. Diebner K (1962) Fusionsprozesse mit hilfe konvergenter stosswellen—einige aeltere und neuere versuche und ueberlegungen. Kerntechnik 3:89–93
16. Derjaguin BV et al (1989) Titanium fracture yields neutrons? Nature 34:492
17. Fujii MF et al. (2002) Neutron emission from fracture of piezoelectric materials in deuterium atmosphere. Jpn J Appl Phys Pt.1 41:2115–2119
18. Sobolev GA, Shestopalov IP, Kharin EP (1998) Implications of solar flares for the seismic activity of the Earth. Izvestiya, Phys Solid Earth 34:603–607
19. Volodichev NN et al. (1999) Lunar periodicity of the neutron radiation burst and seismic activity on the Earth. In: Proceeding of the 26th International cosmic ray conference, Salt Lake City
20. Kuzhevskij M, Nechaev OY, Sigaeva EA (2003) Distribution of neutrons near the Earth's surface. Nat Hazard 3:255–262
21. Kuzhevskij M (2003) Neutron flux variations near the Earth's crust. A possible tectonic activity detection. Nat Hazard 3:637–645
22. Sigaeva EA et al (2006) Thermal neutrons' observations before the Sumatra earthquake. Geophys Res Abstr 8:00435
23. Borla O, Lacidogna G, Carpinteri A (2015) Piezonuclear neutron emissions from earthquakes and volcanic eruptions. Acoustic, electromagnetic, neutron emissions from fracture and earthquakes, Chapter 10, Springer
24. Cardone F, Mignani R (2007) Deformed spacetime, Chapters 16 and 17. Springer, Dordrecht
25. Cardone F, Cherubini G, Petrucci A (2009) Piezonuclear neutrons. Phys Rev Lett A 373:862–866
26. Carpinteri A, Cardone F, Lacidogna G (2010) Energy emissions from failure phenomena: mechanical, electromagnetic, nuclear. Exp Mech 50:1235–1243
27. Carpinteri A, Lacidogna G, Manuello A, Borla O (2013) Piezonuclear fission reactions from earthquakes and brittle rocks failure: evidence of neutron emission and nonradioactive product elements. Exp Mech 53(3):345–365
28. Carpinteri A, Cardone F, Lacidogna G (2009) Piezonuclear neutrons from brittle fracture: early results of mechanical compression tests. Strain 45:332–339
29. Carpinteri A, Chiodoni A, Manuello A, Sandrone R (2011) Compositional and microchemical evidence of piezonuclear fission reactions in rock specimens subjected to compression tests. Strain 47(2):267–281
30. Carpinteri A, Manuello A (2011) Geomechanical and geochemical evidence of piezonuclear fission reactions in the Earth's crust. Strain 47(2):282–292
31. Carpinteri A, Borla O, Lacidogna G, Manuello A (2010) Neutron emissions in brittle rocks during compression tests: monotonic versus cyclic loading. Phys Mesomech 13:264–274
32. Carpinteri A, Manuello A (2012) An indirect evidence of piezonuclear fission reactions: geomechanical and geochemical evolution in the Earth's crust. Phys Mesomech 15:14–23
33. Cardone F, Carpinteri A, Lacidogna G (2009) Piezonuclear neutrons from fracturing of inert solids. Phys Lett A 373:4158–4163
34. Carpinteri A, Lacidogna G, Manuello A, Borla O (2011) Energy emissions from brittle fracture: neutron measurements and geological evidences of piezonuclear reactions. Strength, Fract Complex 7:13–31
35. Carpinteri A, Lacidogna G, Manuello A, Borla O (2012) Piezonuclear fission reactions in rocks: evidences from microchemical analysis, neutron emission, and geological transformation. Rock Mech Rock Eng 45(4):445–459

36. Carpinteri A, Lacidogna G, Borla O, Manuello A, Niccolini G (2012) Electromagnetic and neutron emissions from brittle rocks failure: Experimental evidence and geological implications. Sadhana 37(1):59–78
37. Manuello A et al. (2015) Neutron emissions and compositional changes at the compression failure of iron-rich natural rocks. Acoustic, electromagnetic, neutron emissions from fracture and earthquakes, Chapter 3, Springer
38. Carpinteri A et al. (2015) Frequency-dependent neutron emissions during fatigue tests on iron-rich natural rocks. Acoustic, electromagnetic, neutron emissions from fracture and earthquakes, Chapter 4, Springer
39. Carpinteri A, Lacidogna G, Borla O (2015) Alpha particle emissions from Carrara marble specimens crushed in compression and X-ray photoelectron spectroscopy of correlated nuclear transmutations. Acoustic, electromagnetic, neutron emissions from fracture and earthquakes, Chapter 5, Springer
40. Carpinteri A, Borla O, Lacidogna G (2015) Elemental content variations in crushed mortar specimens measured by instrumental neutron activation analysis (INAA). Acoustic, electromagnetic, neutron emissions from fracture and earthquakes, Chapter 6, Springer
41. Invernizzi S et al. (2015) Piezonuclear evidences from tensile and compression tests on steel. Acoustic, electromagnetic, neutron emissions from fracture and earthquakes, Chapter 7, Springer
42. Hudson JA, Crouch SL, Fairhurst C (1972) Soft, stiff and servo-controlled testing machines: a review with reference to rock failure. Eng Geol 6:155–189
43. Carpinteri A (1989) Cusp catastrophe interpretation of fracture instability. J Mech Phys Solids 37:567–582
44. Carpinteri A, Pugno N (2005) Are scaling laws of strength of solids related to mechanics or to geometry? Nat Mater 4:421–423
45. Carpinteri A, Corrado M (2009) An extended (fractal) overlapping crack model to describe crushing size-scale effects in compression. Eng Fail Anal 16:2530–2540
46. Liu L (2004) The inception of the oceans and CO_2-atmosphere in the early history of the Earth. Earth Planet Sci Lett 227:179–184
47. Padron E et al (2008) Changes on diffuse CO_2 emission and relation to seismic activity in and around El Hierro, Canary Islands. Pure Appl Geophys 165:95–114
48. Taylor SR, McLennan SM (2009) Planetary crusts: their composition, origin and evolution. Cambridge University Press, Cambridge
49. Carpinteri A, Manuello A (2015) Evolution and fate of chemical elements in the Earth's crust, ocean, and atmosphere. Acoustic, electromagnetic, neutron emissions from fracture and earthquakes, Chapter 12, Springer
50. Carpinteri A, Manuello A, Negri L (2015) Chemical evolution in the Earth's mantle and its explanation based on piezonuclear fission reactions. Acoustic, electromagnetic, neutron emissions from fracture and earthquakes, Chapter 13, Springer
51. Carpinteri A, Manuello A, Negri L (2015) Piezonuclear fission reactions triggered by fracture and turbulence in the rocky and gaseous planets of the solar system. Acoustic, electromagnetic, neutron emissions from fracture and earthquakes, Chapter 14, Springer
52. Fleischmann M, Pons S (1989) Electrochemically induced nuclear fusion of deuterium. J Electroanal Chem 261:301–308
53. Preparata G (1991) A new look at solid-state fractures, particle emissions and «cold» nuclear fusion. Il Nuovo Cimento 104A:1259–1263
54. Mizuno T (1998) Nuclear transmutation: the reality of cold fusion. Cold fusion technology, Concord, New Hampshire
55. Carpinteri A et al (2013) Mehanical conjectures explaining cold nuclear fusion. In: Conference and exposition on experimental and applied mechanics (SEM), Lombard, Illinois 3:353–367
56. Carpinteri A et al. (2015) Cold nuclear fusion explained by hydrogen embrittlement and piezonuclear fissions at the metallic electrodes—Part I: Ni-Fe and Co-Cr electrodes. Acoustic, electromagnetic, neutron emissions from fracture and earthquakes, Chapter 8, Springer

57. Carpinteri A et al (2014) Hydrogen embrittlement and "cold fusion" effects in palladium during electrolysis experiments. In: Conference and exposition on experimental and applied mechanics (SEM), Greenville, South Carolina 6:37–47
58. Carpinteri A et al (2015) Cold nuclear fusion explained by hydrogen embrittlement and piezonuclear fissions at the metallic electrodes—Part II: Pd and Ni electrodes. Acoustic, electromagnetic, neutron emissions from fracture and earthquakes, Chapter 9, Springer
59. Carpinteri A, Lacidogna G, Manuello A, Borla O (2012) Piezonuclear neutrons from earthquakes as a hypothesis for the image formation and the radiocarbon dating of the Turin Shroud. Sci Res Essays 7(29):2603–2612
60. Carpinteri A, Lacidogna G, Borla O (2015) Is the Shroud of Turin in relation to the Old Jerusalem historical earthquake? Acoustic, electromagnetic, neutron emissions from fracture and earthquakes, Chapter 11, Springer
61. Phillips TJ, Hedges REM (1989) Shroud irradiated with neutrons? Nature 337:594
62. Acbas G et al (2014) Optical measurements of long-range protein vibrations. Nat Commun 5:3076
63. http://www.buffalo.edu/news/releases/2014/01/012.html
64. Mofard MRK, Kamm RD (2010) Cellular mechanotransduction. Cambridge University Press, Cambridge
65. Carpinteri A, Lacidogna G, Piana G, Bassani A (2017) TeraHertz mechanical vibrations in lysozyme: Raman spectroscopy versus modal analysis. J Mol Struct 1139:222–230
66. Carpinteri A, Piana G, Bassani A, Lacidogna G (2019) TeraHertz vibration modes in Na/K-ATPase. J Biomol Struct Dyn 37:256–264
67. Lacidogna G, Piana G, Bassani A, Carpinteri A (2017) Raman spectroscopy of Na/K-ATPase with special focus on low frequency vibrations. Vib Spectrosc 92:298–301
68. Vysotskii VI, Kornilova AA (2010) Nuclear transmutation of stable and radioactive isotopes in biological systems. Pentagon Press, New Delhi

Chapter 2
Experimental Evidence of Fracto-Emissions: Acoustic, Electromagnetic, Subatomic Particle

Abstract In the present chapter, acoustic (AE), electromagnetic (EME), and neutron (NE) emissions are measured during laboratory compression tests on natural rock specimens loaded up to failure. All the signals are acquired by a National Instruments Digitizer with eight channels simultaneously sampling. The main purpose is to give experimental evidence to the three different forms of energy emission from rocks under compression. The tests are performed on magnetite and basalt specimens at a constant displacement rate. AE signals are detected by applying to the specimen surface a piezoelectric (PZT) transducer with resonance frequency of about 150 kHz. EME signals are revealed by the current induced in a closed circuit by the change in the magnetic flux during specimen compression. The specimens are also monitored by means of a He^3 proportional neutron detector. During the tests, the AE signals are firstly detected and then the EME. All the recorded signals are correlated to the load vs time diagrams. The EME signals are obtained, in particular, during the typical snap-back instabilities, which characterize the load versus displacement diagrams of brittle materials. Neutron emission signals are generally identified at the end of the tests. As a matter of fact, neutron bursts usually occur only when the behavior of the specimen in compression is particularly brittle. Applications of these monitoring techniques to earthquake forecasting will be proved to be possible in the Part V of this book, where we will see that the temporal sequence of the events results to be totally different from that of laboratory tests: first the neutron emission (one week before the earthquake), then the electromagnetic emission (3–4 days before the earthquake), and eventually the acoustic emission (one day before the earthquake).

Keywords Iron-rich natural rocks · Compression tests · Brittle crushing failure · Acoustic emission · Electromagnetic emission · Neutron emission · Phono-fission reactions

A. Carpinteri, *Terahertz Phonons and Nanomechanical Instabilities*,
https://doi.org/10.1007/978-3-032-14692-2_2

2.1 Preliminary Remarks

It is possible to demonstrate experimentally that the failure phenomena, in particular when they occur in a brittle way, i.e., with an external mechanical energy release, emit additional forms of energy related to the fundamental natural forces. Experimental evidence and confirmation are found that energy emission of different forms occurs from solid-state fractures. The tests were carried out at the Laboratory of Fracture Mechanics of Politecnico di Torino, Italy. By subjecting quasi-brittle materials such as natural rocks to compression tests, for the first time bursts of neutron emission (NE) during the failure process were observed [1–5], necessarily involving nuclear reactions. In addition, the well-known acoustic emission (AE) [6–13] and the phenomenon of electromagnetic radiation (EME) [14–19] were detected. The latter is highly suggestive of charge redistribution during material failure and, at present, under investigation in the scientific community.

The phenomenon of EME is regarded as an important precursor of critical phenomena in Geophysics, such as rock fractures, volcanic eruptions, and earthquakes [19, 20]. For example, anomalous radiations of geoelectromagnetic waves were observed before major earthquakes. At the laboratory scale, rocks and concrete under compression generate AE and EME nearly simultaneously. This evidence suggests that also NE is generated during crack growth, reinforcing the idea that also NE can be applied as a forecasting tool for earthquakes.

While the mechanism of AE is fully understood, being provided by transient elastic waves due to stress redistribution following fracture propagation [6–13], the origin of EME from fracture is still not completely clear and different attempts have been made to explain it.

An explanation to the EME origin in metallic materials is related to dislocation phenomena [16], which however are not able to explain EME from fracture in brittle rocks, where the motion of dislocations is absent [17]. Frid et al. [17] and Rabinovitch et al. [21] have recently proposed a model of the EME origin where, following the rupture of bonds during the crack growth, the mechanical and electrical equilibria are broken at the fracture surfaces with the creation of ions moving collectively as a surface wave on both faces. Lines of positive ions on both newly created faces (which maintain their charge neutrality unlike in the capacitor model) oscillate collectively around their equilibrium positions in opposite phase to the negative ones. The oscillating dipoles created on both faces of the propagating fracture act as the source of emitting antenna (EME).

As regards the neutron emissions, in this chapter we present experimental tests performed on brittle rocks (magnetite and basalt), using a He^3 neutron device and a bubble type BD thermodynamic neutron detector. For brittle specimens of sufficiently large dimensions, neutron emissions, detected by He^3, were found to be up to three orders of magnitude higher than the natural background level at the time of the catastrophic failure. These emissions fully confirm the previous tests on granite [1–5] and are due to phono-fission reactions, which depend on the different modalities of energy emission during the crushing tests. For specimens with sufficiently large size and slenderness, a relatively high mechanical energy emission is expected and hence a higher probability of neutron emissions at the time of failure.

The experimental investigation carried out by the authors may open a new possible scenario, in which the stress state of the elements firstly involves the generation of microcracks, accompanied by mechanical energy release in the field of ultrasonic vibrations that can be measured using suitable AE equipments. Hence, the formation of coherent EME fields occurs over a wide range of frequencies, from few Hz to MHz, and even up to microwave frequencies. This excited state of the matter could be a cause of subsequent resonance phenomena of nuclei able to produce neutron bursts. This hypothesis was proposed by Widom et al. [22, 23]. The microcracking elastic energy release ultimately yields the acoustic vibrations, which are converted into electromagnetic oscillations. The electromagnetic waves, generated during microcracking, accelerate the condensed matter electrons, which then collide with protons producing neutrons and neutrinos [22, 23]. At the end of the book (Part VII), we will propose a simpler explanation based on TeraHertz phonons and plasmons emitted from fracture phenomena occurring at the nanoscale in condensed matter.

2.2 Experimental Set-Up

Experimental compression tests are performed on brittle rock specimens under monotonic displacement control. The materials used for the tests are non-radioactive magnetite and basalt. In this testing program, a total of 29 cylindrical specimens with different size and slenderness are utilized (Fig. 2.1). The compression tests are performed at the Fracture Mechanics Laboratory of Politecnico di Torino. In Table 2.1, the experimental data concerning the tested specimens are summarized.

All the specimens are subjected to uniaxial compression using a MTS servo-controlled hydraulic testing machine with a maximum capacity of 1,000 kN. Each test is performed in piston travel displacement control by setting a constant piston

Fig. 2.1 Magnetite (left) and basalt (right) cylindrical specimens, by varying size-scale and slenderness

Table 2.1 Tested specimens and their mechanical characteristics

Specimen type	Number of specimens	Dimensions		Piston velocity	Volume	Average peak load
		Diameter [mm]	Slenderness	[m/s]	[mm^3]	[kN]
Magnetite						
M-20-0.5	5	20	0.5	5×10^{-7}	3,140	67.46
M-20-1	2	20	1	5×10^{-7}	6,280	48.20
M-20-2	4	20	2	5×10^{-7}	12,560	45.88
M-40-0.5	2	40	0.5	1×10^{-6}	25,120	159.25
M-40-1	6	40	1	1×10^{-6}	50,240	146.87
M-40-2	4	40	2	1×10^{-6}	100,480	109.40
M-90-1	4	90	1	2×10^{-6}	572,265	849.89
Basalt						
B-50-2	2	50	2	1×10^{-6}	196,250	177.64
Specimen type	AE		EME		NE	
	Average frequency [Hz]	Average highest frequency [Hz]	Average frequency [Hz]	Average highest frequency [Hz]	Average neutron background [10^{-2} cps]	Average count rate at neutron emission [10^{-2} cps]
Magnetite						
M-20-0.5	20,975	56,990	19,622	> 46,000	4.84 ± 1.21	Background
M-20-1	25,889	49,488	–	–	5.90 ± 1.48	Background
M-20-2	23,569	42,984	32,312	> 55,000	5.70 ± 1.42	Background
M-40-0.5	53,354	133,133	28,278	> 49,000	5.50 ± 1.38	Background
M-40-1	60,055	90,997	37,787	> 77,000	5.06 ± 1.26	18.75 ± 4.69
M-40-2	33,520	73,385	35,274	> 98,000	5.60 ± 1.40	25.96 ± 6.49
M-90-1	52,350	132,062	110,089	> 1 MHz	4.95 ± 1.24	901.20 ± 225.30
Basalt						
B-50-2	76,668	165,363	204,417	> 335,000	5.60 ± 1.40	14.22 ± 3.55

velocity. The specimens are arranged in contact with the press platens without any coupling material, according to the testing modalities known as "test by means of rigid platens with friction".

The AE activity emerging from the compressed specimens is detected by attaching to the specimen surface a piezoelectric (PZT) transducer, resonant at about 150 kHz, which is able to convert the high-frequency surface movements due to the acoustic wave into an electric signal (the AE signal). The sensitivity of the transducer in the low-frequency range is measured by placing it on a shaker excited with all frequencies in the range 0–10 kHz (white noise). The result of this calibration at low frequencies is 1.2 μV/(mms^{-2}). Resonant sensors are more sensitive than broadband sensors, which are characterized by a flat frequency response in their working range, and then they can be successfully used in monitoring large-sized structures.

The EME detecting device, realized at the National Research Institute of Metrology (INRIM), is constituted by three pickup coils with a different number of turns, made of a 0.2 mm copper wire, that are positioned around the monitored specimen. This instrument, which acquires data in the frequency range from few Hz up to 4 MHz, exploits the induction Faraday's law: the induced voltage in a closed circuit (loop) is proportional to the change in the magnetic flux throughout the circuit. The first coil, constituted by 5 turns, works in a frequency range from 300 kHz to 4 MHz. The other two coils, constituted by 125 turns and 500 turns, work in the frequency range from few kHz to 20 MHz and from few Hz to 1 kHz, respectively.

Due to the difficulties in neutron measurements in the presence of electromagnetic disturbances, EME measurement carries out both the validation of NE signals and the monitoring of EME from fracturing; the simultaneous presence of sharply peaked EME signals (well characterized and far from the continuous magnetic noise) and NE signals can be regarded as the global effect of an ongoing damage process. As a further check on NE signals, a set of passive neutron detectors, based on the superheated bubble detection technique and insensitive to electromagnetic noise, are employed.

2.3 He^3 Neutron Proportional Counter

The He^3 detector, which is used in both the compression tests under monotonic displacement control and in ultrasonic vibration, is a He^3 type (Xeram, France) with electronics of preamplification, amplification, and discrimination directly connected to the detector tube. The detector is powered with 1.3 kV, supplied via a high-voltage Nuclear Instrument Module (NIM). The logic output producing the transistor-transistor logic (TTL) pulses is connected to a NIM counter. The device is calibrated for the measurement of thermal neutrons; its sensitivity is 65 cps/$n_{thermal}$ (a precision of 10% declared by the factory), i.e., a thermal neutron flux of 1 thermal neutron/s cm^2 corresponds to a count rate of 65 cps.

Considering that the fracture of dielectric materials, such as rocks, can lead to the emission of charged and neutral particles (electrons, photons, hard X-rays), in order to avoid possible false neutron measurements, the output of the detector is enabled for detecting signals only exceeding a fixed amplitude. This threshold value is determined by measuring the analog signal of the detector by means of a Co-60 gamma source (half-life: 5.271 years; type decay: beta$^-$; beta maximum energy: 317.8 keV; gammas: 1,173.2 keV and 1,332.5 keV). The presence of an interfering capacity on the charge preamplifier input increases the electronic noise and consequently the probability of spurious counts. For this reason, the coaxial cable used for connecting detector and charge preamplifier presents a low capacity (36 pF/m) and a short length (about 50 cm). Moreover, during the experimental measurements, the front-end electronics is screened with aluminum foils, and the He^3 tube is immersed in a sound-absorbing substance such as polystyrene in order to avoid possible accidental impacts and vibrations.

2.4 Neutron Bubble Detectors

A set of passive neutron detectors insensitive to electromagnetic noise and with zero gamma sensitivity is used in compression tests under cyclic loading. The dosimeters, based on superheated bubble detectors (BTI, Ontario, Canada, Bubble Technology Industries (1992)) [24], are calibrated at the factory against an Am-Be source in terms of NCRP38 (NCRP report 38 (1971)) [25]. Bubble detectors provide instant visible detection and measurement of neutron dose. Each detector is composed of a polycarbonate vial filled with elastic tissue-equivalent polymer, in which droplets of a superheated gas (Freon) are dispersed. When a neutron strikes a droplet, the latter immediately vaporizes, forming a visible gas bubble trapped in the gel. The number of droplets provides a direct measurement of the equivalent neutron dose. These detectors are suitable for neutron integral dose measurements, in the energy ranges of thermal neutrons ($E = 0.025$ eV) and fast neutrons ($E > 100$ keV).

All the signals (AE, EME, and NE) are acquired by a National Instruments Digitizer with eight channels simultaneously sampling at 1 MSa/s. The trigger is set to the AE channel with a detection threshold fixed at 20 mV to filter out the background noise.

For all the specimens, the recorded AE, EME, and NE time series are related to the time history of the applied load.

2.5 Test Results

2.5.1 AE and EME Measurements

In the present investigation, among all the 29 tested samples, the experimental results of four specimens (three of magnetite and one of basalt) are described and examined in detail. All specimens are tested in compression up to failure, showing a brittle response with a rapid decrement in load-carrying capacity when deformed beyond the peak load (Figs. 2.2 and 2.3). Experimental evidence indicates the presence of AE, EME, and NE activities. In Figs. 2.2 and 2.3, the AE and EME bursts, i.e., the signals received by the devices, are reported as accumulated number and as rate in time, whereas the NE bursts are reported as counts per second (cps). It is interesting to note that in this experimental campaign on basalt—unlike those carried out by the authors on other materials, such as concrete, Luserna stone, Carrara marble, and Syracuse limestone [14, 15]—the EME activity is much more widespread during the fatigue loading process and not just concentrated at the moment of the final collapse.

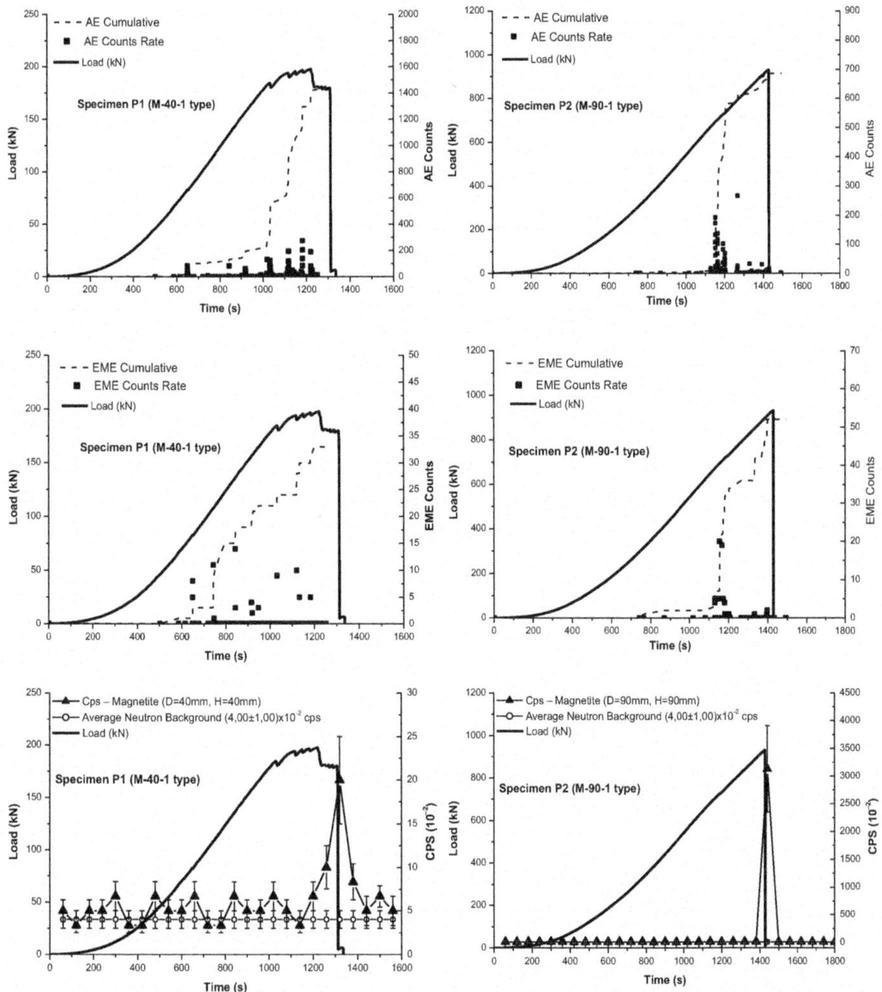

Fig. 2.2 Load versus time diagram of the magnetite specimen P1 (M-40-1 type) (left): accumulated number and rate of AE (upper left); accumulated number and rate of EME (middle left); NE count rate (lower left). Load vs. time diagram of the magnetite specimen P2 (M-90-1 type) (right): accumulated number and rate of AE (upper right); accumulated number and rate of EME (middle right); NE count rate (lower right)

As a matter of fact, in the magnetite specimen P1 (M-40-1 type), whose behavior is described by the load vs. time diagram in Fig. 2.2 (left), the observed bursts of AE and EME activity can be clearly correlated to the stress drops occurring before the final collapse, Fig. 2.2 (upper and middle left). In the magnetite specimen P2 (M-90-1 type), characterized by a perfectly brittle behavior without evident stress drops before the final collapse (note the linearity till failure of the load vs. time diagram in Fig. 2.2 (right)), the specimen failure is preceded by two closely correlated bursts

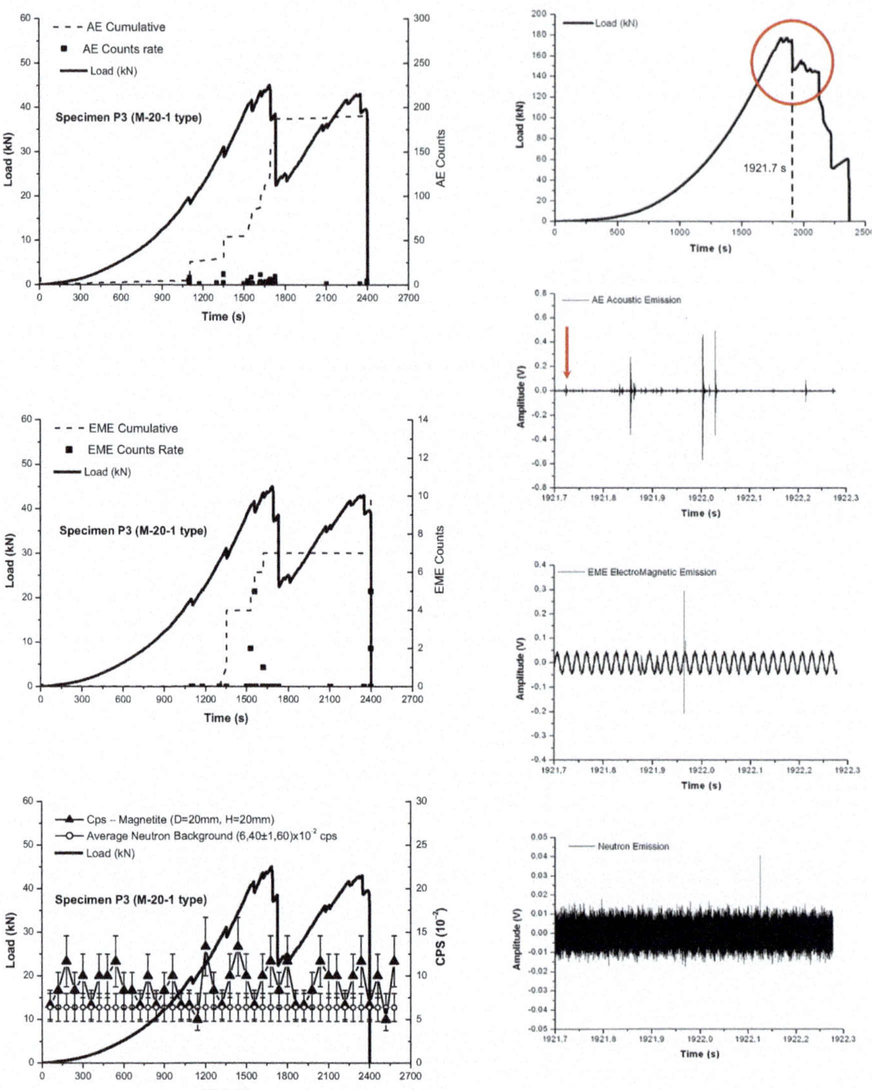

Fig. 2.3 Load versus time diagram of the magnetite specimen P3 (M-20-1 type) (left): accumulated number and rate of AE (upper left); accumulated number and rate of EME (middle left); NE count rate (lower left). Load vs. time diagram of the basalt specimen (upper right); AE, EME (middle right), and NE (lower right) detected in a time window of 0.6 s, starting at 1,921.7 s from the beginning of the test

of AE and EME activity at nearly 80% of the peak load, Fig. 2.2 (upper and middle right). This activity, particularly as regards electromagnetic emission, can be due to the behavior under loading of magnetite that—being rich in iron, about 65% in weight—determines the formation of magnetic charges generated by friction during the loading process, and their spontaneous release independently of the formation of macro-cracks at the time of final collapse.

As a particular case, the load vs. time diagram of the magnetite specimen P3 (M-20-1 type) is double-peaked with a significant stress drop at about 60% of the test duration, followed by a drop in the AE rate, Fig. 2.3 (left). This temporary relaxation in the AE activity describes the well-known Kaiser effect [26], which states that, after stress drops, AE activity is very low during the reloading of the material until the stress exceeds the previously reached values, Fig. 2.3 (upper left). This relaxation is observed also in the EME activity, Fig. 2.3 (middle left), confirming the high correlation between these two phenomena.

2.5.2 *NE Measurements*

As regards the NE measurements, the He^3 neutron detector is switched on at least one hour before the beginning of each compression test, in order to reach the thermal equilibrium of electronics, and to make sure that the behavior of the device is stable with respect to intrinsic thermal effects. For the considered specimens P1, P2, and P3, the average measured natural background level ranges from $(4.00 \pm 1.00) \times 10^{-2}$ to $(6.40 \pm 1.60) \times 10^{-2}$ cps. In general, neutron measurements of specimens M-20-1 yielded values comparable to the natural background, whereas in specimens M-40-1, the experimental data exceeded the background value by approximately four times. For specimens M-90-1, the neutron emissions achieved values up to three orders of magnitude higher than the natural background.

Moreover, the volumes of all the tested specimens are shown in Table 2.1. As reported by the authors in a previous work [27], a volume approximately exceeding 200,000 mm^3, combined with the extreme brittleness of the tested material, represents a threshold value for a neutron emission of about one order of magnitude higher than the natural background.

In this case, it is interesting to highlight how neutron measurements of specimens M-20-1 (6,280 mm^3) yield values comparable to the natural background, whereas in specimens M-40-1 (50,240 mm^3), the experimental data exceed the natural background by about four times. For specimens M-90-1 (572,265 mm^3), the neutron emissions achieve values up to three orders of magnitude higher than the natural background.

As regards the expected energy spectrum, it extends from thermal neutrons (0.025 eV) up to the fast component (few MeV). Also this behavior has already been measured by the authors [1–5] by using specific devices such as proportional counters (He^3 devices) and passive bubble dosimeters.

In Figs. 2.2 and 2.3 (left), the load vs. time diagram and the neutron count rate evolution for each specimen are shown. Moreover, bursts of NE activity are observed at the failure time of specimens P1 (M-40-1 type) and P2 (M-90-1 type), Fig. 2.2 (lower), confirming the need of catastrophic ruptures, i.e., characterized by sudden release of the stored strain energy, to obtain such anomalous neutron emissions.

Furthermore, during the compression tests, a rise in the thermal equivalent neutron dose, analyzed by neutron bubble detectors, is measured, consistently with the increment in the neutron level measured by the He^3 device. In particular, for the specimen P2 (M-90-1 type), a value more than 1,000 times higher than the natural background is found at the end of the test.

2.5.3 AE, EME, and NE Time Correlation

Test results on the basalt specimen P4 (B-50-2 type) present a high degree of correlation among the three emission time series and the load time history. As an example, considering a time window of 0.6 s starting at 1,921.7 s from the beginning of the test, AE bursts followed by an EME pulse in the kHz frequency range are shown in Fig. 2.3 (right). Similar simultaneous EME pulses are observed in the Hz and MHz ranges that, for reasons of space, are not included in the same figure. The time window is related to the evident stress drop indicated by a circle in the load vs. time diagram, Fig. 2.3 (right).

As already discussed in [14], AE and EME signals from a growing fracture present a time delay consistent with their propagation velocities. Being d the distance between source (fracture) and AE transducer, v_{AE} and v_{EME} the average propagation velocities of AE and EME waves with $v_{AE} < < v_{EME}$, the time delay can be estimated by $\Delta t = d/ v_{AE}$. Inserting $d = 10^{-1}$ m and $v_{AE} = 10^3$ m s^{-1}, an estimation of the time delay for the considered event is $\Delta t = 10^{-4}$ s = 100 μs. If then we consider that the main crack propagation, in the specified time window, takes place (begins its first motion, indicated with an arrow in Fig. 2.3 (right)) when the first AE peak of great amplitude is recorded, the main EME pulse follows the AE burst, in spite of the different average propagation velocities of AE and EME signals. Therefore, the EME signal seems to spread anisotropically during the mechanical vibration generated by fracture. Eventually, the NE event is observed at the time of the catastrophic failure of specimen, Fig. 2.3 (upper and lower right).

Considering the behavior of brittle specimens under mechanical loading, further interesting discussions on the variation in the AE vibration frequencies and in the signal peak distributions were reported [14, 15, 28, 29].

2.6 Conclusions

The experimental evidence presented in this chapter confirms the previous investigations on AE and EME signals as collapse precursors in natural materials like rocks. Bursts of AE and EME activity are always observed when significant stress drops occur. This suggests the use of electromagnetic measurements to enhance monitoring systems based on the AE technique. In addition, NE activity is observed when the specimen fails in a sudden, catastrophic way. In particular, for magnetite specimens of sufficiently large size, the neutron flux is found to be up to three orders of magnitude higher than the natural background level at the time of catastrophic failure. Therefore, the observed acoustic, electromagnetic, and neutron activity from laboratory experiments looks promising for effective applications also at the geophysical scale.

Based on the analogy between AE and seismic activity [30–32], on the anomalous radiation of geoelectromagnetic waves observed before major earthquakes [33], and on recent experimental studies that measured neutron components exceeding the usual background in correspondence to seismic activity [34], it could be possible to set up a sort of alarm system based on a regional warning network.

The results obtained from this analysis show how the crack generation is accompanied by mechanical energy emission in the field of ultrasonic vibrations detected by AE sensors. It is also observed that, for constant specimen diameters, the AE signals reach high-frequency peaks for low slenderness values, whereas the EME frequencies increase with the sample size. The highest neutron emissions occur from specimens with EME detected in the field of MHz. This shows that the formation of coherent EME fields (i.e., characterized by evident pulses generated by specific phase relationship between the electric field values at different times) occurs over a wide range of frequencies, from few Hz to MHz and even up to microwaves, during the fracture propagation. This excited state of matter could be the cause of subsequent resonance phenomena of nuclei able to produce neutron bursts in the presence of stress drops or sudden catastrophic fractures. On the other hand, we will see, later in the volume, that the fracto-emission dynamics during the different phases of an earthquake is totally different from that of a crushing test. Neutron emission tends to anticipate AE and EME, when the developing cracks are still in their original stage at the microscale. On the contrary, acoustic emission tends to follow NE and EME, when the developing cracks are already in their final stage at the macro-scale.

References

1. Carpinteri A, Cardone F, Lacidogna G (2009) Piezonuclear neutrons from brittle fracture: early results of mechanical compression tests. Strain 45:332–339
2. Cardone F, Carpinteri A, Lacidogna G (2009) Piezonuclear neutrons from fracturing of inert solids. Phys Lett A 373:4158–4163

3. Carpinteri A, Cardone F, Lacidogna G (2010) Energy emissions from failure phenomena: mechanical, electromagnetic, nuclear. Exp Mech 50:1235–1243
4. Carpinteri A, Borla O, Lacidogna G, Manuello A (2010) Neutron emissions in brittle rocks during compression tests: monotonic versus cyclic loading. Phys Mesomech 13:268–274
5. Carpinteri A, Lacidogna G, Manuello A, Borla O (2011) Energy emissions from brittle fracture: neutron measurements and geological evidences of piezonuclear reactions. Strength, Fract Complex 7:13–31
6. Mogi K (1962) Study of elastic shocks caused by the fracture of heterogeneous materials and its relation to earthquake phenomena. Bull Earthq Res Inst 40:125–173
7. Lockner DA, Byerlee JD, Kuksenko V, Ponomarev A, Sidorin A (1991) Quasi static fault growth and shear fracture energy in granite. Nature 350:39–42
8. Ohtsu M (1996) The history and development of acoustic emission in concrete engineering. Mag Concr Res 48:321–330
9. Rundle JB, Turcotte DL, Shcherbakov R, Klein W, Sammis C (2003) Statistical physics approach to understanding the multiscale dynamics of earthquake fault systems. Rev Geophys 41:1019–1049
10. Niccolini G, Schiavi A, Tarizzo P, Carpinteri A, Lacidogna G, Manuello A (2010) Scaling in temporal occurrence of quasi-rigid-body vibration pulses due to macrofractures. Phys Rev E 82:46115/1–46115/5
11. Carpinteri A, Lacidogna G (2006) Damage monitoring of an historical masonry building by the acoustic emission technique. Mater Struct (RILEM) 39:161–167
12. Carpinteri A, Lacidogna G (2006) Structural monitoring and integrity assessment of medieval towers. J Struct Eng (ASCE) 132:1681–1690
13. Carpinteri A, Lacidogna G (2007) Damage evaluation of three masonry towers by acoustic emission. Eng Struct 29:1569–1579
14. Lacidogna G, Carpinteri A, Manuello A, Durin G, Schiavi A, Niccolini G, Agosto A (2010) Acoustic and electromagnetic emissions as precursor phenomena in failure processes. Strain 47(2):144–152
15. Carpinteri A, Lacidogna G, Manuello A, Niccolini A, Schiavi A, Agosto A (2010) Mechanical and electromagnetic emissions related to stress-induced cracks. Exp Tech 36(3):53–64
16. Misra A (1977) Theoretical study of the fracture-induced magnetic effect in ferromagnetic materials. Phys Lett A 62:234–236
17. Frid V, Rabinovitch A, Bahat D (2003) Fracture induced electromagnetic radiation. J Phys D 36:1620–1628
18. Hadjicontis V, Mavromatou C, Nonos D (2004) Stress induced polarization currents and electromagnetic emission from rocks and ionic crystals, accompanying their deformation. Nat Hazards Earth Syst Sci 4:633–639
19. Warwick JW, Stoker C, Meyer TR (1982) Radio emission associated with rock fracture: possible application to the great Chilean earthquake of May 22, 1960. J Geophys Res 87:2851–2859
20. Nagao T, Enomoto Y, Fujinawa Y et al (2002) Electromagnetic anomalies associated with 1995 Kobe earthquake. J Geodyn 33:401–411
21. Rabinovitch A, Frid V, Bahat D (2007) Surface oscillations. A possible source of fracture induced electromagnetic oscillations. Tectonophysics 431:15–21
22. Widom A, Swain J, Srivastava YN (2013) Neutron production from the fracture of piezoelectric rocks. J Phys G: Nucl Part Phys 40:15006(8pp)
23. Widom A, Swain J, Srivastava YN (2014) Photo-disintegration of the Iron nucleus in fractured magnetite rocks with magnetostriction. Meccanica 50:1205–1216
24. Bubble Technology Industries (1992) Instruction manual for the bubble detector, Chalk River, Ontario, Canada
25. National Council on Radiation Protection and Measurements (1971) Protection against neutron radiation, NCRP Report 38
26. Kaiser J (1950) Ph. D. dissertation, Munich (FRG), Technische Hochschule München, Germany
27. Carpinteri A, Lacidogna G, Manuello A, Borla O (2013) Piezonuclear fission reactions from earthquakes and brittle rocks failure: evidence of neutron emission and non-radioactive product elements. Exp Mech 53:345–365

28. Aggelis DG, Soulioti DV, Sapouridis N, Barkoula NM, Paipetis AS, Matikas TE (2011) Acoustic emission characterization of the fracture process in fibre reinforced concrete. Constr Build Mater 25:4126–4131
29. Aggelis DG, Mpalaskas AC, Matikas TE (2013) Acoustic signature of different fracture modes in marble and cementitious materials under flexural load. Mech Res Commun 47:39–43
30. Scholz CH (1968) The frequency-magnitude relation of microfracturing in rock and its relation to earthquakes. Bull Seismol Soc Am 58:399–415
31. Carpinteri A, Lacidogna G, Pugno N (2006) Richter's laws at the laboratory scale interpreted by acoustic emission. Mag Concr Res 58:619–625
32. Niccolini G, Carpinteri A, Lacidogna G, Manuello A (2011) Acoustic emission monitoring of the Syracuse Athena temple: scale invariance in the timing of ruptures. Phys Rev Lett 14:108503
33. Carpinteri A, Lacidogna G, Borla O, Manuello A, Niccolini G (2012) Electromagnetic and neutron emissions from brittle rocks failure: experimental evidence and geological implications. Sadhana 37:59–78
34. Volodichev NN, Kuzhevskij BM, Nechaev OY, Panasyuk MI, Podorolsky AN, Shavrin PI (2000) Sun-moon-earth connections: the neutron intensity splashes and seismic activity. Astron Vestnik 34:188–190

Part II
Subatomic Particle Emissions and Chemical Composition Changes in Compression Failure of Natural and Artificial Rocks

Chapter 3
Iron-Rich Natural Rocks: Crushing Tests, Neutron Emissions, and Balanced Chemical Composition Changes

Abstract Neutron emissions (NEs) are measured during laboratory experiments conducted on iron-bearing and iron-rich natural rocks. In particular, magnetite specimens are loaded up to the final failure under monotonic displacement control. Also basalt rocks are tested under cyclic loading conditions (2 Hz) up to the final failure. In order to detect neutron emissions, the tests are monitored by two different neutron measurement devices: He^3 proportional counter and thermodynamic (bubble) detectors. After the experiments, Energy Dispersive X-ray Spectroscopy (EDS) analyses are carried out to detect possible indirect evidence of phono-fission reactions on the fracture surfaces. In particular, quantitative evidence of nuclear reactions, involving an iron decrement and the corresponding increment in lighter elements, is observed in olivine, a crystalline mineral phase widely diffused in the basalt matrix, and in magnetite. These results confirm the data observed for Luserna stone (granitic orthogneiss) and that phono-fission reactions take place in iron-bearing natural rocks subjected to damage accumulation and cracking. This could constitute the major explanation to the transformation of our planet from basaltic (15% in Fe) to sialic or granitic (4% in Fe).

Keywords Basalt · Magnetite · Crushing failure · Neutron emissions · Energy dispersive X-ray spectroscopy (EDS) · Iron decrement · Geochemical evolution · Basaltic-to-sialic transformation

3.1 Preliminary Remarks

It is possible to demonstrate experimentally that brittle fracture in solid materials can be accompanied by the release of different forms of energy [1–3]. In recent studies, it has been observed that quasi-brittle materials such as granitic orthogneiss (Luserna stone) subjected to compression tests under monotonic displacement control, by cyclic loading, or by ultrasonic vibration, are characterized by neutron emissions up to one order of magnitude greater than the natural background level [1–11]. These tests were conducted on Luserna stone specimens with different shapes and

A. Carpinteri, *Terahertz Phonons and Nanomechanical Instabilities*,
https://doi.org/10.1007/978-3-032-14692-2_3

dimensions and characterized by an iron oxide content of approximately 1.5%. In the case of this rock, the iron oxides are prevalently concentrated within two minerals: phengite and biotite. These two minerals, rather common in such a stone (up to 20% and 2%in volume, respectively), have shown important changes in the mineral chemistry of the fracture surfaces after the experiments [8–11]. The reduction in Fe content (~25%) is almost perfectly compensated by increments in Al, Si, and Mg [8–11].

In the present investigation, neutron emission measurements, by means of a He^3 proportional counter and thermodynamic bubble detectors, are performed during compression tests on magnetite specimens and during cyclic loading tests carried out on basalt rocks. The employed materials are chosen in order to correlate the iron contents (~15% for basalt and ~72.5% for magnetite) to the neutron emission levels measured during crushing and fatigue tests.

The crushing tests on magnetite are performed using cylindrical specimens of different diameters coming from the San Leone mine. This deposit is located at about 30 km Southwest of Cagliari, Sardinia (Italy), near the Basso-Sulcis batholit, a granodioritic intrusive of the Hercynian age [12]. This mine is rather recent—it was discovered in 1860 during the industrial development [12]. In 1892, the mine started its activity, which ended in 1963. The mineralization of the deposit is mainly represented by magnetite. The ore bodies within skarn were generated by the thermometamorphism of previous Paleozoic limestones in contact with granitic intrusions of Variscan age (300×10^6 years ago). The mean iron concentrations in the San Leone magnetite are between 72.5% and 75% [12].

Another iron-rich material used for the cyclic loading experiments is the basalt coming from Mount Etna. Basalt, a very common extrusive igneous rock, is the dominant material making up the oceanic Earth's Crust and represents the principal product of volcanic eruptions. It is characterized by a mafic chemistry (with a high content of iron and magnesium), is dark gray, and shows porphyritic texture with few mm-sized phenochrists of plagioclase, clinopyroxene, and olivine [13].

After the experiments, the basalt and magnetite fracture surfaces are analyzed by Energy Dispersive X-ray Spectroscopy (EDS) in order to obtain a direct evidence of the phono-fission reactions that take place during the tests. The results confirm the data observed in the case of Luserna stone and permit to recognize different phono-fission reactions induced in natural non-radioactive rocks characterized by a certain iron concentration.

As far as magnetite is concerned, the EDS analysis shows an impressive direct evidence of phono-fission reactions. After the magnetite experiments, macroscopical changes are observed, comparing external and fracture surfaces. Among these changes, the appearance of appreciable quantities of Al, previously absent, is particularly significant. The chemical changes in basalt and magnetite give a valid interpretation to the different chemical compositions between the oceanic and the continental crust. In other terms, we can explain the transition from basaltic to sialic chemical composition that characterized the geochemical evolution of the Earth's Crust [14, 15].

3.2 Granitic Orthogneiss (Luserna Stone): Early Experiments

Tests on prismatic specimens were presented in early contributions [1–3] and are related to phono-fission reactions occurring in solids containing iron (samples of Luserna stone in compression). In such preliminary experiment, four specimens were tested, two made of Carrara marble and two made of Luserna stone (Fig. 3.1). All of them were of the same shape and size, measuring $6 \times 6 \times 10$ cm^3. The neutron measurements obtained on the two Luserna stone specimens exceeded the background level by approximately one order of magnitude, when a catastrophic failure occurred. The first specimen reached at time T = 32 min a peak load of ca 400 kN, corresponding to an average pressure on the bases of 111.1 MPa. When failure occurred, the count rate was found to be $(28.3 \pm 0.2)\ 10^{-2}$ cps, corresponding to an equivalent flux of thermal neutrons of $(43.6 \pm 0.3)\ 10^{-4}\ n_{thermal}\ cm^{-2}\ s^{-1}$ (Fig. 3.2). The second specimen reached at time T = 29 min a peak load of ca 340 kN, corresponding to an average pressure on the bases of 94.4 MPa. When failure occurred, the count rate was found to be $(27.2 \pm 0.2)\ 10^{-2}$ cps, corresponding to an equivalent flux of thermal neutrons of $(41.9 \pm 0.3)\ 10^{-4}\ n_{thermal}\ cm^{-2}\ s^{-1}$.

In more recent experiments, cylindrical specimens with different size and slenderness were selected (Fig. 3.3), instead of prismatic specimens as in the preliminary tests. For the specimens of larger dimension and slenderness, neutron emissions, detected by He^3, were found to be of about one order of magnitude higher than the natural background level at the time of the catastrophic failure. These emissions fully confirmed the preliminary tests [1–10]. For specimens with sufficiently large size and/or slenderness, a relatively high mechanical energy emission is expected and hence a higher probability of neutron emissions at the time of failure. Furthermore, during compression tests with cyclic loading, an equivalent neutron dose was found at the end of the test by neutron bubble detectors, about twice higher than the natural

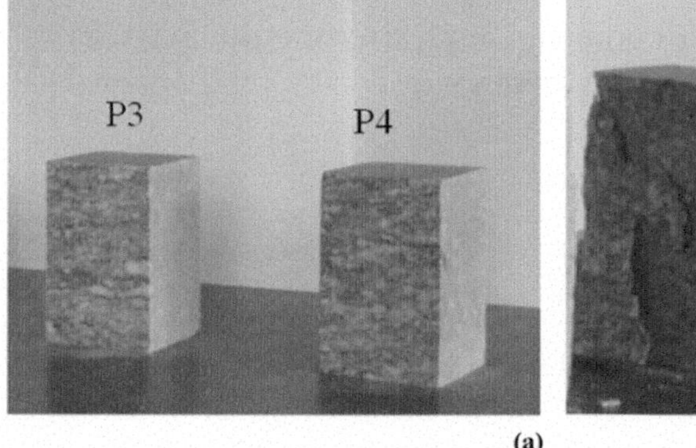

(a) (b)

Fig. 3.1 Granitic orthogneiss (Luserna stone) specimens before (**a**) and after (**b**) the crushing test

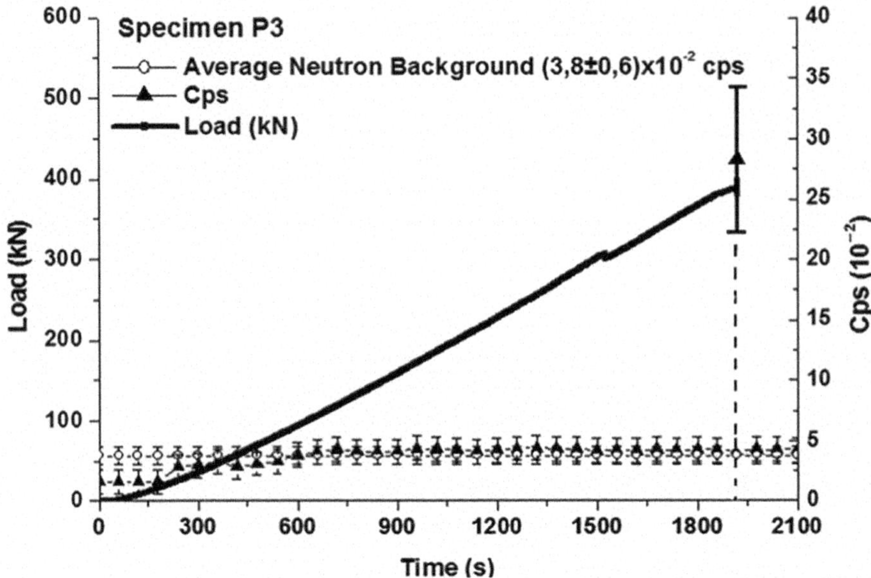

Fig. 3.2 Load versus time diagram and cps values for test specimen P3 of granitic orthogneiss (Luserna stone)

background level [4, 8, 9]. In addition, using an ultrasonic horn suitably joined to the specimen, ultrasonic tests were carried out on Luserna stone producing a continuous vibration at 20 kHz. At the end of the test, an equivalent neutron dose about twice higher than the background level was detected [4, 5]. In the present chapter, in order to evaluate the correlation between iron content and neutron emission, specific tests are described on basalt and magnetite.

3.3 Basalt and Magnetite Specimens: Experimental Set-Up and Neutron Emission Evidence

Experimental compression tests are performed on brittle magnetite specimens under monotonic displacement control and on basalt specimens under cyclic loading. All the experimental tests are realized at the Fracture Mechanics Laboratory of Politecnico di Torino. In particular, the experimental results of the four specimens P1, P2, P3 (magnetite) and P4 (basalt) are reported. The tested materials, the shapes and sizes of specimens, and the characteristics of the testing procedure are summarized in Table 3.1.

The crushing experiments on magnetite are performed in uniaxial compression, utilizing a MTS servo-controlled hydraulic testing machine with a maximum capacity of 1,000 kN. Each test is performed in piston travel displacement control by setting a

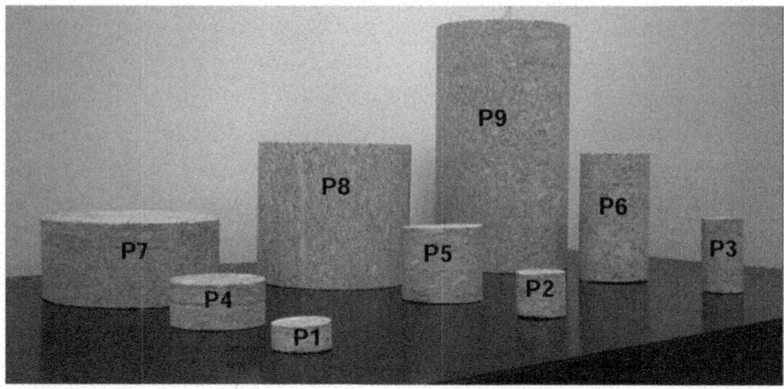

Fig. 3.3 Granite (Luserna stone) cylindrical specimens of different size-scale (Diameter = 28, 53, 112 mm) and slenderness (0.5, 1.0, 2.0), to emphasize the related brittlemess effects. Specimens P1, P2, and P4 show neutron emissions close to the natural background level, whereas Specimens P9, P8, and P6 show neutron emissions of an order of magnitude higher than that of the natural background level

Table 3.1 Tested materials and test modalities

Monotonic displacement control							
Specimen	Material	D (mm)	H (mm)	H/D	Velocity (mm/s)	Peak load (kN)	
P1	Magnetite	40	40	1	0.001	197.97	
P2	Magnetite	90	90	1	0.002	932.38	
P3	Magnetite	20	20	1	0.0005	45.05	
Cyclic loading							
Specimen	Material	D (mm)	H (mm)	H/D	Frequency (Hz)	Maximum load (kN)	Minimum load (kN)
P4	Basalt	80	160	2	2	350	30

constant piston velocity between 5×10^{-4} and 2×10^{-3} mm/s. The test specimens are arranged in contact with the press platens without any coupling material, according to the testing modalities known as “test by means of rigid platens with friction”. The fatigue test on the basalt specimen is performed up to the final failure at a frequency of 2 Hz, with a maximum load of 350 kN and a minimum load of 30 kN (Table 3.1).

The neutron emission measurements on magnetite specimens are performed using an He^3 neutron detector switched on at least one hour before the beginning of each compression test, in order to reach the thermal equilibrium of electronics, and to make sure that the behavior of the device is stable with respect to intrinsic thermal effects. The average measured background level is ranging from $(4.00 \pm 1.00) \times 10^{-2}$ to $(6.40 \pm 1.60) \times 10^{-2}$ cps. Neutron measurements of specimen P3 yield values comparable to the natural background, whereas in specimen P1 the experimental data exceed the background level by approximately five times. For specimen P2, the neutron emission achieves values of about three orders of magnitude higher than the background level

(Fig. 3.4a). Specimen P4, made of basalt, is subjected to fatigue cycles up to the final failure as summarized in Table 3.1. Droplet counting is performed every 12 h and the equivalent neutron dose is calculated [8, 9, 16]. In the same way, the natural background is estimated by means of the two bubble dosimeters. During this test, the natural background is found to be (53.76 ± 13.44) nSv/h. An increment of about twenty times with respect to the background level is detected at specimen failure (Fig. 3.4b). No significant variations in neutron emissions are observed before the failure. The equivalent neutron dose, at the end of the test, is (935.49 ± 233.87) nSv/h.

3.4 Chemical Composition Changes: EDS Analysis on Basalt

After the mechanical loading experiments, Energy Dispersive X-ray Spectroscopy (EDS) is performed on different samples of external and fracture surfaces, belonging to the same specimen used during the cyclic loading test on basalt. The analysis is conducted in order to correlate the neutron emission from the specimen with the variations in rock composition and to detect possible anomalous transformations from iron to lighter elements. The quantitative elemental analyses are performed by a ZEISS Supra 40 Field Emission Scanning Electron Microscope (FESEM) equipped with an Oxford X-rays microanalysis [8–11].

The first analysis is performed on fracture surfaces of basalt specimen P4 after the fatigue test and the consequent failure. In this case, similarly to the case of Luserna stone [11], taking into account the heterogeneity of the material, the samples are carefully chosen to investigate and compare the same minerals before and after the failure. In particular, olivine is considered due to its high iron content (~24%) and because it is rather abundant within this type of rock [11]. In the case of basalt, two different kinds of samples are examined: (i) polished thin sections, finished with a standard petrographic procedure and covered by Cr, for what concerns the external surface; (ii) small portions from the fracture surface without any kind of preparation, apart from the Cr covering. A semi-quantitative nonstandard analysis is performed on the collected spectra, fixing the stoichiometry of the oxides, to correlate the oxides content to the specific crystalline phase. The Cr lines are excluded from the semi-quantitative evaluation [11].

In Fig. 3.5a–c, a polished thin section from the external surface of an integer and un-cracked portion of the basalt specimen P4 is shown together with two fragments from the fracture surface. The polished thin section presents a rectangular geometry (45 × 27 mm) and is 30 μm thick. This kind of analysis involves millions of cubic microns for each acquisition area. From this point of view, there exists a substantial difference between this analysis and the spot analysis reported for the Luserna stone samples [11]. In this case, in fact, a larger portion of material is involved and each analysis is indicative of an investigated volume of $60 \times 20 \times 4\ \mu m^3$.

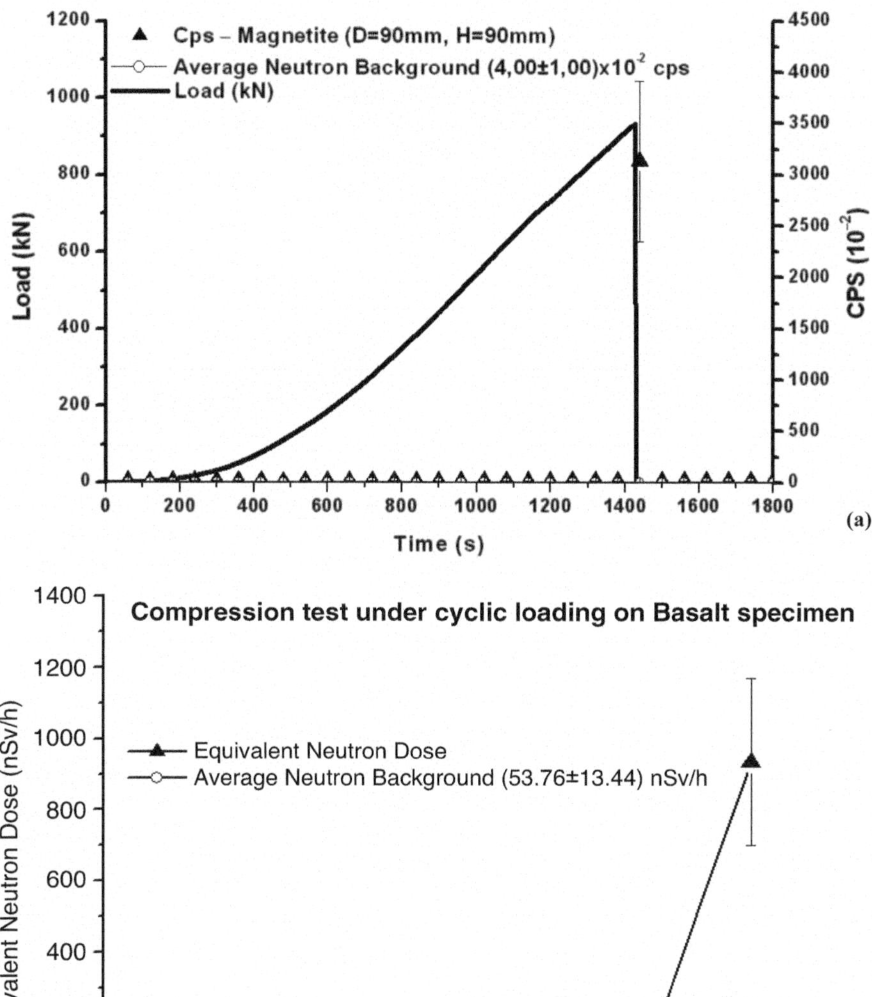

Fig. 3.4 **a** Magnetite, for specimen P2 the neutron emission achieves values of three orders of magnitude higher than the background level. **b** Basalt, a neutron dose increment of about twenty times the background level is detected at the final failure of specimen P4 (fatigue test)

Fig. 3.5 **a** Polished thin sections, finished with a standard petrographic procedure, covered by Cr, are examined to evaluate olivine composition on the external surface. **b**, **c** Small fragments from the fracture surface without any kind of preparation, apart from the Cr covering, are analyzed to evaluate the chemical changes in olivine

In Fig. 3.6a–c, the distributions of Fe, Si, and Mg concentrations in olivine are reported for external and fracture surfaces of specimen P4. It can be observed that the distribution of Fe content for the external surface shows an average value of 18.4% (Fig. 3.6a). In the same graph, the distribution of Fe concentration on the fracture sample shows a significant variation. It can be seen that the mean value of the distribution of measurements performed on the fracture surface is equal to 14.4%, considerably lower than the mean value of external surface measurements (18.4%). Similarly to Fig. 3.6a, in Fig. 3.6b, the Si mass percentage concentrations are considered. For Si contents, the observed variations show a mass percentage increment approximately equal to 2.2%. The average value of Si concentration changes from 18.3% on the external surface to 20.5% on the fracture surface. In Fig. 3.6c, it is shown that, in the case of olivine, also Mg content presents considerable variations. Figure 3.6c shows that the mass percentage concentration of Mg changes from a mean value of 21.2% (external surface) to a mean value of 22.8% (fracture surface), with an increment of 1.6%. Therefore, the iron decrement (−4.0%) in olivine is almost perfectly balanced by increments in silicon (+2.2%) and magnesium (+1.6%). A further analysis conducted on olivine and localized on the fracture surface shows the appearance of Al_2O_3 in the chemical composition of this crystalline phase (Table 3.2). This evidence is particularly important because no Al traces are observed in the olivine sample localized on the external surface. This fact represents a further confirmation that in basaltic olivine, similarly to the case of phengite in Luserna stone [11], the following phono-fission reaction occurs.

$$Fe^{56}_{26} \rightarrow 2Al^{27}_{13} + 2\text{neutrons} \tag{3.1}$$

At the same time, the results involving the Fe decrement and the consequent increments in Si and Mg, discussed above and reported in Fig. 3.6, lead to the conclusion that in olivine, as well as in the biotite of Luserna stone [11], the following phono-fission reaction occurs during the mechanical loading:

$$\mathrm{Fe}_{26}^{56} \rightarrow \mathrm{Mg}_{12}^{24} + \mathrm{Si}_{14}^{28} + 4\mathrm{neutrons} \tag{3.2}$$

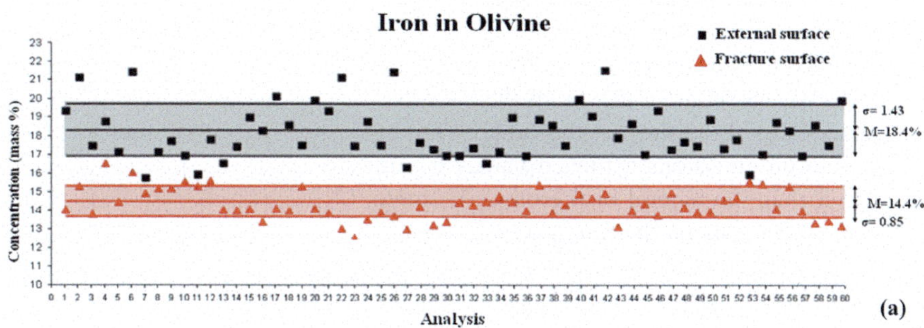

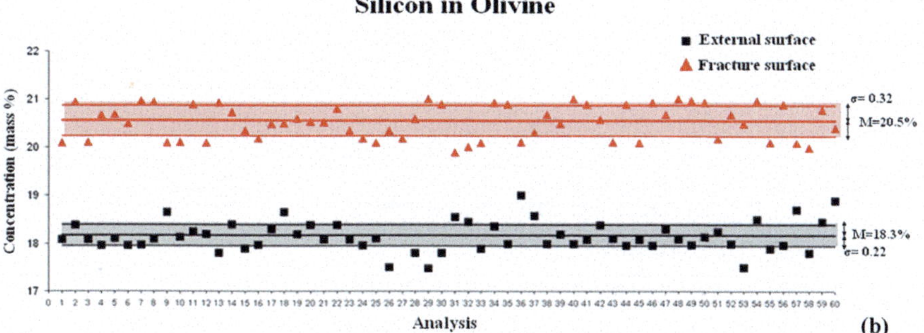

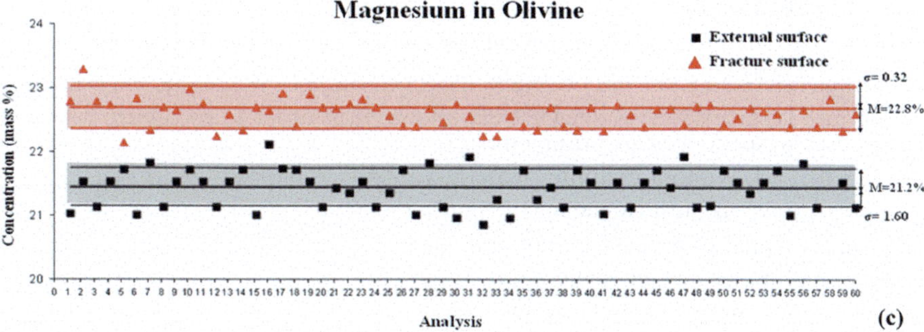

Fig. 3.6 Olivine chemical changes after crushing: The Fe decrement **a** (−4.0%) is almost perfectly balanced by the increments in Si **b** (+2.2%) and Mg **c** (+1.6%)

Table 3.2 Olivine: Fe, Si, Mg, and Ca weight percentage mean values on external and fracture surfaces. Variations with respect to the mineral and to the element itself

	External surface mean value (wt%)	Fracture surface mean value (wt%)	Increment/Decrement with respect to the mineral (%)	Increment/Decrement with respect to the element itself (%)
Fe	18.4	14.4	− 4.0	− 21
Si	18.3	20.5	+ 2.2	+ 12
Mg	21.2	22.8	+ 1.6	+ 7
Ca	0.5	0.5	No changes	No changes

The results reported in Fig. 3.6 represent also an important significance from a geophysical point of view. In fact, at the scale of the Earth's Crust, the non-homogeneous composition of oceanic and continental crusts can be explained by the transition from basaltic to sialic compositions. Comparing the data presented in the literature concerning the composition of the two different types of terrestrial crust, it can be noted that the iron concentration changes from ~8 %, in the oceanic crust, to ~4 % in the continental one [17–22]. Ni changes from ~0.03 %, in the oceanic crust, to ~0.01 % in the continental one (about a threefold decrement). *Vice versa*, Al, Si, Mg, and Na vary from ~7%, ~24%, ~3.6%, and ~1%, in the oceanic crust, to ~8 %, ~28 %, ~1.3%, and ~2.9% in the continental crust, respectively. Considering that approximately 50% of the continental crust has originated over the last 3.8 Gyrs, as the result of oceanic crust subduction [3–5, 14, 15, 18–22], the results presented herein offer a further confirmation that phono-fission reactions are a possible explanation for the chemical changes in the crust in correspondence of mechanical phenomena of fracture, crushing, fragmentation, comminution, erosion, friction, due to seismic and tectonic events [1–5, 9, 14, 15].

3.5 Chemical Composition Changes: EDS Analysis on Magnetite

Similar quantitative results are obtained also in the case of magnetite experiments. The fracture surfaces of specimen P2 are analyzed in order to recognize possible evidence of phono-fission reactions. Taking into account the homogeneity of the chemical composition of this material (Fe-oxides content ~95%), the analysis on the external sample surface is conducted in a different manner with respect to granitic orthogneiss and basalt. In this case, in fact, taking into account the homogeneity of the rock, both map and spot analyses are used to evaluate the changes in element distribution and, eventually, the appearance of other elements previously absent. As mentioned before, the typical composition of magnetite is given by Fe and O, these elements representing about the 95% of this kind of rock. The remaining part is represented by traces of other elements such as Na, Si, Cl, and K. Maps of dimension

250 μm × 200 μm are analyzed to localize the presence of significant changes in the chemical composition.

In Figs. 3.7 and 3.8, the Fe and Al concentrations together with the concentrations of the other elements are reported indicating the mass percentage concentrations in the case of both external and fracture surfaces. At a first glance, it can be noted that, for the fracture surface, a consistent Al content appears after the specimen failure (Fig. 3.8e). The presence of Al and Mn, which are absent in the analysis performed on the external surface, is observed in several points located on the fracture surface (Fig. 3.8e,f).

At the same time, on the fracture surface, we observe a significant Fe decrement together with the increment in elements such as Si and O. In order to evaluate these changes from a statistical point of view, 15 analyses on the external surface and 15 on the fracture surface are carried out and reported in Fig. 3.9 and in Fig. 10a–e. In these diagrams, similarly to the results obtained for olivine and reported in the previous section, the mass percentage concentrations for Fe (Fig. 3.9), Al, Mn, Si, and O (Fig. 3.10) are reported making a comparison between external and fracture surfaces. It is interesting to observe a decrement in Fe concentration of 27.9%, starting from a value of 64.8% (external surface) and going down to a concentration of 36.9% after the experiment (fracture surface, Fig. 3.9). At the same time, Al concentration increases from zero to a mean value of 10.1%. It is interesting to consider that approximately one third of the Fe decrement may be balanced by the Al increment, according to reaction (3.1). The remaining part of the Fe decrement is almost perfectly balanced by the increment in the other elements. In fact, the appearance of Mn (+2.2%), the increment in Si (+8.7%), and the increment in O

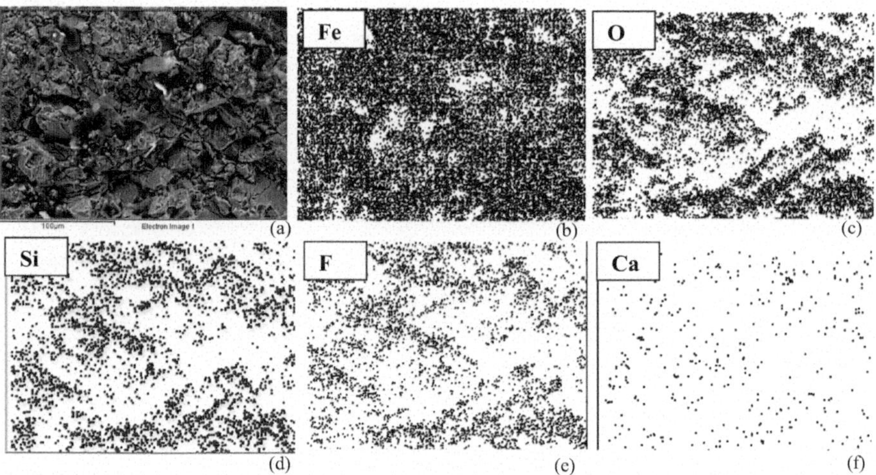

Fig. 3.7 Maps obtained from EDS analysis of magnetite (**a**). Maps, with dimensions of 250 μm × 200 μm, are analyzed before the loading tests. The typical concentration on the external surface shows high percentages of Fe (~65%) (**b**) and O (~30%) (**c**). Remaining concentration is represented by minor contents of Si (**d**), F (**e**), and Ca (**f**)

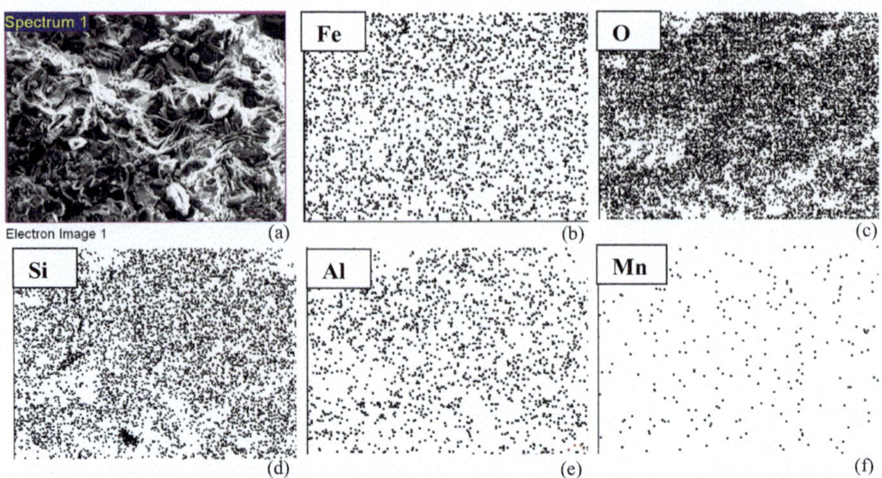

Fig. 3.8 Maps obtained from EDS analysis of the basalt fracture surface (**a**). Maps, with dimensions of 250 μm × 200 μm, are analyzed to localize the presence of significant changes in the chemical composition. The concentration on the fracture surface shows a lower percentage of Fe (~ 35%) (**b**). Concentration of O reaches a content of ~38% (**c**). At the same time, an appreciable Si concentration (~10%) is observed after the test (**d**). Most impressive evidence regards the appearance of Al (**e**) and Mn (**f**) after the experiment with a mass percentage of about 10% and 2.2%, respectively

(+7.6%) may be considered in light of the following reactions (Table 3.3):

$$Fe_{26}^{56} \rightarrow Mn_{25}^{55} + H_1^1 \tag{3.3}$$

$$Fe_{26}^{56} \rightarrow Si_{14}^{28} + O_8^{16} + 2He_2^4 + 4 \text{ neutrons} \tag{3.4}$$

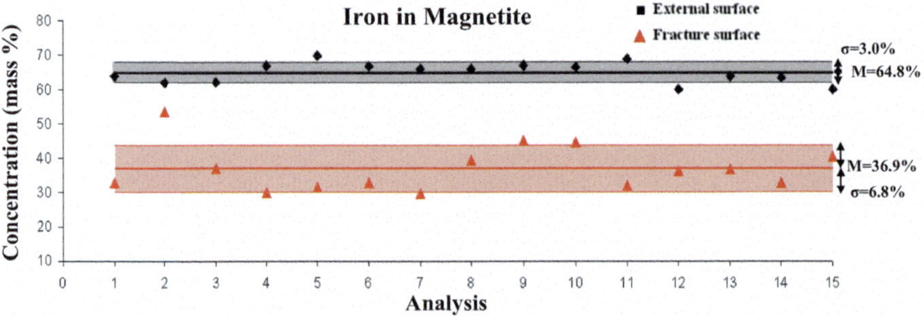

Fig. 3.9 Fe decreases by 27.9% in magnetite after the brittle crushing of the specimen. The Fe concentration on the external surface is 64.8% and the concentration on the fracture surface, after the test, is 36.9%

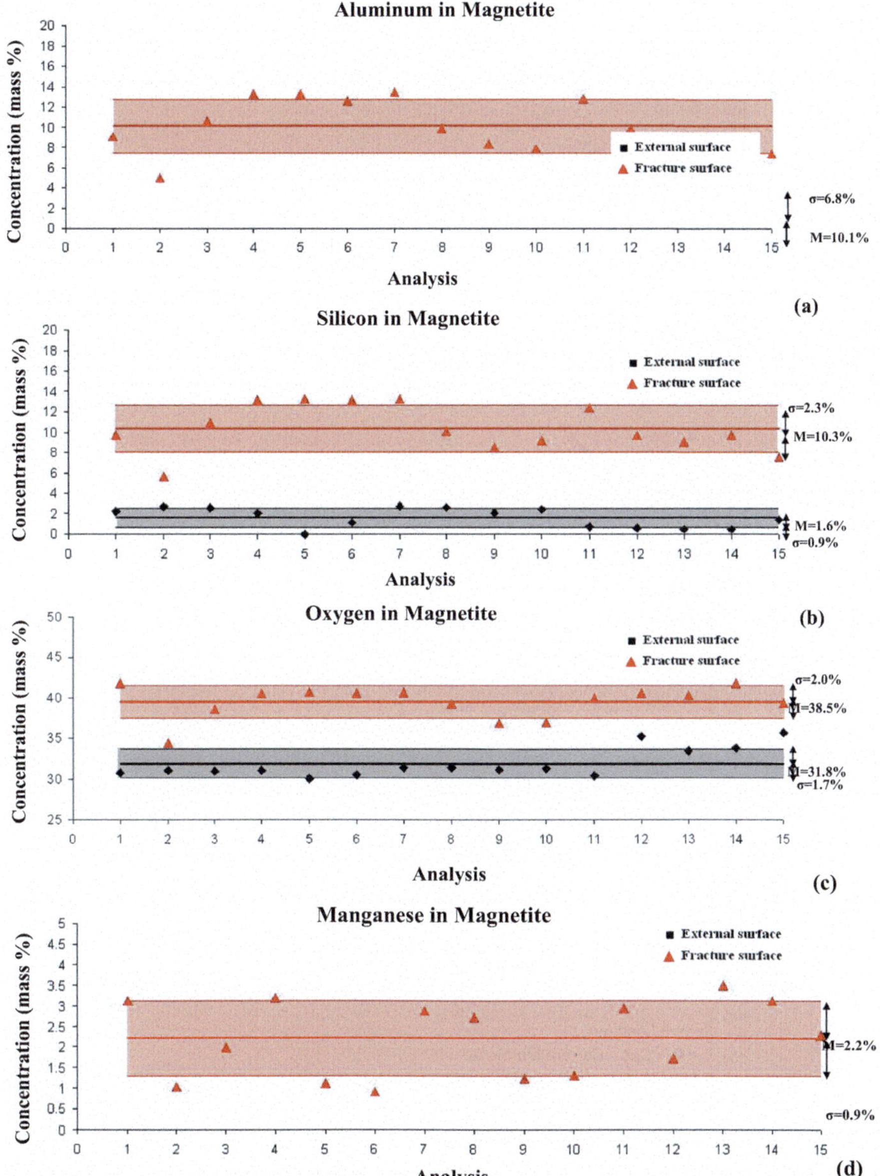

Fig. 3.10 Al, Si, O, and Mn increasing on magnetite fracture surface after the experiment. In particular, Al and Mn, with concentrations of 10.1% and 2.2% after the experiment, are absent before the test

Table 3.3 Magnetite: Fe, Al, Mn, Si, and O weight percentage mean values on external and fracture surfaces. Variations with respect to the mineral and to the element itself

	External surface mean value (wt%)	Fracture surface mean value (wt%)	Increment/Decrement with respect to the mineral (%)	Increment/Decrement with respect to the element itself
Fe	64.8	36.9	− 27.9	− 56%
Al	0.0	10.1	+ 10.1	Before absent
Mn	0.0	2.2	+ 2.2	Before absent
Si	1.6	10.3	+ 8.7	+ 540%
O	31.8	38.5	+ 6.7	+ 21%

The appearance of Al and Mn, previously absent, and the almost perfect balance between the Fe decrement (−27.9%) and the increment in the other elements (+27.7%), well-matched by phono-fission reactions (3.1), (3.3), and (3.4), can be considered as a direct evidence that nuclear reactions of a new type take place.

3.6 Conclusions

The results presented in this chapter about basalt and magnetite are strictly related to other experimental results obtained from similar tests on Luserna stone (granite). In the cases reported herein, the phono-fission reactions regarding iron as the starting element are even more evident. These investigations confirm that high-frequency pressure waves, suitably exerted on inert and stable nuclides, generate nuclear reactions of a new type (phono-fission reactions), with substantial changes in the chemical composition of iron-rich minerals. EDS analyses are performed on different samples of external and fracture surfaces belonging to basalt specimens, in order to get a statistical information about possible changes in the chemical composition of olivine. Considering the results for this crystalline phase, a significant decrement in the iron content (−4%) is balanced by the increment in lighter elements such as Si and Mg.

The results reported for magnetite are even more evident, with the appearance of elements lighter than iron, previously absent. In this case, the decrement in Fe concentration is nearly equal to 30%, and the appearance of Al and Mn, previously absent, represents a very impressive result. In particular for Al, the concentration on the fracture surface reaches, starting from zero, a mean value of over 10%. The remaining part of the Fe decrement is almost perfectly balanced by the increments in Si and O.

The EDS results for magnetite confirm the theoretical explanations to these phenomena by Widom et al. [23, 24]. They considered neutron emissions as a consequence of nuclear reactions taking place in iron-rich rocks during brittle fracture. The same authors argued that neutron emissions may be related to piezoelectric effects and that the fission of iron may be a consequence of the photodisintegration of the nucleus [23].

As a conclusion and according to the experimental results, the hypothesis of phono-fission reactions regarding iron depletion finds a very important evidence and confirmation at the Earth's Crust scale. The phono-fission reactions have thus been considered in order to interpret the most significant geophysical and geological transformations, today still unexplained. The laboratory results herein reported for basalt and magnetite give a new and original interpretation to the evolution from Hadean (i.e., basaltic) to Archean (i.e., granitic) lithosphere in our planet and contribute to explain the difference between oceanic and continental crust compositions due to subductive and seismic events.

Acknowledgements Special thanks are due to Prof. R. Ciccu and Mr F. Argiolas for providing the magnetite specimens.

References

1. Carpinteri A, Cardone F, Lacidogna G (2009) Piezonuclear neutrons from brittle fracture: early results of mechanical compression tests. Strain 45: 332–339; Atti dell'Accademia delle Scienze di Torino, Torino, Italy 33:27–42
2. Cardone F, Carpinteri A, Lacidogna G (2009) Piezonuclear neutrons from fracturing of inert solids. Phys Lett A 373:4158–4163
3. Carpinteri A, Cardone F, Lacidogna G (2010) Energy emissions from failure phenomena: mechanical, electromagnetic, nuclear. Exp Mech 50:1235–1243
4. Carpinteri A, Borla O, Lacidogna G, Manuello A (2010) Neutron emissions in brittle rocks during compression tests: monotonic versus cyclic loading. Phys Mesomech 13:268–274
5. Carpinteri A, Lacidogna G, Manuello A, Borla O (2011) Energy emissions from brittle fracture: neutron measurements and geological evidences of piezonuclear reactions. Strength, Fract Complex 7:13–31
6. Carpinteri A, Lacidogna G, Manuello A, Borla O (2010) Piezonuclear transmutations in brittle rocks under mechanical loading: microchemical analysis and geological confirmations. Recent advances in mechanics, Springer, pp 361–382
7. Carpinteri A, Lacidogna G, Borla O, Manuello A, Niccolini G (2012) Electromagnetic and neutron emissions from brittle rocks failure: experimental evidence and geological implications. Sadhana 37:59–78
8. Carpinteri A, Lacidogna G, Manuello A, Borla O (2012) Piezonuclear fission reactions: evidences from microchemical analysis, neutron emission, and geological transformation. Rock Mech Rock Eng 45:445–459
9. Carpinteri A, Lacidogna G, Manuello A, Borla O (2013) Piezonuclear fission reactions from earthquakes and brittle rocks failure: evidence of neutron emission and nonradioactive product elements. Exp Mech 53(3):345–365
10. Carpinteri A, Borla O, Lacidogna G, Manuello A (2012) Piezonuclear reactions produced by brittle fracture: from laboratory to planetary scale. In: Proceedings of the 19th European conference of fracture, Kazan, Russia
11. Carpinteri A, Chiodoni A, Manuello A, Sandrone R (2011) Compositional and microchemical evidence of piezonuclear fission reactions in rock specimens subjected to compression tests. Strain 47(2):267–281

12. Verhaeren J, Bartholomè P (1979) Petrology of the San Leone magnetite skarn deposit (S. W. Sardinia). Econ Geol 74:53–66
13. Tanguy JC (1978) Tholeiitic basalt magmatism of Mount Etna and its relations with the alkaline series. Contrib Miner Petrol 66:51–67
14. Carpinteri A, Manuello A (2011) Geomechanical and geochemical evidence of piezonuclear fission reactions in the Earth's crust. Strain 47:282–292
15. Carpinteri A, Manuello A (2012) An indirect evidence of piezonuclear fission reactions: geomechanical and geochemical evolution in the Earth's crust. Phys Mesomech 15:14–23
16. Bubble Technology Industries (1992) Chalk River, Ontario, Canada
17. Catling CD, Zahnle KJ (2009) The planetary air leak. Sci Am 300:24–31
18. Taylor SR, McLennan SM (1995) The geochemical evolution of the continental crust. Rev Geophys 33:241–265
19. Taylor SR, McLennan SM (2009) Planetary crusts: their composition, origin and evolution. Cambridge University Press, Cambridge
20. Fowler CMR (2005) The solid earth: an introduction to global geophysics. Cambridge University Press, Cambridge
21. Doglioni C (2007) Interno della Terra. Treccani, Enciclopedia di Scienza e Tecnica: 595–605
22. Rudnick RL, Fountain DM (1995) Nature and composition of the continental crust: a lower crustal perspective. Rev Geophys 33:267–309
23. Widom A, Swain J, Srivastava YN (2013) Photo-disintegration of the iron nucleus in fractured magnetite rocks with magnetostriction. Meccanica 50:1205–1216
24. Widom A, Swain J, Srivastava YN (2013) Neutron production from the fracture of piezoelectric rocks. J Phys G: Nucl Part Phys 40(15006):1–8

Chapter 4
Iron-Rich Natural Rocks: Fatigue Tests and Frequency-Dependent Neutron Emissions

Abstract The results are reported of neutron emission measurements during fatigue experiments at low (2 Hz), intermediate (200 Hz), and high (20 kHz) frequencies. These results confirm that appreciable neutron emissions, without any doubt greater than the natural background level, are observed during damage accumulation in iron-bearing rocks: granite (Fe ~1.5%), basalt (Fe ~15%), and magnetite (Fe ~75%). The neutron detection, together with temperature measurements obtained by infrared techniques, leads to the conclusion that fatigue tests performed at 200 Hz represent the condition for which the neutron emission is the highest. This evidence is corroborated by the seismological observations at Jacinto fault in Southern California, as well as by the neutron emissions detected by the author during seismic activity. On the other hand, as will be explained later in the book, it is only a component of the frequency spectrum that is directly triggering the phono-fission reactions, i.e., the TeraHertz range component. This component is represented by phonons produced by micro and nanocracking and not directly by an instrument or a seismic vibration.

Keywords Fatigue experiments · Granite · Basalt · Magnetite · Neutron emissions · Frequency dependence

4.1 Preliminary Remarks

In the last few years, neutron emissions have been observed several times in experiments characterized by static or repeated (fatigue) loading conditions on inert iron-rich rocks [1–8]. The large amount and the high repeatability of these data suggest that pressure, suitably exerted on inert and stable nuclides, generates nuclear reactions of a new type with subatomic particle emissions [1–3]. In particular, neutron emission measurements, by means of He^3 devices and thermodynamic bubble detectors, were performed during different kinds of experiments: monotonic compression and tensile tests, cyclic loading, ultrasonic vibrations.

A theoretical interpretation was proposed by Widom et al. explaining neutron emissions as a consequence of nuclear fission reactions taking place in iron-rich rocks

A. Carpinteri, *Terahertz Phonons and Nanomechanical Instabilities*,
https://doi.org/10.1007/978-3-032-14692-2_4

during brittle microcracking or fracture [9, 10]. Iron nuclear disintegration takes place when rocks containing such nuclei are crushed. The resulting nuclear transmutations are particularly relevant in the case of magnetite and iron-rich materials in general. The same authors argued that neutron emissions may be related to piezoelectric effects and that the fission of iron may be a consequence of photodisintegration of the nuclei [9]. A different explanation will be provided in the Part VII of this book.

In the experiments reported in previous publications, granitic orthogneiss specimens showed a maximum neutron emission approximately 10 times higher than the measured natural background level. The emission was $28.0 \pm 0.2 \times 10^{-2}$ cps, against a measured background noise of $3.8 \pm 0.2 \times 10^{-2}$ cps [3–8]. At the same time, Etna basalt, used during similar tests, produced neutron bursts with a dose up to 10^2 times the background equivalent dose, in correspondence to the catastrophic compression failure.

The most impressive result is that of magnetite specimens, characterized by a very high content of iron (~75%). They were subjected to crushing experiments and reached a neutron emission dose of (935.49 ± 233.87 nSv/h), up to 10^3 times the background equivalent dose of (5.76 ± 13.44 nSv/h). The results obtained in the last few years from fatigue tests performed at low and high frequencies are herein extended to an intermediate frequency, around 200 Hz, by means of an electromechanical Amsler Vibrofore. This equipment allows to apply fatigue loading with a frequency range between 100 and 260 Hz, a mean load up to 50 kN, and an alternate load up to 50 kN. The cyclic tests, using 2 Hz or 20 kHz as working frequency, were conducted by a servo-hydraulic press for the lower frequency and by a Bandelin HD 2200 ultrasonic sonotrode for the higher frequency [4, 5]. From the experiments reported in the literature, it is possible to correlate neutron emissions, nuclear transmutations, and compositional changes [3–8]. By working on the results coming from the experiments conducted on natural rocks in static and cyclic-fatigue loading, it can be observed that the greater the iron content, the higher the neutron emission [1–8]. From the same experiments, the neutron level may be also correlated to the size-scale of the specimen. As a matter of fact, the greater the size-scale and/or the slenderness, the higher the neutron bursts.

Luserna stone (a granitic rock), basalt, and magnetite, containing different iron concentrations, are tested in fatigue experiments using low (2 Hz), intermediate (200 Hz), and high (20 kHz) frequencies. We investigate different cyclic conditions to confirm that energy emission in the form of neutrons takes place also in the case of fatigue tests under different frequency regimes. The experiments have also the purpose to give first indications about a possible correlation between neutron emission and loading frequency.

4.2 Experimental Set-Up

4.2.1 Materials

The specimens are obtained from different iron-bearing rocks. The tests are conducted on Luserna stone (granitic orthogneiss) that is characterized by a content in iron of about 1.5%, on basalt, with an iron content of ~15%, and on magnetite, with an iron content of ~75%. The employed materials are chosen in order to confirm the correlation between iron concentration and neutron emission level. During the experimental program, the specimens are subjected to fatigue cycles with different working frequencies. More precisely, the test frequencies are 2 Hz, 200 Hz, and 20 kHz. The main purpose is that of detecting neutron emission activity during the tests in order to correlate it to the frequency range, as well as to the iron content and to the specimen size.

From a petrographic point of view, the Luserna stone used for the experiments is a leucogranitic orthogneiss, probably from the Lower Permian Age, that outcrops in the Luserna-Infernotto basin (Cottian Alps, Piedmont), at the border between Turin and Cuneo provinces (North-western Italy) [11–14]. The rock is characterized by a micro "Augen" texture, it is gray-greenish or locally pale blue in color. Geologically, Luserna stone pertains to the Dora-Maira Massif [12, 13], which represents a part of the ancient European continent during Alpine orogenesis [11, 13].

The second material employed during the experimental study is a basalt originating from the Mount Etna. Mount Etna is composed for the most part of intermediate alkaline products, that may be defined as sodic trachybasalts or trachyandesites. The strato-volcanio itself overlies tholeiitic basalts [15].

Basalt is a very common extrusive igneous rock, the dominant material making up the Earth's oceanic crust, and represents the principal product from volcanic eruptions. The basalt used in the tests reported in this chapter is characterized by a mafic composition (with a high content of iron and magnesium), is dark gray and shows porphyritic texture with few mm-sized phenocrysts of plagioclase, clinopyroxene, and olivine [15].

Magnetite specimens come from the San Leone mine. This deposit is located at about 30 km Southwest of Cagliari, Sardinia (Italy), near the Basso-Sulcis batholith, a granodioritic intrusive of the Hercynian Age [16]. The mineralization of the deposit is mainly represented by magnetite. The ore bodies within skarn generated by the thermometamorphism of previous Palaeozoic limestones in contact with granitic intrusions of Variscan Age. The mean iron concentration of the San Leone magnetite is between 72.5 and 75.0% [16]. All the prepared specimens are cylindrical, with a diameter of 50 mm and a height of 100 mm. Figure 4.1a–c shows three different specimens representing the three types of tested rock. The geometrical and mechanical characteristics of Luserna stone, basalt, and magnetite specimens are summarized in Table 4.1.

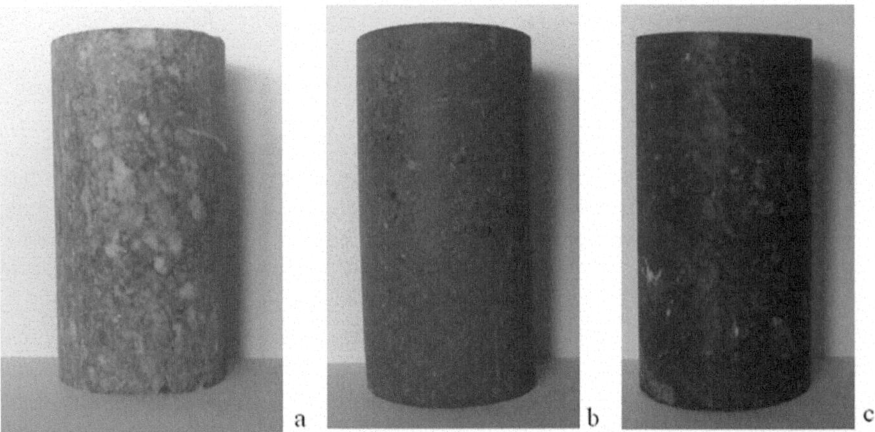

Fig. 4.1 Luserna stone (**a**), basalt (**b**), and magnetite (**c**) specimens adopted during the fatigue tests. The specimens present cylindrical shape, the diameter is 50 mm, and the height 100 mm. All the specimens are drilled from blocks of quarries located in Piedmont (granite), Mount Etna-Sicily (basalt), and Sardinia (magnetite)

Table 4.1 Tested materials

Materials	Number of specimens	D (mm)	H (mm)	$\lambda = H/D$	Compression strength (MPa)	Elastic modulus (GPa)
Granite	2	50	100	2	162	50
Basalt	2	50	100	2	170	60
Magnetite	2	50	100	2	250	220

4.2.2 *Testing Procedure and Devices*

For the lowest frequency (2 Hz), six specimens, two for each kind of rock (D = 50 mm, H = 100 mm, $\lambda = 2$) are considered. The tests are carried out by means of a servo-hydraulic press, with a maximum capacity of 1,800 kN, working by a digital-type electronic control unit. The management software is TESTXPERTII by Zwick/Roel (Zwick/Roel Group, Ulm, Germany), whereas the mechanical parts are manufactured by Baldwin (Instron Industrial Products Group, Grove City, PA, USA) (Fig. 4.2a). The applied force is determined by measuring the pressure in the loading cylinder by means of a transducer. The margin of error in the determination of the force is 1%, which makes it a class 1 mechanical press.

The cyclic loading is programmed at a frequency of 2 Hz and with a load excursion from a minimum load of 10 kN to a maximum of 60 kN. Differently from the tests performed under monotonic displacement control, neutron emissions from tests under cyclic compression loading are measured by using neutron bubble detectors. Due to their isotropic angular response, three BDT and three BD-PND detectors are located, at a distance of about 5 cm, all around the specimen. The detectors are

(a)

(b)

(c)

Fig. 4.2 Servo-hydraulic press employed during the low-frequency fatigue tests (**a**). Electromechanical Amsler Vibrofore 10 HPF 422 used during the experimental tests at the frequency of 200 Hertz (**b**). Bandelin HD 2200 sonotrode used during ultrasonic tests at the frequency of 20 kHz (**c**)

previously activated, unscrewing the protection cap, to reach the suitable thermal equilibrium, and they are kept active throughout the test duration. Furthermore, a BDT and a BD-PND detectors are used as background control during the test. Similar test configurations are adopted for basalt and magnetite.

As far as the intermediate frequency (200 Hz) is concerned, the cylindrical specimens are subjected to compression–compression fatigue tests by means of an electromechanical Amsler Vibrofore 10 HPF 422 (Fig. 4.2b). This equipment allows to apply fatigue loading with a frequency range between 100 and 260 Hz, a mean load up to 50 kN, and an alternate load up to 50 kN. The machine working conditions are related to resonance conditions of the system, which is composed by two masses (a seismic huge mass and the specimen mass) and two springs (a huge machine spring element and the specimen stiffness) disposed in series (Fig. 4.2b). Working condition difficulties are often related to specimen damping. In the case of rocks, a tuning activity is needed to reach resonance conditions. The specimens are set between two compression platens, a preload (mean load) is set and then alternate loading applied.

Also in this case, the neutron emission is detected by using neutron bubble detectors. After 5×10^7 cycles, the tests are interrupted. During the fatigue tests, two sets of measurements are performed: the specimen-surface thermal infrared measurements and the neutron emission measurements by thermodynamic detectors.

The ultrasonic tests are conducted at a frequency of 20 kHz. The ultrasonic oscillation is generated by a high-intensity ultrasonic horn (Bandelin HD 2200) (Fig. 4.2c). The device guarantees constant amplitude independently of changing conditions within the sample. The apparatus consists of a generator that converts electrical energy to 20 kHz ultrasounds and of a transducer that switches this energy into longitudinal mechanical vibration at the same frequency. The specimens are connected to the ultrasonic horn by a glued screw inserted into a 5-mm deep hole (Fig. 4.2). This kind of connection is realized to achieve a resonance condition, considering the speed of sound in Luserna stone and the length of the specimen. Ultrasonic irradiation of the specimen is carried out for 3 h. After the switching on of the transducer, 10% of the maximum power is reached in 20 min. Successively, the transducer power increases to 20% after 1 h and then reaches a maximum level of about 30% after 2 h. Afterward, the transducer works at the same power condition up to the end of the test.

4.3 Neutron and Temperature Measurements

The thermal emission is acquired by means of a NEC TH7100WX infrared thermocamera (Figs. 4.3a and 4.4). The specimens are black painted to maximize thermal emission. Thermal acquisition frame is set to 2 frame/min. The thermal acquisition system acquires the thermal contour of the surface of the element set in front of the IR lens. In particular, the camera is focused on the specimen cylindrical surface, as the radiation of the surface is perpendicular to the same surface. Then, the correct temperature measurement is obtained if the camera is set parallel to cylinder tangent surface. To avoid measurement errors, only the maximum measured temperatures on the specimens are recorded for the data processing. It can be verified that the maximum temperature is measured in correspondence to the line that lies in the tangent plane, which is parallel to the lens. The surface temperature values are processed by subtracting room temperature thus obtaining the surface thermal increment DeltaT. It is well-known [17–20] that fatigue cyclic loading produces a thermal increment in rubber, ceramics, and metals. Different causes contribute to

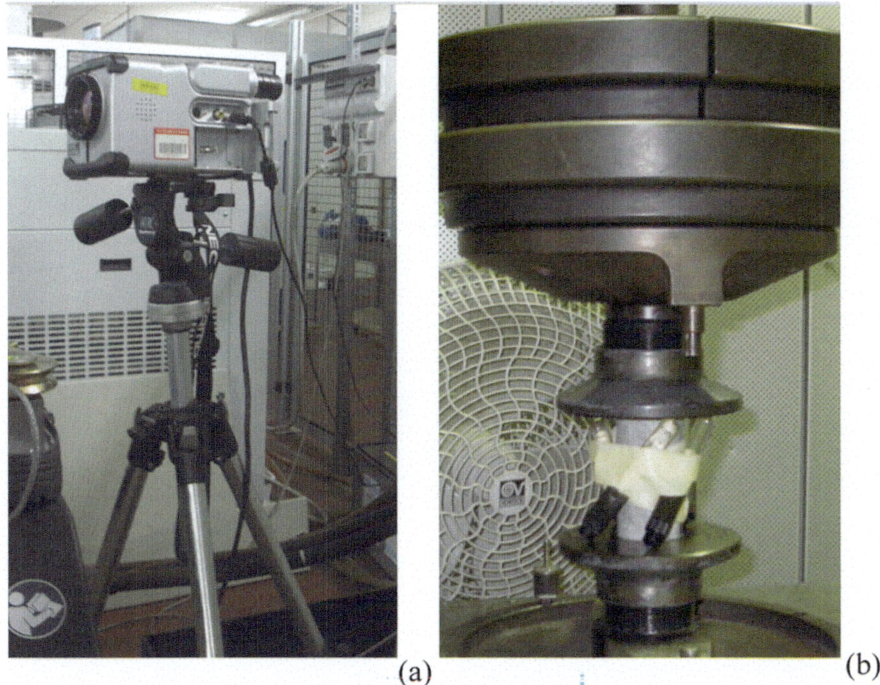

Fig. 4.3 NEC TH7100WX infrared thermo-camera used for the temperature monitoring of the external surface of the specimen (**a**). Set of passive neutron detectors insensitive to electromagnetic noise used for neutron emission detection (**b**)

this phenomenon in the elastic conditions [20, 21], the principal being thermoelastic effects and internal friction dissipations. The latter phenomenon consists in a structural crystal reorganization implying a change in structural damping and in vibration frequencies, as well as in microplastic deformation [20–25].

The neutron activity in fatigue experiments is detected by a thermodynamic Neutron Detection Technique. Since neutrons are electrically neutral particles, they cannot directly produce ionization in a detector and therefore cannot be directly detected. This means that neutron detectors must rely upon a conversion process where an incident neutron interacts with a nucleus to produce a secondary charged particle. These charged particles are then detected, and from them, the neutron's presence is deduced. For an accurate neutron rate evaluation, a set of passive neutron detectors, based on the superheated bubble detection technique, are employed. A set of passive neutron detectors insensitive to electromagnetic noise and with zero gamma sensitivity is used (Fig. 4.3b). The dosimeters, based on superheated bubble detectors (BTI, ON, Canada) (Bubble Technology Industries 1992), are calibrated at the factory against an AmBe (Americium-Beryllium) source in terms of National Council on Radiation Protection and Measurements 1971 (NCRP38). Bubble detectors are the most sensitive and accurate available neutron dosimeters that provide

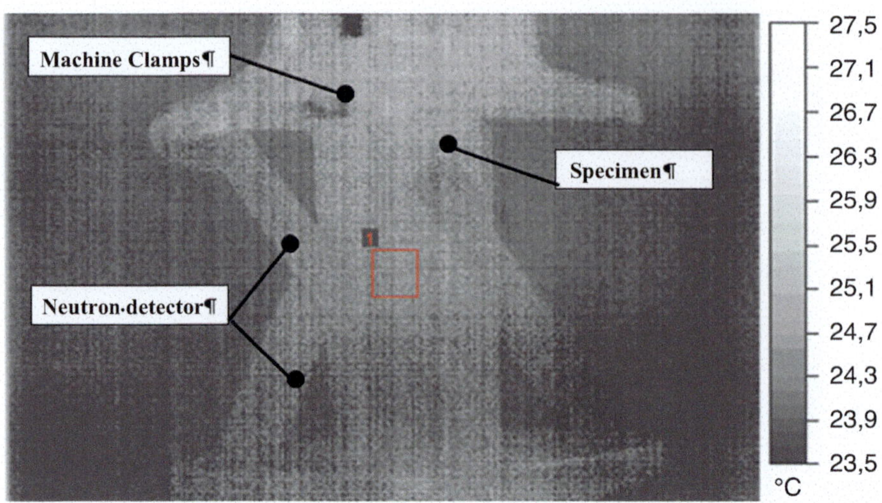

Fig. 4.4 Thermal contour in a basalt specimen during compression fatigue loading. The specimen, the neutron dosimeters, and the machine clamps are indicated. The detected temperatures range between 23.5 and 27.5 °C

instant visible detection and measurement of neutron dose. Each detector is composed of a polycarbonate vial filled with elastic tissue equivalent polymer, in which droplets of a superheated gas (Freon) are dispersed. When a neutron strikes a droplet, the latter immediately vaporizes, forming a visible gas bubble trapped in the gel. The number of droplets provides a direct measurement of the equivalent neutron dose with an efficiency of about 20%. These detectors are suitable for neutron dose measurements in the energy range of thermal neutrons (E = 0.025 eV, BDT type) and fast neutrons (E = 100 keV, BD-PND type).

4.4 Experimental Results

The preliminary experimental results reported in the literature [1–8] are confirmed by those obtained from the fatigue tests carried out on the cylindrical specimens. In addition, the experimental results demonstrate that neutron emissions follow an anisotropic and impulsive distribution from a specific zone of the specimen. As a matter of fact, the detected neutron flux, and the consequent neutron dose, is inversely proportional to the square of the distance from the source. For these reasons, to avoid underestimating data acquisition, different bubble dosimeters are placed around the test specimen. For the different materials, bubbles counting is performed every 12 h and the equivalent neutron dose is calculated during low-frequency fatigue experiments. In the same way, the natural background equivalent dose is estimated by means of the two bubble dosimeters used for assessment. For all the tests, the

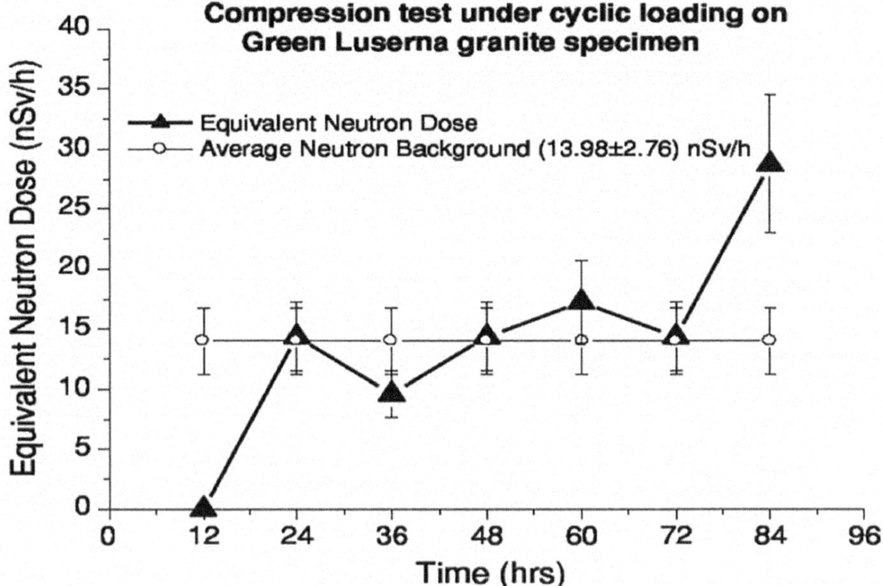

Fig. 4.5 Equivalent neutron dose variation from a Luserna stone specimen during the test. Only at final rupture, the dose is twice the natural background

natural background was found to be (13.98 ± 2.76) nSv/h. In Fig. 4.5, the equivalent neutron dose variation, evaluated during the cyclic (2 Hz) compression test of granitic orthogneiss, is reported. An increment of more than twice the background level is detected at the specimen failure. No significant variations in neutron emission are observed before failure. The equivalent neutron dose, at the end of the test, is (28.74 ± 5.75) nSv/h.

As previously, in Fig. 4.6a,b, the results obtained by thermodynamic detectors and the equivalent neutron dose for a basalt specimen subjected to fatigue test at low frequency is reported. Also in this case, bubbles counting is performed every 12 h and the equivalent neutron dose is calculated. During the test, the natural background is found to be (53.76 ± 13.44) nSv/h. The neutron equivalent dose variation, evaluated during the cyclic loading, is reported in Fig. 4.6b. An increment of about twenty times with respect to the background level is detected at the specimen failure. No significant variations in neutron emission are observed before failure. The equivalent neutron dose, at the end of the test, is (935.49 ± 233.87) nSv/h.

As far as the magnetite specimen is concerned, the natural background was found to be (15.05 ± 3.76) nSv/h. For this very iron-rich rock, the equivalent neutron dose variation, evaluated during the cyclic loading test at low frequency (2 Hz), is reported in Fig. 4.7. An increment of about one hundred times with respect to the background level is detected at failure. No significant variations in neutron emission are observed before failure. The equivalent neutron dose, at the end of the test, is 1,036.78 ± 259.19 nSv/h.

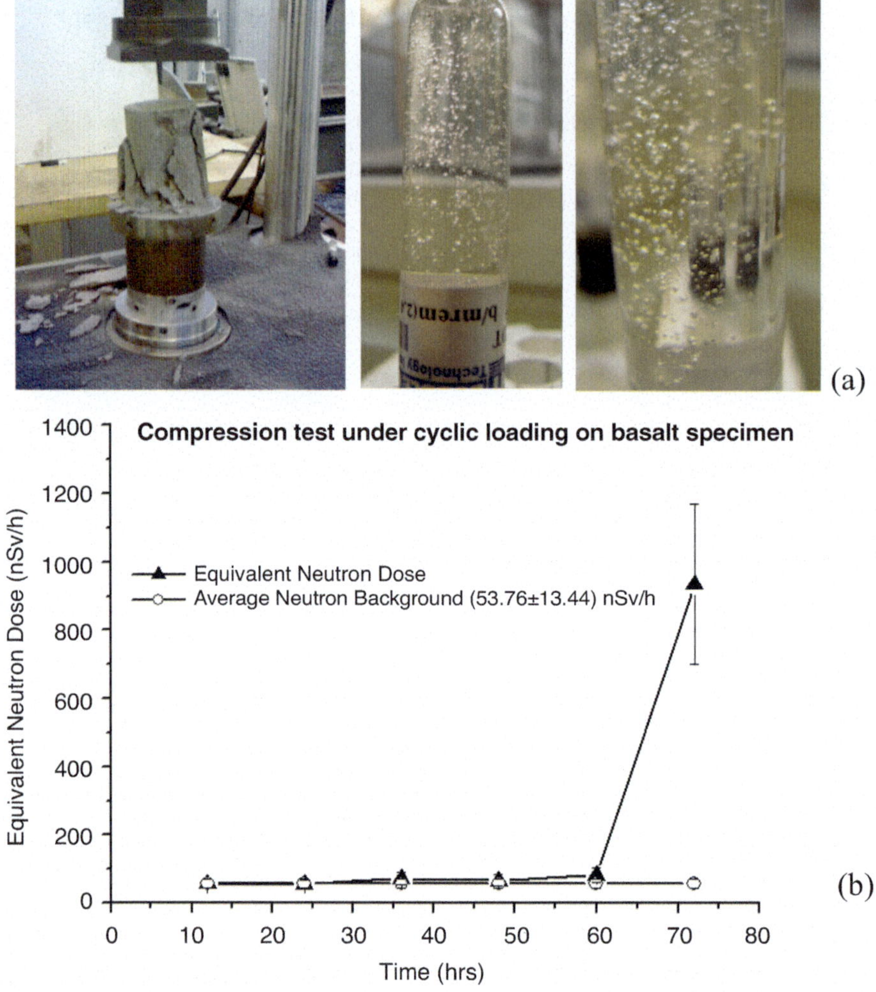

Fig. 4.6 Results from bubble thermodynamic dosimeters obtained at the end of a low-frequency (2 Hz) test on basalt (**a**). Equivalent neutron dose variation during the same test (**b**). Only at final rupture, the dose is approximately 20 times the natural background

In Fig. 4.8, the neutron emission level is reported for a comparison between granitic orthogneiss, basalt, and magnetite, when subjected to fatigue loading with intermediate frequency (200 Hz). It is interesting to note that the emission trends are conserved considering the behavior obtained in the case of low-frequency. Also in this case, in fact, the granitic specimens show a small but appreciable increment. The basalt specimens show an increment up to one order of magnitude greater than the background level, and the magnetite specimens show a neutron emission up to two orders of magnitude greater than the background level.

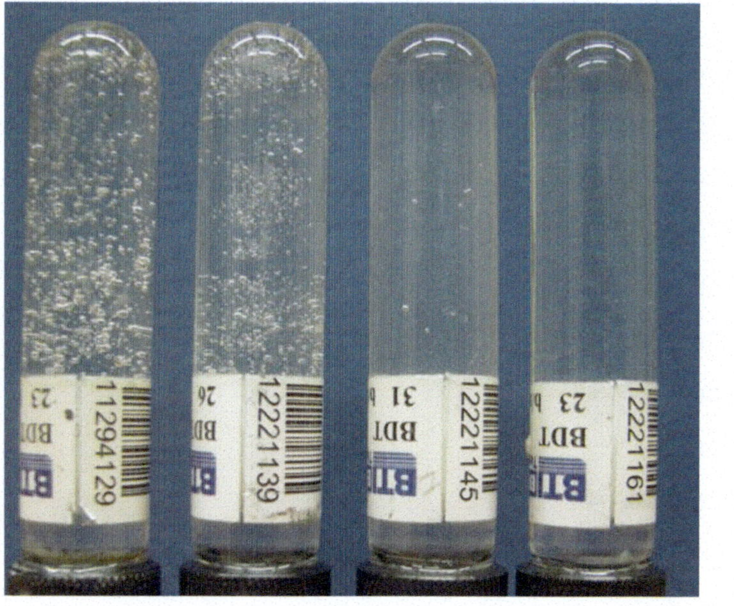

(a)

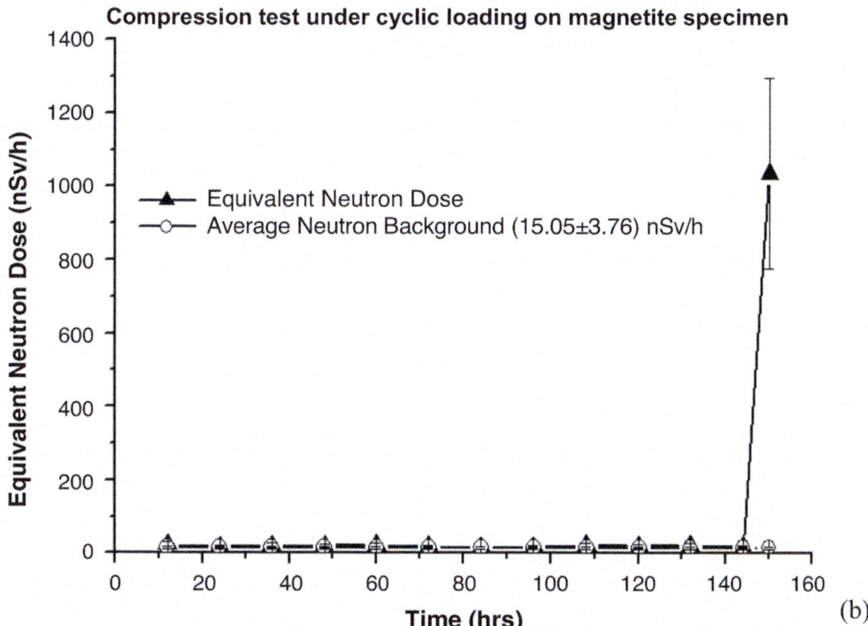

(b)

Fig. 4.7 Results from bubble thermodynamic dosimeters obtained at the end of a low-frequency (2 Hz) test on magnetite (**a**). Equivalent neutron dose variation during the same test (**b**). Only at final rupture, the dose is two orders of magnitude higher than the natural background

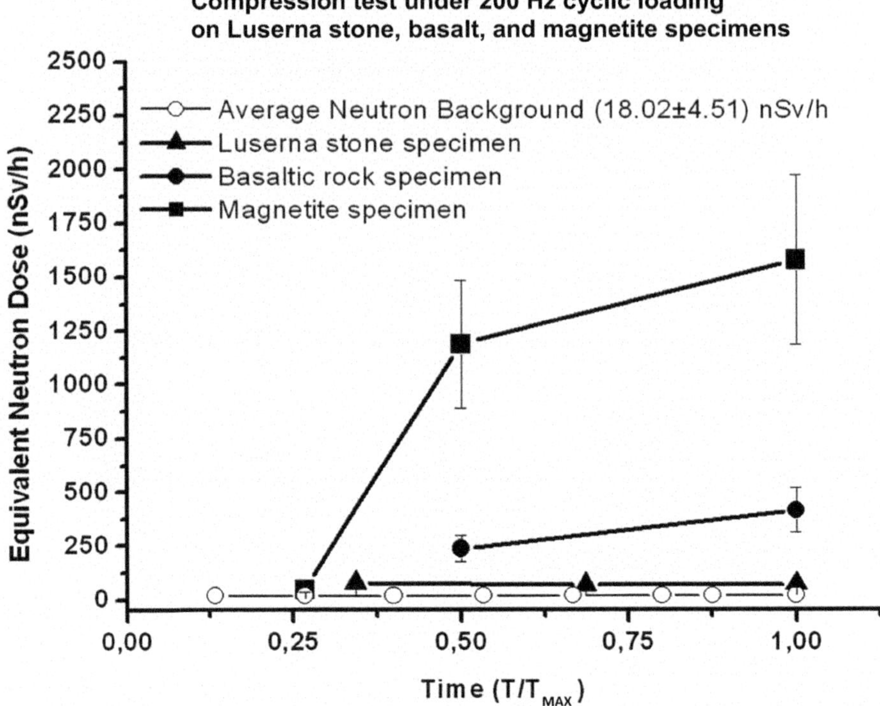

Fig. 4.8 Comparison between the equivalent neutron doses during intermediate-frequency (200 Hz) tests on granite, basalt, and magnetite

The last kind of experiments is conducted using ultrasound vibration. The ultrasonic test on the Luserna stone specimen (D = 53 mm, H = 100 mm) is carried out at the Medical and Environmental Physics Laboratory in the Experimental Physics Department of the University of Turin. A relative natural background measurement is performed by means of the He^3 detector for more than 6 h. The average natural background is $(6.50 \pm 0.85)\ 10^{-3}$ cps, for a corresponding thermal neutron flux of $(1.00 \pm 0.13)\ 10^{-4}$ thermal neutrons $cm^{-2}\ s^{-1}$. This natural background level, lower than the one calculated during the monotonic compression tests at the Fracture Mechanics Laboratory of Politecnico di Torino, is related to the location of the Experimental Physics Laboratory, which is three floors below the ground level.

Similar experiments are conducted using a basalt specimen. In this case, the thermal neutron flux is about two orders of magnitude greater than the natural background level. Ultrasonic experiments are not conducted on magnetite due to the brittleness of this rock. For magnetite, in fact, the connection to the ultrasonic horn by a glued screw inserted in a 5-mm deep hole is not possible, due to the sudden failure of the specimen after few minutes from the beginning of the test.

During the ultrasonic test, the specimen temperature is monitored by using a multimeter/thermometer (Tektronix mod. S3910). The temperature reaches 50° C

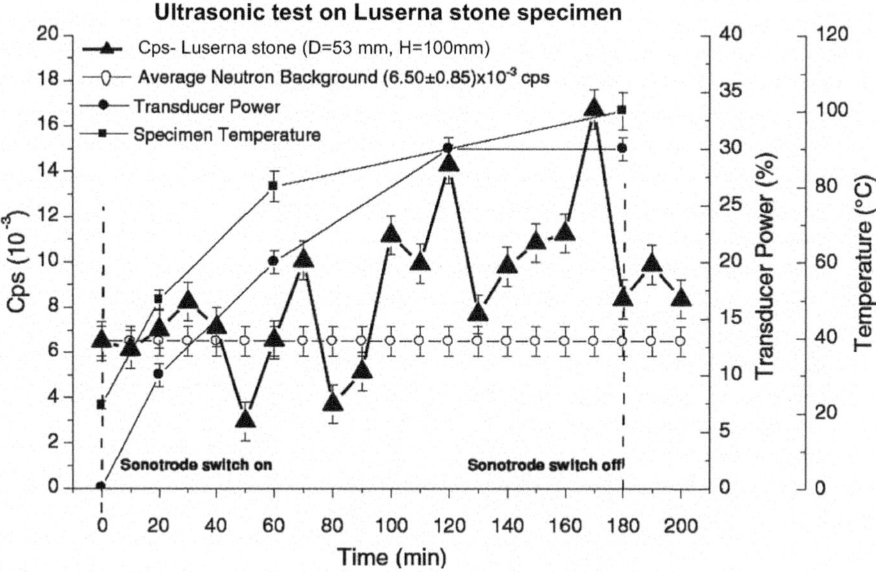

Fig. 4.9 Ultrasonic-frequency (20 kHz) test on Luserna stone: variations of neutron flux (cps) and temperature

after 20 min, and then increases up to a maximum level of 100° C at the end of the ultrasonic test. In Fig. 4.8, the detected neutron emissions are compared to the transducer power trend and to the specimen temperature. A significant increment in neutron activity after 130 min from the beginning of the test is measured (Fig. 4.9). At this time, the transducer power reaches 30% of the maximum, with a specimen temperature of about 90° C. Some neutron variations are detected during the first hour of the test, but they may be due to usual fluctuations in the natural background. At the switching off of the sonotrode, the neutron activity decreases to the typical background level.

In Fig. 4.10, the ratios of neutron emission to neutron background for the different kinds of test and rock are reported, in order to compare frequency range, iron content, and neutron flux. The neutron emissions are considered also in the case of monotonic loading condition, assuming the results obtained from the static tests [3–8]. From the results shown in Fig. 4.10, it is possible to observe that the maximum neutron emission level starts to decrease from the monotonic loading condition, in which the neutron emission is maximum for all the materials, down to the ultrasonic range (20 kHz), for which the neutron emission (Luserna stone and basalt) is only some times greater than the background level. A particular consideration has to be given to the case of fatigue loading at the intermediate frequency (200 Hz). For this frequency range, an higher value of the neutron emission can be observed for all the tested materials (Fig. 4.10). This frequency appears to generate the highest neutron emissions. It is also interesting to consider that the frequency range around 200 Hz, for which the neutron emission is maximized, can be recorded in the preparation zone of the earthquakes. In particular, this kind of microseismic activity was observed in the monitoring of the Jacinto fault propagation (Southern California) [26]. In the preparation zone of an impending earthquake, collapsing pores, grain boundary slippage, and microcracking may cause acoustic emissions in a frequency range up to 300 kHz [27–29]. These results may be correlated to the neutron emissions repeatedly observed before and during earthquakes, leading to consider also the Earth's Crust, in addition to cosmic rays, as being a relevant source of neutron flux variations [30–34]. Neutron emissions exceeded the neutron background up to 1,000 times in correspondence to a seismic event with a Richter magnitude of the 4th degree [5, 29]. It is interesting to note that, from the laboratory to the Earth' Crust (tectonic) scale, the frequency of 200 Hz, produced by the testing machine or by the microseismic activity, may be considered as a catalyzing mechanism able to maximize the neutron emissions during cyclic solicitations. On the other hand, as will be explained later in the book, it is only a component of the frequency spectrum that is directly triggering the phono-fission reactions, i.e., the TeraHertz range. This component is represented by phonons produced by micro and nanocracking and not directly by the vibrating instrument.

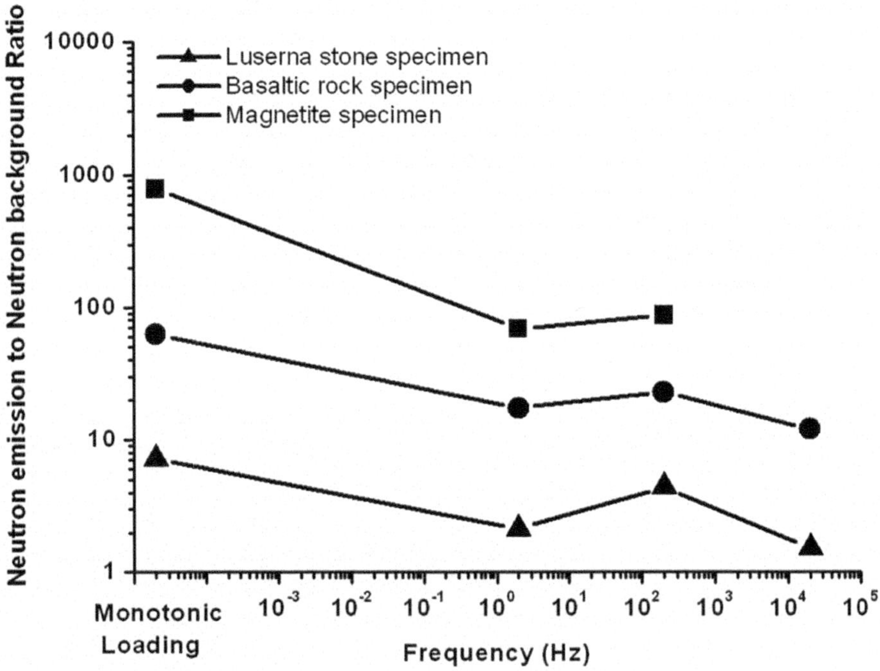

Fig. 4.10 Comparison between the different neutron emission to neutron background ratios for the monotonic loading condition and the fatigue conditions ranging from 2 Hz to 20 kHz

4.5 Conclusions

Fatigue experiments at low (2 Hz), intermediate (200 Hz), and high (20 kHz) frequencies are performed with different iron-bearing natural rocks. The results confirm that neutron emissions greater than the natural background level may be observed during damage accumulation in granitic orthogneiss, basalt, and magnetite. The neutron detection, together with temperature measurements obtained by infrared revelation, leads to the conclusion that fatigue tests performed at 200 Hz represent the condition for which the neutron emission is the highest for specimens subjected to cyclic loading.

As summarized in Fig. 4.10, the emission of neutrons can be related to the iron content in the different materials, and to the testing conditions adopted during the experiments (monotonic or cyclic at different working frequencies). It can be observed that the higher the iron content, the greater the neutron emission. The high neutron flux emitted from magnetite during brittle fracture is confirmed in the case of fatigue tests performed at the intermediate frequency (200 Hz). The Etna basalt shows an average emission about one order of magnitude greater than those found in Luserna stone. Etna basalt shows, at the same time, the most relevant thermal increment measured by infrared waves.

As regards granite in the different testing conditions, it is possible to confirm that the neutron emission is a few times the natural background level for cyclic loads, and up to 10 times the background in the case of crushing.

Particular attention has to be drawn in the case of fatigue at the intermediate frequency of 200 Hz. At this frequency, a higher value of the neutron emission to neutron background ratio can be observed for all the tested materials (Fig. 4.10). This frequency seems to generate higher neutron emissions than the other investigated loading frequencies. It is also interesting to consider that the frequency range around 200 Hz, for which the neutron emission is maximized, can be also recorded in the preparation zone of earthquakes and during fault propagation. In particular, this kind of microseismic activity was observed in the recent monitoring carried out during the Jacinto fault propagation in Southern California. In the preparation zone of an impending earthquake, collapsing pores, grain boundary slippage, and micro-cracking may cause acoustic emissions in a frequency range up to 300 kHz [26–29]. These results may be correlated to the neutron emissions recursively observed during earthquakes, leading to consider also the Earth's Crust, in addition to cosmic rays, as being a relevant source of neutron flux variations [30–34]. Neutron emissions exceeded the natural background up to 1,000 times during seismic events with a Richter magnitude equal to the 4th degree [5, 29].

References

1. Carpinteri A, Cardone F, Lacidogna G (2009) Piezonuclear neutrons from brittle fracture: early results of mechanical compression tests. Strain 45:332–339

2. Cardone F, Carpinteri A, Lacidogna G (2009) Piezonuclear neutrons from fracturing of inert solids. Phys Lett A 373:4158–4163
3. Carpinteri A, Cardone F, Lacidogna G (2010) Energy emissions from failure phenomena: mechanical, electromagnetic, nuclear. Exp Mech 50:1235–1243
4. Carpinteri A, Borla O, Lacidogna G, Manuello A (2010) Neutron emissions in brittle rocks during compression tests: monotonic versus cyclic loading. Phys Mesomech 13:268–274
5. Carpinteri A, Lacidogna G, Manuello A, Borla O (2012) Piezonuclear fission reactions: evidences from microchemical analysis, neutron emission, and geological transformation. Rock Mech Rock Eng 45:445–459
6. Carpinteri A, Lacidogna G, Manuello A, Borla O (2013) Piezonuclear fission reactions from earthquakes and brittle rocks failure: evidence of neutron emission and nonradioactive product elements. Exp Mech 53(3):345–365
7. Carpinteri A, Lacidogna G, Manuello A, Borla O (2013) Energy emissions from brittle fracture: neutron measurements and geological evidences of piezonuclear reactions. Strength, Fract Complex 7:13–31
8. Carpinteri A, Lacidogna G, Borla O, Manuello A, Niccolini G (2012) Electromagnetic and neutron emissions from brittle rocks failure: experimental evidence and geological implications. Sadhana 37:59–78
9. Widom A, Swain J, Srivastava YN (2013) Photo-disintegration of the iron nucleus in fractured magnetite rocks with magnetostriction. Meccanica 50:1205–1216
10. Widom A, Swain J, Srivastava YN (2013) Neutron production from the fracture of piezoelectric rocks. J Phys G: Nucl Part Phys 40(150006):1–8
11. Vola G, Marchi M (2010) Quanitative phase analysis (QPA) of the Luserna stone. Period Miner 79(2):45–60
12. Sandrone R, Cadoppi P, Sacchi R, Vialon P (1993) The Dora-Maira Massif. In: Von Raumer JF, Neubauer F (eds) Pre-Mesozoic geology in the Alps. Springer, Berlin, pp 317–325
13. Sandrone R, Borghi A (1992) Zoned garnets in the northern Dora-Maira Massif and their contribution to a reconstruction of the regional metamorphic evolution. Eur J Minerals 4:465–474
14. Sandrone R, Colombo A, Fiora L, Fornaro M, Lovera E, Tunesi A, Cavallo A (2004) Contemporary natural stones from the Italian western Alps (Piedmont and Aosta Valley Regions). Period Miner 73:211–226
15. Tanguy JC (1978) Tholeiitic basalt magmatism of Mount Etna and its relations with the alkaline series. Contrib Miner Petrol 66:51–67
16. Verkaeren J, Bartholomè P (1979) Petrology of the San Leone magnetite skarn deposit (S. W. Sardinia). Econ Geol 74:53–66
17. Chrysochoos A, Louche H (2000) An infrared image processing to analyse the calorific effects accompanying strain localisation. Int J Eng Sci 28:1759–1788
18. Curà F, Gallinatti AE, Sesana R (2012) Dissipative aspects in thermographic methods. Fatigue Fract Eng Mater Struct 5:1133–1147
19. Kim J, Jeong HY (2010) A study on the hysteresis, surface temperature change and fatigue life of SM490A, SM490A-weld and FC250 metal materials. Int J Fatigue 32:1159–1166
20. Crupi V (2008) An unifying approach to assess the structural strength. Int J Fatigue 30:1150–1159
21. Luong MP (1998) Fatigue limit evaluation of metals using an infrared thermographic technique. Mech Mater 28:155–163
22. Curà F, Curti G, Gallinatti AE, Sesana R (2006) Thermomechanical model and experimental analysis of progressive fatigue damage in steel specimens. In: 8th Biennial ASME conference on engineering systems design and analysis (ESDA): 95114
23. Curti G, Curà F, Sesana R (2006) Thermomechanical model and experimental analysis of progressive fatigue damage in steels specimens. In: Proceedings of ESDA 2006, 8th Biennial ASME conference, July 4–7. Turin, Italy
24. Doudard C, Calloch S, Hild F, Cugy P, Galtier A (2005) A probabilistic two scale model for high cycle fatigue life predictions. Fatigue Fract Eng Mater Struct 28:279–288

25. Curti G, Curà F, Sesana R (2005) A new iteration method for the thermographic determination of fatigue limit in steels. Int J Fatigue 27:453–459
26. Aster RC, Shearer PM (1991) High-frequency borehole seismograms recorded in the San Jacinto Fault zone, Southern California. Part 2; attenuation and site effects. Bull Seism Soc Am 81(4):1081–1100
27. Lockner DA, Byerlee JD, Kuksenko V, Ponomarev A, Sidorin A (1991) Quasi-static fault growth and shear fracture energy in granite. Nature 350:39–42
28. Rabinovitch A, Frid V, Bahat D (2007) Surface oscillations—a possible source of fracture induced electromagnetic radiation. Tectonophysics 431:15–21
29. Amstrong BH, Valdes CM (1991) Acoustic Emission/Microseismic activity at very low strain levels. In: Sachse W, Roget J, Yamaguchi K (eds) Acoustic emission: current practice and future directions, ASTM STP 1077, American society for testing and materials, Philadelphia
30. Kuzhevskij BM, Nechaev OY, Sigaeva EA, Zakharov VA (2003) Neutron flux variations near the Earth's Crust. A possible tectonic activity detection. Nat Hazards Earth Syst Sci 3:637–645
31. Kuzhevskij BM, Nechaev OY, Sigaeva EA (2003) Distribution of neutrons near the Earth's surface. Nat Hazards Earth Syst Sci 3:255–262
32. Antonova VP, Volodichev NN, Kryukov SV, Chubenko AP, Shchepetov AL (2009) Results of detecting thermal neutrons at Tien Shan High Altitude station. Geomag Aeron 49:761–767
33. Volodichev NN, Kuzhevskij BM, Nechaev OY, Panasyuk MI, Podorolsky AN, Shavrin PI (2000) Sun-moon-earth connections: the neutron intensity splashes and seismic activity. Astron Vestnik 34:188–190
34. Sigaeva E et al (2006) Thermal neutrons' observations before the Sumatra earthquake. Geophys Research Abs 8:00435

Chapter 5
Carrara Marble: Crushing Tests, Alpha Particle Emissions, and Correlated Nuclear Transmutations

Abstract Neutron emission measurements are carried out by means of a He^3 detector on Carrara marble specimens under compression. While iron-rich minerals generate neutrons—due to phono-fission reactions involving the transformation of one iron atom into two atoms of aluminum or into other lighter elements—this phenomenon does not appear in marble crushing tests. On the other hand, significant alpha particle emissions are detected by a 6150AD-k probe during the same compression tests on marble. The external and fracture surfaces belonging to Carrara marble specimens crushed during the compression tests are analyzed by X-ray Photoelectron Spectroscopy (XPS). Such quantitative compositional analyses are carried out in order to detect any variation in Carrara marble chemical composition due to brittle failure. A total decrement in Ca, Mg, and O by 13% is exactly equivalent to the increment in C, which is observed on the fracture surface, with respect to the external surface. The assumed transmutations involve elements with an equal number of protons and neutrons. Due to this peculiar reason, the microchemical analyses suggest phono-fission reactions with alpha particle emission, and not excess neutron emission.

Keywords Carrara marble · X-ray photoelectron spectroscopy (XPS) · Alpha particle emissions · Nuclear transmutations · Alkaline-earth element decrement · Carbon increment

5.1 Preliminary Remarks

After summarizing the preliminary results already presented in [1–3], involving compression tests on Luserna stone, further experiments performed on Carrara marble specimens under mechanical compression loading are described. The neutron natural background monitoring, by means of He^3 devices and bubble detectors, was performed during the compression tests under monotonic displacement control. The compression tests on specimens with different sizes and shapes were carried out at the Fracture Mechanics Laboratory of Politecnico di Torino.

A. Carpinteri, *Terahertz Phonons and Nanomechanical Instabilities*,
https://doi.org/10.1007/978-3-032-14692-2_5

In papers [1–3], that neutron emission from Luserna stone specimens of larger dimensions exceeded the natural background level by one order of magnitude at failure. On the other hand, in the compression tests of Carrara marble specimens, the neutron measurements yield values comparable to the natural background level, even at the time of failure. While Luserna stone generates neutrons—due to nuclear reactions involving fission of iron into aluminum or other lighter elements—this phenomenon does not appear in marble crushing. However, even if no relevant neutron emissions are detected during compression loading of marble, important phono-fission reactions are observed, substantiated by a significant alpha particle flux. A similar evidence of alpha particle emission was also observed during cyclic loading tests on cylindrical steel bars [4].

The external and fracture surfaces belonging to Carrara marble specimens crushed during the compression tests are analyzed by X-ray Photoelectron Spectroscopy (XPS). A decrement in Ca, Mg, and O as well as an increment in C contents is observed on the fracture surface with respect to the external surface. These analyses suggest phono-fission reactions with alpha particle emissions and without neutron emissions.

A theoretical explanation about phono-fission reactions was firstly provided based on classical Nuclear Physics [5, 6]. An alternative explanation based on TeraHertz phonon and plasmon resonance with the atomic lattice will be provided in Part VII.

5.2 Neutron and Alpha Particle Emission Detection Techniques

Since neutrons are electrically neutral particles, they cannot directly produce ionization in a detector, and therefore cannot be directly detected. This means that neutron detectors must rely upon a conversion process where an incident neutron interacts with a nucleus to produce a secondary charged particle. These charged particles are then detected, and from them the neutron presence is deduced. For an accurate neutron evaluation, a He^3 proportional detector is used.

Moreover, a set of passive neutron detectors, based on the superheated bubble detection technique and insensitive to electromagnetic noise, are employed.

5.2.1 *He^3 Neutron Proportional Counter*

The He^3 detector used in the compression tests under monotonic displacement control is a He^3 type (Xeram, France), with electronics of preamplification, amplification, and discrimination directly connected to the detector tube. The detector, filled with 4 bars of helium-3 gas, is powered with 1.3 kV, supplied via a high voltage Nuclear Instrument Module (NIM). The logic output producing the transistor–transistor logic

(TTL) pulses is connected to a NIM counter. The device is calibrated for the measurement of thermal neutrons; its sensitivity is 65 cps/$n_{thermal}$ ($\pm$10% declared by the factory), i.e., a thermal neutron flux of 1 thermal neutron/s cm^2 corresponds to a count rate of 65 cps.

5.2.2 *Neutron Bubble Detectors*

A set of passive neutron detectors, insensitive to electromagnetic noise and with zero gamma sensitivity, is used in compression tests. The dosimeters, based on superheated bubble detection (BTI, Ontario, Canada, Bubble Technology Industries (1992)) [7], are calibrated at the factory against an Am-Be source in terms of NCRP38 (NCRP report 38 (1971)) [8]. Bubble detectors provide instant visible detection and measurement of neutron dose. Each detector is composed of a polycarbonate vial filled with elastic tissue-equivalent polymer, in which droplets of a superheated gas (Freon) are dispersed. When a neutron strikes a droplet, the latter immediately vaporizes, forming a visible gas bubble trapped in the gel. The number of bubbles provides a direct measurement of the equivalent neutron dose. These detectors are suitable for neutron integral dose measurements, in the energy ranges of thermal neutrons ($E = 0.025$ eV) and fast neutrons ($E > 100$ keV).

5.2.3 *The 6150AD-K Probe for Alpha Particle Measurements*

For alpha particle emissions, a 6150AD-k probe with a sealed proportional counter is used. A peculiarity of this device is that it does not require refilling or flushing from external gas reservoirs. The probe is sensitive to alpha and/or beta particles, and/or to gamma radiation, by means of a removable discriminator plate (stainless steel, 1 mm thick). An electronic switch allows for the operating mode “alpha” to detect alpha radiation only. In this operational mode, the radiation detection is very sensitive because the background level is much lower. The 6150AD-k probe is used in the operating mode “alpha” also to monitor the background level before and after the test.

5.3 Tests on Luserna Stone Specimens

During previous experimental compression tests, neutron emissions were measured from nine Luserna stone cylindrical specimens, of different size and shape (Table 5.1, Fig. 5.1), denoted by P1, P2,..., P9 [1–3]. In the following, the main results are briefly presented. The He^3 neutron detector was switched on at least one hour before the beginning of each compression test, in order to reach the thermal equilibrium of electronics, and to make sure that the behavior of the device was stable with respect to intrinsic thermal effects. The detector was placed in front of the test specimen at a distance of 10 cm and it was enclosed in a polystyrene case in order to avoid "spurious" signals coming from impacts and vibrations.

Neutron measurements for specimens P2, P3, P4, P7 yielded values comparable to the natural background, whereas in specimens P1 and P5, the experimental data exceeded the background level by approximately four times. For specimen P6, neutron emissions of about five times the background level were observed concomitant to the sharp stress drop at the time of failure, whereas for specimens P8 and P9, the neutron emissions achieved values by approximately one order of magnitude higher than the natural background. In Table 5.1, the experimental data from the compression tests on the nine Luserna stone specimens are summarized. These phenomena were assumed to be caused by reactions occurring in iron atoms [9, 10].

Considering the experimental results from the previous work [11], a volume approximately exceeding 200,000 mm^3, combined with the extreme brittleness of the tested material, represents a threshold value for a neutron emission of about one order of magnitude higher than the natural background. As regards the expected energy spectrum, it extends from thermal neutrons (0.025 eV) up to the fast components (few MeV).

5.4 Tests on Carrara Marble Specimens: Experimental Set-Up and Results

Neutron emissions are measured on 27 Carrara marble cylindrical specimens, three for each size and shape (see Table 5.2), denoted by M1, M2,..., M9. In Fig. 5.2a, Carrara marble specimens with the same size but different slenderness are shown. The tests are carried out by means of a MTS servo-hydraulic press, with a maximum capacity of 1,000 kN, working with an electronic control unit. The applied force is determined by measuring the pressure in the loading cylinder by means of a transducer. The specimens are arranged with the two smaller surfaces in contact with the press platens, without coupling materials in-between, according to the testing modalities known as "test by means of rigid platens with friction". The platens are controlled by means of a wire-type potentiometric displacement transducer. The tests are performed under displacement control, with the planned displacement velocity equal to 0.0005 mm/s for the specimens with a slenderness 0.5, 0.001 mm/s for slenderness 1, and 0.002 mm/s for slenderness 2.

A comparative measurement of natural neutron background is performed in order to assess the average background affecting data acquisition in experimental room conditions. The He^3 device is located in the same position as the experimental set-up and the background measurements are performed fixing at 60 s the acquisition time, during a preliminary period of more than three hours, for a total number of 200

Table 5.1 Luserna stone specimens under compression test with monotonic displacement control: specimen size-scale/slenderness and neutron emissions

Specimens	Dimensions		Piston velocity	Volume	Average peak load	Average neutron background	Count rate at the peak load
	Diameter (mm)	Slenderness	(m/s)	(mm^3)	(kN)	(10^{-2} cps)	(10^{-2} cps)
P1	28	0.5	1×10^{-6}	8,616	52.19	3.17 ± 0.32	8.33 ± 3.73
P2	28	1	1×10^{-6}	17,232	33.46	3.17 ± 0.32	Background
P3	28	2	1×10^{-6}	34,464	41.28	3.17 ± 0.32	Background
P4	53	0.5	1×10^{-6}	58,434	129.00	3.83 ± 0.37	Background
P5	53	1	1×10^{-6}	116,868	139.10	3.84 ± 0.37	11.67 ± 4.08
P6	53	2	1×10^{-6}	233,736	206.50	4.74 ± 0.46	25.00 ± 6.01
P7	112	0.5	1×10^{-5}	551,434	1,099.30	4.20 ± 0.80	background
P8	112	1	1×10^{-5}	1,102,868	1,077.10	4.20 ± 0.80	30.00 ± 11.10
P9	112	2	1×10^{-5}	2,205,736	897.80	4.20 ± 0.80	30.00 ± 10.00

Fig. 5.1 Granite (Luserna stone) cylindrical specimens of different size-scale (Diameter = 28, 53, 112 mm) and slenderness (0.5, 1.0, 2.0), to emphasize the related brittleness effects. Specimens P1, P2, and P4 show neutron emissions close to the natural background level, whereas Specimens P9, P8, and P6 show neutron emissions approximately of one order of magnitude higher than that of the natural background level

Table 5.2 Carrara marble specimens under compression test with monotonic displacement control: specimen size-scale/slenderness and neutron emissions

Specimens	Dimensions		Piston velocity	Volume	Average peak load	Average neutron background	Count rate at the peak load
	Diameter (mm)	Slenderness	(m/s)	(mm^3)	(kN)	(10^{-2} cps)	(10^{-2} cps)
M1	25	0.5	5×10^{-7}	6,133	69.0	5.64 ± 1.41	Background
M2	25	1	1×10^{-6}	12,266	47.4	5.64 ± 1.41	Background
M3	25	2	2×10^{-6}	24,532	36.1	5.64 ± 1.41	Background
M4	50	0.5	5×10^{-7}	49,062	223.7	7.29 ± 1.83	Background
M5	50	1	1×10^{-6}	98,125	164.1	7.29 ± 1.83	Background
M6	50	2	2×10^{-6}	196,250	144.5	7.29 ± 1.83	Background
M7	100	0.5	5×10^{-7}	392,500	909.0	6.25 ± 1.55	Background
M8	100	1	1×10^{-6}	785,000	880.6	6.25 ± 1.55	Background
M9	100	2	2×10^{-6}	1,570,000	714.0	6.25 ± 1.55	Background

counts. The average measured background level is ranging from $(5.64 \pm 1.41) \times 10^{-2}$ to $(7.29 \pm 1.83) \times 10^{-2}$ cps (Table 5.2).

Additional background measurements are repeated before each test, fixing an acquisition time of 60 s in order to check a possible variation in natural background. Neutron measurements for all specimens yield values comparable to the natural background. In Table 5.2, the experimental data concerning compression tests on the

(a) (b)

Fig. 5.2 **a** Carrara marble cylindrical specimens of constant diameter and different slenderness. **b** Carrara marble specimen M9 after compression failure. The bubble detectors are placed all around the monitored specimen. On the left, the He^3 neutron detector is located at a distance of about 20 cm

Carrara marble specimens are summarized. In Fig. 5.2b, Carrara marble specimen M9 is shown after its crushing failure.

As an example, in Fig. 5.3, the load vs. time diagram and the neutron count rate evolution for specimens M3, M6, and M9 are shown. We can observe that the specimens of smaller dimension (M3) present a ductile behavior in compression, whereas, by increasing the sample size, the behavior becomes more brittle. As a matter of fact, specimens M9 show a catastrophic failure [11].

Bubble counting in the bubble detectors are performed at the end of each compression test and the equivalent neutron dose is calculated. In the same way, the natural background is estimated by means of two bubble dosimeters used for assessment. The natural background is found to be (20.45 ± 4.09) nSv/h. Similarly to the He^3 results, no relevant neutron emissions were detected.

5.5 Microchemical-Composition Evidence of Phono-Fission Reactions in Marble Specimens

X-ray Photoelectron Spectroscopy (XPS) is performed on different samples of external and fracture surfaces belonging to one of the three M9 marble specimens crushed during the compression tests. XPS quantitative compositional analyses are carried out in order to detect any possible variation in Carrara marble composition. The XPS analysis is developed with a VersaProbe5000 Physical Electronics instrument, a monochromatic Al source (1486.6 eV), and a hemispherical analyzer.

Survey scans as well as narrow scans are recorded with a 100 μm spot depth over an area of 500 μm $\times$ 500 μm. An electron flood gun combined with an Argon ion

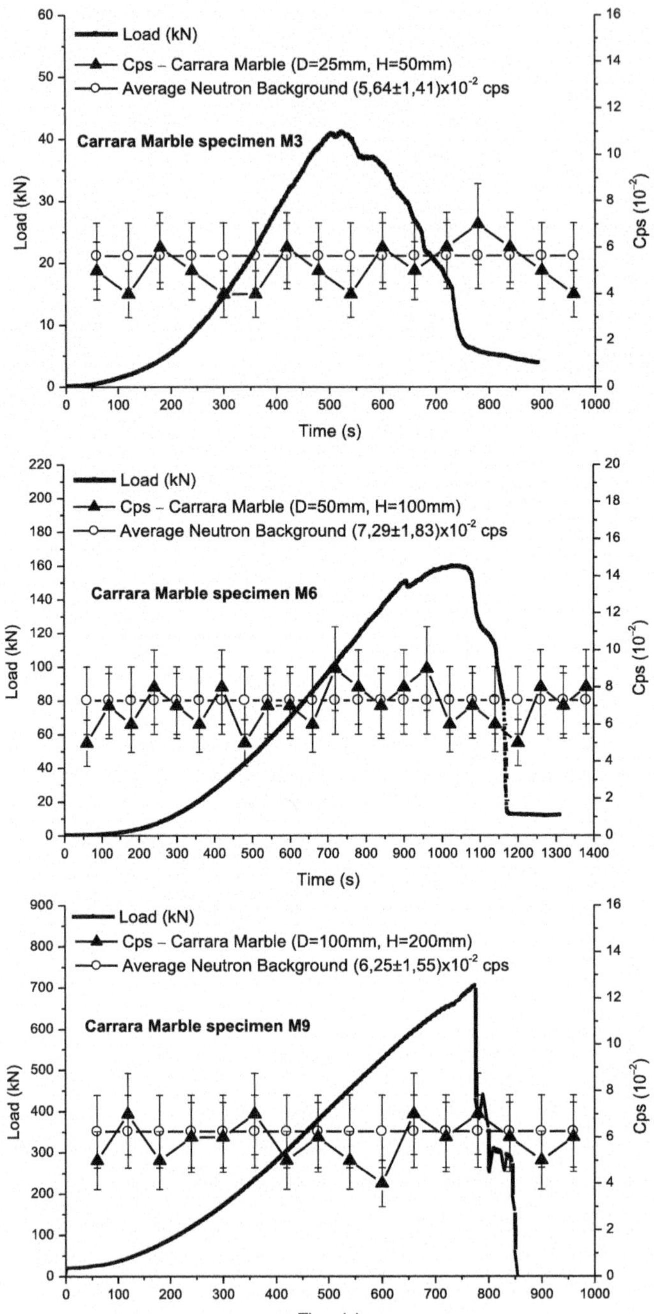

Fig. 5.3 Specimens M3, M6, M9: load versus time diagrams and neutron emission count rates

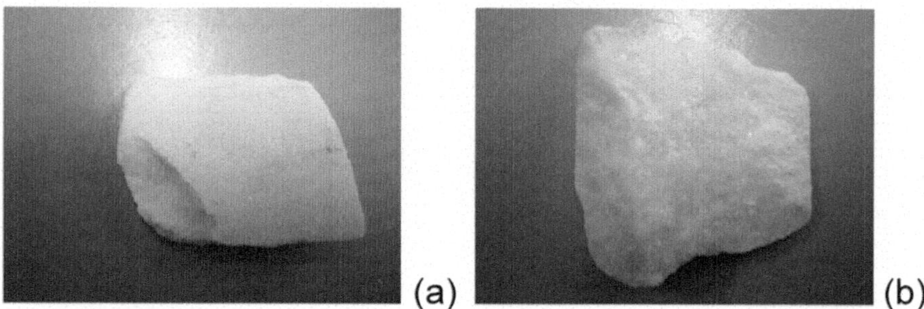

Fig. 5.4 **a** Portion from the external specimen surface. **b** Fracture surface in a fragment belonging to the same specimen

gun is used to control surface charging during the measurements. XPS survey scan (pass energy of 187.85 eV) of Carrara marble surfaces show that Ca, Mg, O, and C are present. The high-resolution C(1 s), Ca(2p), Mg (1 s), and O(1 s) spectra are obtained at a pass energy of 23.50 eV.

In Fig. 5.4a, the external surface of a portion of one of the three M9 specimens is shown. In Fig. 5.4b, the fracture surface of a fragment taken from the same specimen is shown. For the XPS analyses, several spots are localized on the surface of the thin section and on the surface of the fragment. Thirty spots on the external surface and twenty on the fracture surface are selected and analyzed. In Fig. 5.5a–d, the results for the Ca, Mg, O, and C concentrations are shown.

The distributions of O, Ca, and Mg on the external surface, represented in the graphs by black squares, show average values (calculated as the arithmetic mean values) respectively equal to 45.8, 13.4, and 0.7%. In the same graphs, the distributions of O, Ca, and Mg concentrations on the fracture surface (indicated by red triangles) show significant variations. It can be seen that the mean values in this case are respectively equal to 36.8, 9.8, and 0.3%. They are considerably lower than the mean values related to the external surface.

Similarly to Fig. 5.5a–c, in Fig. 5.5d the C mass percentage concentrations are considered in both cases of external and fracture surfaces. The observed average variation shows a mass percentage increment approximately equal to the percentage decrement in O, Ca, and Mg. The average increment in the distributions corresponding to the fracture surface (indicated by red triangles) is approximately equal to 13.0%. The average value of C concentration changes from 40.1% on the external surface to 53.1% on the fracture surface. The evidence emerging from the XPS analyses is that the global value of oxygen, calcium, and magnesium decrement (−13%), and that of C increment (+13.0%) are of the same amount (Table 5.3).

From the results previously shown, it can be clearly seen that phono-fission reactions are possible in the inert and non-radioactive Carrara marble specimens. From the XPS results on crushed samples, the evidence of O, Ca, Mg, and C variations leads to the conclusion that the following phono-fission reactions are occurring:

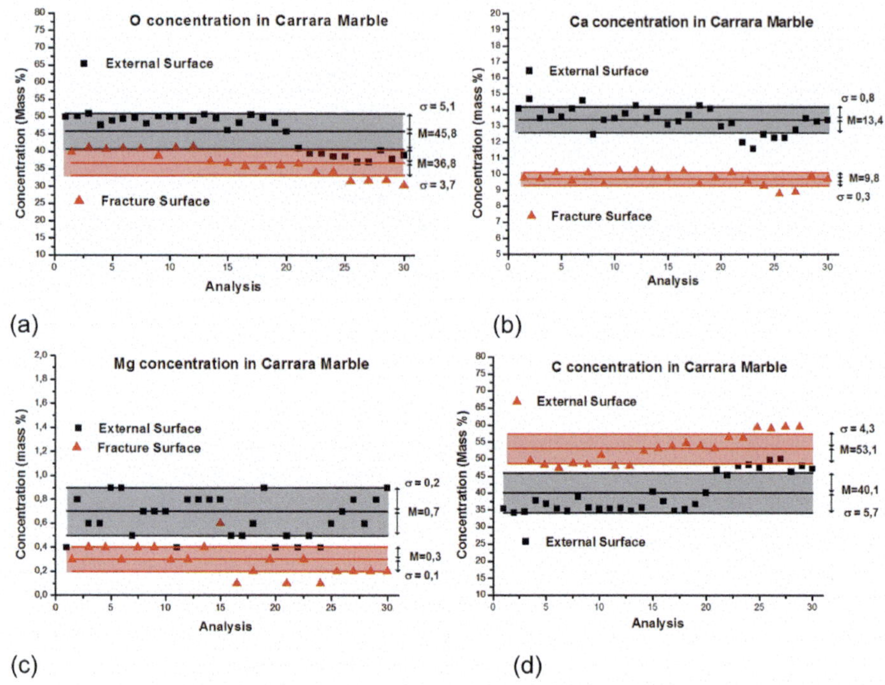

Fig. 5.5 O, Ca, Mg, and C concentrations in a Carrara marble specimen: **a** O on the external surface (black squares) and on the fracture surface (red triangles); **b** Ca on the external surface (black squares) and on the fracture surface (red triangles); **c** Mg on the external surface (black squares) and on the fracture surface (red triangles); **d** C on the external surface (black squares) and on the fracture surface (red triangles)

Table 5.3 Mass percentage concentrations of the involved chemical elements on the external and fracture surfaces, i.e., before and after the experiment (Carrara marble)

	Mass percentage concentrations (%)		Difference of mass percentage concentrations (%)
	External surface	Fracture surface	
Oxygen	45.8 ± 5.1	36.8 ± 3.7	–9.0
Calcium	13.4 ± 0.8	9.8 ± 0.3	–3.6
Magnesium	0.7 ± 0.2	0.3 ± 0.1	–0.4
Carbon	40.1 ± 5.7	53.1 ± 4.3	+ 13.0

$$O_8^{16} \rightarrow C_6^{12} + He_2^4 \tag{5.1}$$

$$Ca_{20}^{40} \rightarrow 3C_6^{12} + He_2^4 \tag{5.2}$$

$$Mg_{12}^{24} \rightarrow 2C_{6}^{12} \tag{5.3}$$

These transmutations involve elements (O, Ca, and Mg) with an equal number of protons and neutrons, like their reaction products (C and He). For this peculiar reason, the microchemical analyses suggest phono-fission reactions without emission of excess neutrons from marble specimens. The potential problem of hydrocarbon impurity [12] cannot bring into question the previous conclusions.

5.6 Alpha Particle Monitoring

Alpha particle emissions are measured on three additional Carrara marble cylindrical specimens (A1, A2, and A3) with different slenderness, from 0.5 to 2. Similarly to Luserna stone, the tests are carried out by means of a MTS servo-hydraulic press, with a maximum capacity of 1,000 kN. The tests are performed under displacement control, with a planned displacement velocity equal to 0.0005 mm/s for the specimen of slenderness 0.5, 0.001 mm/s for slenderness1, and 0.002 mm/s for slenderness 2.

A comparative measurement of alpha natural background is performed in order to assess the average background affecting data acquisition in experimental room conditions. The 6150AD-k probe is located in the same position as the experimental set-up during a preliminary period of 24 h, to detect the typical daily fluctuation of alpha field mostly due to radon concentration variability [13]. The average measured background level is approximately of $(12.00 \pm 3.00) \times 10^{-2}$ cps.

In Table 5.4, the experimental data concerning the compression tests on the Carrara marble specimens are summarized.

In Fig. 5.6, the load versus time diagram and the alpha count rate evolution for specimen A3 are reported. The specimen shows a catastrophic behavior at the final collapse, with a maximum alpha emission of $(110.00 \pm 28.00) \times 10^{-2}$ cps, about 10 times higher than the environmental level.

For all the specimens, a considerable increment in alpha counting rate (counts per second) is observed. The counting rate decreases to the typical background level at the end of each experiment.

Table 5.4 Carrara marble compression tests under monotonic displacement control: Alpha particle emissions

Specimens	Dimensions		Piston velocity	Volume	Average peak load	Average alpha background	Maximum count rate at the peak load
	Diameter (mm)	Slenderness	(m/s)	(mm³)	(kN)	(10^{-2} cps)	(10^{-2} cps)
A1	50	2	2×10^{-6}	196,250	154.9	12.00 ± 3.00	46.00 ± 12.00
A2	100	0.5	5×10^{-7}	392,500	810.8	12.00 ± 3.00	61.00 ± 15.00
A3	100	1	1×10^{-6}	785,000	737.9	12.00 ± 3.00	110.00 ± 28.00

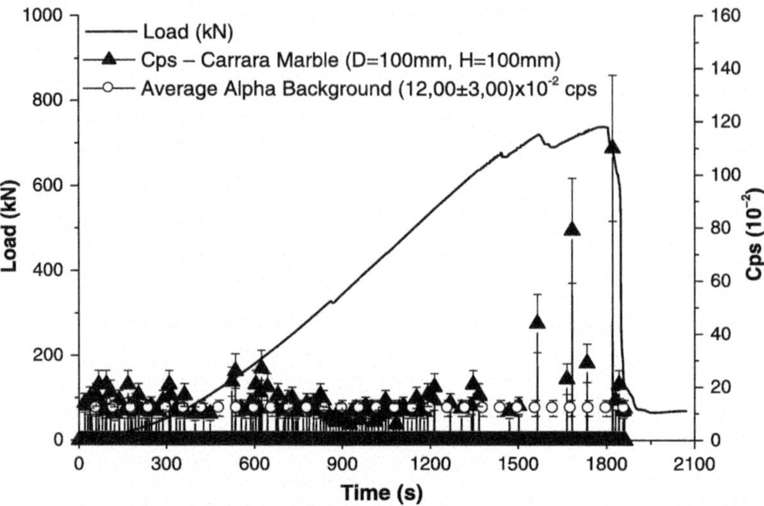

Fig. 5.6 Marble specimen A3: Load versus time diagram and alpha emission counting rate

Recent works by Moiser-Boss [14, 15] also demonstrate the presence of high-energy alpha particles emitted during experiments of Pd/D codeposition.

5.7 Conclusions

In the two previous chapters, neutron emission measurements performed on iron-rich natural rocks in compression were shown. From these experiments, it can clearly be assumed that phono-fission reactions giving rise to excess neutron emissions are possible in inert non-radioactive solids. In particular, during compression tests on specimens of sufficiently large size, the neutron flux was found to be up to orders of magnitude higher than the background level at the time of the catastrophic failure. Further neutron emission measurements were carried out also on Carrara marble specimens. Even if no significant neutron emissions are detected, important phono-fission reactions are observed, also confirmed by considerable alpha particle emissions up to one order of magnitude higher than the natural background level. The compositional analysis by X-ray Photoelectron Spectroscopy (XPS) confirms the photo-fission reactions involving the transmutation of calcium, magnesium, and oxygen into carbon. As in the case of Luserna stone, this conjecture finds a surprising evidence and confirmation at the Earth crust scale. As a matter of fact, the phono-fission reactions could interpret the most significant geophysical and geological transformations, today still unexplained, like those regarding calcareous rocks.

References

1. Carpinteri A, Cardone F, Lacidogna G (2010) Energy emissions from failure phenomena: mechanical, electromagnetic, nuclear. Exp Mech 50:1235–1243
2. Carpinteri A, Borla O, Lacidogna G, Manuello A (2010) Neutron emissions in brittle rocks during compression tests: monotonic versus cyclic loading. Phys Mesomech 13:268–274
3. Carpinteri A, Lacidogna G, Manuello A, Borla O (2011) Energy emissions from brittle fracture: neutron measurements and geological evidences of piezonuclear reactions. Strength, Fract Complex 7:13–31
4. Albertini G, Calbucci V, Cardone F, Fattorini G, Magnani R, Petrucci A, Ridolfi F, Rotili A (2013) Evidence of alpha emission from compressed steel bars. Int J Mod Phys B 27(23):1350124
5. Widom A, Swain J, Srivastava YN (2013) Neutron production from the fracture of piezoelectric rocks. J Phys G: Nucl Part Phys 40(015006):1–8
6. Widom A, Swain J, Srivastava YN (2014) Photo-disintegration of the iron nucleus in fractured magnetite rocks with magnetostriction. Meccanica 50:1205–1216
7. Bubble Technology Industries (1992) Instruction manual for the bubble detector. Chalk River, Ontario, Canada
8. National Council on Radiation Protection and Measurements (1971) Protection against neutron radiation, NCRP Report 38
9. Cardone F, Cherubini G, Petrucci A (2009) Piezonuclear neutrons. Phys Lett A 373:862–866
10. Cardone F, Mignani R, Petrucci A (2009) Piezonuclear decay of thorium. Phys Lett A 373:1956–1958
11. Carpinteri A, Lacidogna G, Manuello A, Borla O (2013) Piezonuclear fission reactions from earthquakes and brittle rocks failure: evidence of neutron emission and non-radioactive product elements. Exp Mech 53:345–365
12. Ni M, Ratner BD (2008) Differentiating calcium carbonate polymorphs by surface analysis techniques—an XPS and TOF-SIMS study. Surf Interface Anal 40:1356–1361
13. Postendörfer J, Butterweck G, Reineking A (1994) Daily variation of the radon concentration indoors and outdoors and the influence of meteorological parameters. Health Phys 67:283–287
14. Mosier-Boss PA et al (2007) Use of CR-39 in Pd/D co-deposition experiments. Eur Phys J Appl Phys 40:293–303
15. Mosier-Boss PA et al (2010) Comparison of Pd/D co-deposition and DT neutron generated triple tracks observed in CR-39 detectors. Eur Phys J Appl Phys 51:20901–20911

Chapter 6
Iron-Enriched Mortar: Crushing Tests and Chemical Composition Instrumental Neutron Activation Analysis (INAA)

Abstract The previous chapters concerning neutron emission measurements highlight phono-fission reactions during mechanical tests on iron-rich natural materials. Based on this experimental evidence, iron can be considered as a chemical element subject to fission into aluminum or into magnesium and silicon. In the present chapter, the Instrumental Neutron Activation Analysis (INAA) is applied in order to provide an additional experimental evidence to elemental content variations in mortar specimens subjected to compression tests up to crushing failure. To emphasize the occurrence of such a phenomenon also in artificial materials, the specimens are highly enriched of iron oxide. Twenty-four chemical elements, including iron, aluminum, magnesium, and silicon, are quantified before and after the mechanical tests by means of both chemical and INAA analyses. Our intention is mainly that of confirming the presence of low-energy nuclear reactions involving fission of iron into aluminum. To this purpose, the volumetric concentrations of iron and aluminum before and after the compression tests of the mortar specimens are presented and discussed.

Keywords Compression tests · Iron-enriched mortar · Phono-fission reactions · Instrumental neutron activation analysis (INAA) · Iron decrement · Aluminum increment

6.1 Preliminary Remarks

The results presented in [1–3] and related to compression tests on Luserna stone samples demonstrate that neutron emissions detected from specimens of sufficiently large dimensions exceed the natural background level by approximately one order of magnitude at the moment of specimen failure.

These emissions are due to phono-fission reactions of iron into aluminum, or into magnesium and silicon. The assumed fissions are supported by spectroscopical analyses of the fracture surface. The results of Energy Dispersive X-ray Spectroscopy (EDS), performed on samples coming from the Luserna stone specimens of the earlier experiments [1–3], show that on the fracture surfaces a considerable reduction in the

A. Carpinteri, *Terahertz Phonons and Nanomechanical Instabilities*,
https://doi.org/10.1007/978-3-032-14692-2_6

iron content (~25%) is very consistently balanced by an increment in Al, Si, and Mg concentrations [4]. A theoretical explanation to such reactions was provided in [5, 6], whereas an alternative explanation based on the resonance of phonons and plasmons with atomic lattices will be given in Part VII.

In the present chapter—after reporting an interesting experimental result on a single prismatic specimen ($4 \times 4 \times 16$ cm^3) of cementitious mortar enriched with iron oxides—a new experimental evidence of phono-fission reactions is obtained with cubic mortar specimens (of 1.00 cm side) enriched with iron oxides and subjected to mechanical loading. The purpose is mainly that of analyzing the nuclear reaction, $Fe^{56} \rightarrow 2Al^{27} + 2n$, involving the symmetrical fission of iron into two atoms of aluminum, during the compression tests. The last are carried out at the Fracture Mechanics Laboratory of Politecnico di Torino. During the compression test of the single prismatic specimen ($4 \times 4 \times 16$ cm^3), a neutron emission approximately three times higher than the average natural background is observed at the moment of brittle failure. On the other hand, due to the small dimensions of the cubic specimens, no relevant neutron emissions are detected during mechanical loading. In any event, an important phono-fission evidence is observed in the crushed mortar samples. They are analyzed by the Instrumental Neutron Activation Analysis (INAA) [7] measuring the γ radiation emitted during the subsequent radioactive decay of the produced nuclei. This analytical technique offers positive features such as high accuracy, sensitivity, and the possibility of multi-element analysis.

Compared to the EDS technique that is able to analyze only the fracture surface, the INAA technique allows a detailed volumetric analysis for the determination of elements even in minimal amounts. For this reason, the two compositional analysis methods are considered as complementary techniques. Moreover, we are dealing with a non-destructive procedure: no complex operations like dissolution, digestion, or other chemical treatments of the samples are required, reducing the possibility of sample contamination. The irradiations with neutrons are carried out in the Laboratory for Applied Nuclear Energy (LENA) of the University of Pavia, in a TRIGA MARKII reactor (General Atomics) [8]. The samples are irradiated for different times with different neutron fluencies depending on matrix type and on the elements to be determined. The experimental data collected during the tests highlight a significant increment in Al^{27}, up to 66% of the initial concentration. Since the measurement method based on INAA is validated with the Certified Reference Materials (CRMs) (NIST 2709 San Joaquin Soil and NIST 2704 Buffalo River Sediment), the detected variations can be associated to the phono-fission reactions occurring during the failure test.

6.2 Instrumental Neutron Activation Analysis (INAA)

The Instrumental Neutron Activation Analysis (INAA) is a nuclear analytical technique useful for the qualitative and quantitative determination of trace elements and macro-constituents in different matrices.

A sample is subjected to a neutron flux and radioactive nuclides are produced. During the radioactive decay, the nuclides emit gamma rays, whose energies are characteristic for each nuclide. The intensity of these gamma rays compared to those emitted by a standard nuclide allows a measure of the concentrations of the various elements.

The neutron-gamma reaction is the fundamental one in neutron activation analysis. For example, consider the following reaction:

$$\mathrm{Al}^{27}(\mathrm{n},\gamma) \rightarrow \mathrm{Al}^{28} + \beta^{-} + \gamma \tag{6.1}$$

where Al^{27} is a stable isotope of aluminum, whereas Al^{28} is a radioactive isotope (with a half-life t½ of 139 s). The gamma ray emitted during the decay of the Al^{28} nucleus has an energy of 1,778.9 keV and this gamma ray is characteristic for this nuclide.

On the other hand, the activity of a particular radionuclide, at any time t during an irradiation, can be calculated from the following equation:

$$A_t = \sigma_{\mathrm{act}}\varphi N\left(1-\mathrm{e}^{-\lambda t}\right) \tag{6.2}$$

where A_t is the activity in number of decays per unit time, σ_{act} is the activation cross-section, φ is the neutron flux (usually given in number of neutrons cm^{-2} s^{-1}), N is the number of parent atoms, λ is the decay constant (number of decays per unit time), and t is the irradiation time. After the sample has been activated and the resulting gamma ray energies and intensities have been determined using a solid-state detector (usually Germanium), gamma rays passing through the detector generate free-electrons. The number of electrons (current) is related to the energy of the gamma rays.

Each radioactive nuclide is also decaying during the counting interval and corrections must be considered for this decay. The standard form of the radioactive decay correction is

$$A = A_0\mathrm{e}^{-\lambda t} \tag{6.3}$$

where A is the activity at any time t, A_0 is the initial activity, λ is the decay constant, and t is time.

6.3 Preliminary Test on a Single Prismatic Mortar Specimen

As anticipated in Sect. 6.1, a prismatic cementitious mortar specimen enriched with iron, measuring $4 \times 4 \times 16$ cm^3 (Fig. 6.1, left), is tested under compression by an electronic controlled servo-hydraulic press with a maximum capacity of 1,000 kN. This machine makes it possible to carry out tests in either load control or displacement control. The tests are performed by piston travel displacement control, setting a velocity of 0.001 mm/s. Neutron emission measurements are made by means of a He^3 proportional counter [2, 3] placed at a distance of 10 cm from the specimen and enclosed in a polystyrene case, to prevent impacts and vibrations. The detector relies on a conversion process where an incident neutron interacts with a nucleus to produce a secondary charged particle. These charged particles are then detected, and from them the neutron presence is deduced. In particular, the device is calibrated for the measurement of thermal neutrons; its sensitivity is 65 cps/$n_{thermal}$ (±10% declared by the factory), i.e., a thermal neutron flux of 1 thermal neutron/s cm^2 corresponds to a count rate of 65 cps. The monitored neutron emissions (Fig. 6.1, right) exceed the natural background level by approximately three times, when the brittle failure occurs.

Although the neutron activation analysis considers only the iron isotope 58, whereas the investigated phono-fission reaction regards Fe^{56}, important considerations can be nevertheless made for these iron nuclei.

It is well-known that the natural abundance of Fe^{58} is 0.3% (compared to 91.7% of Fe^{56}). From the point of view of neutron activation, it is the isotope on which the concentration analysis of iron is based. Once subjected to a suitable neutron flux, the radioactive nuclide Fe^{59} is produced with a half-life of about 45 days. This isotope is accompanied by three different gamma ray energy emissions, typical of this nuclide, and then iron concentration is calculated.

In particular, the INAA results for the fragments coming from the fracture surface show a lower concentration of iron (–14%) if compared to the ones contained on the external surface of the specimen. This iron decrement appears to be balanced by a rather close increment in aluminum (+19%). Moreover, being the tested artificial material characterized by a sufficiently precise and homogeneous elemental distribution, the results of neutron activation analysis are even more accurate and reliable.

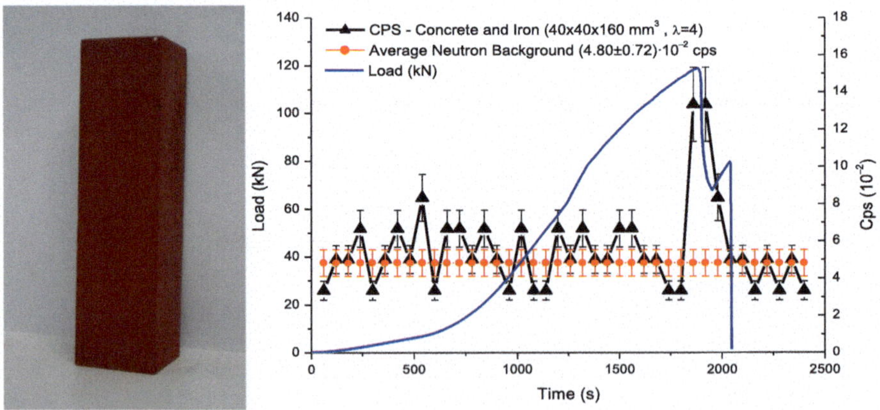

Fig. 6.1 Prismatic mortar specimen (4 × 4 × 16 cm^3) enriched with iron oxide (left). Load versus time diagram and neutron emission count rate (right)

The results obtained in this preliminary test lead us to consider specimens of smaller dimensions (cubes of side 1.00 cm) to verify, by means of the INAA (Instrumental Neutron Activation Analysis) technique, whether the neutron emissions are linked to compositional changes in the elements constituting the samples.

6.4 Tests on Small Cubic Mortar Specimens: Experimental Set-Up

Phono-fission evidence is observed in five of the seven cubic mortar specimens (of 1.00 cm side) enriched with iron oxides. In Fig. 6.2(left), the specimens are shown. The tests are carried out by a MTS servo-hydraulic press, with a maximum capacity of 250 kN, working by a digital type electronic control unit. The force applied is determined by measuring the pressure in the loading cylinder by means of a transducer. The specimens are arranged (Fig. 6.2, right) with two surfaces in contact with the press platens, without coupling materials in-between, according to the testing modalities known as "test by means of rigid platens with friction". The platens are controlled with a wire-type potentiometric displacement transducer. Finally, the tests are performed under displacement control, with a planned displacement velocity equal to 0.001 mm/s.

During the experimental tests, the full stress–strain curves are considered. The complete curve for specimen P6 is plotted in Fig. 6.3(left). After the specimen crushing, down to a heap of fragments (Fig. 6.3, right), the resistance to further deformation reaches a minimum after about 500 s from the beginning of the compression test. After that, the load begins to increase once more. The load increment continues until all the fragments are pulverized. The slope of the load versus displacement

Fig. 6.2 Cubic mortar specimens (1.00 cm side) enriched with iron oxide (left). Mortar specimen during the compression test: The He^3 neutron detector is positioned at a distance of about 25 cm from the specimen (right)

curve then coincides with the testing machine stiffness. The test is stopped when a load five times higher than the peak load is reached.

In Table 6.1, experimental data concerning the compression tests on the seven mortar specimens are summarized.

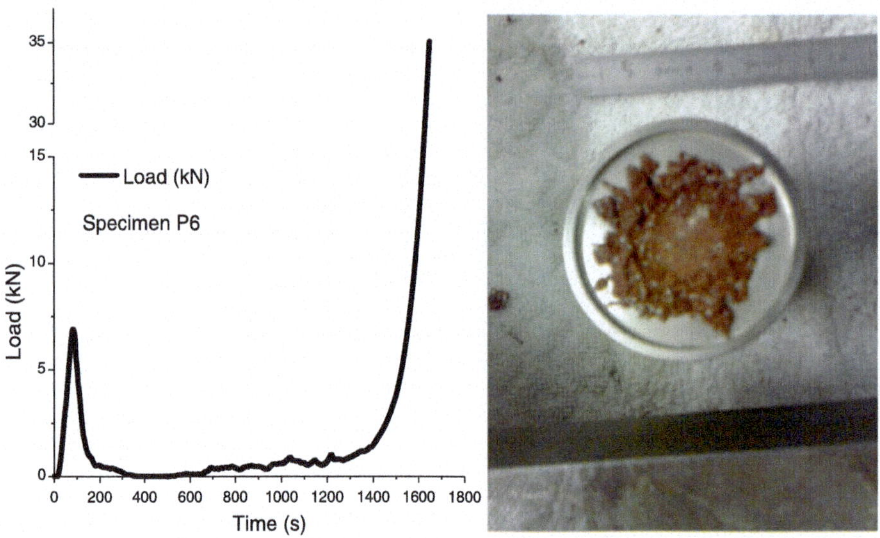

Fig. 6.3 Specimen P6, load versus time diagram (left). Specimen P6 after the crushing test (right)

Table 6.1 Compression tests on iron-enriched mortar specimens. Dimensions and neutron emissions

	Dimension	Piston velocity	Peak load	Load at compaction	Neutron background	Neutron count rate at the peak load
Specimens	Side (mm)	(m/s)	(kN)	(kN)	(10^{-2} cps)	(10^{-2} cps)
P1	10	1×10^{-6}	2.69	33.45	3.73 ± 0.37	Background
P2	10	1×10^{-6}	7.11	52.66	3.73 ± 0.37	Background
P3	10	1×10^{-6}	6.50	44.94	3.73 ± 0.37	Background
P4	10	1×10^{-6}	7.55	30.23	3.73 ± 0.37	Background
P5	10	1×10^{-6}	7.89	34.66	3.73 ± 0.37	Background
P6	10	1×10^{-6}	6.89	35.27	3.73 ± 0.37	Background
P7	10	1×10^{-6}	5.71	31.14	3.73 ± 0.37	Background

6.5 Tests on Small Cubic Mortar Specimens: Experimental Results

In order to verify possible phono-fission reactions, an analysis of the variations in the concentration of the isotope Al^{27} by means of INAA is carried out on the mortar specimens before and after the compression tests.

In Table 6.2, the chemical composition analysis before the compression test on three of the seven tested specimens is reported.

As explained in great detail in [1–3], we intend to detect the nuclear reaction, $Fe^{56} \rightarrow 2Al^{27} + 2n$, involving the symmetrical fission of iron into two atoms of aluminum, during the compression test. Iron oxide was added to the samples in

Table 6.2 Mortar specimens: Chemical composition analysis before the compression test 1

Concentration (μg/g)						
Element	Sample 1	Sample 2	Sample 3	Mean value	Standard deviation %	Mass %
Mg	3,690	3,798	3,960	3,906	6.0	0.39
Ca	148,987	139,929	136,780	138,400	5.8	14
V	7.4	9.2	5.5	7.1	16	0.0007
Al	14,771	14,928	15,260	14,720	5.3	1.5
Mn	179	170	173	172	6.2	0.02
Si	244,616	270,000	255,663	254,424	3.4	25
Sm	0.72	0.62	0.66	0.67	7.1	6.7E-05
Mo	0.11	0.22	0.25	0.20	37	2.0E-05
U	0.51	0.43	0.65	0.53	21	5.3E-05
Br	0.25	0.21	0.21	0.23	9.3	2.3E-05
As	3.6	3.1	5.1	4.0	26	0.0004
K	6,601	6,689	5,032	6,107	15	0.61
La	5.0	4.7	4.4	4.7	6.0	0.0005
Ce	5.8	6.3	6.6	6.2	6.7	0.0006
Se	0.26	0.14	0.27	0.22	32	2.2E-05
Th	1.6	1.9	2.2	1.9	15	0.0002
Cr	32	30	34	32	6.5	0.003
Hf	0.74	0.88	0.96	0.86	13	8.6E-05
Cs	0.64	0.64	0.54	0.60	9.0	6.0E-05
Ni	12	12	12	12	0.34	0.001
Sc	0.68	0.69	0.92	0.76	18	7.6E-05
Rb	23	29	24	25	12	0.003
Fe	45,271	51,168	53,951	50,130	8.8	5.0
Ta	0.11	0.11	0.08	0.10	16	1.0E-05

order to emphasize such a nuclear reaction. An increment in the isotope Al^{27} is therefore expected.

In this framework, the seven cubic specimens of 1.00 cm side are prepared. The elemental content of Al^{27} in each specimen is measured by INAA. After a cooling time of 6 months, during which the induced radioactivity became negligible, the seven specimens are subjected to failure. After the failure test, the measurement of the elemental content of Al^{27} in each specimen is repeated.

The INAA feature of being a non-destructive technique allows to measure the elemental content in the same specimen before and after the failure test. To quantify the Al^{27} content, the nuclear reaction Al^{27} $(n, \gamma) \rightarrow Al^{28}$ (t½ 139s) is considered. The corresponding gamma emission occurs at 1,778.9 keV. Samples, standards, and CRMs are irradiated for 30 s with a neutron flux of $1 \times 10^{13} cm^{-2} s^{-1}$ in the pneumatic fast transfer thimble (Rabbit) of the reactor. The gamma spectrometry is carried out by counting the gamma emission for 300 s after a cooling time of 300 s. A home-made standard solution is prepared as a comparator starting from a certified primary solution (Inorganic Ventures) of aluminum with a concentration of $9{,}996 \pm 56$ μg/ml. The analytical method is checked by measuring the Al^{27} content of two certified reference materials (NIST 2709 San Joaquin Soil and NIST 2704 Buffalo River Sediment).

The aluminum solution is prepared gravimetrically by pipetting aliquots of the certified primary solution onto a filter paper rolled up as a cylinder and inserted in polyethylene vials (Kartell). The same vials are also used for the mortar samples and the Certified Reference Materials (CRM). Samples, standards, CRMs, and blanks (polyethylene vials) are inserted in containers for neutron irradiation.

The gamma counting facility consists of an HPGe detector coupled to a multi-channel acquisition system (DSPEC from ORTEC-USA). The collected spectra are analyzed and processed with the Gammavision (ORTEC-USA) software package.

The results obtained during the INAA analyses are summarized in Table 6.3.

The experimental data collected highlight an average increment in the Al^{27} content within the samples. In particular, the results obtained show a significant increment in Al^{27} in three of the seven samples. The concentration of Al^{27} in Sample 1 is more than 50% higher than the concentration of the same isotope before the failure test, whereas Samples 4 and 7 show an increment of 15% and 8%, respectively. It is interesting to observe that no sample shows a significant decrement in Al^{27} concentration. Since the measurement method based on INAA is validated with the CRMs, the detected

Table 6.3 Concentrations of Al^{27} before and after the failure tests. The INAA relative concentration uncertainty due to the device sensitivity is to be considered equal to 4%

	Concentration (μg/g) of Al^{27} before and after the failure test						
	Sample 1	Sample 2	Sample 3	Sample 4	Sample 5	Sample 6	Sample7
Before	14,490	14,500	14,880	13,230	15,000	15,090	13,440
After	24,110	14,590	15,570	15,160	14,880	14,630	14,380
After/before	1.66	1.01	1.05	1.15	0.99	0.97	1.08

variations can be associated to the assumed phono-fission reactions occurring during the failure test.

6.6 Conclusions

Neutron emission measurements performed on Luserna stone specimens in compression are shown in [1–3]. From these experiments, it can be clearly seen that phono-fission reactions giving rise to neutron emissions are possible in inert non-radioactive solids. In particular, during compression tests on specimens of sufficiently large size, the neutron flux is found to be of one order of magnitude higher than the natural background level at the time of catastrophic failure.

About the sophisticated INAA technique, a preliminary compression test is conducted on a prismatic specimen ($4 \times 4 \times 16\,\text{cm}^3$) realized by cementitious mortar enriched with iron. During the test, a neutron emission by about three times higher than the average natural background is observed at the moment of final rupture. Based on the same technique, further neutron emission measurements are carried out on small cubic mortar specimens (of 1.00 cm side) enriched with iron. In these samples, even if no significant neutron emissions were detected, important compositional changes are observed.

As a matter of fact, our intention is mainly that of analyzing the concentration of aluminum before and after the compression test, to demonstrate the occurrence of low-energy nuclear reactions involving the symmetrical fission of iron into two atoms of aluminum.

Differently from [3, 4], in the present investigation the compositional analysis of the cementitious mortar specimens is conducted with the INAA method that allows elemental analyses within the entire volume of the samples, and not only on the fracture surfaces.

The Instrumental Neutron Activation Analysis confirms the phono-fission reaction $Fe^{56} \rightarrow 2Al^{27} + 2n$, involving the transmutation of iron into aluminum, not only in natural rocks like Luserna stone but also in artificial materials such as cementitious mortar.

Acknowledgements Dr. L. Bergamaschi and Dr. L. Giordani from the National Research Institute of Metrology (INRIM) are gratefully acknowledged for their assistance with the Instrumental Neutron Activation Analysis (INAA) during the period 2009–2011.

The author is also grateful to Prof. S. L. Pagliolico from Politecnico di Torino for the suggestions about the suitable composition of the cementitious mortar specimens enriched by iron oxide, and to Dr. F. Canonico from Buzzi Unicem factory for taking care of manufacturing the mortar specimens.

References

1. Carpinteri A, Cardone F, Lacidogna G (2010) Energy emissions from failure phenomena: mechanical, electromagnetic, nuclear. Exp Mech 50:1235–1243
2. Carpinteri A, Borla O, Lacidogna G, Manuello A (2010) Neutron emissions in brittle rocks during compression tests: monotonic versus cyclic loading. Phys Mesomech 13:268–274
3. Carpinteri A, Lacidogna G, Manuello A, Borla O (2011) Energy emissions from brittle fracture: neutron measurements and geological evidences of piezonuclear reactions. Strength Fracture Complexity 7:13–31
4. Carpinteri A, Chiodoni A, Manuello A, Sandrone R (2011) Compositional and microchemical evidence of piezonuclear fission reactions in rock specimens subjected to compression tests. Strain 47(s2):282–292
5. Widom A, Swain J, Srivastava YN (2013) Neutron production from the fracture of piezoelectric rocks. J Phys G: Nucl Part Phys 40(015006):1–8
6. Widom A, Swain J, Srivastava YN (2014) Photo-disintegration of the Iron nucleus in fractured magnetite rocks with magnetostriction. Meccanica 50:1205–1216
7. Alfassi ZB (1994) Chemical analysis by nuclear methods. Wiley
8. Laboratory of Applied Nuclear Energy (LENA): http://www.unipv-lena.it/english-version/irradiation-service-a-analysis.html. Accessed June 2013

Chapter 7
Gypsum and Quartz: Crushing Tests, Fracto-Emissions, and Phono-Fission/Fusion Nuclear Reactions

Abstract Extensive experimental investigations are conducted on Gypsum and Quartz specimens of different sizes. They are brought to a complete compression failure, showing two different failure modalities: (1) Very brittle loading drop for micro-crystalline Gypsum and Quartz; (2) Post-peak strain-softening behavior for macro-crystalline Gypsum. All the tested specimens emit acoustic and electromagnetic waves, and the single events are cumulated up to the peak load. On the other hand, neutron emissions are evident only for the largest specimens. The significant chemical composition changes occurring on the fracture surfaces are consistently explained by the assumption of Low-Energy Nuclear Reactions (LENR), both fusion and fission reactions. It is the first time that fusion reactions appear due to fracture, whereas fission reactions have already consistently explained the results related to iron-rich natural and artificial rocks.

Keywords Gypsum · Quartz · Crushing failure · Size effects · Fracto-emissions · Phono-fission · Phono-fusion · LENR

7.1 Preliminary Remarks

Crushing tests on Gypsum and Quartz are carried out at the Laboratory of Fracture Mechanics of Politecnico di Torino, Italy. By subjecting such quasi-brittle materials to compression tests, bursts of neutron emission (NE) during the failure process are observed [1–5], necessarily involving nuclear reactions, besides the well-known acoustic emission (AE) [6–13], and the phenomenon of electromagnetic emission (EME) [14–19]. EME is regarded as an important precursor of critical phenomena in geophysics, such as rock fractures, volcanic eruptions, and earthquakes [19, 20]. For example, anomalous radiations of geo-electromagnetic waves are observed before major earthquakes. At the laboratory scale, rocks and concrete under compression generate AE and EME nearly simultaneously. Also NE are generated during crack growth, reinforcing the idea that also NE can be applied to forecast earthquakes.

A. Carpinteri, *Terahertz Phonons and Nanomechanical Instabilities*,
https://doi.org/10.1007/978-3-032-14692-2_7

While the mechanism of AE is fully understood, being provided by transient elastic waves due to stress redistribution after fracture propagation [6–13], the origin of EME from fracture is not completely clear and different attempts have been made to explain it.

An explanation of the EME origin in metals is related to dislocation phenomena [16], which however are not able to explain EME from fracture in brittle materials, where the motion of dislocations is negligible. Frid et al. [17] and Rabinovitch et al. [21] proposed a model of the EME origin where, following the rupture of bonds during the crack growth, mechanical and electrical equilibria are broken at the fracture surfaces with the creation of ions moving collectively as a surface wave on both crack faces. The resulting oscillating dipoles created on both faces of the propagating fracture act as the source of EME.

As regards the neutron emissions, experimental tests performed on Gypsum and Quartz are presented, using a He^3 neutron device and a bubble type BD thermodynamic neutron detector. For brittle specimens of sufficiently large dimensions and/or slenderness, neutron emissions, detected by He^3, are found to be up to three/four times higher than the natural background at the time of failure. These emissions fully confirm the previous tests [1–5] and are due to phono-fission reactions, which depend on the different modalities of mechanical energy release during the tests. For specimens with sufficiently large size and slenderness, a relatively high mechanical energy emission is expected, and hence a higher probability of neutron emissions at the time of failure. The formation of coherent EME fields occurs over a wide range of frequencies, from few Hz to MHz, and even up to microwave frequencies. This excited state of the matter could be one of the causes of subsequent resonance phenomena of the nuclei able to produce neutron bursts. This hypothesis was also confirmed by Widom et al. [22, 23]. Carpinteri and Lucia [24, 25] have recently proposed an alternative model where THz frequencies are reached by phonons and plasmons with the possibility of coupling interactions with atomic lattices and then the disintegration of some nuclei.

Carpinteri and Borla [26–28] explained through phono-fission reactions the geochemical evolution of our planet, which is due to the most critical tectonic events. As in geochemistry so even in electrochemistry, nano and microcracking of the electrodes produce THz vibrations and phono-fission reactions with sub-atomic particle emissions [29].

7.2 Experimental Set-Up

Compression tests are performed on rock specimens (nine of micro-crystalline Gypsum, six of macro-crystalline Gypsum, and five of Quartz) under monotonic displacement control. The materials used for the tests are non-radioactive Gypsum (in the micro- and macro-crystalline form) and Quartz. In these tests, a total of 20 specimens with different size and slenderness are used (Fig. 7.1). In Tables 7.1 and 7.2, the experimental data concerning the tested specimens are summarized.

Fig. 7.1 Gypsum specimens varying the slenderness (left) and Quartz specimens with irregular shapes (right)

Table 7.1 Tested specimens and their geometrical and mechanical characteristics

Specimen type	Number of specimens	Dimensions		Piston velocity	Volume	Average peak load
		Diameter (mm)	Slenderness λ	(m/s)	(mm^3)	(kN)
Gypsum (micro-crystalline)						
G-50-0.5-micro	3	50	0.5	1×10^{-6}	49.062	72.52 ± 3.65
G-50-1.0-micro	3	50	1.0	1×10^{-6}	98.125	55.35 ± 2.20
G-50-2.0-micro	3	50	2.0	1×10^{-6}	196.250	44.80 ± 3.83
Gypsum (macro-crystalline)						
G-50-1.0-macro	3	50	1.0	1×10^{-6}	98.125	52.19 ± 4.47
G-50-2.0-macro	3	50	2.0	1×10^{-6}	196.250	49.03 ± 3.61
Quartz						
Q-40-1.5	5	40	1.5	1×10^{-6}	75.360	168.07 ± 18.74

All the specimens are subjected to uniaxial compression using a MTS servo-controlled hydraulic testing machine with a maximum capacity of 600 kN. Each test is performed in piston travel displacement control by setting a constant piston velocity. The specimens are arranged in contact with the press platens without any coupling material, according to the testing modalities known as "test by means of rigid platens with friction".

The AE activity emerging from the compressed specimens is detected by attaching to the specimen surface a piezoelectric (PZT) transducer, which is able to convert the high-frequency surface vibrations due to the acoustic waves into an electric signal (the AE signal). The sensitivity of the transducer in the low-frequency range is measured by placing it on a shaker excited with all frequencies in the range 0–10 kHz (*white noise*). The result of this calibration at low frequencies is 1.2 μV/(mm s^{-2}). Resonant sensors are more sensitive than broadband sensors, which are characterized by a flat

Table 7.2 Tested specimens and their fracto-emission average characteristics

Specimen type	AE		EME		NE	
	Average frequency (kHz)	Average highest frequency (kHz)	Average frequency (MHz)	Average highest frequency (MHz)	Average neutron background (10^{-2} cps)	Average count rate at the neutron emission peak (10^{-2} cps)
Gypsum (micro-crystalline)						
G-50-0.5-micro	31.84 ± 6.35	65.46 ± 7.59	37.59 ± 6.84	72.01 ± 15.25	5.20 ± 1.30	Background
G-50-1.0-micro	38.61 ± 7.64	63.01 ± 4.27	29.64 ± 10.91	76.49 ± 15.47	5.20 ± 1.30	Background
G-50-2.0-micro	35.92 ± 6.75	62.79 ± 5.85	38.66 ± 9.07	71.25 ± 12.64	5.20 ± 1.30	15.85 ± 3.96
Gypsum (macro-crystalline)						
G-50-1.0-macro	33.79 ± 6.89	79.27 ± 6.42	40.24 ± 8.46	75.69 ± 13.55	5.41 ± 1.35	Background
G-50-2.0-macro	30.96 ± 5.76	81.58 ± 7.14	38.99 ± 9.33	74.98 ± 14.99	5.41 ± 1.35	18.75 ± 4.69
Quartz						
Q-40-1.5	36.63 ± 7.72	76.22 ± 8.09	44.05 ± 10.96	77.62 ± 11.04	5.13 ± 1.28	21.11 ± 5.27

frequency response in their working range, and then they can be successfully used in monitoring large-sized structures.

The EME equipment consists of a telescopic antenna, having a maximum length of 125 cm. The telescopic antenna can be considered as a sort of length-adjustable monopole. By pulling it to the right length, the antenna can be tuned to operate at different frequencies. For this reason, it represents a "wide band" device, in the sense that it is possible to adjust its length according to the frequency/wavelength that the operator wants to receive. Telescopic antennas are generally used for frequencies $\leq$500 MHz. Moreover, the antenna is coupled with an Agilent DSO1052B oscilloscope (50 MHz, 2 channels) that allows appropriate monitoring of EME signals with frequencies up to tens of MHz.

The neutron emission (NE) detector used in the compression tests under monotonic displacement control is of the He^3 type with electronics of pre-amplification, amplification, and discrimination directly connected to the detector tube. The detector is powered with 1.3 kV, supplied via a high-voltage NIM (Nuclear Instrument Module). The logic output producing the TTL (transistor-transistor logic) pulses is connected to a NIM counter. The device is calibrated for the measurement of thermal neutrons. Its sensitivity is 65 cps/$n_{thermal}$ ($\pm$10% declared by the factory), i.e., a thermal neutron flux of 1 thermal neutron/cm^2 s corresponds to a count rate of 65 cps.

Furthermore, Energy Dispersive X-ray Spectroscopy (EDS) is carried out in order to identify a possible evidence of phono-fission reactions that can take place during the compression failure. The elemental analyses are performed by a ZEISS Auriga field emission scanning electron microscope (FESEM) equipped with an Oxford INCA energy-dispersive X-ray detector (EDX), with a resolution of 124 eV @ MnKa. The energy used for the analyses is equal to 18 keV.

7.3 Test Results

The tested specimens (Fig. 7.1), with their geometrical and mechanical characteristics, are described in Table 7.1.

The experimental diagrams reported in the following and describing load and fracto-emission measurement versus time are related to three of the total 20 specimens (one for each material: micro-crystalline Gypsum, macro-crystalline Gypsum, and Quartz). All specimens are tested in compression up to complete failure, showing either a catastrophic drop (*snap-back*) in the load carrying capacity, when deformed beyond the peak load (Figs. 7.2 and 7.4), or a post-peak softening behavior (Fig. 7.3). The post-peak branch of micro-crystalline Gypsum and Quartz mechanical responses is characterized by a sharp strain localization. Experimental evidence emerges of AE, EME, and NE activities. In Figs. 7.2, 7.3, and 7.4, AE and EME are reported as cumulative event numbers, whereas the NE bursts are reported as *counts per second* (cps). For all the specimens, the observed bursts can be clearly correlated to the loading peaks.

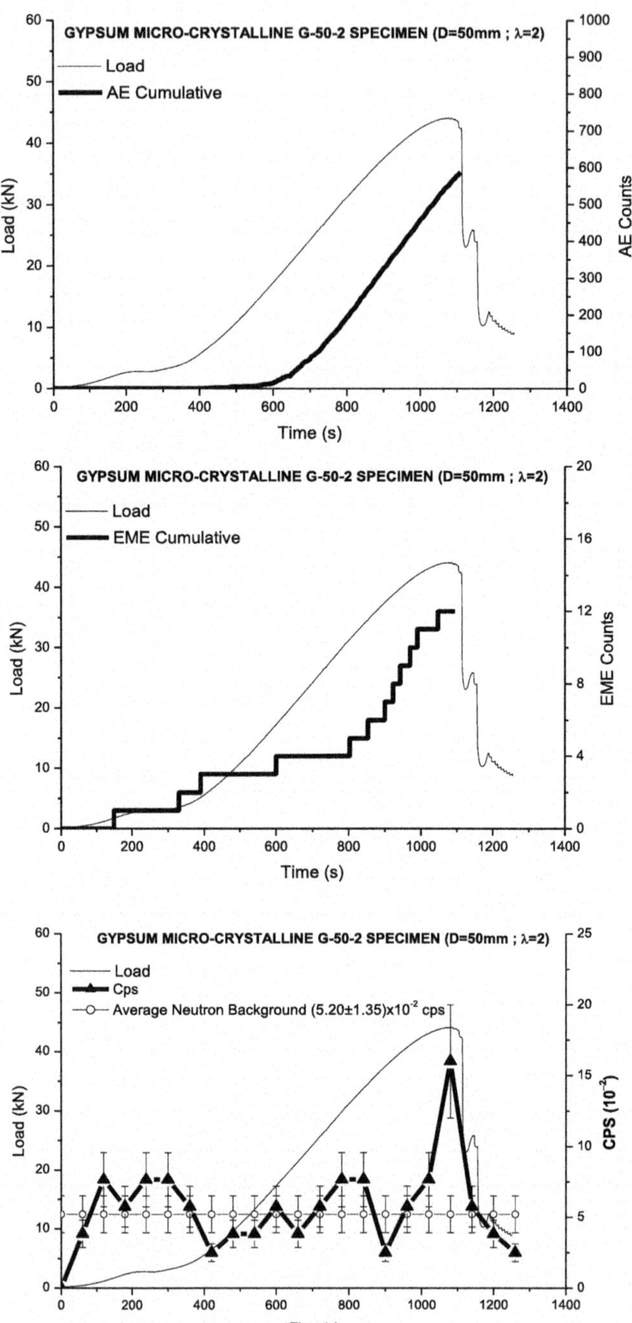

Fig. 7.2 Load versus time diagram of the Gypsum micro-crystalline specimen (G-50-2.0): cumulated number of AE (up); cumulated number of EME (middle); NE count rate (down)

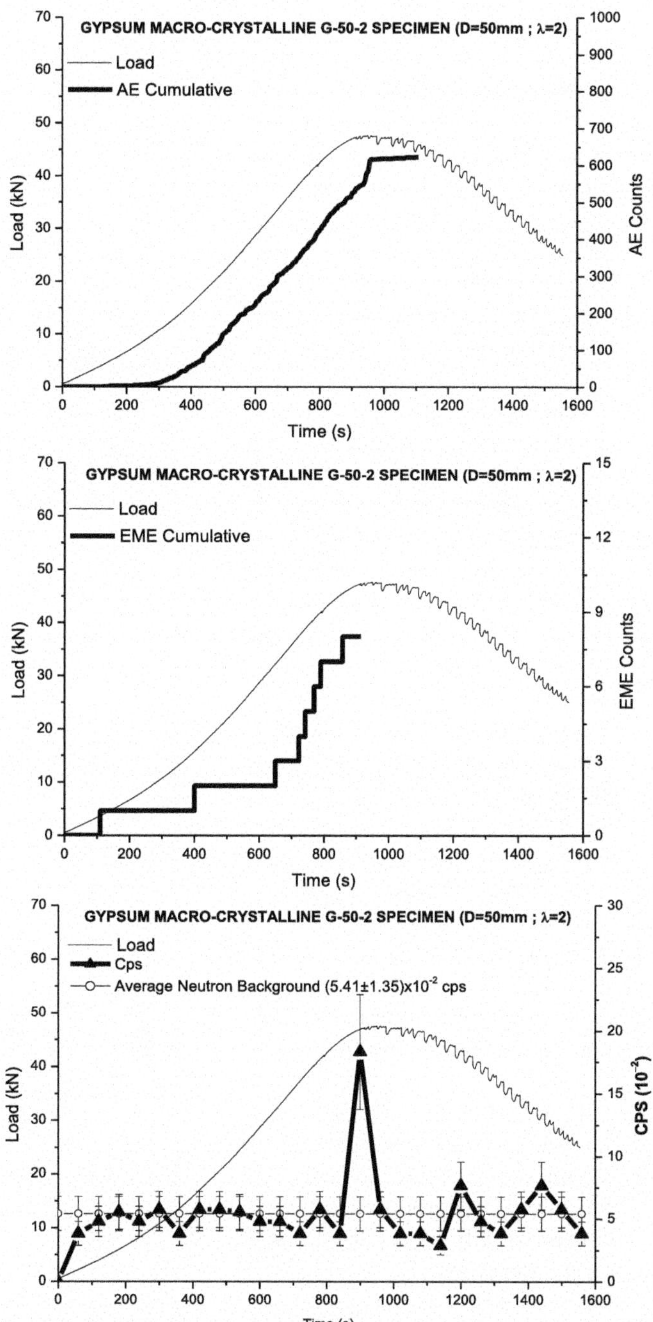

Fig. 7.3 Load versus time diagram of the Gypsum macro-crystalline specimen (G-50-2.0): cumulated number of AE (up); cumulated number of EME (middle); NE count rate (down)

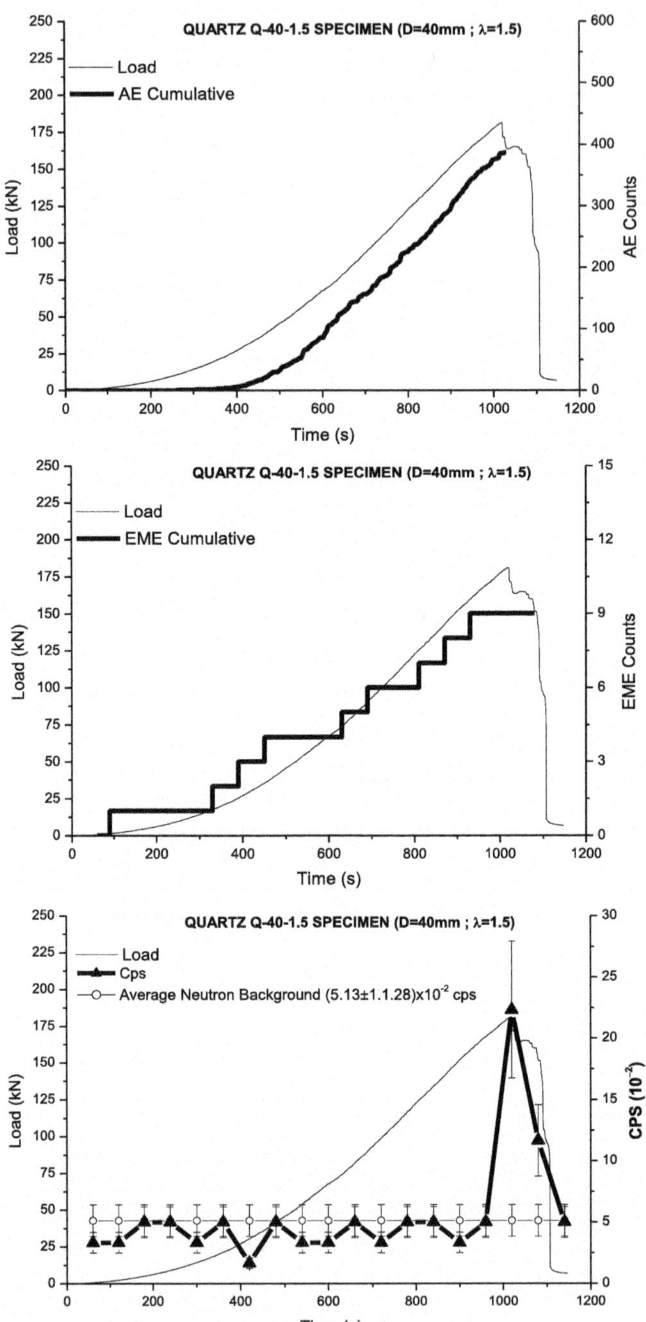

Fig. 7.4 Load versus time diagram of the Quartz specimen (Q-50-1.5): cumulated number of AE (up); cumulated number of EME (middle); NE count rate (down)

As regards the NE measurements, the He^3 neutron detector is switched on at least one hour before the beginning of each compression test, in order to reach the thermal equilibrium in electronics and to make sure that the behavior of the device is stable with respect to intrinsic thermal effects. For the considered specimens, the average measured natural background level ranges from $(5.13 \pm 1.28) \times 10^{-2}$ to $(5.41 \pm 1.35) \times 10^{-2}$ cps. In general, neutron measurements for specimens G-50-0.5 and G-50-1.0 yield values comparable to the natural background, whereas in specimens G-50-2.0 and in Quartz specimens the experimental data exceeded the natural background level by three/four times, approximately. In Table 7.2, the frequencies of AE and EME are reported together with the neutron flux at the peak load for three slendernesses of micro-crystalline Gypsum (three identical specimens per each slenderness), for two slendernesses of macro-crystalline Gypsum (three identical specimens per each slenderness), and for five irregularly shaped specimens of Quartz.

7.4 Nuclear and Stoichiometric Balances

In Table 7.3 and Fig. 7.5, EDS results for C, O, Si, and Mg concentrations on the external and fracture surfaces of micro-crystalline Gypsum specimens are shown. Considering the mean values of the two distributions over the fifteen investigated spots, the increments and decrements are reported. The stoichiometric balances, which are based on the assumed nuclear balances (phono-fissions), reproduce the experimental data very accurately. The only significant divergence regards the carbon

Table 7.3 Test performed on a micro-crystalline gypsum specimen. Element concentrations on external and fracture surfaces; nuclear and stoichiometric balances

Element	External surface mean value	Fracture surface mean value	Increment or decrement
	(wt%)	(wt%)	(wt%)
C	14.7	20.6	+5.9
O	60.2	55.4	−4.8
Si	1.5	0.2	−1.3
Mg	1.3	0.5	−0.8
Ca	11.2	11.2	0.0
S	11.7	11.9	+0.2
Nuclear balances			
$O_8^{16} \rightarrow C_6^{12} + He_2^4$			
$Si_{14}^{28} \rightarrow 2C_6^{12} + He_2^4$			
$Mg_{12}^{24} \rightarrow 2C_6^{12}$			
Stoichiometric balances			
O(−4.8%) = C(+3.6%) + He(+1.2%)			
Si(−1.3%) = C(+1.1%) + He(+0.2%)			
Mg(−0.8%) = C(+0.8%)			

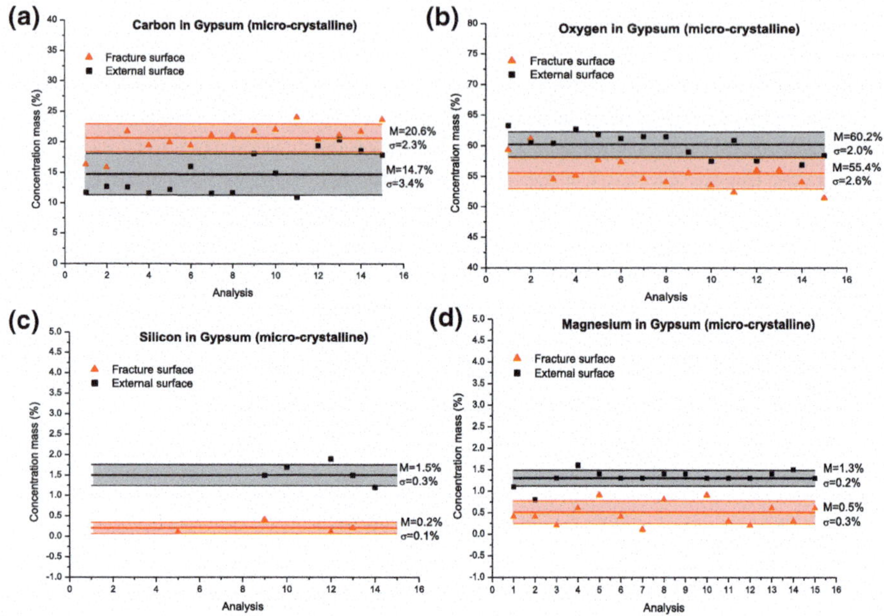

Fig. 7.5 EDS results for C, O, Si, and Mg concentrations in micro-crystalline gypsum: **a** C concentration on external surfaces (black squares) and on fracture surfaces (red triangles); **b** O concentration on external surfaces (black squares) and on fracture surfaces (red triangles); **c** Si concentration on external surfaces (black squares) and on fracture surfaces (red triangles); **d** Mg concentration on external surfaces (black squares) and on fracture surfaces (red triangles)

increment, which experimentally appears to be equal to 5.9% against 5.5% of the theoretical prediction.

In Table 7.4 and Fig. 7.6, EDS results for C, O, S, and Ca concentrations on the external and fracture surfaces of macro-crystalline Gypsum specimens are shown. Considering the mean values of the two distributions over the fifteen investigated spots, the increments and decrements are reported. The stoichiometric balances, which are based on the assumed nuclear balances (phono-fissions and -fusions), reproduce the experimental data very accurately. The only significant divergence regards the oxygen increment, which experimentally appears to be equal to 8.3% against 3.7% of the theoretical prediction.

Finally, in Table 7.5 and Fig. 7.7, EDS results for C, O, Si, N, S, and Ca concentrations on the external and fracture surfaces of Quartz specimens are shown. Considering the mean values of the two distributions over the fifteen investigated spots, the increments and decrements are reported. The stoichiometric balances, which are based on the assumed nuclear balances (prevalently, phono-fissions), reproduce the experimental data very accurately. The only significant divergence regards the carbon increment, which experimentally appears to be equal to 6.5% against 6.2% of the theoretical prediction.

Table 7.4 Test performed on a macro-crystalline gypsum specimen. Element concentrations on external and fracture surfaces; nuclear and stoichiometric balances

Element	External surface mean value (wt%)	Fracture surface mean value (wt%)	Increment or decrement (wt%)
C	32.7	16.1	−16.6
O	47.8	56.1	+8.3
N	1.0	0.8	−0.2
Si	0.7	0.9	+0.2
S	8.7	13.0	+4.3
Mg	0.8	1.0	+0.2
Ca	8.3	12.1	+3.8
Nuclear balances			
$4C_6^{12} \rightarrow Ca_{20}^{40} + 2He_2^4$			
$Ca_{20}^{40} \rightarrow S_{16}^{32} + 2He_2^4$			
$S_{16}^{32} \rightarrow 2O_8^{16}$			
Stoichiometric balances			
$C(-16.6\%) = Ca(+13.8\%) + He(+2.8\%)$			
$Ca(-10.0\%) = S(+8.0\%) + He(+2.0\%)$			
$S(-3.7\%) = O(+3.7\%)$			

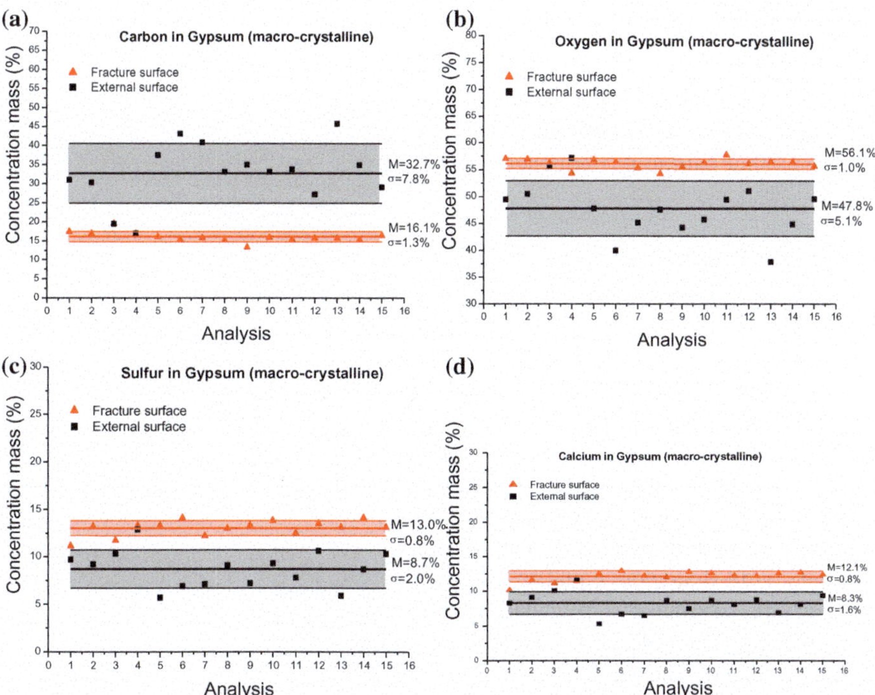

Fig. 7.6 EDS results for C, O, S, and Ca concentrations in macro-crystalline gypsum: **a** C concentration on external surfaces (black squares) and on fracture surfaces (red triangles); **b** O concentration on external surfaces (black squares) and on fracture surfaces (red triangles); **c** S concentration on external surfaces (black squares) and on fracture surfaces (red triangles); **d** Ca concentration on external surfaces (black squares) and on fracture surfaces (red triangles)

Table 7.5 Test performed on a quartz specimen. Element concentrations on external and fracture surfaces; nuclear and stoichiometric balances

Element	External surface mean value	Fracture surface mean value	Increment or decrement
	(wt%)	(wt%)	(wt%)
C	55.6	62.1	+6.5
O	31.0	24.1	−6.9
Si	9.6	10.9	+1.3
N	1.3	0.1	−1.2
Mg	0.1	0.1	0.0
Na	0.5	0.4	−0.1
S	1.2	0.4	−0.8
Ca	0.5	0.1	−0.4
Nuclear balances			
$O_8^{16} \rightarrow C_6^{12} + He_2^4$			
$S_{16}^{32} \rightarrow 2C_6^{12} + 2He_2^4$			
$Ca_{20}^{40} \rightarrow 3C_6^{12} + He_2^4$			
$2N_7^{14} \rightarrow Si_{14}^{28}$			
Stoichiometric balances			
O(−6.9%) = C(+5.2%) + He(+1.7%)			
S(−0.8%) = C(+0.6%) + He(+0.2%)			
Ca(−0.4%) = C(+0.4%) + He(+0.0%)			
N(−1.2%) = Si(+1.2%)			

7.5 Conclusions

Extensive experimental investigations are conducted on Gypsum and Quartz compression specimens of different sizes. They are brought to complete failure, showing two different failure modalities: (1) Very brittle loading drop for micro-crystalline Gypsum and Quartz; (2) Post-peak strain-softening behavior for macro-crystalline Gypsum. All the tested specimens emit acoustic and electromagnetic waves, and the single events are cumulated up to the peak load. On the other hand, neutron emissions are evident only for the largest specimens, which are more brittle than the smaller ones. The significant chemical composition changes occurring on the fracture surfaces are consistently explained by the assumption of phono-fission/fusion reactions. It is the first time that fusion reactions emerge from crushing tests, whereas fission reactions have already successfully explained the results related to other materials like the iron-rich natural and artificial rocks. Let us observe that, in the case of macro-crystalline Gypsum, an original correlation seems to appear between mechanical behavior (post-peak strain-softening) and LENR (Low-Energy Nuclear Reactions) modalities (multi-body phono-fusion reactions).

The content of this chapter was preliminarily published in [30].

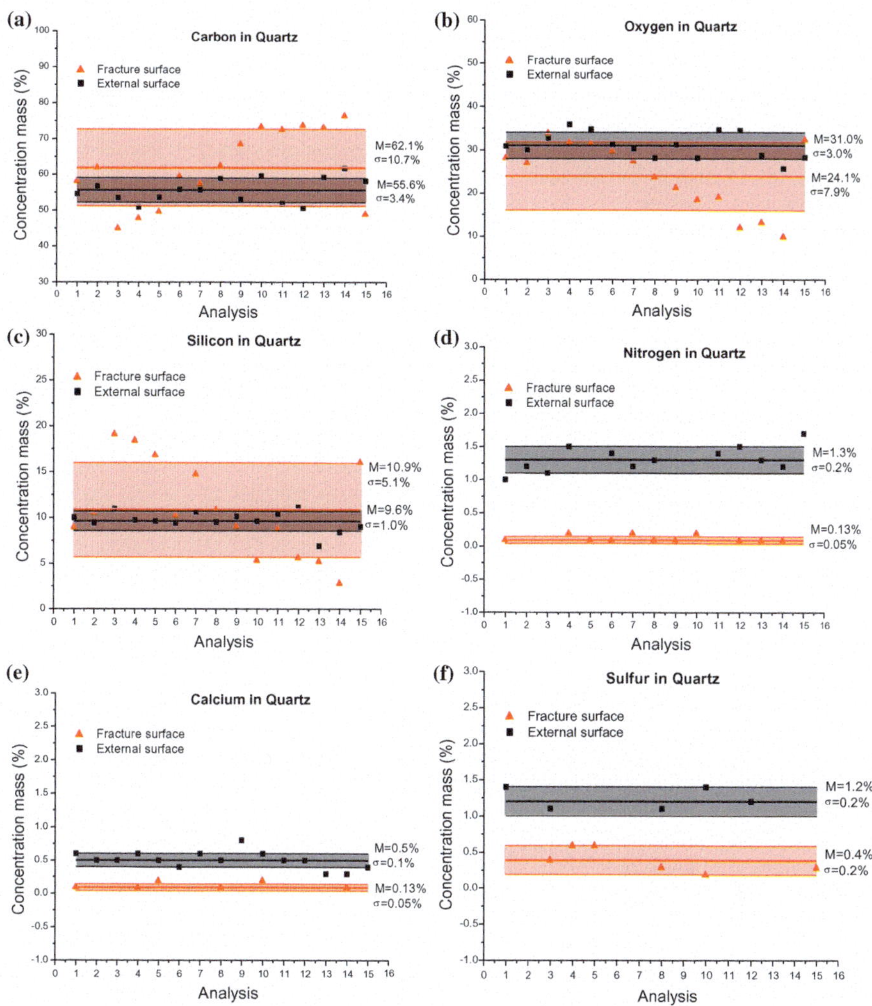

Fig. 7.7 EDS results for C, O, Si, N, Ca, and S concentrations in Quartz: **a** C concentration on external surfaces (black squares) and on fracture surfaces (red triangles); **b** O concentration on external surfaces (black squares) and on fracture surfaces (red triangles); **c** Si concentration on external surfaces (black squares) and on fracture surfaces (red triangles); **d** N concentration on external surfaces (black squares) and on fracture surfaces (red triangles); **e** S concentration on external surfaces (black squares) and on fracture surfaces (red triangles); **f** Ca concentration on external surfaces (black squares) and on fracture surfaces (red triangles)

References

1. Carpinteri A, Cardone F, Lacidogna G (2009) Piezonuclear neutrons from brittle fracture: early results of mechanical compression tests. Strain 45:332–339
2. Cardone F, Carpinteri A, Lacidogna G (2009) Piezonuclear neutrons from fracturing of inert solids. Phys Lett A 373:4158–4163
3. Carpinteri A, Cardone F, Lacidogna G (2010) Energy emissions from failure phenomena: mechanical, electromagnetic, nuclear. Exp Mech 50:1235–1243
4. Carpinteri A, Borla O, Lacidogna G, Manuello A (2010) Neutron emissions in brittle rocks during compression tests: monotonic versus cyclic loading. Phys Mesomech 13:268–274
5. Carpinteri A, Lacidogna G, Manuello A, Borla O (2011) Energy emissions from brittle fracture: neutron measurements and geological evidences of piezonuclear reactions. Strength Fracture Complex 7:13–31
6. Mogi K (1962) Study of elastic shocks caused by the fracture of heterogeneous materials and its relation to earthquake phenomena. Bull Earthq Res Inst 40:125–173
7. Lockner DA, Byerlee JD, Kuksenko V, Ponomarev A, Sidorin A (1991) Quasi static fault growth and shear fracture energy in granite. Nature 350:39–42
8. Ohtsu M (1996) The history and development of acoustic emission in concrete engineering. Mag Concr Res 48:321–330
9. Rundle JB, Turcotte DL, Shcherbakov R, Klein W, Sammis C (2003) Statistical physics approach to understanding the multiscale dynamics of earthquake fault systems. Rev Geophys 41:1019–1049
10. Niccolini G, Schiavi A, Tarizzo P, Carpinteri A, Lacidogna G, Manuello A (2010) Scaling in temporal occurrence of quasi-rigid-body vibration pulses due to macrofractures. Phys Rev E 82(46115):1–8
11. Carpinteri A, Lacidogna G (2006a) Damage monitoring of an historical masonry building by the acoustic emission technique. Mater Struct (RILEM) 39:161–167
12. Carpinteri A, Lacidogna G (2006b) Structural monitoring and integrity assessment of medieval towers. J Struct Eng (ASCE) 132:1681–1690
13. Carpinteri A, Lacidogna G (2007) Damage evaluation of three masonry towers by acoustic emission. Eng Struct 29:1569–1579
14. Lacidogna G, Carpinteri A, Manuello A, Durin G, Schiavi A, Niccolini G, Agosto A (2010) Acoustic and electromagnetic emissions as precursor phenomena in failure processes. Strain 47(2):144–152
15. Carpinteri A, Lacidogna G, Manuello A, Niccolini G, Schiavi A, Agosto A (2010) Mechanical and electromagnetic emissions related to stress-induced cracks. Exp Tech 36(3):53–64
16. Misra A (1977) Theoretical study of the fracture-induced magnetic effect in ferromagnetic materials. Phys Lett A 62:234–236
17. Frid V, Rabinovitch A, Bahat D (2003) Fracture induced electromagnetic radiation. J Phys D 36:1620–1628
18. Hadjicontis V, Mavromatou C, Nonos D (2004) Stress induced polarization currents and electromagnetic emission from rocks and ionic crystals, accompanying their deformation. Nat Hazards Earth Syst Sci 4:633–639
19. Warwick JW, Stoker C, Meyer TR (1982) Radio emission associated with rock fracture: possible application to the great Chilean earthquake of May 22, 1960. J Geophys Res 87:2851–2859
20. Nagao T, Enomoto Y, Fujinawa Y et al (2002) Electromagnetic anomalies associated with 1995 Kobe earthquake. J Geodyn 33:401–411
21. Rabinovitch A, Frid V, Bahat D (2007) Surface oscillations. A possible source of fracture induced electromagnetic oscillations. Tectonophysics 431:15–21
22. Widom A, Swain J, Srivastava YN (2013) Neutron production from the fracture of piezoelectric rocks. J Phys G: Nucl Part Phys 40(15006):1–8
23. Widom A, Swain J, Srivastava YN (2015) Photo-disintegration of the iron nucleus in fractured magnetite rocks with magnetostriction. Meccanica 50:1205–1216

24. Lucia U, Carpinteri A (2015) GeV plasmons and spalling neutrons from crushing of iron-rich natural rocks. Chem Phys Lett 640:112–114
25. Carpinteri A, Lucia U, Zucchetti M, Borla O (2023a) THz vibrations and phono-fission reactions from crushing of iron-rich natural rocks. J Condensed Matter Nucl Sci 37:84–97
26. Carpinteri A, Borla O (2018) Nano-scale fracture phenomena and TeraHertz pressure waves as the fundamental reasons for geochemical evolution. Strength Fracture Complex 11:149–168
27. Carpinteri A, Borla O (2017) Fracto-emissions as seismic precursors. Eng Fract Mech 177:239–250
28. Carpinteri A, Borla O (2019) Acoustic, electromagnetic, and neutron emissions as seismic precursors: the lunar periodicity of low-magnitude seismic swarms. Eng Fract Mech 210:29–41
29. Carpinteri A, Borla O, Manuello A (2023b) Hydrogen embrittlement, microcracking, and THz vibrations in the metal electrodes of cold-fusion electrolysis experiments: repeatability of nuclear and stoichiometric balances. J Condens Matter Nucl Sci 37:68–83
30. Carpinteri A, Borla O, Accornero F (2023c) Gypsum and quartz specimens in compression failure: Fracto-emissions and related stoichiometric balances. Eng Fract Mech 284:109202

Part III
Neutron Emissions from Tensile Failure and Fatigue of Metallic Materials

Chapter 8
Steel Bars Under Tension, Compression, and Impact Loading

Abstract Phono-fission reactions and neutron emissions are triggered by high-frequency pressure waves in inert, non-radioactive materials. This phenomenon has previously been treated for brittle rocks. In the present chapter, the investigation is extended to a more ductile material, like steel, subjected to different loading conditions. Although phono-fission reactions are more likely to take place during the failure of brittle materials, the phenomenon is revealed also with the more ductile failure of metallic materials.

Keywords Steel bars · Phono-fission reactions · Neutron emissions · Tensile and compression tests · Impact tests

8.1 Preliminary Remarks

Neutron emission measurements were successfully performed on brittle solid specimens during crushing test failure. From those experiments, it can be clearly seen that phono-fission reactions giving rise to neutron emissions are possible in inert, non-radioactive brittle materials like Luserna stone or mortar [1–5]. A theoretical explanation has been also provided in recent works [6, 7], whereas an alternative explanation will be given in Chap. 22, based on phonons and plasmons rising from nanomechanics instabilities.

The term "phono-fission" reaction refers to low-energy nuclear reactions (LERN), which are triggered by the fundamental action of pressure waves. The phenomenon of neutron emission was firstly recognized in stable iron nuclides contained in aqueous solutions of iron chloride or nitrate subjected to cavitation induced by ultrasounds [8]. It has recently and repeatedly been detected also in solids during mechanical tests on laboratory specimens. On the other hand, quantitative Scanning Electron Microscope (SEM) analyses of the fracture surfaces of broken rock specimens showed that different kinds of element transmutation could be the results of fission reactions generating excess neutron emission [9, 10].

A. Carpinteri, *Terahertz Phonons and Nanomechanical Instabilities*,
https://doi.org/10.1007/978-3-032-14692-2_8

In the present chapter, the investigation is extended to a more ductile metallic material, like steel, subjected to different testing conditions.

Steel bars subjected to uniform tensile or compression loading, as well as notched specimens subjected to Charpy impact test, are described. From these tests, evidence of neutron emission is obtained, which depends on loading condition, specimen size, and, therefore, on the brittleness of the mechanical response [11].

For an accurate neutron evaluation, a He^3 proportional counter (Xeram, France) and a He^3 radiation monitor AT1117M (ATOMTEX, Minsk, Republic of Belarus) are used. Since neutrons are electrically neutral particles, they cannot directly produce ionization in a detector, and therefore cannot be directly detected. This means that neutron detectors must rely upon a conversion process where an incident neutron interacts with a nucleus to produce a secondary charged particle. These charged particles are then detected, and from them the neutron presence is deduced. In addition, a set of passive neutron detectors (BDT Bubble Dosimeter and BD-PND Bubble Dosimeter models) insensitive to electromagnetic noise and with a zero gamma sensitivity is used. The dosimeters, based on superheated bubble detectors (BTI, Ontario, Canada) (BUBBLE TECHNOLOGY INDUSTRIES (1992)) [12], are calibrated at the factory against an AmBe source in terms of NCRP38 (NCRP report 38 (1971)) [13].

Semi-quantitative results from XRF (X-Ray Fluorescence Spectroscopy) and SEM analyses are also described. Due to the size of the steel bars, a portable XRF analysis is carried out to characterize the chemical composition of the fracture surface of the broken specimens. On the other hand, since the Charpy test specimens are smaller enough, they are analyzed by SEM. Although phono-fission reactions are more likely to take place during the failure of brittle materials, the phenomenon is revealed also with the more ductile failure of metallic materials.

8.2 Steel Specimens Under Tensile and Compression Loading

Specific tests are conducted on eighteen steel bars of different diameter subjected to tensile stress condition, and on two bars subjected to compression stress condition [11].

Regarding the eighteen specimens tested in tensile condition, nine have variable diameter ranging from 14 to 28 mm, whereas the remaining nine have the same diameter equal to 40 mm and a length of 1.0 m.

In Table 8.1, the applied loads, as well as the geometrical and mechanical characteristics for each of the first nine specimens are summarized. The specimens are subjected to tensile loading up to failure according to the EN ISO 6892 recommendation [14].

To carry out these experiments, a hydraulic press, Walter Bai type, with electronic control is used. Each test is conducted in three subsequent stages. In the first stage,

Table 8.1 Tensile loading tests on steel bars of different diameters

Tensile loading tests							
Bar type	Section S_0 (mm^2)	Yield load P_y (kN)	Ultimate load P_u (kN)	Yielding strength (MPa)	Stress peak (MPa)	Elongation at P_u (%)	Elongation at failure (%)
D14_1	153.94	84.68	96.81	550.11	628.88	25.56	37.71
D20_1	314.16	160.36	194.17	510.45	618.06	34.39	53.72
D20_2	314.16	166.61	193.24	530.33	615.10	39.28	54.81
D22_1	380.13	148.52	182.35	390.71	479.70	27.71	46.07
D22_2	380.13	155.39	188.08	408.78	494.78	33.72	52.26
D24_1	452.39	230.00	306.41	508.66	677.65	38.18	58.99
D24_2	452.39	290.05	337.10	641.15	745.15	31.36	50.74
D28_1	615.44	351.42	437.58	571.02	711.00	37.96	61.27
D28_2	615.44	362.30	441.62	588.69	717.57	36.69	55.77

it is controlled by stress increments of 15 MPa/s up to the value of about 500 MPa, which corresponds to the yield strength of the material. Subsequently, the test is controlled by an imposed velocity of 0.16 mm/s up to an elongation equal to 10% of the initial length. In the last stage, which ends with the specimen failure [11], the imposed deformation is applied by displacement increments of 0.33 mm/s.

The additional tests conducted on the nine steel specimens with constant diameter and length (40 mm and 1,000 mm, respectively) are carried out in the framework of the usual procedures, according to which only the ultimate strength is registered.

In Table 8.2, the applied loads, as well as the specimen geometrical characteristics, are summarized. An average loading ramp of 323 kN/min is applied and a mean ultimate strength of 808.81 kN is observed.

Table 8.2 Tensile loading tests on steel bars with a diameter of 40 mm

Tensile loading tests			
Bar type	Section S_0 (mm^2)	Ultimate load P_u (kN)	Stress peak (MPa)
D40	1,256	797.0	634.55
D40	1,256	799.8	636.78
D40	1,256	800.7	637.50
D40	1,256	791.6	630.25
D40	1,256	800.7	637.50
D40	1,256	801.9	638.46
D40	1,256	811.9	646.42
D40	1,256	837.0	666.40
D40	1,256	838.7	667.75

To carry out the tensile test on the specimens with a diameter of 40 mm, a hydraulic press (METRO COM), with manual control and a maximum load of 1,000 kN, is used. Each test is conducted in a single stage up to the specimen failure.

Concerning the trials in compression loading condition, two specimens with diameter of 40 mm are tested using a hydraulic press, Galdabini type, with a maximum load of 5,000 kN (Fig. 8.1a). A loading ramp of 58 kN/min is applied and the tests stop after about 30 min, at a load of 2,000 kN, corresponding to a specimen shortening of approximately 50% (Fig. 8.1b). Considering a symmetric behavior of steel in tensile and compression conditions, the yield strength (638.56 kN) and the ultimate strength (850.31 kN) of specimen C1 are identified considering the yielding and the ultimate strengths obtained during the tensile loading test. Similar results are also obtained for specimen C2 (Table 8.3).

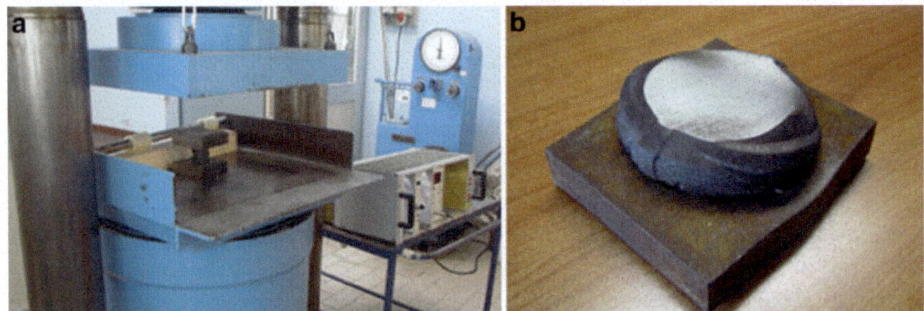

Fig. 8.1 **a** Experimental set-up for the compression test performed on specimen C1. **b** Specimen C1 at the end of the compression test (observe the presence of cracks on the lateral surface of the specimen)

Table 8.3 Compression loading tests on steel bars

Compression tests	
Specimen C1	
Section S_0	1,256 (mm^2)
Yield load, P_y	638.56 (kN)
Ultimate load, P_u	850.31 (kN)
Specimen C2	
Section S_0	1,256 (mm^2)
Yield load, P_y	607.21 (kN)
Ultimate load, P_u	808.69 (kN)

8.3 Neutron Emission Detection

As a first hint, the measurement of natural neutron background, for each test described in the following, is performed. He^3 devices (thermal neutron counter and ATOMTEX) and thermodynamic bubble detectors (BDT and BD-PND types) are employed to monitor the typical daily fluctuation of background level. Additional background measurements, for a period of at least 1 h, are repeated before each test in order to check a possible variation in natural background. Moreover, to provide a precise neutron evaluation in terms of cps (counts per second), a suitable reference time is chosen depending on the experimental condition. It is fixed up to 60 s to reduce the probability to detect zero counts during the neutron emission evaluation.

The tensile tests on the nine steel bars reported in Table 8.1 and the compression tests performed on specimens C1 and C2 are monitored by the He^3 (Xeram, France) neutron detector. The He^3 proportional counter is placed close to the experimental set-up for both tests. In Fig. 8.2, the load vs. time diagram for specimen D24_1 is reported with the neutron emission measurements. The average neutron background level measured before the test is equal to $(7.22 \pm 1.42) \times 10^{-2}$ cps. It can be noted that during the test, and in particular in correspondence to the achievement of the yield strength limit equal to 230 kN (Fig. 8.2), neutron emissions increase up to $(11.67 \pm 2.29) \times 10^{-2}$ cps. The increment is equal to about 50% of the background level. In addition, in correspondence to the ultimate strength (306 kN), a maximum neutron emission of $(16.67 \pm 2.29) \times 10^{-2}$ cps is measured. This last emission level corresponds to an incremented flux more than twice the natural background level measured before the experiment. Finally, after the steel bar failure, the neutron emissions decrease almost instantaneously down to the background level measured before the experiment.

As regards the tensile loading tests on the other steel bars reported in Table 8.1, a similar behavior in neutron emission measurements is evidenced for specimens characterized by diameters equal to or greater than 24 mm. In the following (Fig. 8.3), the load versus time diagrams are reported together with the neutron emission measurements also for specimens D20, D22, and D28. For the largest specimens (D28), neutron emissions reach values up to three times higher than the average natural background.

The remaining nine tensile loading tests on specimens with a diameter of 40 mm are monitored by the ATOMTEX He^3 neutron device and by six bubble neutron detectors, three BDT and three BD-PND. The average neutron background level measured by bubble detectors before the test is equal to (0.063 ± 0.016) microSv/h. The bubble dosimeters are placed all around the specimens, whereas the He^3 detector is positioned at a distance of about 50 cm from the steel bar. A neutron emission up to (0.115 ± 0.029) microSv/h, corresponding to an increased flux more than twice the background level, is monitored at the end of the test, only in the low-energy range. This behavior is also confirmed by the data acquired by the Atomtex device. In fact, no significant increment in the high-energy neutron spectrum is observed.

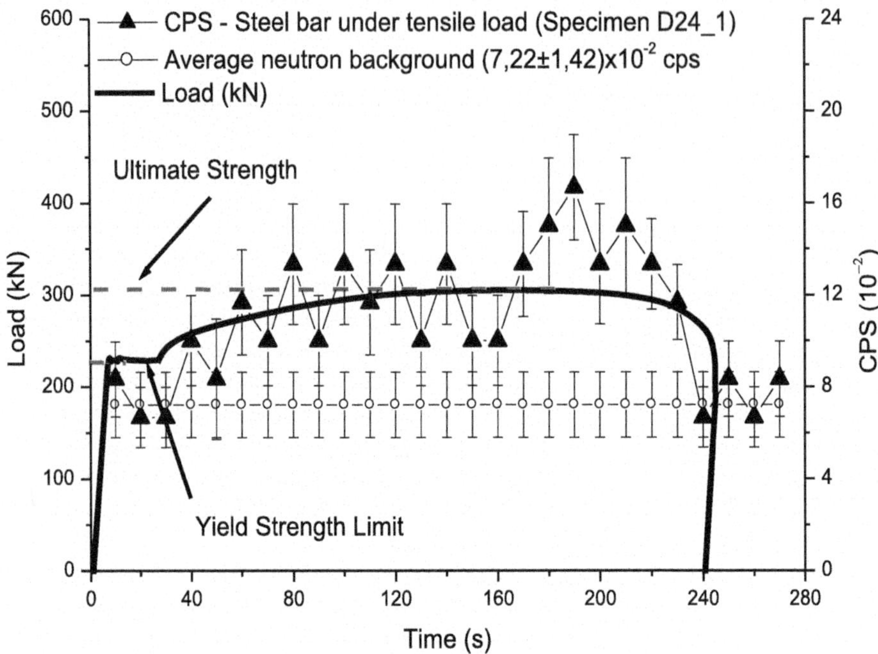

Fig. 8.2 Load versus time diagram and neutron emission measurements for specimen D24_1

For specimen C1, the load versus time diagram is reported in Fig. 8.4. Also in this case, the neutron emissions show an appreciable increment immediately after the achievement of the ultimate load (850 kN). At this point, the maximum neutron emission level ($(19.99 \pm 2.96) \times 10^{-2}$ cps) corresponds to an increment of about three times the background level. As can be seen from Fig. 8.1b, the final section area of specimen C1 is sensibly larger than the initial nominal area, so that the theoretical ultimate strength is widely overcome.

For sample C1, similarly to specimen D24_1, it is also possible to observe that the maximum neutron emission level, reached at the ultimate strength and equal to $(19.10 \pm 2.29) \times 10^{-2}$ cps, corresponds to an increment of about three times with respect to the natural background level [11].

It is important to emphasize that the emission of neutrons is a burst-like phenomenon due to three main reasons. The first is that local propagation (or redistribution) of defects is discontinuous and takes place similarly to the stick–slip phenomenon in friction. The second is due to the directionality of the neutron emission, that influences the probability of being acquired by the instrument. Finally, it depends on the experimental condition and on the reference time window over which the neutron counting rate is measured. If the reference time is not sufficiently long, the probability exists to detect zero counts in the chosen time window. For this reason, if the experimental conditions allow it (i.e., the duration of the test), it

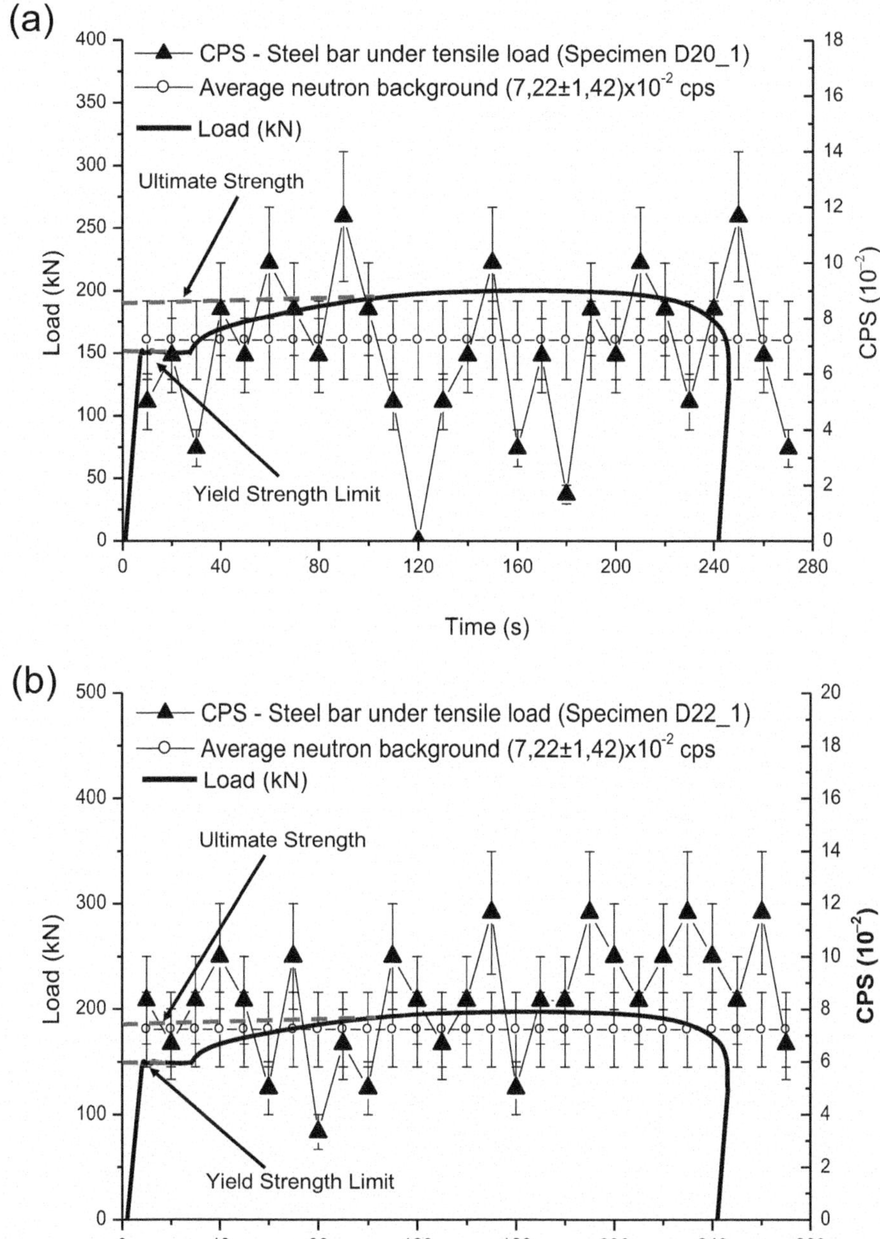

Fig. 8.3 Load vs. time diagrams and neutron emission measurements for specimens D20 (**a**), D22 (**b**), and D28 (**c**), under tensile loading tests

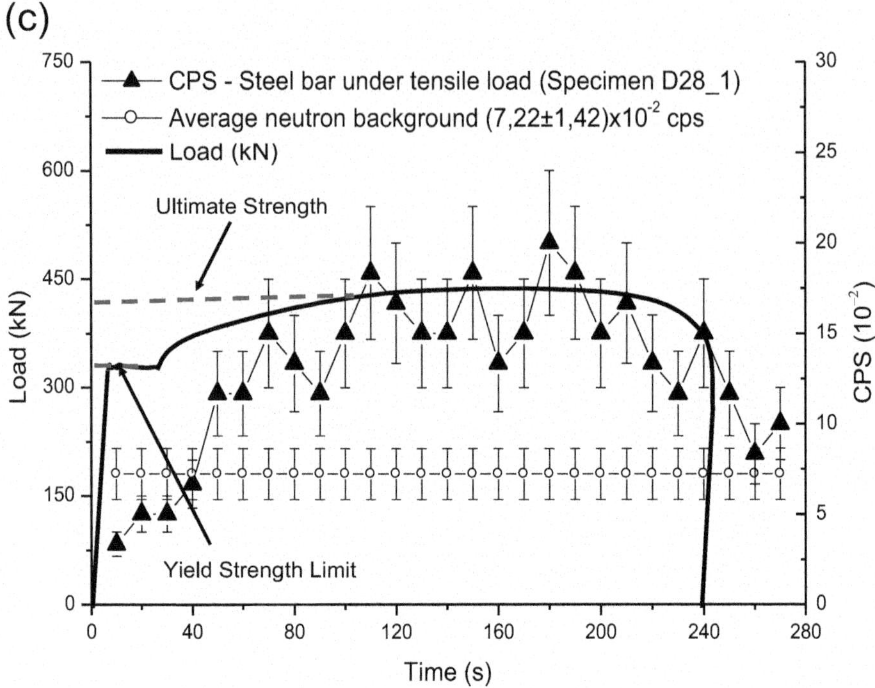

Fig. 8.3 (continued)

is better to set a reference time window sufficiently long to monitor an appreciable number of events.

8.4 XRF Chemical Analysis of Steel Bars

The emission of neutrons is usually due to nuclear phenomena, such as fission and spallation, which yield products different from the atomic elements that are present prior to the test. Those elements can be both radioactive or not, like in the case of the so-called low-energy fission reactions [15].

Due to the size of the steel bars, one of the most affordable techniques is the portable XRF (X-ray fluorescence), which is commonly adopted for the characterization of corrosion resistant steel bars [16]. Unfortunately, the XRF technique is not able to detect light elements, with atomic number lower than 15.

XRF analyses are performed with an Assing LITHOS 3000 portable spectrometer equipped with a molybdenum tube and a Peltier cooled Si-PIN detector with a sensitive area of about 7 mm^2, and a beryllium (Be) window 12.5 μm thick. For the measurements on the Graduale the X-ray tube voltage is 25 kV, the current is 50 μA,

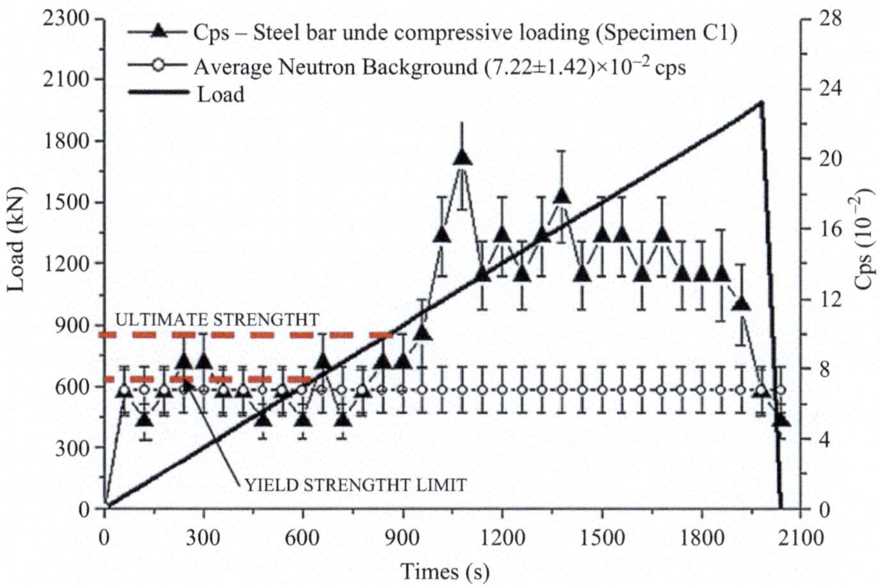

Fig. 8.4 Load versus time diagram and neutron emission measurements for compression specimen C1

and the acquisition time is 720 s. A 10 mm collimator is used for an investigated area of approximately 5 mm in diameter.

The acquisition is carried out adopting a calibration profile for metals, both on the cutting surface and on the fracture surface. Several acquisition points are selected, close to the bar axis as well as close to the lateral surface. Finally, acquisitions on the lateral surface are also performed.

Figure 8.5 shows the details of the acquisition point locations.

Figure 8.6 shows the scattered X-ray spectrum, for the different acquisition points, which allows for the element tracing. As expected, iron (Fe) is by a great extent the most abundant element.

The spectral analysis allows for the determination of additional elements, such as Mn, Cr, Mo, Ni, and Cu, which are frequent impurities in common steel bars. In addition, a rather well-detectable peak in the spectrum is present in correspondence to scandium (Sc) with an approximate concentration of 50–100 ppm. Scandium is a rather rare element on the Earth's crust; therefore, the presence of the peak is rather intriguing. Moreover, the magnitude of the peak seems greater on the fracture surface, especially close to the lateral surface, rather than in the inner part, as shown in the comparison spectrum of Fig. 8.6c. This suggests that the presence of scandium could be somehow linked to the level of strain experienced by the material.

It is worth noting that the adopted portable XRF analyzer is not suitable to detect light elements as aluminum (Al), which is expected to be a phono-fission product from iron. Nevertheless, scandium is detected from photo-fission experiments on iron [17],

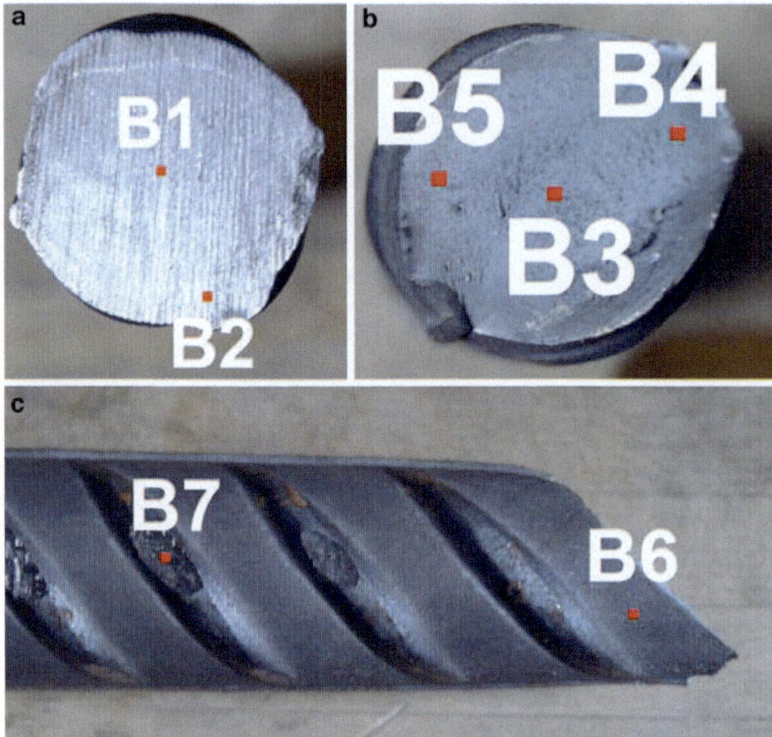

Fig. 8.5 XRF acquisition points on the steel bar: cutting section (**a**), fracture surface (**b**), lateral surface (**c**)

as well as in iron meteorites, due also to the effect of exposition to cosmic radiation [18]. Therefore, although further analyses are necessary, the presence of scandium could be a clue of the presence of other lighter fission products. For example, the phono-fission transmutation of Fe into Sc should be accompanied by the presence of boron (B).

The acquisition on the lateral surface (points B6 and B7) did not reveal any anomalous element, a part from a much higher presence of Ni and Cu, which is likely due to surface passivation treatments.

8.5 Charpy Impact Test

The Charpy impact test is carried out according to the European standard code EN 10045/1-1990 [19]. The nominal energy of the adopted Charpy pendulum is 300 J. Five standard steel specimens with a V-notch and a 10×8 mm^2 section are analyzed. The second specimen is preventively refrigerated in liquid nitrogen to increase its

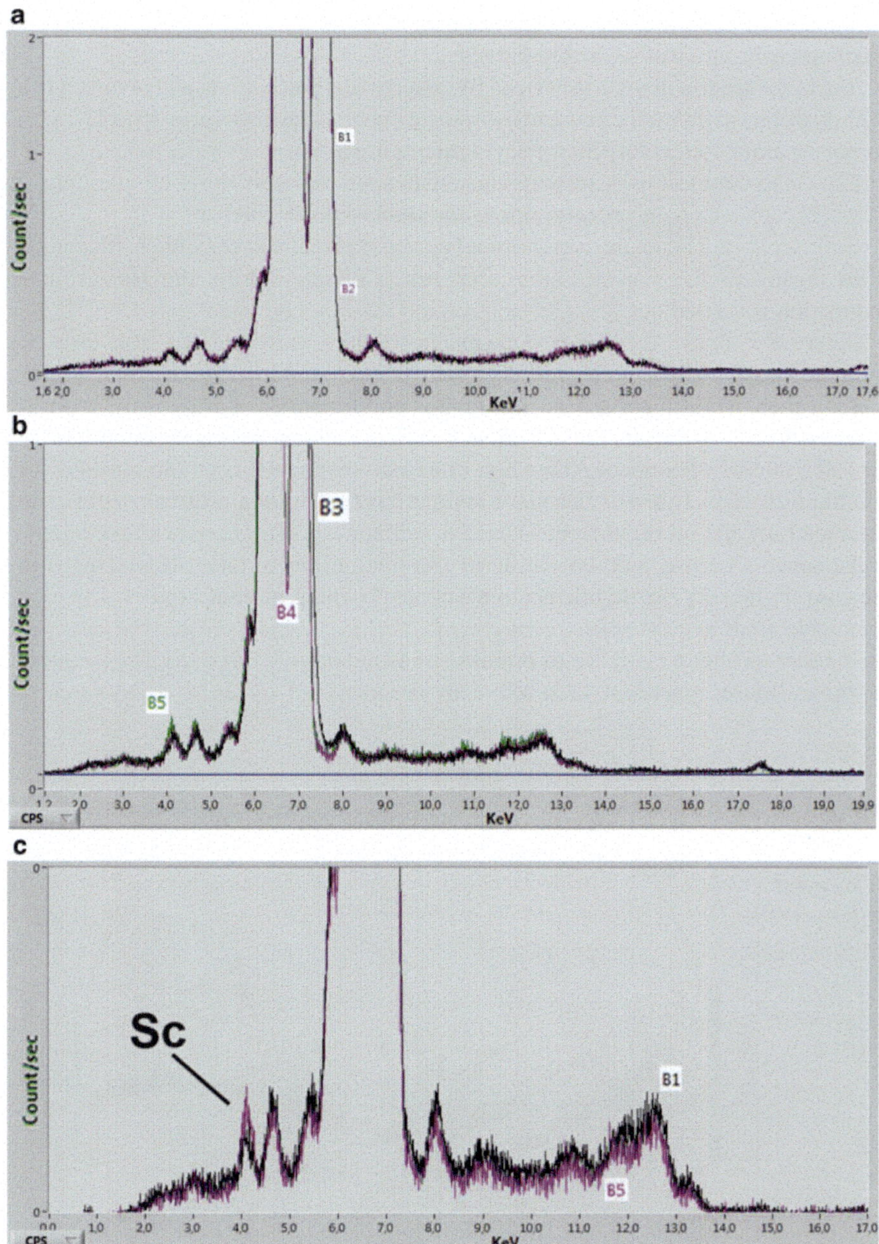

Fig. 8.6 XRF spectra: acquisition on the cutting section points B1 and B2 (**a**); acquisition on the fracture surface points B3, B4, and B5 (**b**); comparison between points B1 and B5 (**c**)

brittleness. In this case, the dissipated energy during the test is about 6 J, and the specimen split into two separated halves.

The other specimens, characterized by different carbon content and obtained from slightly different thermal treatments, dissipate energies ranging from 76 to 175 J, and do not separate completely after the pendulum impact.

The Charpy impact tests are performed with approximately thirty minutes interval between each other, in order to allow for stable bubble nucleation in the bubble detector sensors. The analysis performed does not reveal any sensible bubble nucleation in any of the sensors, after each test. Contemporarily, the He^3 Atomtex acquisition is carried out.

Figure 8.7 shows that, in correspondence to the third and the fifth tests, the recorded neutron emission is well above the background mean level. Nevertheless, such difference is still comprised within the confidence bar.

In particular, it appears that the greater variations from the background level are recorded in correspondence to the first (150 J dissipated energy), the third (175 J), and the fifth (135 J) tests. This suggests that brittleness is not the only governing parameter. In fact, in order to be able to detect neutron emissions with the adopted instruments, a certain amount of energy dissipation has to take place. This is not the case of the very brittle refrigerated sample 2, which provides only 6 J, with no detectable neutron emissions.

In order to obtain evidence of possible transmutations of iron, as a consequence of phono-fission reactions, the specimens are analyzed by SEM (EDAX), at the

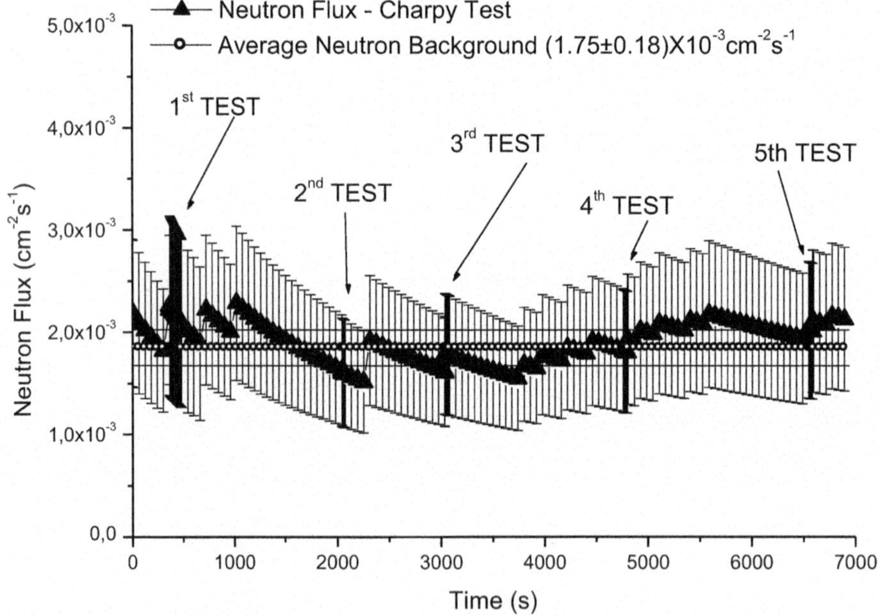

Fig. 8.7 Diagram of the Atomtex neutron acquisitions during the Charpy impact tests

Department of Material Science of Politecnico di Torino. The instrument can perform the Energy Dispersive X-ray Spectroscopy. The volume of material involved in the present measurement can be estimated to be equal to 1 cubic millimeter, based on the energy of the laser beam and on the area of the acquisition spot (30 × 30 μm^2).

Different areas of each specimen are analyzed, respectively in correspondence to the surface of the crack growing from the tip of the V-notch (brittle propagation), or rather far from it (more stable propagation). In addition, the lateral surfaces of the specimens and the area in correspondence to the impact are investigated. Most of the analyzed surfaces are rough, thus the acquisition is not in the ideal instrumental condition.

The quantitative element composition measured for the third specimen is reported in Table 8.4. An example of the element spectrum is shown in Fig. 8.8. No clear evidence of anomalous elements is recorded in correspondence to the fracture surface, to the lateral surface, or to the impact region. On the other hand, Fig. 8.9a shows the spectrum when the acquisition is performed in the vicinity of an inclusion, corresponding to the acquisition area shown in Fig. 8.9b. In this case, the composition percentage is different, but this could be due to the original composition of the inclusion.

A low neutron emission is recorded, which is not completely distinguishable from the background. The metallic crystalline lattice structure does not appear to be particularly prone to phono-fission phenomena. In addition, the small volumes and dissipated energies, combined with some difficulties in the positioning of the sensors, negatively affect the chance of detecting neutron emissions.

Table 8.4 Quantitative spectrography for the third Charpy impact specimen

Location	Weight %				
	Fe	Si	Mn	Al	Ca
Fracture surface	98.50	0.39	1.11	–	–
	98.73	0.29	0.99	–	–
	98.57	0.24	1.19	–	–
	98.62	0.36	1.02	–	–
	98.42	0.41	1.17	–	–
	98.09	0.35	1.57	–	–
V-notch surface	98.37	0.47	1.16	–	–
	98.23	0.47	1.30	–	–
Impact area	98.36	0.63	1.02	–	–
	97.53	0.96	1.52	–	–
	97.59	1.32	1.10	–	–
Inclusion	70.60	11.07	–	10.27	8.06

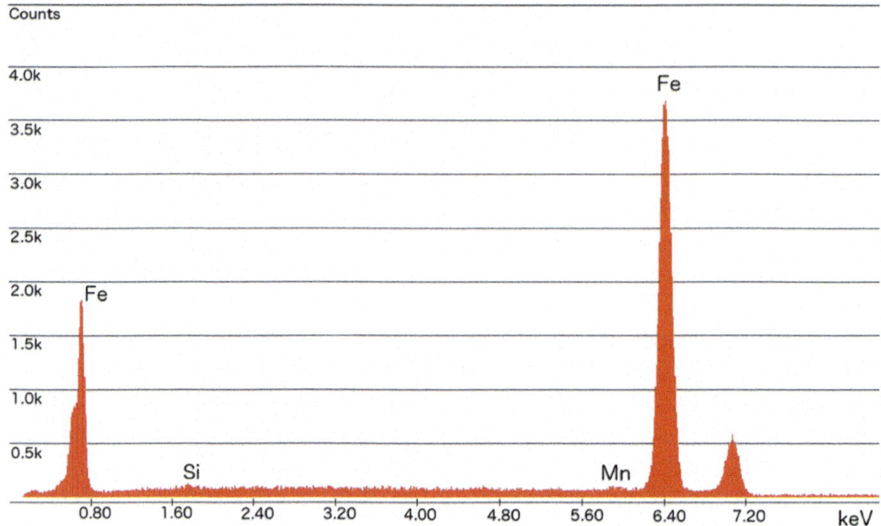

Fig. 8.8 Typical isotopic spectrum, where no anomalous elements are detected

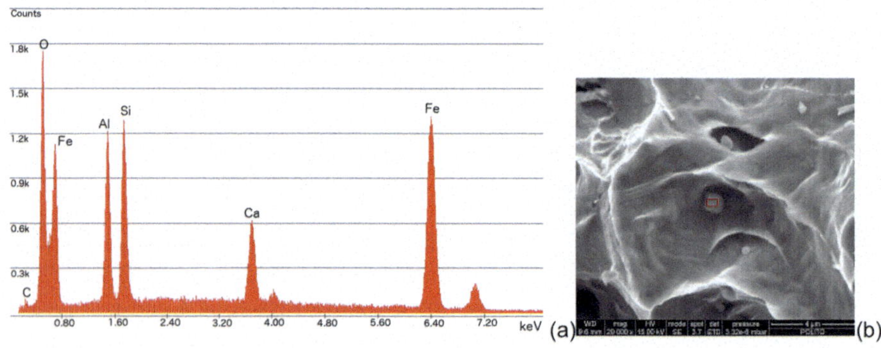

Fig. 8.9 Acquisition in correspondence to an inclusion: element spectrum (**a**); micrography (**b**)

8.6 Conclusions

In the present chapter, an investigation about phono-fission neutron emissions from ductile materials, like steel, subjected to different loading conditions, is presented. Steel bars subjected to uniform tensile or compression loading tests, as well as notched specimens subjected to the Charpy impact test, are considered. Some positive evidence of neutron emission is reported, which depends on the loading condition, the size and the brittleness of the specimen, and also on the amount of the dissipated mechanical energy. The XRF analysis of the fracture surfaces of steel bars broken in tension reveals the presence of an anomalous element, i.e., Scandium (Atomic Number 21), which is very rare on the Earth's crust, whereas it is often found on iron surfaces subjected to photodisintegration (artificial irradiation or cosmic radiation). In the case of meteorites, the cause of the atomic disintegration or fission of iron nuclei could be represented also by the very energetic impacts and not only by the cosmic rays.

Acknowledgements We deeply acknowledge the collaboration by Prof. Jean Marc Tulliani (Politecnico di Torino) for SEM analysis and Charpy impact tests, as well as by Dr. Marco Nicola (Adamantio srl, c/o Incubatore d'Impresa dell'Università di Torino, via Quarello 11/a, Torino, Italy) for the semi-quantitative XRF analyses of the steel bars.

References

1. Carpinteri A, Cardone F, Lacidogna G (2009) Piezonuclear neutrons from brittle fracture: early results of mechanical compression tests. Strain 45:332–339
2. Cardone F, Carpinteri A, Lacidogna G (2009) Piezonuclear neutrons from fracturing of inert solids. Phys Lett A 373:4158–4163
3. Carpinteri A, Cardone F, Lacidogna G (2010) Energy emissions from failure phenomena: mechanical, electromagnetic, nuclear. Exp Mech 50:1235–1243
4. Carpinteri A, Borla O, Lacidogna G, Manuello A (2010) Neutron emissions in brittle rocks during compression tests: monotonic vs cyclic loading. Phys Mesomech 13:268–274
5. Carpinteri A, Lacidogna G, Manuello A, Borla O (2011) Energy emissions from brittle fracture: neutron measurements and geological evidences of piezonuclear reactions. Strength Fracture Complex 7:13–31
6. Widom A, Swain J, Srivastava YN (2013) Neutron production from the fracture of piezoelectric rocks. J Phys G: Nucl Part Phys 40(015006):1–8
7. Widom A, Swain J, Srivastava YN (2014) Photo-disintegration of the iron nucleus in fractured magnetite rocks with magnetostriction. Meccanica 50:1205–1216
8. Cardone F, Cherubini G, Petrucci A (2009) Piezonuclear neutrons. Phys Lett A 373:862–866
9. Carpinteri A, Chiodoni A, Manuello A, Sandrone R (2011) Compositional and microchemical evidence of piezonuclear fission reactions in rock specimens subjected to compression tests. Strain 47(s2):282–292
10. Carpinteri A, Manuello A (2010) Geomechanical and geochemical evidence of piezonuclear fission reactions in the Earth's crust. Strain 47(2):267–281
11. Carpinteri A, Lacidogna G, Manuello A, Borla O (2013) The phenomenon of neutron emission from earthquakes and brittle rocks failure: mechanical tests, microchemical analyses, geological consequences. Exp Mech 53:345–365
12. Bubble Technology Industries (1992) Instruction manual for the bubble detector, Chalk River, Ontario, Canada
13. National Council on Radiation Protection and Measurements (1971) Protection against neutron radiation, NCRP Report 38
14. EN ISO 6892-1:2009 (2009) Metallic materials—Tensile testing—Part 1: method of test at room temperature
15. Gönnenwein F, Börsig B (1991) Tip model of cold fission. Nucl Phys A 530(1):27–57
16. Sharp SR, Lundy LJ, Nair H, Moen CD, Johnson JB, Sarver BE (2011) Acceptance procedures for new and quality control procedures for existing types of corrosion resistant reinforcing steel. Virginia Department of Transportation, Innovation and Research
17. Fulmer CB, Toth KS, Williams IR, Handley TH, Dell GF, Callis EL, Jenkin TM, Wyckoff JM (1970) Photonuclear reactions in iron and aluminum bombarded with high-energy electrons. Phys Rev C 2(4):1371–1378

18. Wanke H (1960) Scandium-45 as reaction product of cosmic radiation in iron meteorites. Zeitschrift fuer Naturforschung (West Germany) Divided into Z. Nautrforsch., A, and Z. Naturforsch., B: Anorg Chem Org Chem Biochem Biophys 15a:953–964
19. EN 10045/1 Metallic Materials—Charpy Impact test Test Method (1990) European committee for standardization. Rue de Stassart 36, Bruxelles, Belgium

Chapter 9
Fatigue in Aluminum Alloys: Early-Stage Diagnosis by Neutron Emissions

Abstract In the last two decades, a large amount of experimental data has been collected about acoustic, electromagnetic, and subatomic particle emissions from solid media subjected to brittle fracture. This experimental evidence can be explained by considering the close relationship between the frequency of pressure waves generated during the fracture process and the size of the forming cracks. Based on the preceding observations, in the present chapter neutron emission measurements are applied for the first time in the early-stage fatigue diagnosis of metallic materials. The attention is focused onto the fatigue behavior of EN-AW6082 aluminum alloy specimens in the Very High Cycle Fatigue (VHCF) regime. In order to investigate the scale effects on both fatigue resistance and neutron emission, five different diameters in the middle cross-section of the specimens are selected between 3 and 30 mm. The VHCF tests are performed by means of an ultrasonic fatigue testing machine able to reach 10^{10} cycles in approximately one week. The experimental results show a decrement in the fatigue resistance by increasing the specimen size. In addition, considerable neutron emissions are revealed during the fatigue tests, which are correlated to micro- and nano-cracking in the samples. Furthermore, an increment in the neutron emission is found by increasing the specimen diameter in the middle cross-section. These results permit to consider the application of neutron emissions as a new and promising structural health monitoring technique for critical components subjected to fatigue damage initiation and accumulation.

Keywords Aluminum alloy · Fatigue initiation · Very high cycle fatigue (VHCF) · Size effects · Micro- and nano-cracking · Neutron emission · Early-stage diagnosis

9.1 Preliminary Remarks

The experimental activities carried out during the last two decades have allowed to collect a great amount of data about the different forms of energy released during the crack formation and propagation in brittle solids, in terms of acoustic, electromagnetic, and subatomic particle emissions [1–4]. In particular, several compression

A. Carpinteri, *Terahertz Phonons and Nanomechanical Instabilities*,
https://doi.org/10.1007/978-3-032-14692-2_9

tests, which were performed on natural rocks characterized by different brittleness properties and chemical compositions, allowed to reveal evident peaks of neutron emission corresponding to the time of specimen failure, even several times higher than the natural neutron background [3, 5–7].

Together with the anomalous neutron and alpha particle emissions [4], sensible chemical composition changes on the fracture surfaces were noticed, with the decrement in the concentration of heavier elements such as iron, nickel, chromium, and others, as well as the contemporary increment in lighter elements, like magnesium, aluminum, silicon, and oxygen, always fulfilling the mass balances [3, 4, 6]. The only possible reason for the surprising results obtained can be recognized in low-energy nuclear reactions (LENR) occurring on the fracture of the material. An explanation of these reactions was found in the interaction between the condensed matter and the frequency of pressure waves produced by cracks at the different scales. In fact, according to the proposed hypothesis, the wavelength of the phonon propagating in the solid body is always of the same order of magnitude as that of the mechanical instability from which it is emitted [4, 7, 8]. Therefore, by considering a wave speed of the order of 10^3 m/s and the simple relation between frequency and wavelength, it is possible to observe that phonons with frequencies in the range between kHz and MHz represent ultrasonic acoustic signals and are caused by cracks at the scale of the millimeter-to-meter. At the same time, phonons with frequencies in the range between MHz and GHz, triggered by microcracking, can produce electromagnetic waves of the same frequency as a secondary effect. Finally, nanocracks can induce phonons with frequencies in the range between GHz and THz, as well as the emission of neutrons as a secondary effect. In addition, since the Debye frequencies of medium-weight elements are of the order of magnitude of TeraHertz (e.g., 7.77 THz for iron), the phonons produced by nanocracking can induce resonance phenomena in the crystal lattices and, as a consequence of energy accumulation (incubation times), the fission of some nuclei with the emission of subatomic fragments and particles [4, 7, 8]. A more complete view on the theoretical model will be provided in the Part VII of this book.

The experimental investigations, carried out on both natural and artificial materials, can give an important confirmation to the proposed explanation for the different types of recorded emissions that are also called "fracto-emissions" [2, 4, 7]. Therefore, in the present chapter, it is proposed to apply the neutron emission monitoring in order to explore the possibility of an early-stage diagnosis of fatigue microcracking (initiation) in metallic materials.

Since the first times of the Industrial Revolution, the collapse of structural elements caused by cyclic loading has become a problem of paramount importance in the structural integrity assessment [9]. In order to investigate the mechanical response of metallic and non-metallic materials subjected to cyclic loading, a huge amount of tests was performed under different loading and environmental conditions. The main output of fatigue tests is generally represented by Wöhler's diagrams, which relate the applied stress range to the number of cycles needed to induce the specimen failure [10]. In addition, due to the statistical dispersion of the experimental data, a large amount of fatigue tests is needed to provide a reliable assessment of the fatigue life of the structural component [11, 12]. On the other hand, it is worth to emphasize that it is not possible to test too many specimens, especially in the ultra-long life regime, since it is very time-consuming with servo-hydraulic and rotating bending machines [13]. In fact, the traditional fatigue testing set-ups are characterized by a working frequency in the range between 1 and 100 Hz, so that it would not be possible to conduct fatigue tests beyond 10^7 cycles within a reasonable time [9, 13, 14]. On the other hand, the development of high-strength metal alloys and the consequent requirement of analyzing their long-term fatigue behavior, induced the Scientific Community to introduce new devices able to increase the loading frequency and, as a consequence, to reduce the testing time [9, 13]. Therefore, thanks to the introduction of ultrasonic fatigue testing machines (UFTM), which are characterized by a working frequency of 20 kHz, it was possible to investigate the mechanical behavior of metal components in the Very High Cycle Fatigue (VHCF) regime [9, 13]. More recently, the concepts of VHCF have been applied for the first time in the civil engineering field to explain the collapse of the Polcevera Bridge (Genoa, Italy) [14–18].

Although the UFTMs allow to reach a very high number of cycles in a relatively short time, an important limit to this kind of fatigue testing devices is the possibility to easily test only specimens with a small diameter [19]. In fact, two important aspects should be considered when performing VHCF tests on larger specimens. The former can be found in the noticeable heat generation produced by internal friction and fast plastic deformation of the samples, although the self-heating could be reduced by adopting a forced-air cooling system and a pulse-pause sequence [13]. The latter reason is due to the difficulty in providing a sufficient stress range in the middle cross-section to ensure their failure under cyclic loading, without increasing the stress concentration factor beyond values higher than 1.15 [13]. In other terms, only numerical models can predict the specimen-size effects on the VHCF resistance [11, 20–24], which cannot be fully validated by experimental results. Actually, the only way to overcome this limitation and to validate the different models proposed in the literature to capture the influence of the specimen size on the gigacycle resistance is to test materials with relatively low fatigue resistance and, at the same time, to adopt optimization procedures for the specimen geometry in order to maximize the nominal stress range in the middle cross-section [11, 14, 16].

Another crucial aspect in the VHCF testing is the possibility to monitor the damage in the specimen subjected to cyclic loading. Nowadays, one of the most exploited non-destructive techniques for VHCF damage is the application of infrared thermography and resonance-based parameters [24]. In-situ thermography allows to detect the crack initiation site and the consequent crack propagation within the specimen [25]. In fact, plastic deformations induced by the application of cyclic loading in ultrasonic fatigue testing produce a considerable increment in the temperature of the sample when it is close to failure [25–27]. In addition, the Scientific Community studied the application of the nonlinear β parameter to investigate the evolution of fatigue damage during VHCF tests. More in detail, the β parameter relates the amplitudes of the fundamental and second harmonics. While the amplitude of the resonance vibration is constant during the ultrasonic fatigue test, since it depends only on loading conditions, a change in the spectral power density of the second harmonic is expected over the time because of fatigue damage accumulated within the specimen. Therefore, this nonlinear parameter, coupled with the resonance frequency of the vibrating system, could provide important information about the fatigue damage evolution [24]. Recently, the acoustic emission technique has been proposed to monitor in-situ damage occurred in aluminum alloy samples during ultrasonic fatigue tests [28].

In the present chapter, a thermal neutron emission-based damage monitoring technique is applied for the first time in the early-stage fatigue diagnosis of EN-AW6082 aluminum alloy hourglass and dog-bone samples subjected to ultrasonic fatigue. In order to investigate the specimen-size effect on the neutron emission, five different diameters of the middle cross-section are selected between 3 and 30 mm (one-order of magnitude range). During VHCF tests, considerable thermal neutron emissions are found, which detect the nanocracking formation in the samples. At the same time, a very peculiar influence of the specimen size on the thermal neutron emissions is revealed. Therefore, such damage monitoring method based on thermal neutron emission could represent in the future an important structural health monitoring technique for critical components subjected to fatigue damage initiation and accumulation.

9.2 Testing Set-Up

9.2.1 Ultrasonic Fatigue Testing Machine

The ultrasonic fatigue testing machines were conceived to produce sinusoidal displacements at the ends of the specimen, which are able to establish resonance at the fundamental frequencies of the specimen itself, the cyclic loading being indirectly applied as a function of the imposed displacement amplitude [9, 13]. In the present investigation, the fatigue tests are performed by means of an ultrasonic testing machine (UFTM MU90) produced by the Italian company Italsigma®. The UFTM MU90 is conceptually composed of four main devices: an ultrasonic generator, a piezoelectric transducer, a booster, and a catenoidal horn that is connected to the specimen by means of a screwed connection [11, 14]. The ultrasonic generator Branson® CR-20 produces an electric signal at 20 kHz that is transmitted to the piezoelectric transducer. The piezoelectric transducer is able to convert the received electric signal into a mechanical vibration of sinusoidal shape, which is characterized by the same original frequency of 20 kHz, by means of the converse piezoelectric effect. The amplitude of the mechanical vibration imposed to the piezoelectric transducer can be changed in the range 5.3–21.7 μm by varying the output voltage settings of the ultrasonic generator. The booster Branson® 2000X Series Gold is screwed to the piezoelectric converter, which provides a first amplification to the mechanical vibration by a factor 1.5 and a fixed support to the whole resonating system. A further amplification of the imposed displacement by a factor 2.0 is finally produced by the catenoidal Branson® 126–192 horn, which is connected to the booster on the upper side and to the specimen by means of a M6 screw [11, 14]. The effect of the two in-series displacement amplifiers is to extend the field of the available displacement amplitudes from the original interval guaranteed only by the piezoelectric transducer (5.3–21.7 μm) to a wider range, i.e., 16–60 μm, where the upper limit is set in order to avoid possible failures of the catenoidal horn [14].

As regards the monitoring systems, the eddy current sensor Micro-Epsilon® eddyNCDT 3300/3301 allows the continuous registration of the actual working frequencies of the specimens and the automatic adjustment of the power provided to the ultrasonic generator, in order to keep a constant amplitude of the mechanical vibration in the middle section of the sample [14]. Furthermore, it is worth noting that the piezoelectric transducer is able to work just in the range of frequencies between 19.5 and 20.5 kHz. Therefore, when the crack size in the specimen is large enough to cause a decrement in the vibration frequency below the limit of 19.5 kHz, the piezoelectric transducer is no longer able to guarantee sinusoidal displacements to the sample and the fatigue test is stopped [11, 14].

As previously stated in Sect. 9.1, one of the main issues related to VHCF tests is the relevant increment in the sample temperature caused by the very high strain rate of the imposed displacement amplitude. Therefore, with the aim at avoiding excessive local temperature peaks, an air-cooling system is provided, as well as it is possible to conduct tests in pulse-pause modality. The continuous control of the temperature of the specimen is made with a pyrometer Optris® CT LT22CF. The pyrometer is characterized by a temperature range between –50 and 975 °C, an accuracy of ±1 °C, and a resolution of 0.1 °C [14]. At the same time, the surfaces of the specimens are covered with a matte-black painting to increase their infrared emissivity [14]. In Figs. 9.1 and 9.2, the main components of the ultrasonic testing machine are portrayed.

In spite that, in the tests reported in the following, no preload is applied, the UFTM is provided by a load cell, which is able to reach a maximum constant vertical load of 1.5 kN. During this experimental program, the fatigue tests are conducted under fully reversed constant stress amplitude tension–compression conditions with a load ratio $R = -1$. Furthermore, T-rosette strain gauges are used in a half-bridge configuration to monitor the stress distributions and validate the results of the numerical simulations

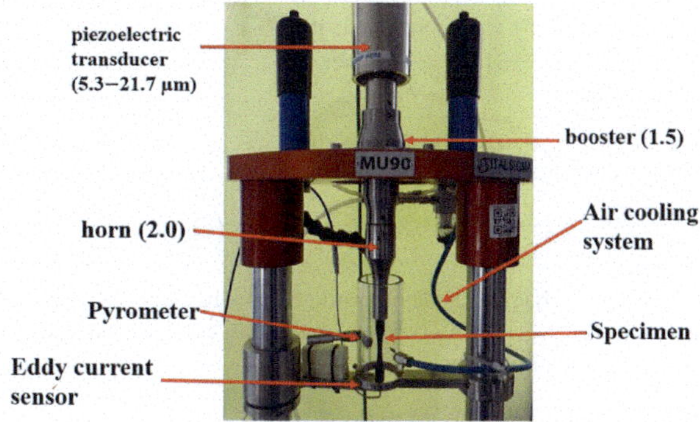

Fig. 9.1 Ultrasonic fatigue testing machine MU90 Italsigma®, operative part

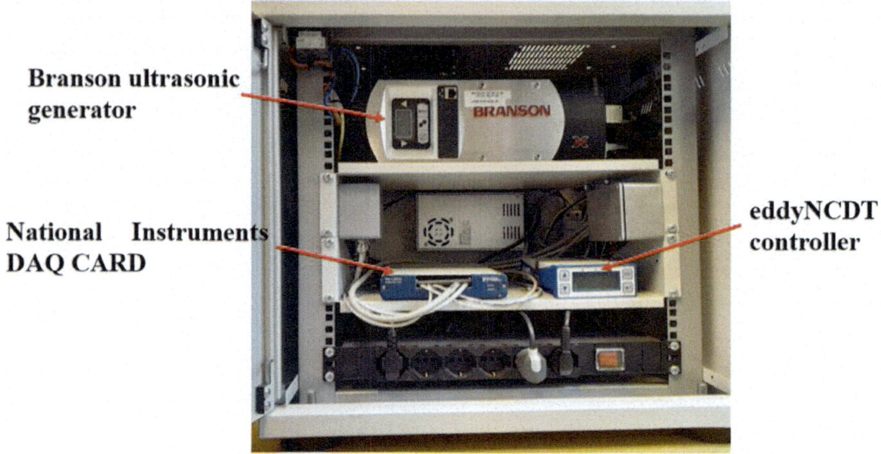

Fig. 9.2 Ultrasonic fatigue testing machine MU90 Italsigma®, control unit

performed before the fatigue tests. The sampling frequency for the strain gauge acquisitions is set to 300 ksamples/s for the specimens of 6 and 12 mm in diameter, whereas higher sampling frequencies are adopted for the specimens of 24 and 30 mm [14].

9.2.2 *Material*

The material adopted for the VHCF tests is the EN-AW6082/AlSi1MgMn wrought aluminum alloy, part of the 6xxx Aluminum-Magnesium-Silicon family, which is received from the producer in the T6 tempered condition [14]. The alloy is known and widely used because of the excellent corrosion resistance and the high tensile strength. Table 9.1 reports the mass percentages of the elements in solid solution, apart from aluminum itself [17]. The specimens are shaped by three cylindrical 4 m long bars, with diameters equal to 20, 30, and 40 mm. The specimens are shaped from the three original bars by means of a Computerized Numerical Control (CNC) machining process [14]. The physical and mechanical properties of the material are thus assessed. The mass density is evaluated by means of an analytical weight balance Kern® ALJ 310-4A, with resolution ± 0.1 mg and linearity ± 0.3 mg. The dynamic elastic modulus, E_{d}, is assessed according to the ASTM Standard E1876-15 on three round samples with diameter of 6 mm and length of 120 mm by means of the non-destructive measurement of the first fundamental longitudinal frequency.

The yielding strength, σ_y, the ultimate strength, σ_u, and the ultimate deformation, ε_u, are obtained by means of quasi-static tensile tests on dog-bone specimens, whose geometry is chosen by following the ASTM Standard B557M-15. To this aim, a servo-hydraulic tensile testing machine with a load cell of 500 kN is utilized. The

Table 9.1 Chemical composition of the EN-AW6082 aluminum alloy (Al excluded)

Element	Si	Mg	Mn	Cu	Fe	Cr	Zn	Ti
Min %	0.70	0.60	0.40	–	–	–	–	–
Max %	1.30	1.20	1.00	0.10	0.50	0.25	0.20	0.10

Table 9.2 Physical and mechanical properties of the EN-AW6082 aluminum alloy

Rod diameter (mm)	ρ (Kg/m^3)	E_d (GPa)	σ_y (MPa)	σ_u (MPa)	ε_u (%)	Brinell hardness (HB)
20	2,713	72.3	–	–	–	105 ± 1.4
30	2,700	70.1	356.8 ± 1.3	375.1 ± 0.2	11.1 ± 1.0	102 ± 1.7
40	2,700	70.4	308.1 ± 0.4	329.8 ± 0.1	14.5 ± 0.2	100 ± 0.7

Brinell hardness, HB, is finally estimated according to the ASTM Standard E10-18 on the original cylindrical bars. The results of these tests are shown in Table 9.2, where the standard deviations are evaluated according to the Law of Propagation of Uncertainty [14].

9.2.3 Specimen Design

In order to evaluate the thermal neutron flux emitted during the VHCF tests on the aluminum alloy EN-AW6082, as well as the specimen-size effect on the cumulative neutron counts compared to the natural background level, it is necessary to define in a proper way the geometry of the specimens. An hourglass shape is adopted for the smaller specimens with diameters in the middle cross-section of 3, 6, and 12 mm, whereas a dog-bone geometry is preferred for the larger specimens of 24 and 30 mm in diameter [11, 14]. More in detail, the specimens of 3 and 6 mm in diameter are shaped from the cylindrical rod of 20 mm in diameter, the specimen of 12 mm from the rod of 30 mm, whereas the rod with diameter equal to 40 mm is used to obtain the two largest specimens [14]. An iterative procedure is needed to obtain the geometry for each dimension, in order to ensure that the fundamental longitudinal frequency of the specimen falls in the range between 19.5 and 20.5 kHz, where the ultrasonic fatigue testing machine is able to work. A preliminary geometry for the hourglass specimens is obtained by adopting the analytical expression reported in Bathias and Paris' book [14, 29], whereas the dog-bone specimens are designed according to the equation proposed by Liu et al. [30]. At this stage, the nominal normal stress in the middle cross-section for a unit displacement amplitude, k_σ, is assessed. Subsequently, the previously obtained specimen geometry is refined by using the finite element analysis software ANSYS Workbench®. An axisymmetric 2D model is adopted to discretize the specimen geometry. Secondly, a modal analysis is carried out for the evaluation of the fundamental longitudinal frequency [14]. Therefore, the obtained

Table 9.3 Mechanical properties and risk volumes of the final specimen shapes

Specimen diameter (mm)	k_σ (MPa/μm)	f_{FEM} (Hz)	k_t (−)	V_{90} (mm^3)
3	7.474	20,213	1.021	42
6	4.233	20,047	1.025	301
12	3.191	19,977	1.046	1,731
24	2.679	19,925	1.071	8,928
30	2.443	20,082	1.086	14,563

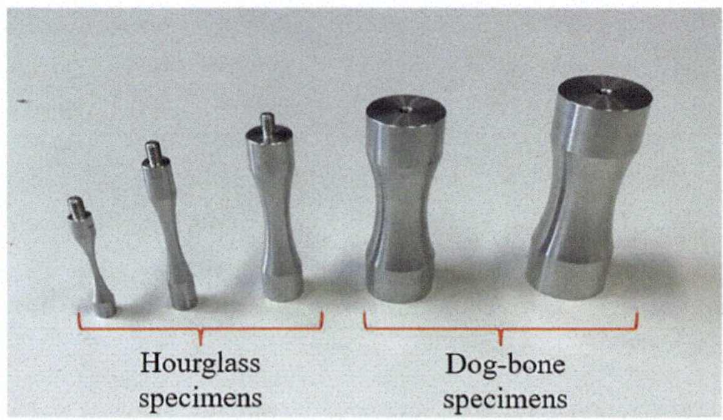

Fig. 9.3 Hourglass and dog-bone specimens. Diameters 3, 6, 12, 24, 30 mm

fundamental frequencies are compared to the limits of the piezoelectric transducer operability. A further harmonic analysis is performed to estimate the normal stress distribution within the specimen volume, so that it is possible to assess the stress concentration factor k_t [31] and the risk volume V_{90}. The original specimen geometry is iteratively modified to ensure a first mode natural frequency close to 20 kHz, a stress concentration factor lower than 1.10, and a nominal normal stress for unit displacement amplitude higher than 2.4 [11, 14]. The final values of the parameter k_σ, the fundamental longitudinal frequency, the stress concentration factor, and the risk volume obtained at the end of the iterative procedure are reported in Table 9.3.

The five specimens are shown in Fig. 9.3. It is worth to highlight that, after the CNC machining process, the specimens are polished with sandpaper to reduce to a final value of 0.4 μm the surface roughness, which is a possible source of unwanted fatigue failures.

9.2.4 Thermal Neutron Counter

The thermal neutron detection during the fatigue tests is accomplished by using a ^3He neutron proportional counter produced by the industrial company Canberra® (France). The operating principle of the counter is based on the fact that the device contains the ^3He gas, whose molecule interacts with the incident thermal neutron to produce a proton, a tritium molecule, and 765 keV of released energy. The proton,

Fig. 9.4 ^{3}He proportional counter

secondary charged particle, can easily be detected as an electromagnetic signal and acquired by the acquisition system [32]. The neutron emission detector is composed of a ^{3}He detector tube and an electronic system of preamplification, amplification, and discrimination of the acquired signal (Canberra® ACHNP97) [3, 5, 6]. It is powered with an electrical tension of 1.25 kV by means of a high voltage Nuclear Instrumentation Module (NIM). The logic output produces the Transistor-Transistor Logic (TTL) pulses, and it is connected to a NIM counter. A threshold value needs to be set for the received signals, in order to avoid possible false neutron measurements, due to the fact that the ^{3}He counter is also sensitive to gamma rays, X rays, and electrons [4–6]. The device is directly calibrated by the company for the detection of thermal neutrons. Its sensitivity is equal to $1\mathrm{n}_{\mathrm{thermal}}/(\mathrm{cm}^2\ \mathrm{s})$, with a declared variability of $\pm 10\%$ [3, 4, 6]. The neutron counter is shown in Fig. 9.4, with its main components.

9.3 Experimental Results

9.3.1 S–N Diagrams

Very high cycle fatigue tests are performed on a total number of 98 specimens (Fig. 9.5). Twenty tests are carried out on hourglass specimens of 3 mm in diameter subjected to a stress amplitude ranging between a minimum value of 150 MPa and a maximum value of 210 MPa. Eighteen of these specimens are led to failure, whereas two of them result to be runouts. Twenty tests are carried out on specimens of 6 mm in diameter, reporting two runouts at 140 MPa. Eighteen failures are obtained for the specimens of 12 mm in diameter between 130 and 180 MPa, whereas two runouts are obtained at 120 MPa. Twenty-one tests are carried out on specimens of 24 mm in diameter, reporting two runouts at 124 MPa. Finally, fifteen failures are obtained

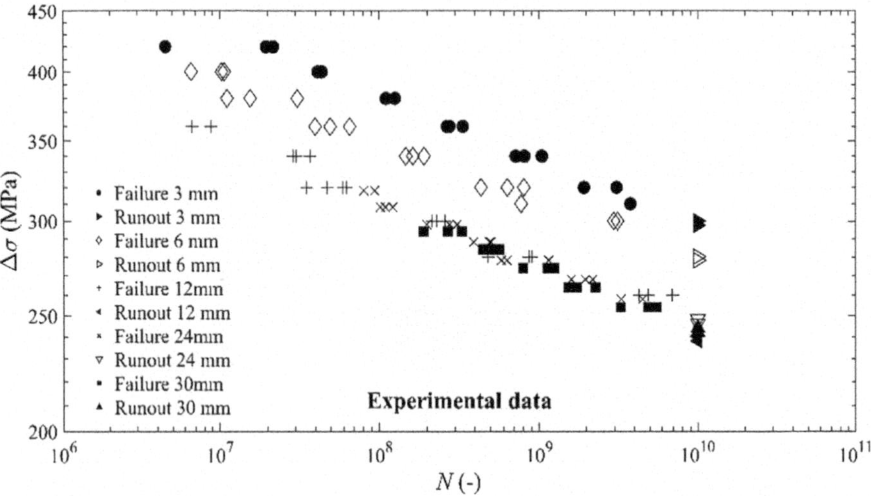

Fig. 9.5 S–N diagrams of the VHCF tests

for the specimens of 30 mm in diameter between 127 and 147 MPa, whereas two runouts are obtained at 122 MPa.

Within the previous complete list, five tests are monitored for the neutron emission in the cases of diameters 3 and 6 mm, four tests for each diameter of 12 and 24 mm, and two tests for the diameter of 30 mm. The main results in terms of neutron emission monitoring are reported in the next sections.

9.3.2 *Neutron Emissions During the VHCF Tests*

Before performing the fatigue tests, background measurements are performed by means of the ^{3}He detector for more than 3 h. The average natural background is of $(6.46 \pm 0.65)\times 10^{-2}$ cps. Subsequently, the specimens are tested with the UFTM and the thermal neutron emissions are detected for the entire duration of the cyclic tests. The detector is placed in front of the UFTM at a distance of 15 cm. In Fig. 9.6a, the neutron emission count rate curve obtained for a specimen of 3 mm in diameter is shown. The specimen, subjected to a stress range of 420 MPa, survived up to 3,533,497 cycles. In the same figure, the previously measured natural background level is reported with white circles. From this figure, neutron emissions about twice the average background level can be observed just before switching off the ultrasonic fatigue testing machine. As an additional supporting evidence, the cumulative count curve is compared to the cumulative average background level (Fig. 9.6b). It is possible to observe a noticeable final difference between the two cumulative curves, with an increment in the neutron emission with respect to the natural background equal to 57.5%.

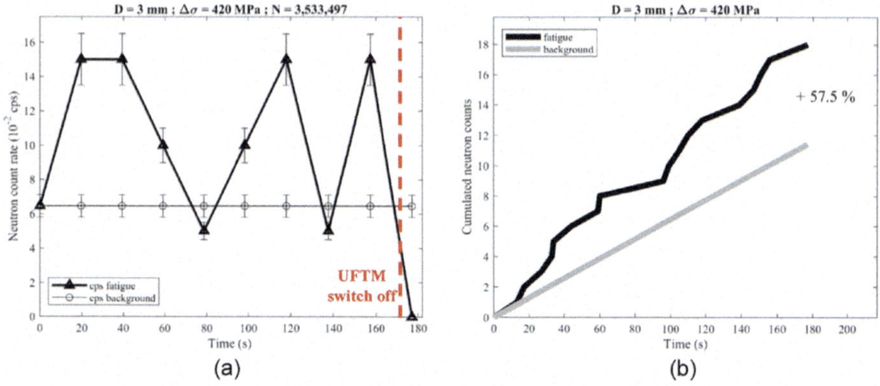

Fig. 9.6 Thermal neutron emissions for the specimen of 3 mm in diameter subjected to a stress range of 420 MPa: **a** neutron count rate; **b** cumulative curve

The neutron emission measurements are also repeated for specimens of 3 mm subjected to different stress range levels. In Fig. 9.7, the deviations from the neutron background of the cumulative curves related to the five tests are shown. From the analysis of the results, it is possible to observe that the revealed neutron emissions show a non-monotonic dependence on the stress range, although cumulative neutron emissions higher than the natural background are always found during VHCF tests, independently of the stress range applied. More in detail, the absolute maximum cumulative neutron emissions are found at a stress range of 420 MPa, whereas an absolute minimum occurs at a stress range of 400 MPa. Analogously, thermal neutron emission measurements are repeated for the other specimen sizes. In Fig. 9.8a, the results obtained for a specimen of 6 mm in diameter, subjected to a mechanical stress range of 380 MPa, are depicted. In this fatigue test, the neutron emissions achieve values about three times the background level. Conversely, immediately after the switching off of the ultrasonic fatigue testing machine, the thermal neutron emissions drop to the natural background level. It is interesting to note that an important increment in the neutron activity is recorded after about 1,600 s from the beginning of the test. This behavior is more evident when analyzing in detail the cumulative neutron emission (Fig. 9.8b), since after 1,600 s an important deviation from the background level can be clearly seen. In this case, a final positive deviation of the cumulative neutron emission with respect to the natural background is found to be equal to 21.8%.

Similarly to what done for the specimens of 3 mm, the neutron emission dependence on the stress range is also investigated for the specimens of 6 mm in diameter. In this case, an absolute minimum of the cumulative neutron emission is found for the stress range of 360 MPa, whereas the maximum deviation from the background level is found for the highest applied stress range, i.e., 400 MPa. More in detail, a neutron emission value about 33% higher than the natural background level is found (Fig. 9.9).

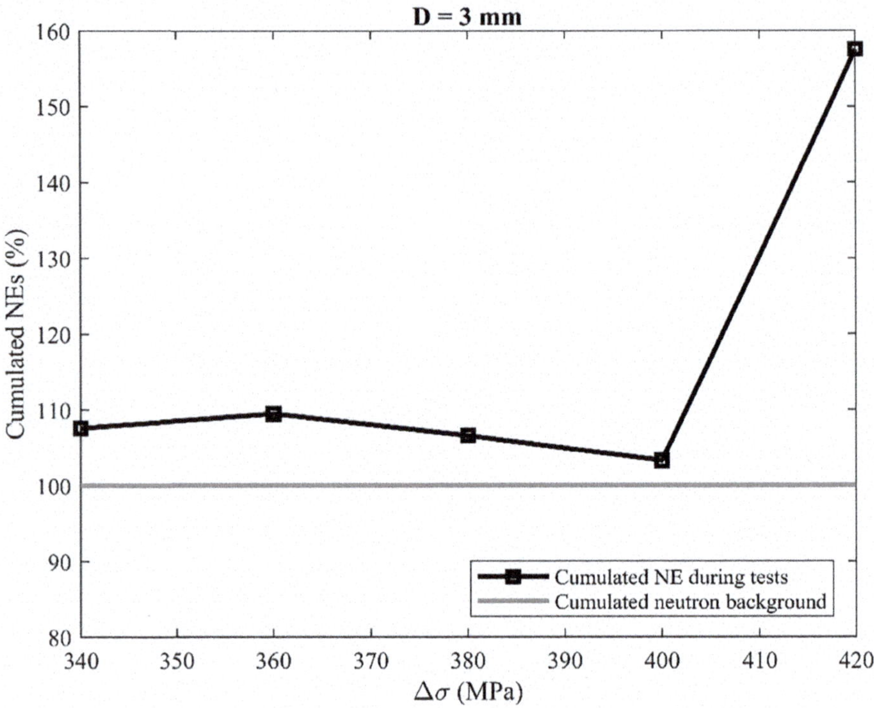

Fig. 9.7 Cumulative neutron emissions against the stress range for the specimens of 3 mm in diameter

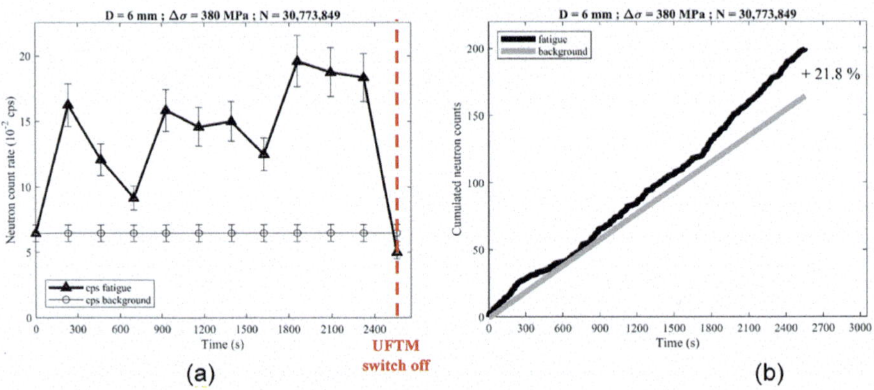

Fig. 9.8 Thermal neutron emissions for the specimen of 6 mm in diameter subjected to a stress range of 380 MPa: **a** neutron count rate; **b** cumulative curve

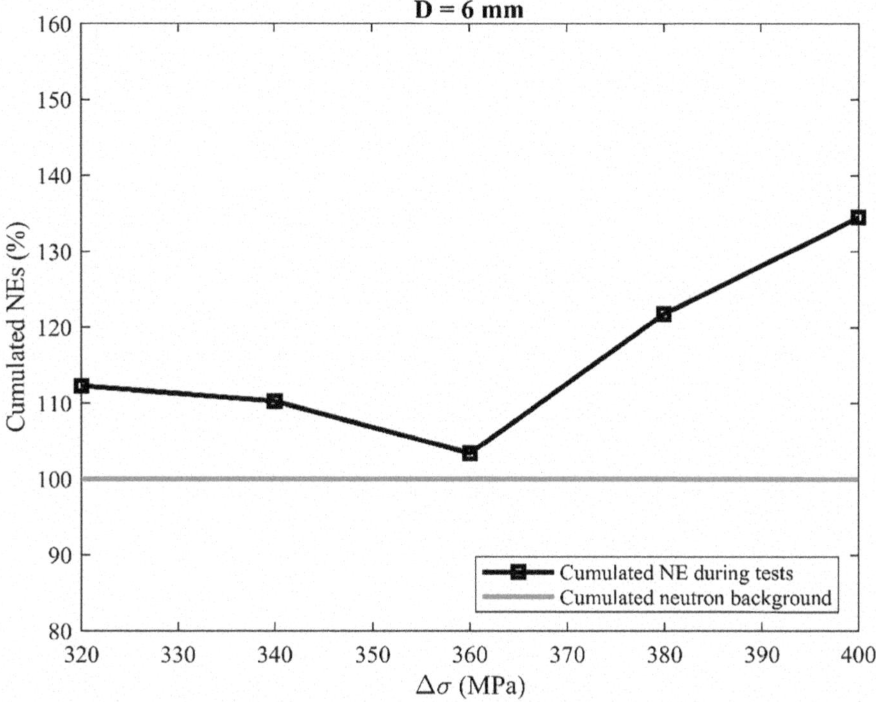

Fig. 9.9 Cumulative neutron emissions against the stress range for the specimens of 6 mm in diameter

The neutron count rates and neutron cumulative curves of a single test performed on a specimen of 12 mm in diameter, which are obtained for an applied stress range of 320 MPa, are depicted in Fig. 9.10a and b, respectively. In this case, the experimental data exceed the background level by approximately 100%. The neutron activity increases significantly after almost 1,200 s from the beginning of the test, with the detection of three main peaks of the neutron emission before the specimen failure. It is interesting to note that, at the switching off of the UFTM, the neutron activity decreases to the background level. In addition, the cumulative neutron emission measured at the end of the test is 18.1% higher than the background level.

In Fig. 9.11, the neutron emission cumulative deviations from the background levels are shown for the four different fatigue tests carried out on the specimens of 12 mm in diameter.

The failures are obtained between 280 and 340 MPa. In this case, the absolute maximum does not occur for the highest stress range applied, although positive deviations from the background level are always found.

The new structural health monitoring technique based on the thermal neutron emissions for the early-stage diagnosis of fatigue damage is also applied to the specimens of 24 mm in diameter. Figure 9.12a and b shows the results obtained for

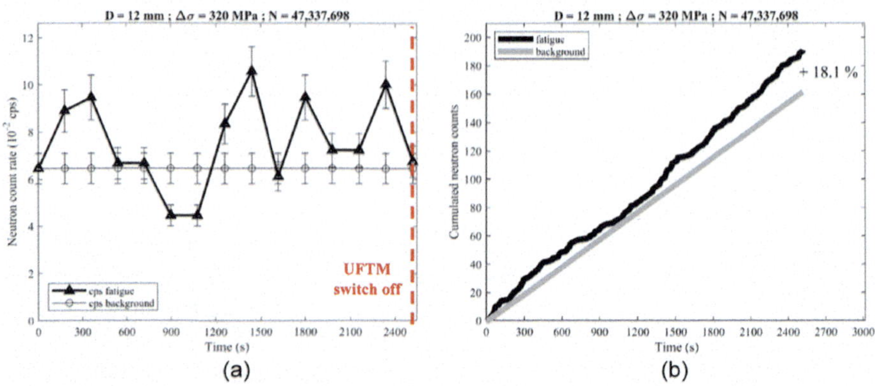

Fig. 9.10 Thermal neutron emissions for the specimen of 12 mm in diameter subjected to a stress range of 320 MPa: **a** neutron count rate; **b** cumulative curve

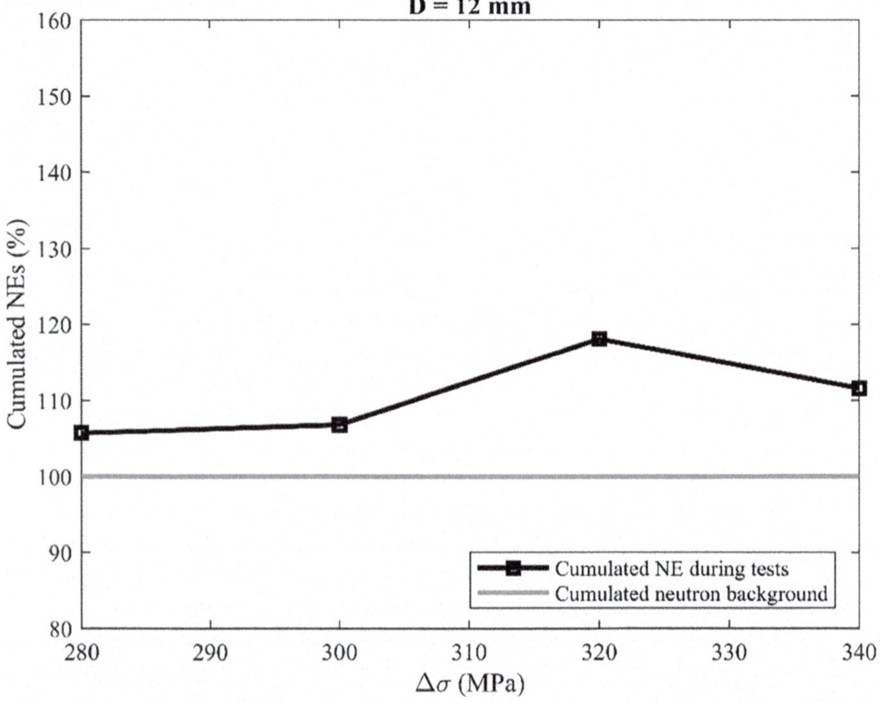

Fig. 9.11 Cumulative neutron emissions against the stress range for the specimens of 12 mm in diameter

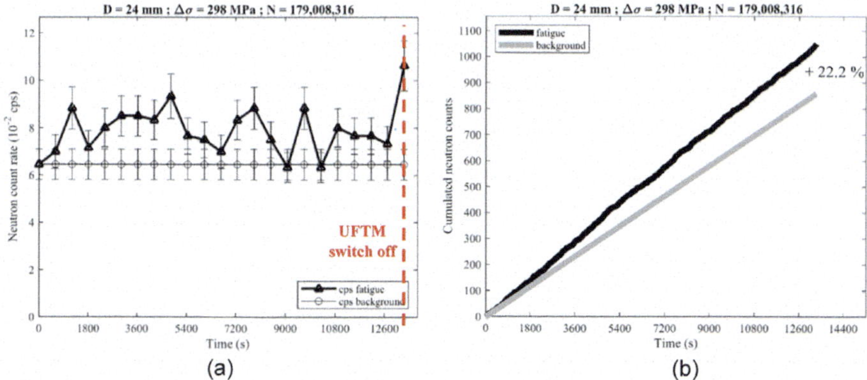

Fig. 9.12 Thermal neutron emissions for the specimen of 24 mm in diameter subjected to a stress range of 298 MPa: **a** neutron count rate; **b** cumulative curve

a stress range of 298 MPa in terms of neutron count rate and cumulative neutron emission during the VHCF test, respectively. The most important observation about this diagram is the existence of an evident peak of the neutron emission just before the specimen failure, with a value approximately twice the background level. The positive final deviation with respect to the average background level assumes in this case the remarkable value of 22.2%.

Figure 9.13 shows the dependency of the neutron emission on the stress range. This figure highlights a little increment in the neutron emission activity with the stress range up to 298 MPa. Conversely, an important final decrement is found for the highest stress range applied, that represents the absolute minimum among the investigated stress ranges. This constitutes a further confirmation to the non-monotonic dependence of the cumulative deviations on this mechanical parameter.

Eventually, the neutron emission measurements are repeated for the specimens of 30 mm in diameter. In this case, the neutron measurements yield values comparable to the natural background and no significant variations in neutron emissions are observed, neither before nor at the specimen failure. As a matter of fact, as can be seen in Fig. 9.14a, the measured neutron count rate (black triangles) is always included in the error band of the background level. This observation means that these small variations may be due only to standard fluctuations of the natural background. Very small variations with respect to the background level are also observed in terms of cumulative neutron emissions (Figs. 9.14b and 9.15), independently of the stress range applied.

A possible explanation could be given by considering that neutron emissions follow an anisotropic and impulsive distribution from a specific zone of the specimen. Therefore, a possible solution to avoid underestimated data acquisition is an experimental measurement by using more than one ^{3}He detector placed around the specimen.

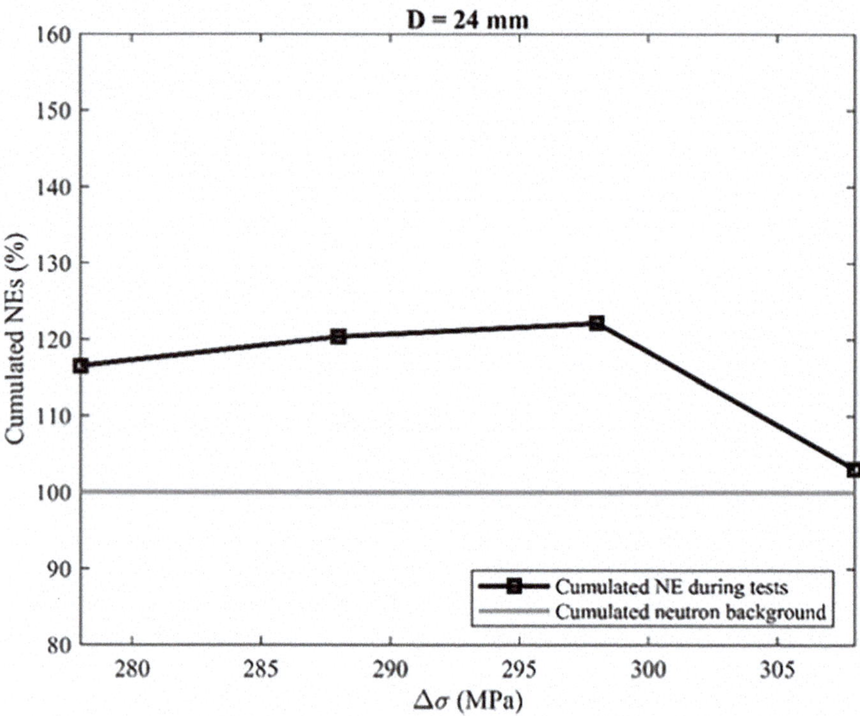

Fig. 9.13 Cumulative neutron emissions against the stress range for the specimens of 24 mm in diameter

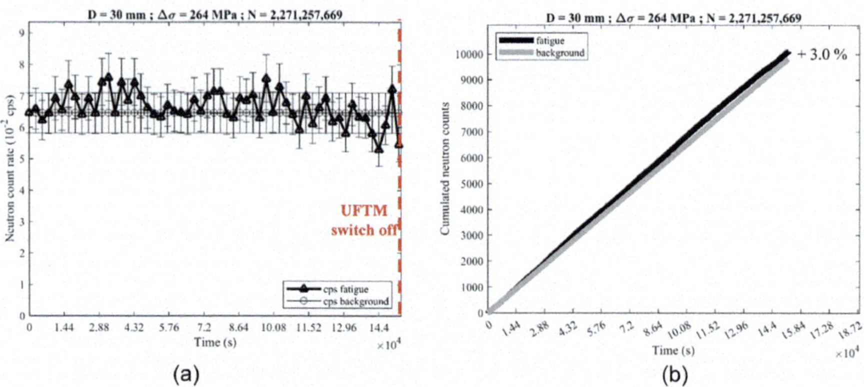

Fig. 9.14 Thermal neutron emissions for the specimen of 30 mm in diameter subjected to a stress range of 264 MPa: **a** neutron count rate; **b** cumulative curve

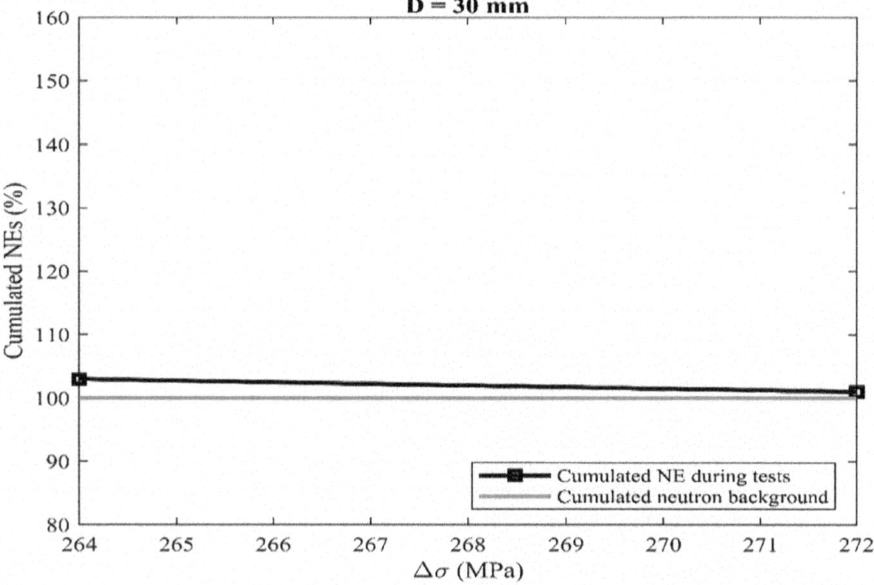

Fig. 9.15 Cumulative neutron emissions against the stress range for the specimens of 30 mm in diameter

9.3.3 Specimen-Size Effects on Neutron Emissions

The cumulated neutron emissions are plotted as a function of specimen size, the stress range being the same. By considering the curve obtained for a stress range equal to 340 MPa, a slight increment in the cumulative neutron emission by increasing the specimen size from 3 up to 12 mm is detected. Analogously, a similar trend is observed for the other applied stress ranges. In particular, a significant and unexpected increment in the cumulative thermal neutron emission is found when the specimen size ranges between 12 and 24 mm. As a consequence, the curves plotted in Fig. 9.16 can be considered as an unambiguous and incontrovertible evidence of the existence of a positive scale effect on the neutron emission activity. In other terms, it can be stated that the larger is the specimen size, the higher results to be the increment in the neutron emission activity with respect to the natural background. This experimental evidence is consistent with the observation that very large specimens are characterized by a very brittle (catastrophic) failure under monotonic loading [33–35], with abrupt acoustic, electromagnetic, and neutron emission bursts. On the other hand, we can justify this increment also by more simple and direct proportional volumetric effects.

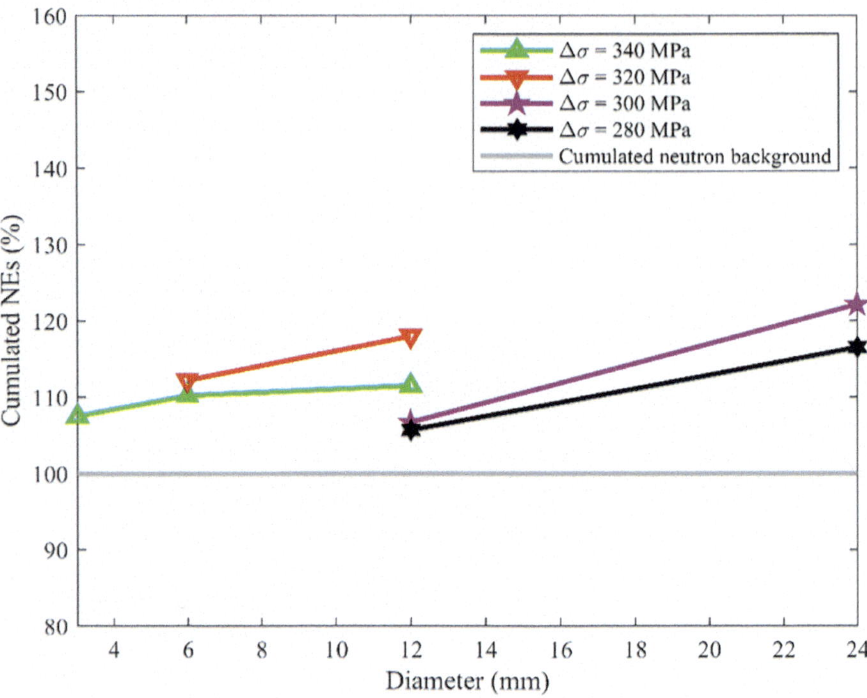

Fig. 9.16 Cumulative neutron emissions as a function of the specimen diameter

9.4 Conclusions

The experimental activity described in the present chapter is focused on the possible application of a new and promising structural health monitoring technique for early-stage fatigue diagnosis of metallic materials based on the neutron emission from the fatigue damage initiation and accumulation. The VHCF tests carried out on aluminum EN-AW6082 specimens, characterized by different diameters in the middle cross-section, spanning from 3 to 30 mm, allow to observe noticeable neutron emissions during the fatigue tests. In particular, it is observed that thermal neutron emissions up top 2–3 times higher than the natural background level take place before the specimen fatigue failure. Moreover, the highest neutron count rate is obtained from the specimen with a diameter in the middle cross-section of 6 mm, subjected to one of the highest stress ranges applied in this experimental program, i.e., 400 MPa. On the other hand, no neutron emission occurs for the largest specimen size (30 mm), although it is worth to note that they are subjected to the lowest stress range. In terms of cumulative neutron emissions, similar results are obtained. In fact, the highest cumulative neutron emission is recorded for one of the smallest specimens of 3 mm in diameter, which is subjected to the highest stress range of 420 MPa, whereas no relevant deviation from the background cumulative curve is found for the largest specimens. As a consequence, it can be stated that the applied stress range certainly plays a fundamental role in the thermal neutron emission from fatigue damage initiation and accumulation, i.e., the lower is the stress range, the lower results to be the thermal neutron emission. At the same time, the influence of the specimen size on the thermal neutron emission is investigated. To this aim, the stress range is kept constant, whereas the other parameter under investigation, i.e., the specimen size, is changed. In this case, the experimental results show an evident influence of the specimen size, e.g., the larger is the specimen size, the higher results to be the thermal neutron emission, the stress range being the same. On the basis of these experimental results, the application of neutron emission during VHCF tests could be considered as a new and promising structural health monitoring technique for critical components subjected to fatigue damage initiation and accumulation.

References

1. Carpinteri A, Cardone F, Lacidogna G (2009) Piezonuclear neutrons from brittle fracture: early results of mechanical compression tests. Strain 45(4):332–339
2. Carpinteri A, Cardone F, Lacidogna G (2010) Energy emissions from failure phenomena: mechanical, electromagnetic, nuclear. Exp Mech 50(8):1235–1243
3. Carpinteri A, Lacidogna G, Borla O, Manuello A, Niccolini G (2012) Electromagnetic and neutron emissions from brittle rocks failure: experimental evidence and geological implications. Sādhāna 37(1):59–78
4. Carpinteri A, Lacidogna G, Manuello A (2015) Acoustic, electromagnetic, neutron emissions from fracture and earthquakes. Springer International Publishing, Switzerland
5. Carpinteri A, Borla O, Lacidogna G, Manuello A (2010) Neutron emissions in brittle rocks during compression tests: monotonic versus cyclic loading. Phys Mesomech 13(5–6):268–274
6. Carpinteri A, Lacidogna G, Manuello A, Borla O (2012) Piezonuclear fission reactions in rocks: evidences from microchemical analysis, neutron emission, and geological transformation. Rock Mech Rock Eng 45(4):445–459
7. Carpinteri A, Borla O (2017) Fracto-emissions as seismic precursors. Eng Fract Mech 177:239–250
8. Carpinteri A, Borla O (2020) Strong correlation between LENR and nano-mechanics instabilities/THz phonons in condensed matter: applications in geophysics, geochemistry, energetics, biology. Infinite Energy 153:32–37
9. Sharma A, Oh M, Ahn B (2020) Recent advances in Very High Cycle Fatigue behavior of metals and alloys—A review. Metals 10(9) 1200:1–23
10. Mairone M, Asso R, Masera D, Invernizzi S, Montagnoli F, Carpinteri A (2022) Fatigue performance analysis of an existing orthotropic steel deck (OSD) bridge. Infrastructures 7(10):135
11. Invernizzi S, Montagnoli F, Carpinteri A (2021) Experimental evidence of specimen-size effects on EN-AW aluminium alloy in VHCF regime. Appl Sci 11(9):4272
12. Tridello A, Boursier Niutta C, Rossetto M, Berto F, Paolino D (2022) Statistical models for estimating the fatigue life, the stress-life relation, and the P-S-N curves of metallic materials in very high cycle fatigue: a review. Fatigue Fract Eng Mater Struct 45(2):332–370
13. Fitzka M, Schönbauer B, Rhein R, Sanaei N, Zekriardehani S, Tekalur S, Carrol J, Mayer H (2021) Usability of ultrasonic frequency testing for rapid generation of high and very high cycle fatigue data. Materials 14(9):2245
14. Montagnoli F (2021) Very high cycle fatigue: size effects and applications in civil engineering. Doctoral Thesis. Politecnico di Torino, Torino, Italy

15. Invernizzi S, Montagnoli F, Carpinteri A (2019) Fatigue assessment of the collapsed XXth century cable-stayed Polcevera Bridge in Genoa. Procedia Struct Integrity 18:237–244
16. Invernizzi S, Montagnoli F, Carpinteri A (2022) Very high cycle corrosion fatigue study of the collapsed Polcevera Bridge, Italy. J Bridge Eng 27(1):04021102
17. Invernizzi S, Montagnoli F, Carpinteri A (2020) The collapse of the Morandi's Bridge: Remarks about fatigue and corrosion. In: IABSE symposium, Wroclaw 2020: Synergy of culture and civil engineering—History and challenges, pp 1040–1047
18. Invernizzi S, Montagnoli F, Carpinteri A (2020) Corrosion fatigue investigation on the possible collapse reasons of Polcevera Bridge in Genoa. In: Lectures notes on mechanical engineering, pp 151–159
19. Tridello A, Paolino D, Rossetto M (2020) Ultrasonic VHCF tests on very large specimens with risk-volume up to 5000 mm^3. Appl Sci 10(7):2210
20. Invernizzi S, Paolino D, Montagnoli F, Tridello A, Carpinteri A (2022) Comparison between fractal and statistical approaches to model size effects in VHCF. Metals 12(9):1499
21. Carpinteri A, Montagnoli F, Invernizzi S (2020) Scaling and fractality in fatigue resistance: specimen-size effect on Wöhler's curve and fatigue limit. Fatigue Fract Eng Mater Struct 43(8):1869–1879
22. Carpinteri A, Montagnoli F (2019) Scaling and fractality in fatigue crack growth: Implications to Paris' law and Wöhler's curve. Procedia Struct Integrity 14:957–963
23. Montagnoli F, Invernizzi S, Carpinteri A (2020) Fractality and size effect in fatigue damage accumulation: comparison between Paris and Wöhler perspectives. In: Lectures notes on mechanical engineering, pp 188–196
24. Mayer H (2016) Recent developments in ultrasonic fatigue. Fatigue Fract Eng Mater Struct 39(1):3–29
25. Ranc N, Messager A, Junet A, Palin-Luc T, Buffière JY, Saintier N, Elmay M, Mancini L, King A, Nadot Y (2022) Internal fatigue crack monitoring during ultrasonic fatigue test using temperature measurements and tomography. Mechan Mater 174:104471
26. Krewerth D, Weidner A, Biermann H (2013) Application of in situ thermography for evaluating the high-cycle and very high-cycle fatigue behaviour of cast aluminium alloy $AlSi_7Mg$ (T6). Ultrasonics 53(8):1441–1449
27. Krewerth D, Lippmann T, Weidner A, Biermann H (2015) Application of full-surface view in situ thermography measurements during ultrasonic fatigue of cast steel G42CrMo4. Int J Fatigue 80:459–467
28. Seleznev M, Weidner A, Biermann H, Vinogradov A (2021) Novel method for in situ damage monitoring during ultrasonic fatigue testing by the advanced acoustic emission technique. Int J Fatigue 142:105918
29. Bathias C, Paris PC (2004) Gigacycle fatigue in mechanical practice. CRC Press
30. Liu Y, Chen S, Tian R, Wang Q (2011) Design of dog-bone-shaped ultrasonic vibrational fatigue specimen and its application in study on VHCF behavior of 6063 aluminium alloy. Adv Mater Res 160–162:783–788
31. Tridello A, Paolino D, Chiandussi G, Rossetto M (2013) Comparison between dog-bone and gaussian specimens for size effect evaluation in gigacycle fatigue. Frattura ed Integrità Strutturale 26:49–56
32. CANBERRA (2017) Neutron detection and counting
33. Bazant Z, Chen E (1997) Scaling of structural failure. Appl Mech Rev 50(10):593–627
34. Carpinteri A (1989) Decrease of apparent tensile and bending strength with specimen size: two different explanations based on fracture mechanics. Int J Solids Struct 25(4):407–429
35. Carpinteri A, Chiaia B (2002) Embrittlement and decrease of apparent strength in large-sized concrete structures. Sādhāna 27(4):425–448

BY

Part IV
Heat Generation, Subatomic Particle Emissions, and Chemical Composition Changes in Electrolysis and Cavitation Experiments

Chapter 10
Electrolysis Experiments with Ni–Fe and Co–Cr Electrodes: Hydrogen Embrittlement, Microcracking, TeraHertz Phonons, and Correlated Nuclear Phenomena

Abstract Remarkable evidence of anomalous nuclear reactions occurring in condensed matter are observed in electrolysis experiments performed worldwide. Despite the great amount of experimental results coming from the so-called "Cold Nuclear Fusion" research activities, the comprehension of these phenomena still remains unsatisfactory and controversial. On the other hand, as reported by several authors, a common feature of the experiments is the appearance of microcracks on the electrode external surfaces after the experiments. In the present chapter, a macromechanical explanation is proposed, considering a new kind of nuclear reactions, the phono-fissions, which are a consequence of hydrogen embrittlement of the electrodes during the electrolysis. The experimental activity is conducted using a Ni–Fe anode and a Co–Cr cathode immersed in a potassium carbonate aqueous solution. Emissions of neutrons and alpha particles are detected during the experiments and the electrode chemical compositions are analyzed both before and after the electrolysis experiment, revealing the effects of phono-fission reactions occurring in the host lattice of the electrodes. The symmetrical fission of Ni appears to be the most evident process. Such reaction produces two Si atoms, or two Mg atoms, with neutrons and alpha particles as additional fragments.

Keywords Electrolysis · Hydrogen embrittlement · Electrode microcracking · TeraHertz phonons · Phono-fission reactions · Neutron emissions · Cold nuclear fusion

10.1 Preliminary Remarks

During the last 35 years, a remarkable evidence of anomalous nuclear reactions occurring in condensed matter has been observed [1–34]. These experimental tests were characterized by significant neutron and alpha particle emissions, as well as by extra-heat generation. In addition, appreciable variations in the chemical composition were detected in the specimens after brittle fracture or fatigue damage accumulation experiments [35–44], see Parts II and III.

A. Carpinteri, *Terahertz Phonons and Nanomechanical Instabilities*,
https://doi.org/10.1007/978-3-032-14692-2_10

Most relevant papers on the so-called Cold Nuclear Fusion describe extensive experimental activities conducted on electrolytic cells powered by direct current and filled with ordinary or heavy water solutions. In particular, in 1989, Fleischmann and Pons proposed the first experiment reproducing Cold Nuclear Fusion by means of electrolysis [6]. They asserted that the Palladium electrode reacts with the Deuterium coming from the heavy water solution [6]. Later works report that Pt and Ti electrodes had also been electrolyzed with D_2O to produce extra-energy and chemical elements previously absent [26, 28, 29]. Extra-energy was also produced from electrolysis with Ni cathodes and H_2O-based electrolyte [13]. Furthermore, it was affirmed that a voltage sufficient to induce plasma can generate a large variety of anomalous nuclear reactions when Pd, W, or C cathodes are adopted [16, 21–25].

In different experiments, the generated heat was calculated to be several times higher than the input energy, and the neutron emission rate, during the electrolysis, was measured to be about three times the natural background level [6].

In 1998, Mizuno presented the results of the measurements conducted by means of neutron emission detectors and by chemical composition analysis techniques in relation to different electrolysis experiments [22]. A relevant heat generation was observed when the cell was supplied with high voltage, with an excess energy 2.6 times higher than the input energy. Remarkable neutron emissions were revealed during these tests, as well as a considerable amount of new elements, i.e., Pb, Fe, Ni, Cr, and C, with the isotopic distribution of Pb deviating greatly from the natural isotopic abundances [22]. These results suggested that nuclear fission reactions had taken place during the electrolysis process [22]. Later, in 2002 Kanarev and Mizuno reported the results obtained from the surface compositional analysis of iron electrodes (99.90% of Fe) immersed in KOH and NaOH solutions [34]. After the experiments, EDX spectroscopy revealed the appearance of several new chemical elements previously absent. Concentrations of Si, K, Cr, and Cu were found on the surfaces of the operating cathode immersed in KOH. Analogously, concentrations of Al, Cl, and Ca were noticed on the iron electrode surfaces operating in the NaOH solution. These findings are an evidence of compositional changes occurring during plasma formation in electrolysis of water solutions [34].

In 2007, Mosier-Boss et al. [31, 33] obtained important proofs of anomalous measurements in experiments conducted by electrolytic co-deposition cells. More in detail, anomalous effects observed in the Pd/D system included heat and helium-4 generation, tritium, neutron, gamma/X-ray emissions, and chemical composition changes [7, 12, 31, 33].

Quoted from the Italian scientist Giuliano Preparata: "Despite the great amount of experimental results observed by a large number of scientists, a unified interpretation and theory of these phenomena has not been accepted and their comprehension still remains unsolved" [6–9, 26, 27]. On the other hand, as shown by most articles devoted to Cold Nuclear Fusion, one of the principal and most frequent experimental features is the appearance of microcracks on the electrode surfaces after the tests [26, 27]. Such evidence should be directly correlated to hydrogen embrittlement of the metal composing the electrodes (Pd, Ni, Fe, Ti, etc.). This phenomenon, well-known in Metallurgy and, particularly, in Fracture Mechanics, characterizes some metals during forming or finishing operations [45]. In the present study, the host metal matrix (in particular, hydrogen adsorption is remarkable in Ni and Pd) is subjected to mechanical damage and cracking due to external atoms (deuterium or hydrogen) penetrating into the atomic lattice and forcing it by an internal compression, during the gas loading. Hydrogen effects are largely studied, especially in metal alloys where the presence of H free atoms in the host lattice causes the metal to become more brittle and less resistant to crack formation and propagation. In particular, hydrogen generates an internal compression stress that lowers the apparent fracture toughness of the metal, so that brittle crack growth can occur under a hydrogen partial pressure below 1 atm [45, 46].

Some experimental evidence shows that neutron emissions may be strictly correlated to fracture of non-radioactive or inert materials. From this point of view, anomalous nuclear emissions and heat generation had been detected during fracture in fissile materials [2–4] and in deuterated solids [5, 8, 30]. The experiments recently proposed by Carpinteri et al. and by Cardone et al. [35–43] represent the first evidence of neutron emissions due to phono-fissions during failure of inert, stable, and non-radioactive solids under compression, as well as from non-radioactive liquids under ultrasonic cavitation [35, 36]. In particular, damaged and even cratered zones were observed in recent experiments reported by Albertini et al., where iron bars are subjected to pressure waves of ultrasonic frequency. In this case, the presence of elements such as O, Cl, K, Cu cannot be attributed to the occurrence of non-metallic inclusions or to contamination occurred during fabrication [42].

In the present chapter, we analyze neutron and alpha particle emissions during tests conducted on an electrolytic cell, where the electrolysis is obtained using Ni–Fe and Co–Cr electrodes in potassium carbonate aqueous solution. Voltage, current intensity, solution conductivity, temperature, alpha and neutron emissions are monitored. The compositions of the electrodes are analyzed both before and after the tests. Strong evidence suggests that the so-called Cold Nuclear Fusion, interpreted under the light of hydrogen embrittlement, may be explained by phono-fission reactions occurring in the host metal, instead of by the nuclear fusion of H isotopes adsorbed in the atomic lattice. These new kind of fission reactions are observed, from the laboratory to the Earth's crust or tectonic scale, when particularly high-frequency stress waves (phonons) originate from fracture or fatigue phenomena in the laboratory, or in correspondence to an impending earthquake [38–40, 44].

10.2 Experimental Set-Up and Measurement Equipments

10.2.1 The Electrolytic Cell and the Power Circuit

Over the last fifteen years, specific experiments have been conducted on an electrolytic reactor (Owners: Mr Alessandro Goi et al.). The aim was to investigate whether the anomalous heat generation may be correlated to a new type of nuclear reactions during electrolysis phenomena. The reactor was built in order to be appropriately filled with a salt solution of water and potassium carbonate (K_2CO_3). The electrolytic phenomenon was obtained using two metal electrodes immersed in the aqueous solution. The solution container, named also reaction chamber in the following, is a cylinder-shaped element of 100 mm in diameter, 150 mm high, and 5 mm thick. For the reaction chamber, two different materials were used during the experiments: Pyrex glass and Inox AISI 316L steel. The two metallic electrodes were connected to a source of direct current: a Ni–Fe-based electrode as the positive pole (anode), and a Co–Cr-based electrode as the negative pole (cathode) (Fig. 10.1b).

With regards to the experiment described in the present chapter, after approximately 10 operating hours, the generation of cracks on the glass container forced the authors to adopt a more resistant container made of steel. Teflon lids are sealed to both upper and lower openings of the chamber. The reaction chamber base consists of a ceramic plate preventing the direct contact between liquid solution and Teflon lid (Fig. 10.1a). Two threaded holes host the electrodes, which are screwed to the bottom of the chamber, which is successively filled with the solution. A valve at the top of the cell allows the vapor to escape from the reactor and condense in an external collector. Externally, two circular Inox steel flanges, fastened by means of four threaded ties, hold the Teflon layers. The inferior steel flange of the reactor is connected to four supports isolated from the ground by means of a rubber-based material. As mentioned before, a direct current passes through the anode and the cathode electrodes, provided by a power circuit connected to the power grid through an electric socket. The components of the circuit are an isolating transformer, an electronic variable transformer (Variac), and a diode bridge linked in series (Fig. 10.2).

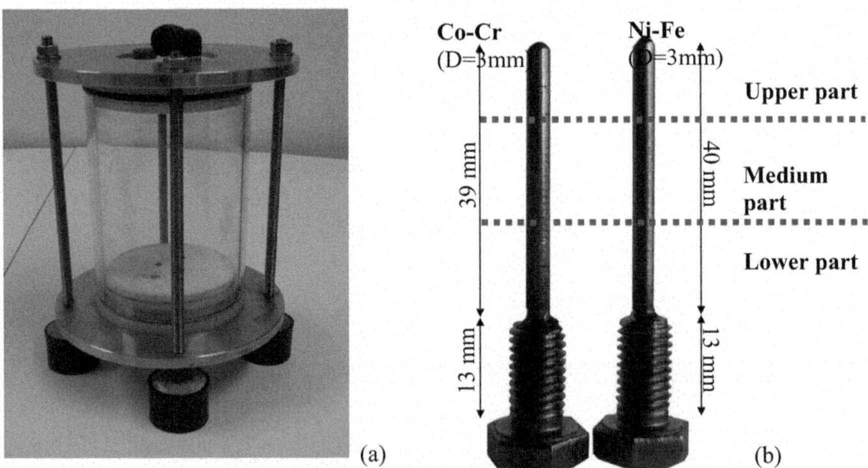

Fig. 10.1 Reaction chamber is a cylinder-shaped element of 100 mm in diameter, 150 mm high, and 5 mm thick (**a**). The two electrodes present a height of about 40 mm for the operating part and a diameter of 3 mm. The threaded portion and the base are 13 mm and 5 mm long, respectively (**b**)

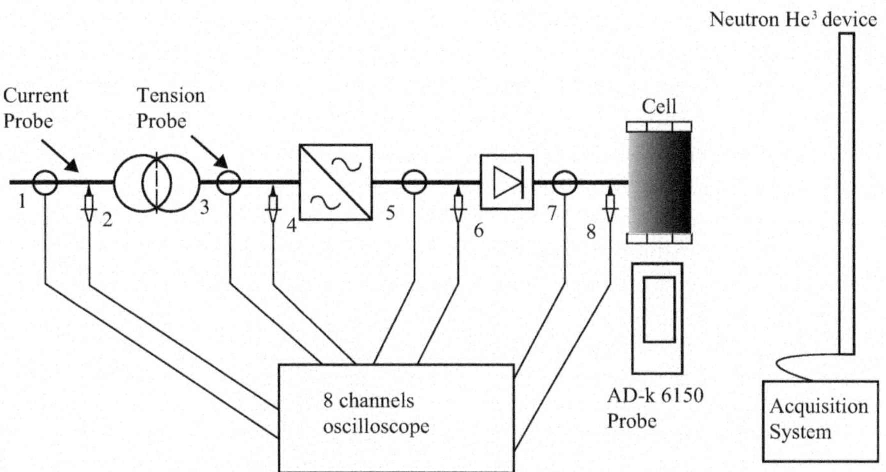

Fig. 10.2 Scheme of the experimental set-up adopted and disposition of the measurement instruments employed during the tests

10.2.2 Measurement Instruments

Different physical quantities are measured during the experiments, such as voltage, current intensity, neutron and alpha particle emissions.

Electric current and voltage probes are positioned in different parts of the circuit as shown in Fig. 10.2. The voltage measurements are performed by a differential voltage probe of 100 MHz with a maximum rated voltage of 1,400 V. The current is measured by a Fluke I 310S probe with a maximum rated current of 30 A. Particular attention is paid to the data obtained from the current and voltage probes positioned at the linear input powering the reaction chamber (probes 7 and 8 in Fig. 10.2), in order to evaluate the power absorbed by the cell. Current intensity and voltage measurements are also taken by means of a multi-meter positioned at the line input. From the turning onto the switching off of the electrolytic cell, current and voltage are found to vary in a range from 3 to 5 A and from 20 to 120 V, respectively. For convenience and clarity, these values are considered as a benchmark to be compared to further measurements.

For what concerns the neutron emission measurements, since neutrons are electrically neutral particles, they cannot directly produce ionization in a detector, and therefore cannot be directly detected. This means that neutron detectors must rely upon a conversion process accounting the interaction between an incident neutron and a nucleus, which produces a secondary charged particle. Such charged particle is then detected and the neutron's presence is revealed from it. For an accurate neutron evaluation, a He^3 proportional counter is employed. The detector used in the tests is a He^3 type (Xeram, France) with pre-amplification, amplification, and discrimination electronics directly connected to the detector tube. The detector is supplied by a high voltage (about 1.3 kV) via NIM (Nuclear Instrument Module). The logic output producing the TTL (transistor–transistor logic) pulses is connected to a NIM counter. The logic output of the detector is enabled for analogue signals exceeding 300 mV. This discrimination threshold is a consequence of the sensitivity of the He^3 detector to the gamma rays ensuing neutron emission in ordinary nuclear processes. This value is determined by measuring the analog signal of the detector by means of a Co-60 gamma source. The detector is also calibrated at the factory for the measurement of thermal neutrons; its sensitivity is 65 cps/$n_{thermal}$ ($\pm$10% declared by the factory), i.e., the flux of thermal neutrons is one thermal neutron/s cm^2, corresponding to a count rate of 65 cps.

For the alpha particle emission, a 6150AD-k probe with a sealed proportional counter is used, which does not require refilling or flushing from external gas reservoirs. The probe is sensitive to alpha, beta, and gamma radiations. An electronic switch allows for the operating mode "alpha" to detect alpha radiation only, such that in this mode the radiation recognition is very sensitive because the background level is much lower. A removable discriminator plate (stainless steel, 1 mm in thickness) distinguishes between beta and gamma radiation detection. An adjustable handle can be locked to the most convenient orientation. The 6150AD-k probe is used in the operating mode "alpha" to monitor the background level before and after the switching on of the cell.

Eventually, before and after the experiments, Energy Dispersive X-ray Spectroscopy is performed in order to recognize a possible direct evidence of phono-fission reactions that can take place during the electrolysis. The elemental analyses are performed by a ZEISS Auriga field emission scanning electron microscope (FESEM) equipped with an Oxford INCA energy-dispersive X-ray detector (EDX) with a resolution of 124 eV @ MnKa. The energy used for the analyses is 18 keV.

10.3 Experimental Results

10.3.1 Preliminary Chemical Composition Analysis of the Electrodes

In Fig. 10.1b, the two electrodes used for the tests are shown. The initial measurement phase implies the use of the Energy Dispersive X-ray spectroscopy (EDX) technique to obtain measurements useful to evaluate the chemical composition of the two electrodes before the experiment. In particular, a series of measurements are repeated in three different regions of interest for each electrode, in order to obtain a sufficient amount of reliable data. Such regions are the upper, the middle, and the lower parts of the single electrode, as reported in Fig. 10.1b.

In Fig. 10.3a and b, the average chemical element concentrations of the electrodes used for the electrolysis are shown. In the initial condition, the Ni–Fe electrode (anode) is composed by approximately 44% in Ni, 30% in Fe, and 23% in O. The remaining percentage includes contents of Si, Mn, Ca, Al, K, Na, Mg, Cl, and S, observable only in traces (Fig. 10.3a). On the other hand, the Co–Cr cathode is composed approximately by 44% in Co, 18% in Cr, 4% in Fe, 25% in O, and traces of other elements such as Si, Al, Mg, Na, W, Cu, and S (Fig. 10.3b). Table 10.1 summarizes the results for the compositional analysis conducted on K_2CO_3, the salt used for the aqueous solution ($K_2CO_3 + H_2O$), where the solute to solvent ratio is approximately 40 g/l.

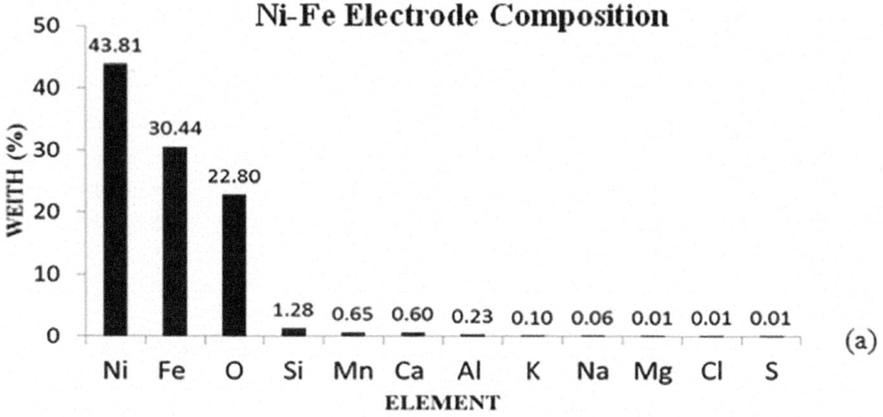

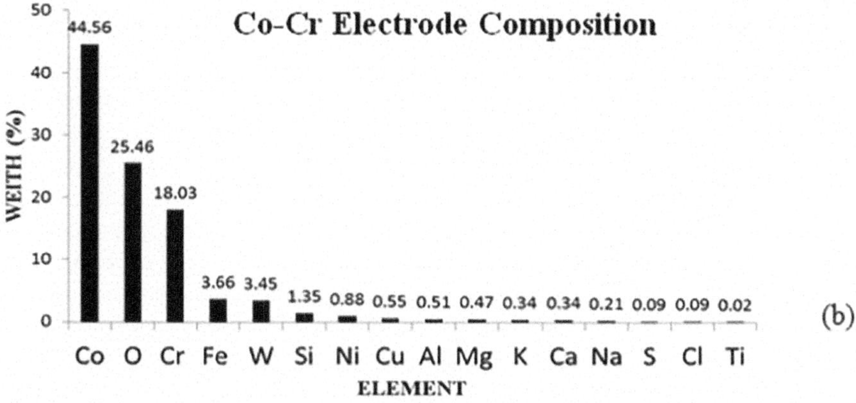

Fig. 10.3 Initial chemical element concentrations in the two electrodes used for the electrolysis

Table 10.1 EDX spectroscopy of the K_2CO_3 salt used for the aqueous solution

Element	Weight%	Atomic%	Compd%	Formula
C	13.02	22.05	47.72	CO_2
K	43.40	22.57	52.28	K_2O
O	43.58	55.38		
Total	100.00			

10.3.2 Neutron and Alpha Particle Detection During the Experiment

Neutron emission measurements performed during the experimental activity are represented in Fig. 10.4. The measurements performed by the He^3 detector are conducted for a total time of about 38 h. The natural background level is measured for different time spans before and after switching on the reaction chamber. These measurements report an average neutron background of $5.17 \pm 1.29 \times 10^{-2}$ cps. Furthermore, when the reactor is active, it is possible to observe that, after a time span of about 3 h (200 min), neutron emissions of about 4 times the background level may be recognized. After 11 h (650 min) from the beginning of the measurements, it is possible to observe a neutron emission level of about one order of magnitude greater than the background level. Similar results are observed after 20 h (1,200 min) and 25 h (1,500 min), when neutron emissions of about 5 times and 10 times the natural background level are measured, respectively.

In Fig. 10.5a and b, the alpha particle emissions are reported. The data are related to an alpha emission level monitored by means of the 6150AD-k probe set to the operating mode “alpha”. The measurements shown in Fig. 10.5a are referred to the data acquired for a time interval equal to 60 min, when the reaction chamber is operating (cell on). The data in Fig. 10.5b represent the alpha particle emissions corresponding to the background level and are obtained by measurements acquired, also in this case, for a 60-min time interval (cell off). From these diagrams, it can be noticed that the number of counts per second acquired by the probe increases considerably when the electrolytic cell is operating (cell on) (Fig. 10.5a). In addition, the mean values of two alpha emission time series are computed, one when the cell is switched on and the other when the cell is off. The first time series show an average alpha emission of about 0.030 c/s (count per second), whereas the second one provides the background emission level in the laboratory with a mean value of about 0.015 c/s. It is evident that the average alpha particle emission during the electrolysis is twice the alpha background level. These results, together with the evidence of neutron emissions reported in Fig. 10.4, are particularly interesting when considering the compositional variation described in the following, and will be useful to corroborate the conjecture of phono-fission for the chemical elements constituting the electrodes. In Fig. 10.5c, the cumulative curves for the alpha emission counts are reported. It is evident that the total count value, monitored when the cell is operating

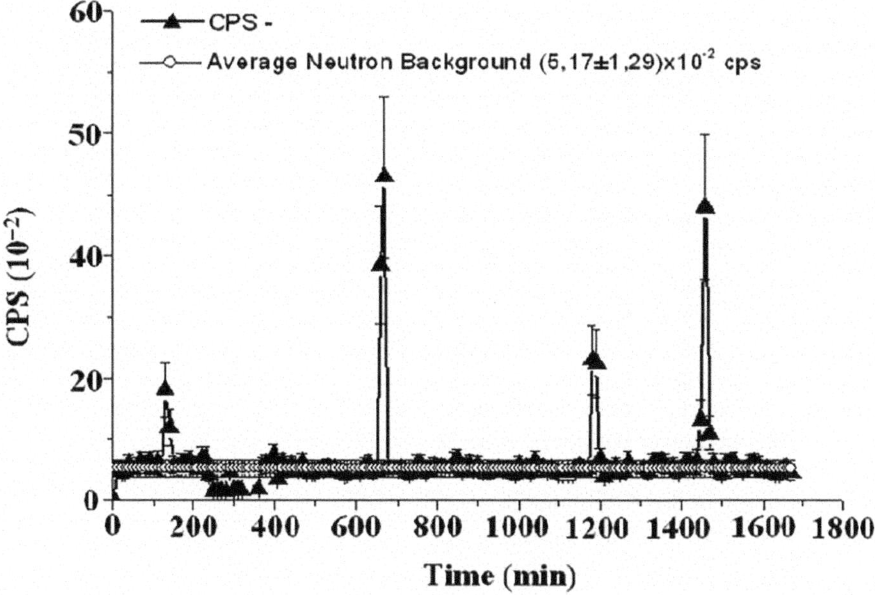

Fig. 10.4 Neutron emission measurements up to 4 and 10 times the natural background level are observed during the experiment

(cell on), is approximately twice the value measured for the background level (cell off).

10.3.3 Chemical Composition Analysis of the Electrodes After the Experiment

As reported previously, Energy Dispersive X-ray Spectroscopy is performed in order to recognize a possible evidence of phono-fission reactions taking place during the electrolysis. In Fig. 10.6a and b, two images of the Co–Cr electrode surface, respectively before the experiment and after 32 h since the beginning of it, are reported. It is shown that the electrode, after several operating hours, presents microcracks and cracks visible on its external surface (Fig. 10.6b).

The experimental activity is developed in three different phases in order to investigate on a possible evolution in the chemical composition of the electrode surface. A first analysis is carried out to evaluate the composition of the electrodes before they undergo the electrolysis experiment (0 h), (Table 10.2). The second analysis is conducted after an initial operating time interval of about 4 h (Tab 10.2). After this, a third- and a fourth-step analyses are performed. For these two final steps, the cell

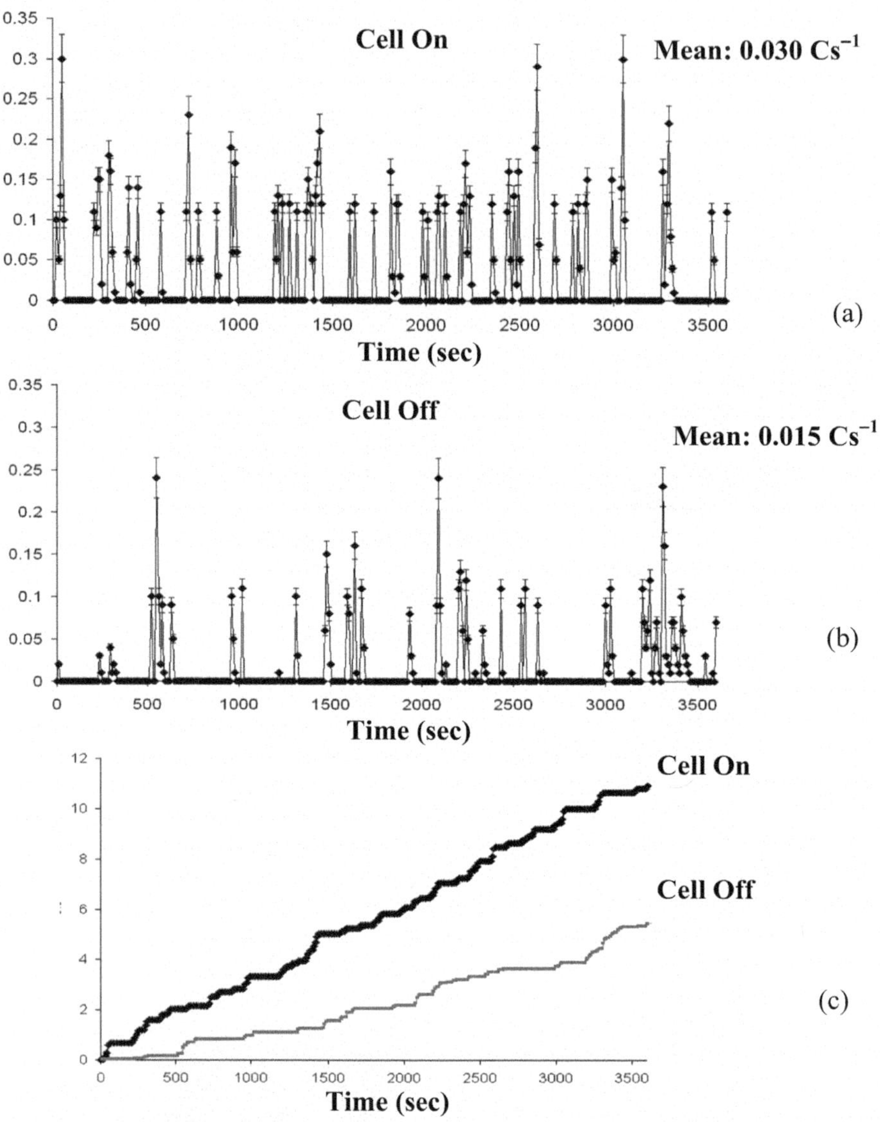

Fig. 10.5 Alpha particle data acquired for a time interval equal to 60 min when the reaction chamber is operating (**a**). Alpha particle emission rates corresponding to the background level (**b**). Cumulative curves for the alpha emissions (**c**)

operated for 28 and 6 h, respectively, corresponding to a cumulative working time of 32 (4 h + 28 h) and 38 h (4 h + 28 h + 6 h) (Table 10.2).

In the case of the Ni–Fe electrode, the resulting mean concentrations of Ni, Si, Mg, Fe, and Cr are reported for each step in Table 10.2. In Figs. 10.7, 10.8, 10.9, 10.10 and 10.11, the EDX measurements for each element are reported considering

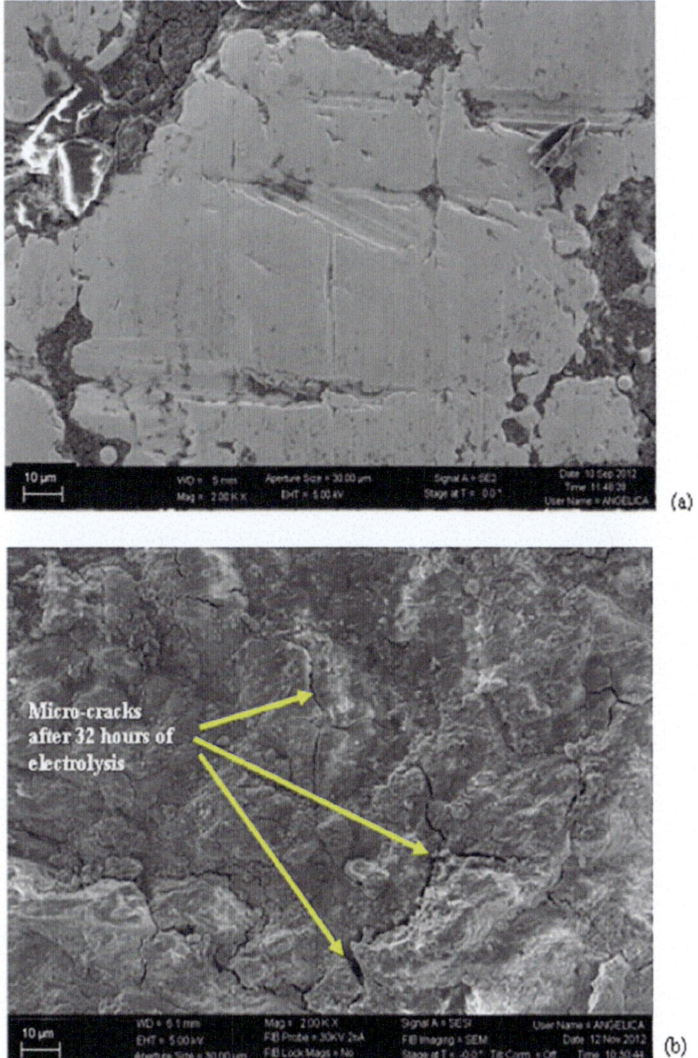

Fig. 10.6 Image of the Co–Cr electrode surface before the experiment (**a**). Electrode after 32 operating hours presents cracks and microcracks visible on the external surface (**b**)

the four steps previously described (Figs. 10.7a, 10.8a, 10.9a, 10.10a and 10.11a). At the same time, the evolution of the mean values of each time series along with their respective standard deviations, corresponding to 0, 4, 32, and 38 h, are reported by histograms (Figs. 10.7b, 10.8b, 10.9b, 10.10b and 10.11b). After 38 h, the appearance of Cr, before absent, is detected as reported in Table 10.2 and in Fig. 11a. In particular, the Ni concentration shows a total average decrement of 8.6%, from 43.9% to 35.3% after 38 h (Tab 10.2 and Fig. 10.7a and b). This Ni depletion is about one-fifth of the

Table 10.2 Ni–Fe electrode. Element concentrations before the experiment, after 4, 32, and 38 h of testing

	Ni (%)	Si (%)	Mg (%)	Fe (%)	Cr (%)
Before the experiment	43.9	1.1	0.1	30.5	–
After 4 h	43.6	1.1	0.4	30.7	–
After 32 h	35.2	5.0	0.2	27.9	–
After 38 h	35.3	1.5	4.8	27.3	3.0

* The values reported for the mass % of each element are referred to the mean value of all the performed measurements

initial Ni content. A mean increment in Si concentration after 32 h of 3.9% and an average increment in Mg concentration after 38 h, starting from 0.1% up to 4.8%, can be observed from the data reported in Table 10.2, as well as in Figs. 10.8 and 10.9. Similar considerations may be done also for Fe and Cr concentrations. The average Fe content decreases of 3.2%, changing from 30.5 to 27.3% at the end of the experiment (Table 10.2 and Fig. 10.10). On the other hand, the Cr concentration appears only in the last phase with an appreciable increment of 3.0% (Table 10.2 and Fig. 10.11). Such decrements in Ni and Fe are almost perfectly balanced by the increments in other elements: Si, Mg, and Cr. In particular, since the analysis excludes Ni content variations on the other electrode, the balance: Ni (−8.6%) = Si (+3.9%) + Mg (+4.7%) can be reasonably explained only by the following symmetrical phono-fissions:

$$Ni^{58}_{28} \rightarrow 2Si^{28}_{14} + 2 \text{ neutrons} \tag{10.1}$$

$$Ni^{58}_{28} \rightarrow 2Mg^{24}_{12} + 2He^{4}_{2} + 2 \text{ neutrons} \tag{10.2}$$

At the same time, the balance Fe (−3.2%) ≅ Cr (+3.0%) may be explained by the reaction:

$$Fe^{56}_{26} \rightarrow Cr^{52}_{24} + He^{4}_{2} \tag{10.3}$$

It is very interesting to observe that reactions (10.1) and (10.2) imply neutron emissions, as well as reactions (10.2) and (10.3) imply the emissions of alpha particles.

As far as the Co–Cr electrode is concerned, it is possible to observe variations even more evident in the concentrations of the most abundant constituting elements. In particular, the average Co concentration decreases by 23.5%, from an initial percentage of 44.1% to a concentration of 20.6% after 32 h (Table 10.3 and Fig. 10.12a and b). At the same time, an increment of 23.2% can be observed in the Fe content after 32 h, changing from 3.1% before the experiment to 26.3% at the end of the second phase (Table 10.3 and Fig. 10.13a and b). It is rather impressive that the decrement in Co and the increment in Fe are almost the same: Co (−23.5%) ≅

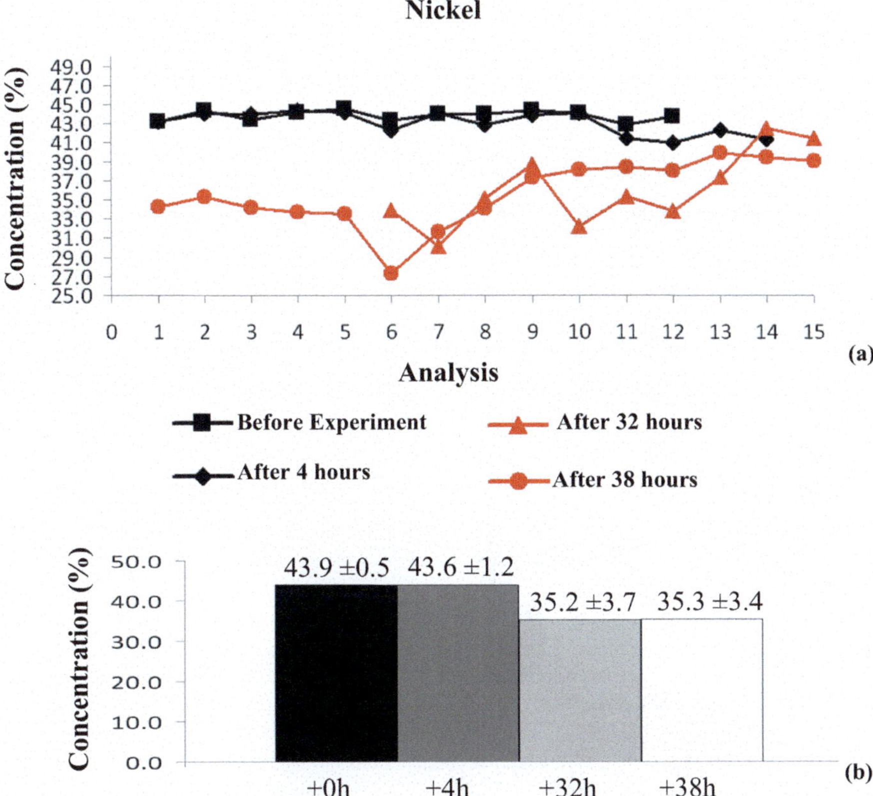

Fig. 10.7 Ni concentration before the experiment, after 4, 32, and 38 h (**a**). The average values of Ni concentration change from a mass percentage of 43.9%, at the beginning of the experiment, to 35.2% and 35.3% after 32 and 38 h, respectively (**b**)

Fe (+23.2%). According to the compositional analysis on the Ni–Fe electrode, no Co concentration is found to lead us to consider the chemical migration to explain its depletion in the Co–Cr electrode, thus the following reaction seems the most reasonable possibility:

$$\mathrm{Co}_{27}^{59} \rightarrow \mathrm{Fe}_{26}^{56} + \mathrm{H}_{1}^{1} + 2 \text{ neutrons} \tag{10.4}$$

In Table 10.3 and in Figs. 10.14 and 10.15, Cr and K concentrations are reported for the different phases of the experiment. The K increment by about 12.4% after 32 h may be only partially balanced by the decrement in Cr (8.1%) according to the next reaction (10.5). The remaining increment in K (4.3%) can be considered as an effect of the K_2CO_3 aqueous solution deposition at the end of the third interval.

Considering these variations, also the following phono-fission reaction, involving Cr as the starting element and K as the resultant, can be considered:

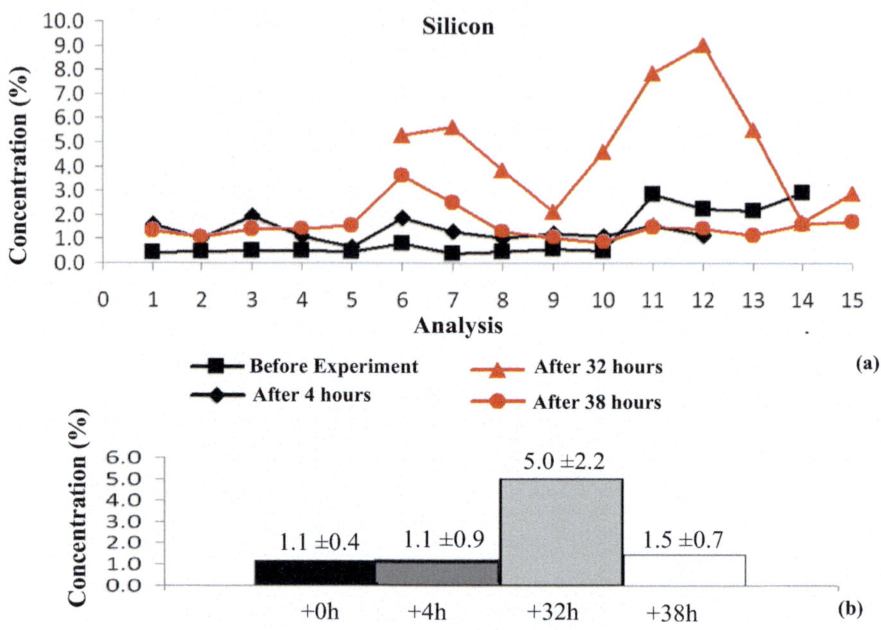

Fig. 10.8 Si concentration before the experiment, after 4, 32, and 38 h (**a**). Mean values of Si concentration change from a mass percentage of 1.1%, at the beginning of the experiment, to 5.0% and 1.5% after 32 and 38 h, respectively (**b**)

$$Cr^{52}_{24} \rightarrow K^{39}_{19} + H^{1}_{1} + 2He^{4}_{2} + 4 \text{ neutrons} \tag{10.5}$$

Also in this case, it is remarkable that both reactions (10.4) and (10.5) imply neutron emissions, whereas reaction (10.5) implies also the emission of alpha particles.

It is important to consider that the balances reported, for the Ni–Fe electrode (reactions 10.1–10.3) and for the Co–Cr electrode (reactions 10.4 and 10.5), are obtained considering the values of the second or third interval corresponding to the largest variation in each element concentration (Tables 10.2 and 10.3).

All these results suggest that, during the gas loading performed by hydrogen or deuterium, the host atomic lattice is subjected to mechanical damage and cracking due to atom adsorption and penetration. Evidence of a diffused cracking is identified on the electrode surface after the experiments (Fig. 10.6b). Accordingly, we argue that the hydrogen, favoring the crack formation and propagation in the metal, comes from the electrolysis of water. In fact, being the electrodes immersed in a liquid solution, their surfaces are exposed to the formation of gaseous hydrogen from the decomposition of water caused by the current passage.

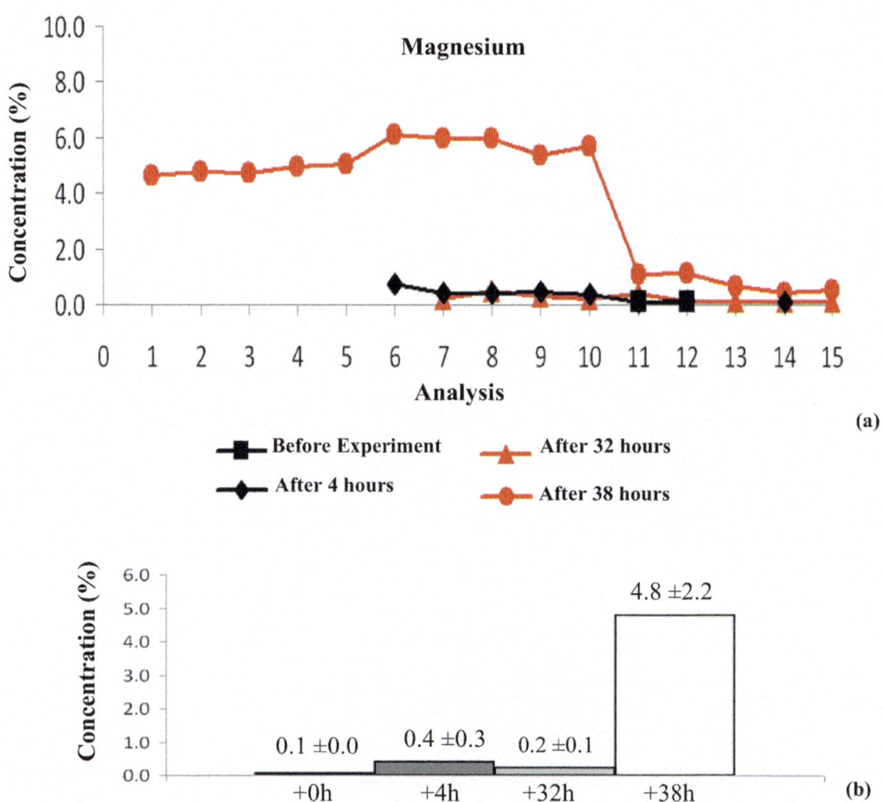

Fig. 10.9 Mg concentration before the experiment, after 4, 32, and 38 h (**a**). Mean values of Mg concentration change from a mass percentage of 0.1 and 0.4%, at the beginning of the experiment and after 32 h, to 4.8% after 38 h (**b**)

10.4 Conclusions

Neutron emissions up to one order of magnitude higher than the natural background level are observed during the operating time of an electrolytic cell. In particular, after a time period of about 3 h, neutron emissions of about 4 times the background level are measured. After 11 h, it is possible to observe neutron emissions of about one order of magnitude greater than the background level. Similar results are observed after 20 and 25 h. To the damage accumulation corresponds an incubation time period for the beginning of the phono-fission reactions.

When the cell is switched on, the average alpha particle emission is about 0.030 c/s for 1 h of measurement; this value corresponds to an alpha emission level of twice the background measured in the laboratory before and after the experiment (0.015 c/s).

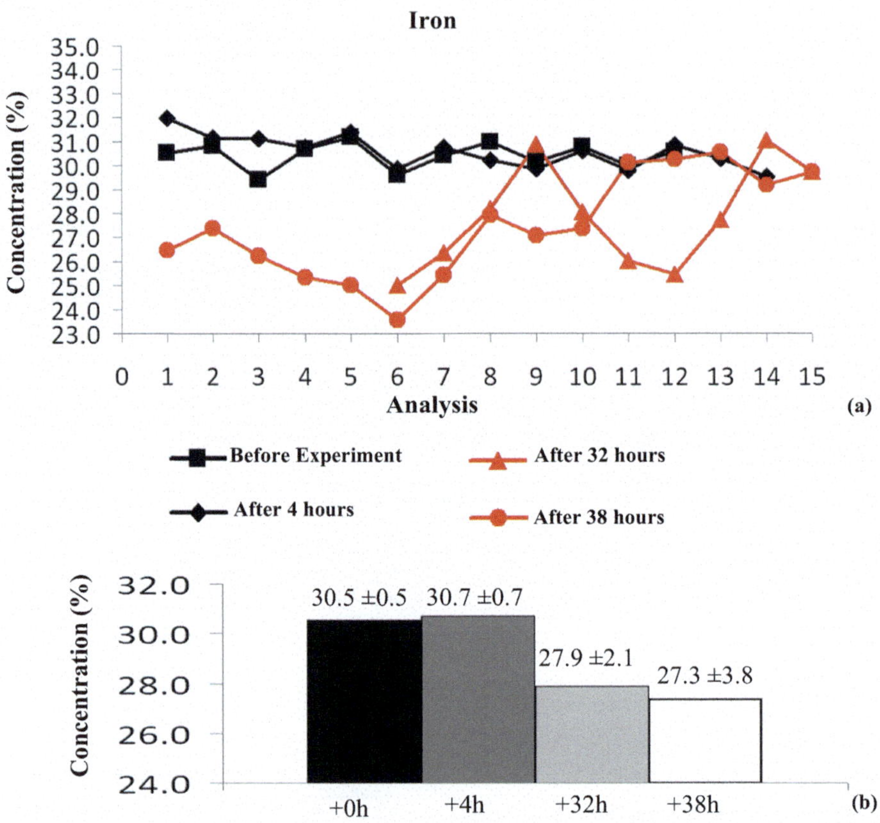

Fig. 10.10 Fe concentration before the experiment, after 4, 32, and 38 h (**a**). Mean values of Fe concentration change from a mass percentage of 30.5%, at the beginning of the experiment, to 27.9% and 27.3% after 32 and 38 h, respectively (**b**)

By the EDX analysis performed on the two electrodes in three subsequent steps, significant chemical composition changes are recorded. In general, the decrements in Ni and Fe in the Ni–Fe electrode are almost perfectly balanced by the increments in lighter elements: Si, Mg, and Cr. In fact, the balance Ni (−8.6%) ≅ Si (+3.9%) + Mg (+4.7%) is satisfied by reactions (10.1) and (10.2). In particular, the Si atoms could have undergone further reactions, which would explain its concentration drop observed at the third step. At the same time, the balance Fe (−3.2%) ≅ Cr (+3.0%) may be explained considering reaction (10.3).

As far as the Co–Cr electrode is concerned, the Co decrement is almost perfectly balanced by the Fe increment: Co (−23.5%) ≅ Fe (+23.2%). This last evidence, which is really impressive considering the mass percentage involved, is explainable only considering reaction (10.4). Eventually, the Cr decrement and the K increment are explained taking into account reaction (10.5) and the solution deposition. In particular, the K increment by about 12.4% may be only partially balanced by the

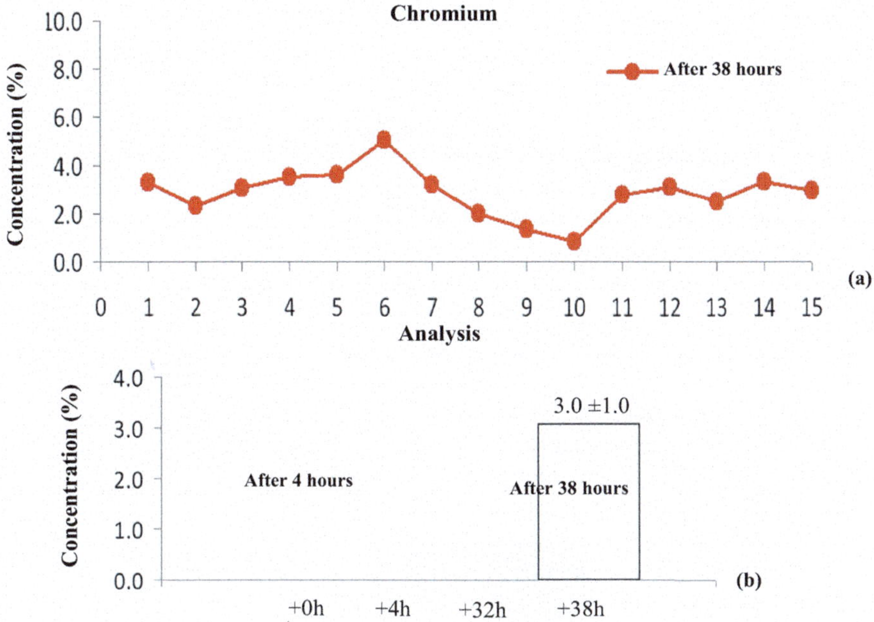

Fig. 10.11 Cr concentration after 38 h (**a**). The mean value of Cr concentration changes from 0% (initial absence of this element) to a mass percentage of 3.0% after 38 h (**b**)

Table 10.3 Co–Cr electrode. Element concentrations before the experiment, after 4, 32, and 38 h of testing

	Co (%)	Fe (%)	Cr (%)	K (%)
Before the experiment	44.1	3.1	17.8	0.5
After 4 h	43.7	1.6	17.8	2.2
After 32 h	20.6	26.3	9.7	12.9
After 38 h	34.4	6.6	5.1	4.4

* The values reported for the mass % of each element are referred to the mean value of all the performed measurements

decrement in Cr (8.1%) according to reaction (10.5). The remaining increment in K (4.3%) could be considered as an effect of the K_2CO_3 aqueous solution deposition at the end of the third step.

Electrochemical reasons cannot explain the variations observed. Elements such as Ni and Co show no traces on the surface of the other electrode and cannot be found dissolved in the solution. Also the so-called *electromigration*, having a reduced effect due to the size of the electrode, cannot be considered as an explanation to compositional changes in the order of tens of percentage points.

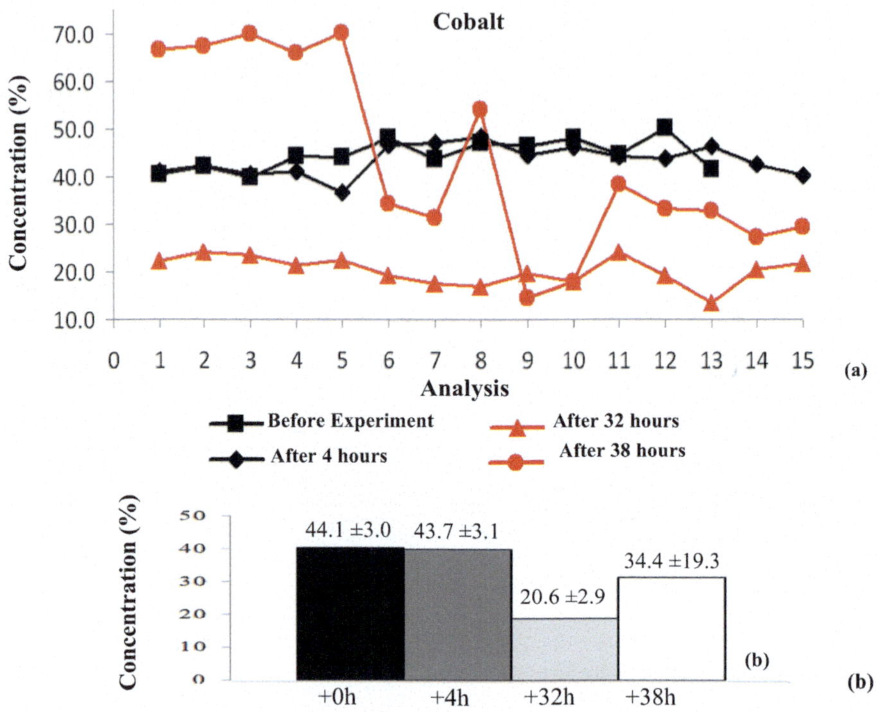

Fig. 10.12 Co concentration before the experiment, after 4, 32, and 38 h (**a**). Mean values of Co concentration change from a mass percentage of 44.1%, at the beginning of the experiment, to 20.6% and 34.4% after 32 and 38 h, respectively (**b**)

Chemical composition changes and subatomic particle emissions may be accounted as direct evidence of phono-fission reactions, which are correlated also to microcrack formation and propagation. The last are due to hydrogen embrittlement. According to the present interpretation of the so-called Cold Nuclear Fusion, hydrogen, which is imputed to favor the crack formation and propagation in the electrodes, comes from the electrolysis of water. Being the electrodes immersed in an aqueous solution, the metal surface is exposed to the formation of gaseous hydrogen from the decomposition of the water molecule caused by the current passage. In addition, the high current density contributes to the formation and penetration of hydrogen into the metal atomic lattice. The so-called Cold Nuclear Fusion, interpreted under the light of hydrogen embrittlement, may be explained by phono-fission reactions occurring in the metallic electrode, rather than by nuclear fusion reactions involving hydrogen isotopes forced into the metal lattice.

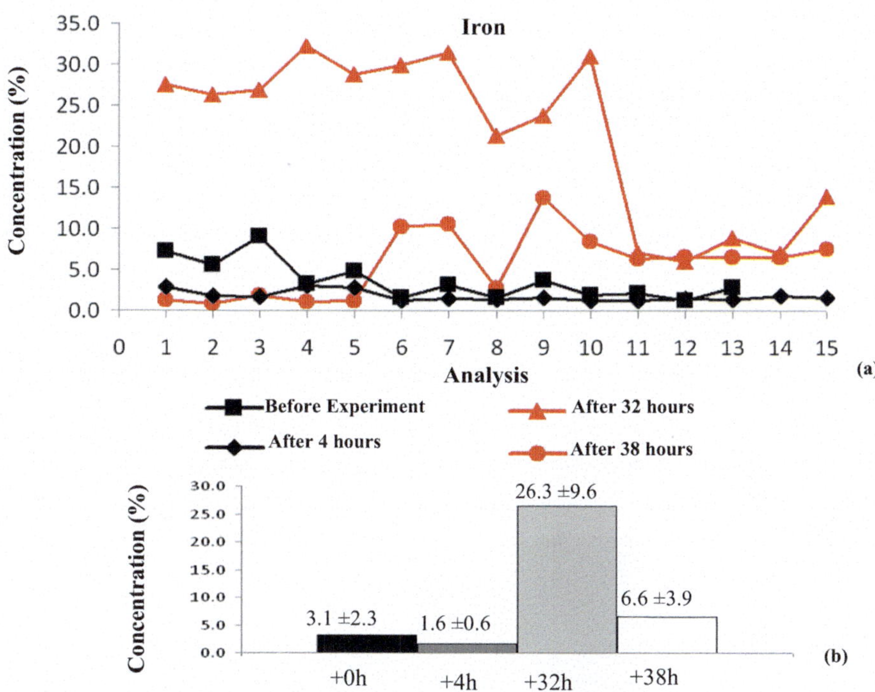

Fig. 10.13 Fe concentration before the experiment, after 4, 32, and 38 h (**a**). Mean values of Fe concentration change from a mass percentage of 3.1%, at the beginning of the experiment, to 26.3% and 6.6% after 32 and 38 h, respectively (**b**)

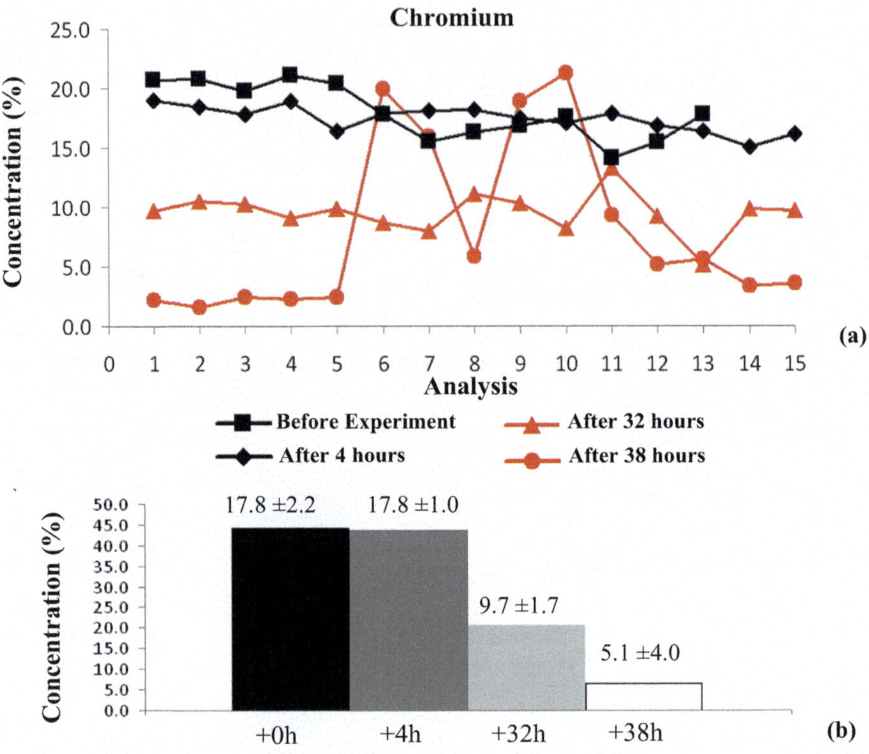

Fig. 10.14 Cr concentration before the experiment, after 4, 32, and 38 h (**a**). Cr concentration changes from a mass percentage of 17.8%, at the beginning of the experiment, to 9.7% and 5.1% after 32 and 38 h, respectively (**b**)

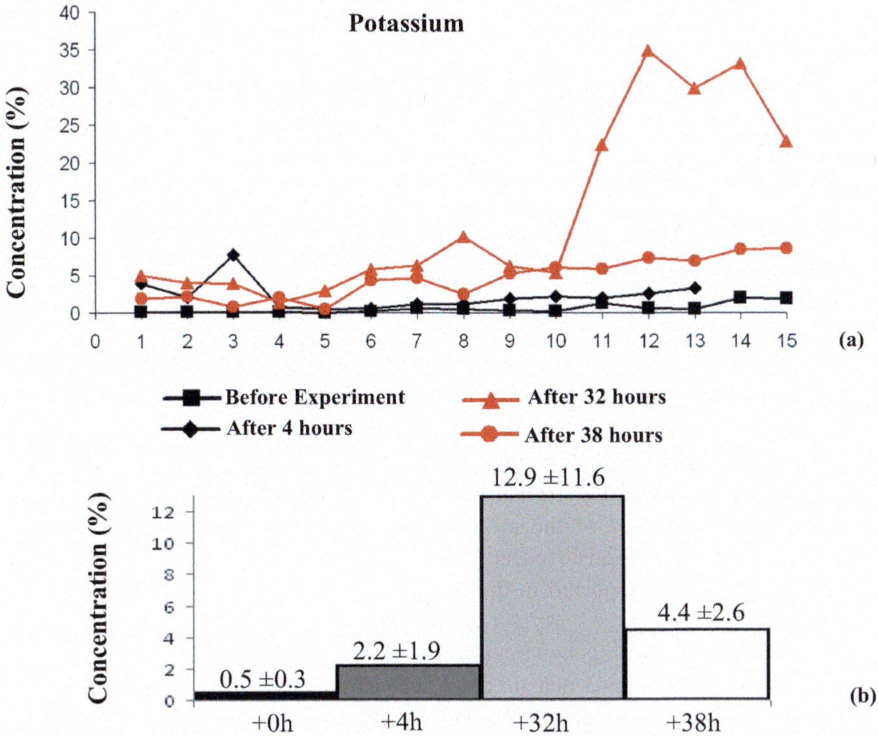

Fig. 10.15 K concentration before the experiment, after 4, 32, and 38 h (**a**). K concentration changes from 0.5%, at the beginning, and 2.2%, after 4 h, to 12.9% and 4.4% after 32 and 38 h, respectively (**b**)

Acknowledgements Dr. A. Sardi and Dr. F. Durbiano (INRIM) are gratefully acknowledged for their kind collaboration during the experimental set-up definition and for the assistance given during the measurements. Also Dr. A. Chiodoni (IIT) is warmly acknowledged for the EDX spectroscopy analyses.

References

1. Borghi DC, Giori DC, Dall'Olio A (1992) Experimental evidence on the emission of neutrons from cold hydrogen plasma. In: Proceedings of the international workshop on few-body problems in low-energy physics, Alma-Ata, Kazakhstan, pp 147–154; Unpublished Communication (1957); Comunicacao n. 25 do CENUFPE, Recife, Brazil (1971), pp 147–154
2. Diebner K (1962) Fusionsprozesse mit Hilfe konvergenter Stosswellen – einige aeltere und neuere Versuche und Ueberlegungen. Kerntechnik 3:89–93
3. Kaliski S (1978) Bi-conical system of concentric explosive compression of D-T. J Tech Phys 19:283–289

4. Winterberg F (1984) Autocatalytic fusion–fission implosions. Atomenergie-Kerntechnik 44:146
5. Derjaguin BV et al (1989) Titanium fracture yields neutrons? Nature 34:492
6. Fleischmann M, Pons S, Hawkins M (1989) Electrochemically induced nuclear fusion of deuterium. J Electroanal Chem 261:301
7. Bockris JM, Lin GH, Kainthla RC, Packham NJC, Velev O (1990) Does tritium form at electrodes by nuclear reactions?. In: The first annual conference on cold fusion. National Cold Fusion Institute, University of Utah Research Park, Salt Lake City
8. Preparata G (1991) Some theories of cold fusion: a review. Fusion Tech 20:82
9. Preparata G (1991) A new look at solid-state fractures, particle emissions and "cold" nuclear fusion. Il Nuovo Cimento 104 A:1259–1263
10. Mills RL, Kneizys P (1991) Excess heat production by the electrolysis of an aqueous potassium carbonate electrolyte and the implications for cold fusion. Fusion Technol 20:65
11. Notoya R, Enyo M (1992) Excess heat production during electrolysis of H_2O on Ni, Au, Ag and Sn electrodes in alkaline media. In: Proceedings of the third international conference on cold fusion. Universal Academy Press, Tokyo, Nagoya Japan
12. Miles MH, Hollins RA, Bush BF, Lagowski JJ, Miles RE (1993) Correlation of excess power and helium production during D_2O and H_2O electrolysis using palladium cathodes. J Electroanal Chem 346:99–117
13. Bush RT, Eagleton RD (1993) Calorimetric studies for several light water electrolytic cells with nickel fibrex cathodes and electrolytes with alkali salts of potassium, rubidium, and cesium. In: Fourth international conference on cold fusion. Lahaina, Maui. Electric Power Research Institute, p 13
14. Fleischmann M, Pons S, Preparata G (1994) Possible theories of cold fusion. Nuovo Cimento Soc Ital Fis A 107:143
15. Szpak S, Mosier-Boss PA, Smith JJ (1994) Deuterium uptake during Pd-D codeposition. J Electroanal Chem 379:121
16. Sundaresan R, Bockris JOM (1994) Anomalous reactions during arcing between carbon rods in water. Fusion Technol 26:261
17. Arata Y, Zhang Y (1995) Achievement of solid-state plasma fusion ("cold-fusion"). Proc Jpn Acad 71(B):304–309
18. Ohmori T, Mizuno T, Enyo M (1996) Isotopic distributions of heavy metal elements produced during the light water electrolysis on gold electrodes. J New Energy 1(3):90–99
19. Monti RA (1996) Low energy nuclear reactions: experimental evidence for the alpha extended model of the atom. J New Energy 1(3):131
20. Monti RA (1998) Nuclear transmutation processes of lead, silver, thorium, uranium. In: The seventh international conference on cold fusion. ENECO Inc., Vancouver Salt Lake City
21. Ohmori T, Mizuno T (1998) Strong excess energy evolution, new element production, and electromagnetic wave and/or neutron emission in light water electrolysis with a tungsten cathode. Infinite Energy 20:14–17
22. Mizuno T (1989) Nuclear transmutation: the reality of cold fusion. Infinite Energy Press, New Hampshire
23. Little SR, Puthoff HE, Little ME (1998) Search for excess heat from a Pt electrode discharge in K_2CO_3–H_2O and K_2CO_3-D_2O electrolytes. Infinite Energy 5:34
24. Ohmori T, Mizuno T (2000) Nuclear transmutation reaction caused by light water electrolysis on tungsten cathode under incandescent conditions. J New Energy 4(4):66–78
25. Ransford HE (1999) Non-stellar nucleosynthesis: transition metal production by DC plasma-discharge electrolysis using carbon electrodes in a non-metallic cell. Infinite Energy 4(23):16–22
26. Storms E (2000) Excess power production from platinum cathodes using the Pons-Fleischmann effect. In: 8th international conference on cold fusion, Lerici (La Spezia). Italian Physical Society, Bologna, Italy, pp 55–61
27. Storms E (2007) Science of low energy nuclear reaction: a comprehensive compilation of evidence and explanations about cold fusion. World Scientific Publishing, Singapore

28. Mizuno T et al (2000) Production of heat during plasma electrolysis. Jpn J Appl Phys A 39:6055
29. Warner J, Dash J, Frantz S (2002) Electrolysis of D2O with titanium cathodes: enhancement of excess heat and further evidence of possible transmutation. In: The ninth international conference on cold fusion. Tsinghua University, Beijing, China, p 404
30. Fujii MF et al (2002) Neutron emission from fracture of piezoelectric materials in deuterium atmosphere. Jpn J Appl Phys 41:2115–2119
31. Mosier-Boss PA et al (2007) Use of CR-39 in Pd/D co-deposition experiments. Eur Phys J Appl Phys 40:293–303
32. Swartz M (2008) Three physical regions of anomalous activity in deuterated palladium. Infinite Energy 14:19–31
33. Mosier-Boss PA et al (2010) Comparison of Pd/D co-deposition and DT neutron generated triple tracks observed in CR-39 detectors. Eur Phys J Appl Phys 51(2):20901–20911
34. Kanarev M, Mizuno T (2002) Cold fusion by plasma electrolysis of water. New Energy Technol 1:5–10
35. Cardone F, Mignani R (2003) Possible observation of transformation of chemical elements in cavitated water. Int J Mod Phys B 17:307–317
36. Cardone F, Cherubini G, Petrucci A (2009) Piezonuclear neutrons. Phys Lett A 373:862–866
37. Carpinteri A, Cardone F, Lacidogna G (2009) Piezonuclear neutrons from brittle fracture: early results of mechanical compression tests. Strain 45: 332–339. Atti dell'Accademia delle Scienze di Torino 33:27–42
38. Cardone F, Carpinteri A, Lacidogna G (2009) Piezonuclear neutrons from fracturing of inert solids. Phys Lett A 373:4158–4163
39. Carpinteri A, Cardone F, Lacidogna G (2010) Energy emissions from failure phenomena: mechanical, electromagnetic, nuclear. Exp Mech 50:1235–1243
40. Carpinteri A, Lacidogna G, Manuello A, Borla O (2012) Piezonuclear fission reactions: evidences from microchemical analysis, neutron emission, and geological transformation. Rock Mech Rock Eng 45:445–459
41. Cardone F, Mignani R, Petrucci A (2012) Piezonuclear reactions. J Adv Phys 1:3
42. Albertini G, Calbucci V, Cardone F, Petrucci A, Ridolfi F (2014) Chemical changes induced by ultrasounds in iron. Appl Phys A 114(4):1233–1246
43. Petrucci A, Mignani R, Cardone F (2011) Comparison between piezonuclear reactions and CMNS phenomenology. In: Violante V, Sarto F (eds) Proceedings of the 15th international conference on condensed matter nuclear science, Part 1, Oct. 5–9, 2009, Rome, Italy, p 246
44. Carpinteri A, Lacidogna G, Manuello A, Borla O (2013) Piezonuclear fission reactions from earthquakes and brittle rocks failure: evidence of neutron emission and nonradioactive product elements. Exp Mech 53:345–365
45. Milne I, Ritchie RO, Karihaloo B (2003) Comprehensive structural integrity: fracture of materials from nano to macro, vol 6. Elsevier, Chapter 6.02, pp 31–33
46. Liebowitz H (1971) Fracture. An advanced treatise. Academic Press, New York, San Francisco, London

Chapter 11
Electrolysis Experiments with Pd and Ni Electrodes: Hydrogen Embrittlement, Microcracking, TeraHertz Phonons, and Correlated Nuclear Phenomena

Abstract Preliminary electrolysis experiments provided evidence of phono-fission reactions occurring in condensed matter (see Chap. 10). These experiments were characterized by significant neutron and alpha particle emissions, together with appreciable variations in the chemical composition on the electrode surfaces. A macromechanical reason for the so-called "Cold Nuclear Fusion" is proposed. The hydrogen embrittlement due to hydrogen atoms produced by the electrolysis plays an essential role in the production of microcracking in the electrode host metals (Ni–Fe and Co–Cr alloys). Our hypothesis is that phono-fission reactions occur in correspondence to the microcrack formation and/or propagation. In order to confirm the early results obtained by the Ni–Fe and Co–Cr electrodes and presented in the previous chapter, electrolytic tests are conducted using 100% Pd at the cathode and 90% Ni at the anode. As a result, relevant chemical composition changes and the appearance of elements previously absent are observed on the Pd and Ni electrodes after the experiment, as well as significant neutron and alpha particle emissions. The most relevant process emerging from the experiment is the primary fission of palladium into iron and calcium. Then, secondary fissions of the products, as well as of nickel on the other electrode, in turn appear producing oxygen atoms, as well as additional alpha particles and neutrons.

Keywords Palladium · Nickel · Electrolysis · Hydrogen embrittlement · Electrode microcracking · TeraHertz phonons · Phono-fission reactions · Neutron emissions · Cold nuclear fusion

11.1 Preliminary Remarks

A remarkable evidence of anomalous nuclear reactions occurring in condensed matter was observed by different authors [1–34]. These experiments are characterized by extra-heat generation, neutron emission, and alpha particle detection. Some of these studies, using electrolytic devices, reported also a significant evidence of chemical composition changes after microcracking of the electrodes [35–40].

A. Carpinteri, *Terahertz Phonons and Nanomechanical Instabilities*,
https://doi.org/10.1007/978-3-032-14692-2_11

As reported also previously (Chap. 10), in 1998 Mizuno presented the results of the measurements conducted by means of neutron detectors and compositional analysis techniques in relation to different electrolytic experiments. On the other hand, as shown by most of the articles devoted to Cold Nuclear Fusion, one of the principal and recurring features of the experiments is the appearance of microcracks on the electrode surfaces after the tests [26, 27]. Such evidence might be directly correlated to hydrogen embrittlement of the metallic material composing the electrodes (Pd, Ni, Fe, Ti, etc.). This phenomenon, well-known in metallurgy and, particularly, in fracture mechanics, characterizes metals during forming or finishing operations [41, 42]. In order to confirm the early results obtained by Ni–Fe and Co–Cr electrodes [43–45], electrolytic tests are conducted using 100% Pd at the cathode and a Ni–Fe alloy (91% of Ni) at the anode. The host metal matrix is subjected to mechanical damage and cracking due to hydrogen atoms penetrating into the metal lattice and forcing it during the gas loading. Hydrogen effects are extensively studied, especially for those metal alloys where the hydrogen adsorption is particularly high. The hydrogen atoms generate an internal compression stress that lowers the apparent fracture toughness of the metal, so that brittle crack formation and/or propagation can occur with a hydrogen partial pressure below 1 atm [41, 42]. The obtained results provide an important confirmation about the hypothesis proposed by the author and reported in previous papers on electrolysis with Ni–Fe and Co–Cr electrodes [43, 44].

11.2 Experimental Set-Up

The experimental set-up adopted during the tests is the same as that described in Chap. 10. Similarly to the previous experiment, the two metallic electrodes are connected to a source of direct current: the Ni electrode as the positive pole (anode), and the Pd electrode as the negative pole (cathode). The reaction chamber is the same as that employed by the author for the previous preliminary tests. The current intensity and the neutron emissions are analyzed according to the same procedures adopted in the case of the Ni–Fe and Co–Cr electrodes.

11.3 Neutron Emission Measurements

Neutron emission measurements performed during the experimental activity are represented in Fig. 11.1. The measurements performed by the ^{3}He detector are conducted for a total time period of 24 h. The natural background level is measured for different time intervals before and after switching on the reaction chamber. These measurements provide an average neutron background of $(3.23 \pm 1.49) \times 10^{-2}$ cps. Furthermore, when the reaction chamber is active, it is possible to observe that, after a time interval of about 7.5 h (460 min), neutron emissions of about 3 times the background level are detected. After 9 h (540 min) from the beginning of the

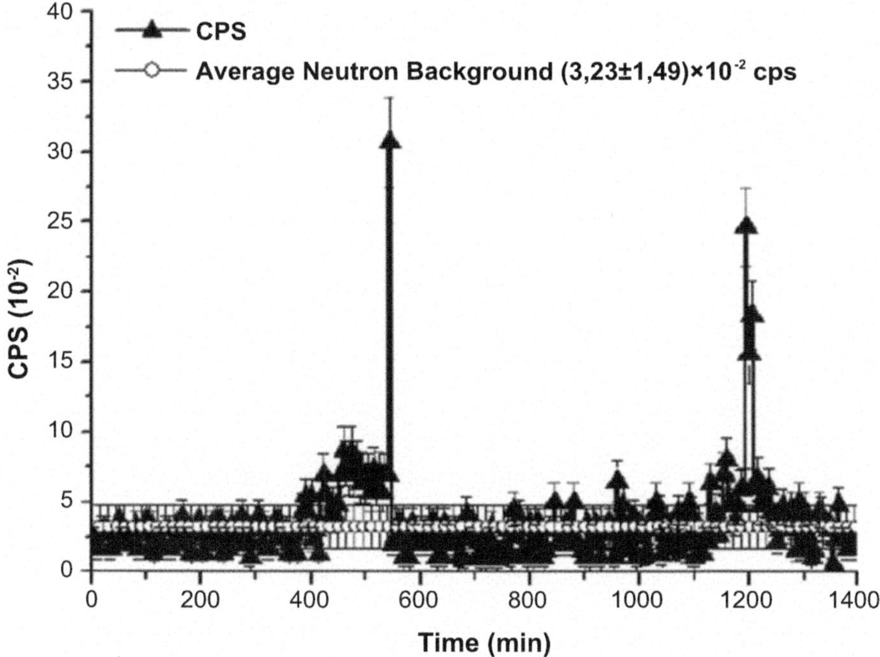

Fig. 11.1 Neutron emission measurements. Emission peaks between 3 and 10 times the natural background level are observed during the experiment

measurements, it is possible to observe neutron emission peaks of about one order of magnitude greater than the background level.

Similar results were observed after 20 h (1,200 min), when neutron emissions up to 7 times the background level were measured.

11.4 Chemical Composition Analysis of the Pd Electrode

The chemical compositions before and after the experiment are taken into account on the two electrodes (Tables 11.1 and 11.2).

Table 11.1 Element concentrations before and after the electrolysis test (Pd electrode)

Mean values							
	Pd	Fe	Ca	O	Mg	K	Si
Before the experiment (%)	100.0	0.0	0.0	0.0	0.0	0.0	0.0
After 20 h (%)	71.3	2.0	0.2	18.5	1.0	1.5	1.1

Table 11.2 Element concentrations before and after the electrolysis (Ni electrode)

Mean values					
	Ni	O	Si	Fe	Al
Before the experiment (%)	91.6	2.0	0.3	2.4	0.0
After 20 h (%)	68.5	21.5	1.1	0.4	1.8

In addition, considering the neutron emission measurements and according to the hydrogen embrittlement hypothesis suggested by Carpinteri et al. [43, 44], the presence of microcracks and macrocracks on the electrode surface (Fig. 11.2a–c) is accounted in the macromechanical interpretation of the phenomenon. This evidence is particularly strong in the case of the Pd electrode, where a macroscopic fracture takes place during the test. The macrocrack presents a width of about 40 μm, observable also at naked eyes (Fig. 11.2).

Considering the average decrement in Pd (−28.6%), reported in Table 11.1, a first-generation fission can be assumed:

$$\mathrm{Pd}_{46}^{106} \rightarrow \mathrm{Ca}_{20}^{40} + \mathrm{Fe}_{26}^{56} + 10 \text{ neutrons} \tag{11.1}$$

According to reaction (11.1), the Pd decrement is balanced by the Ca and Fe increments in the following proportions: 10.8% and 15.1%, respectively. These variations are accompanied by a neutron emission corresponding to the remaining 2.7% of the mass concentration (Figs. 11.3, 11.4, 11.5 and 11.6).

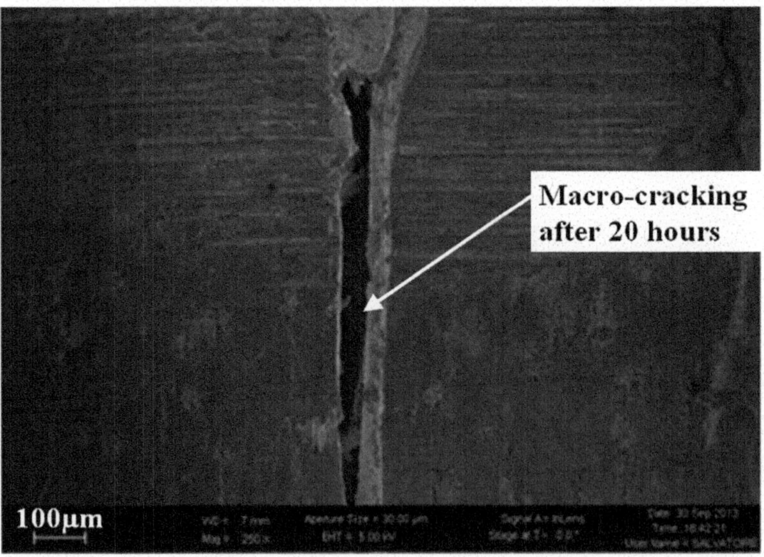

Fig. 11.2 Image of the palladium electrode surface. The crack presents a width of approximately 40 μm

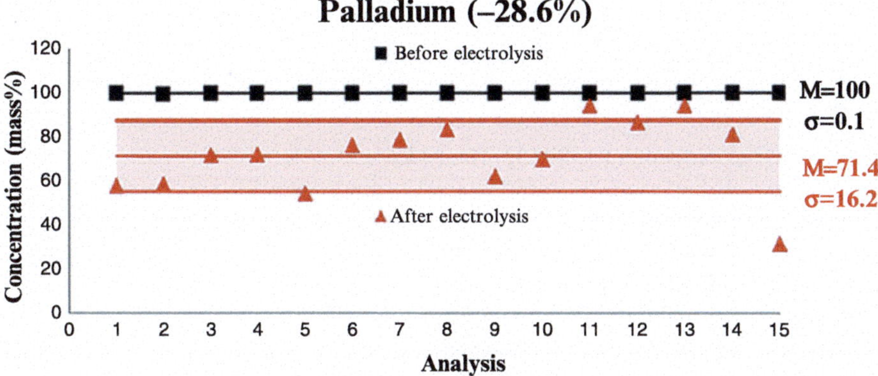

Fig. 11.3 Pd concentrations measured on 15 different spots of the electrode surface, before (black squares) and after (red triangles) the electrolysis experiment. The average concentration M and the corresponding standard deviation σ are also reported

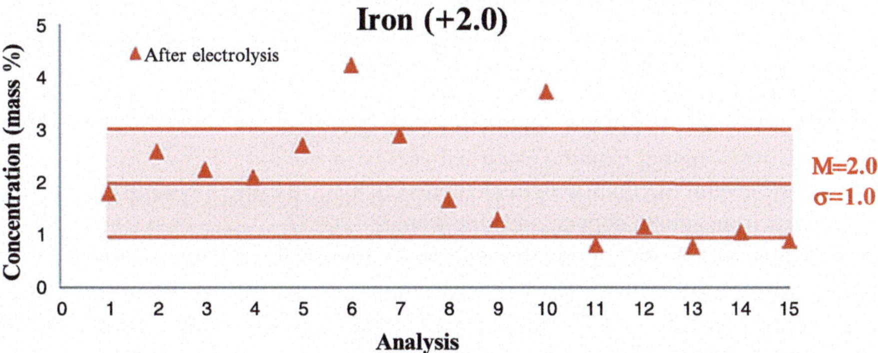

Fig. 11.4 Fe concentrations measured on the electrode surface after the electrolysis. Iron is absent before the experiment

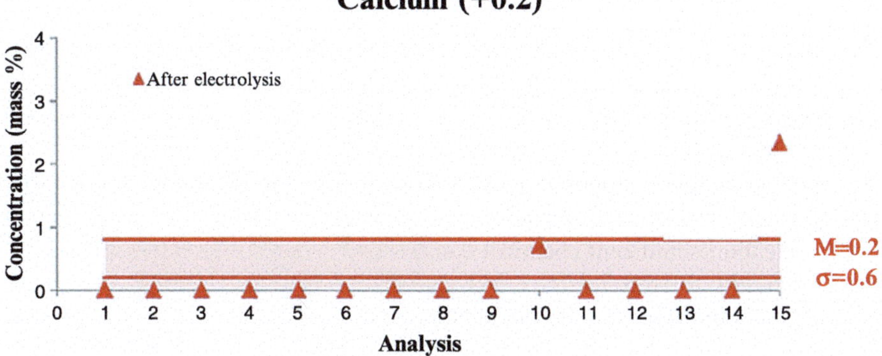

Fig. 11.5 Traces of Ca are detected after the electrolysis

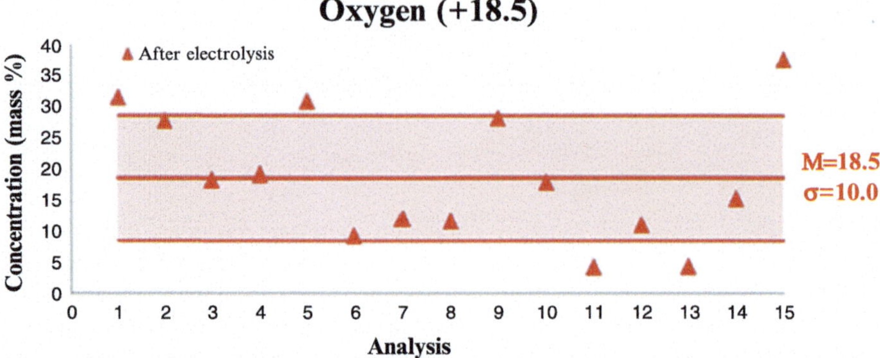

Fig. 11.6 The presence of oxygen is remarkable only after the experiment

The whole iron increment, according to reaction (11.1), could be entirely taken as the starting element for the production of other elements. Hence, a second hypothesis can be considered involving Fe as starting element and O as the product, together with alpha particle and neutron emissions:

$$Fe_{26}^{56} \rightarrow 3O_{8}^{16} + He_{2}^{4} + 4 \text{ neutrons} \tag{11.2}$$

According to reaction (11.2), the iron decrement produces 12.9% of oxygen with alpha particle (He) and neutron emissions. The total measured increment in oxygen after the experiment is equal to 18.5% (Table 11.1). This quantity is only partially explained by reaction (11.2). The remaining 5.6% of O concentration could be explained considering reaction (11.3) involving Ca as the starting element:

$$Ca_{20}^{40} \rightarrow 2O_{8}^{16} + 2He_{2}^{4} \tag{11.3}$$

$$Ca_{20}^{40} \rightarrow O_{8}^{16} + Mg_{12}^{24} \tag{11.4}$$

From reaction (11.3), we can consider a decrement in Ca equal to 5.9%. This decrement gives increments of 4.7% in O and 0.6% in He. On the other hand, from reaction (11.4), we obtain a further decrement in Ca concentration of 1.6%, as well as the formation of 1.0% of Mg and 0.6% of O (Fig. 11.7).

The calculated total increment in O of 18.2% is very close to the experimental value of 18.5%. At the same time, the Mg increment observed after the experiment can be explained by reaction (11.4), see also Table 11.1. According to reactions (11.3) and (11.4), the following balances may be considered: Ca (−5.9%) = O (+4.7%) + He (+1.2%), and Ca (−1.6%) = O (+0.6%) + Mg (+1.0%). Taking into account the same reactions, and considering the Ca increment coming from reaction (11.1) (10.8%), a concentration of 3.3% remains to be matched. To this aim, it is possible to take into account additional reactions involving Ca as starting element and Si, K,

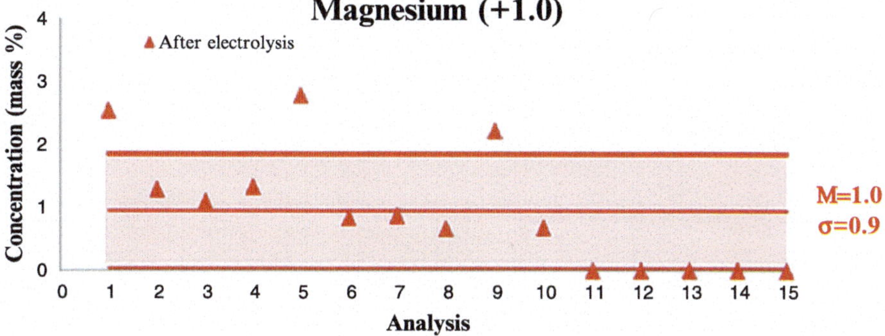

Fig. 11.7 The magnesium presence is evident after the experiment, whereas there is no trace of it before

and C as the products:

$$Ca_{20}^{40} \rightarrow K_{19}^{39} + H_1^1 \tag{11.5}$$

$$Ca_{20}^{40} \rightarrow C_6^{12} + Si_{14}^{28} \tag{11.6}$$

From these reactions, the following balances may be considered: Ca (−1.5%) = K (+1.5%), and Ca (−1.6%) = C (+0.5%) + Si (+1.1%). Considering the experimental residual amount of 0.2% for Ca, and the previously calculated residual amount of 3.3% for Ca, the previous two reactions provide a complete matching (Figs. 11.8 and 11.9).

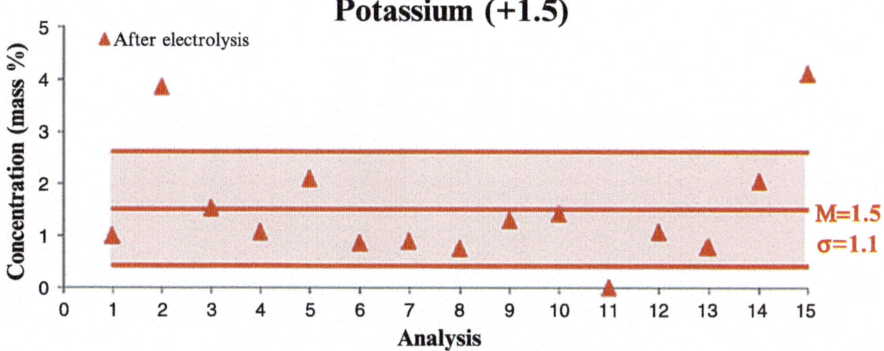

Fig. 11.8 Potassium evidence emerges only after the experiment

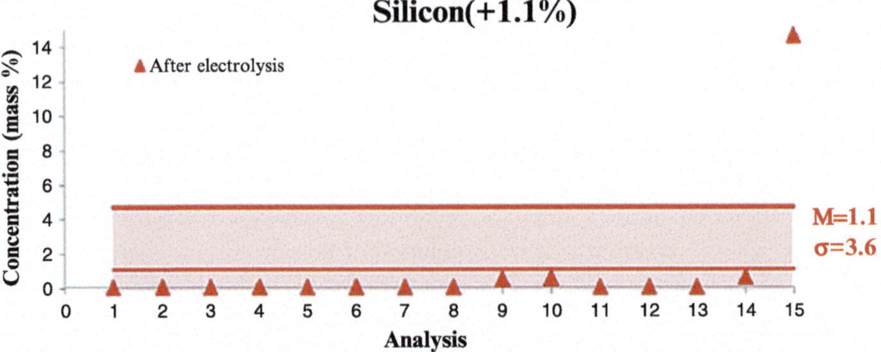

Fig. 11.9 EDS measurements show evident traces of silicon only after the electrolysis experiment

11.5 Chemical Composition Analysis of the Ni Electrode

Let us consider the nickel electrode. Table 11.2 summarizes the concentration changes after the electrolysis experiment. Nickel diminishes by 23.1%, whereas the most relevant positive variation is that of oxygen (+19.5%). It is worth observing that the average concentration decrement in Fe (−2.0%) is comparable to the average increment in Al (Table 11.2). Figures 11.10 and 11.11 show the set of concentrations measured before and after the experiment for Ni and O.

On the basis of the phono-fission reaction conjecture, we can assume the oxygen average variation as a nuclear effect caused by the following reaction:

$$Ni_{28}^{59} \rightarrow 3O_{8}^{16} + 2He_{2}^{4} + 3 \text{ neutrons} \tag{11.7}$$

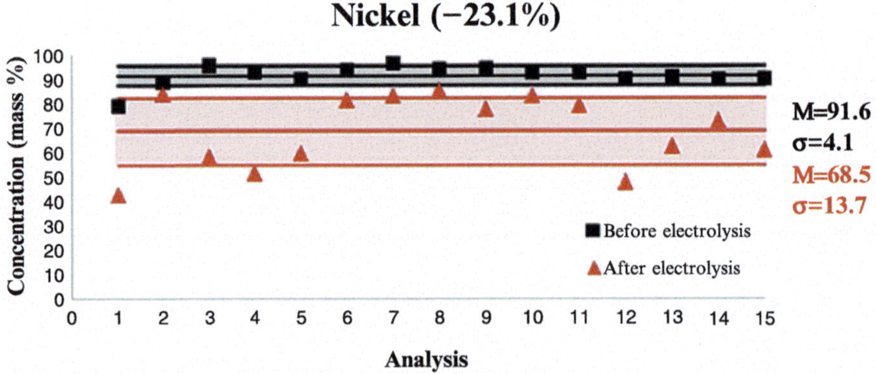

Fig. 11.10 Nickel electrode: the series of the Ni concentrations measured before and after 20 h of electrolysis. The average variation between the two series is 23.1%

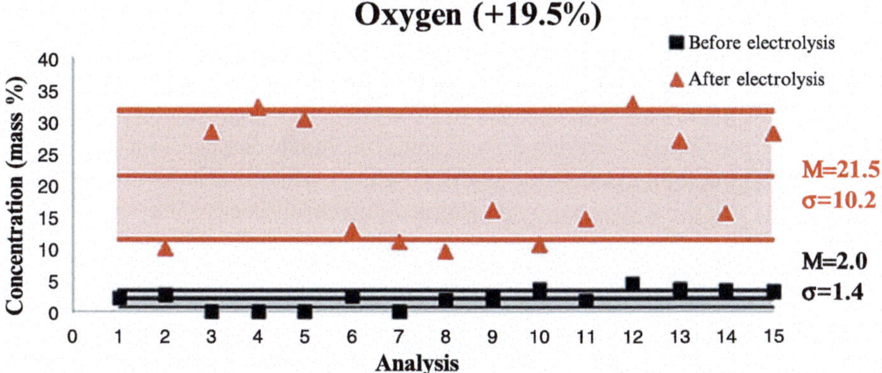

Fig. 11.11 O concentration before and after the experiment. The average values of O concentration change from a mass percentage of 2.0%, at the beginning of the experiment, to 21.5% at the end

A second hypothesis can be considered for the Al average variation, which is consistent with the following reaction:

$$Fe_{26}^{56} \rightarrow 2Al_{13}^{27} + 2 \text{ neutrons} \tag{11.8}$$

A third hypothesis can be made considering the silicon average variation:

$$Ni_{28}^{59} \rightarrow 2Si_{14}^{28} + 3 \text{ neutrons} \tag{11.9}$$

Reactions (11.7), (11.8) and (11.9) imply emissions of neutrons, which provide a great support to the hypothesis based on phono-fission reactions. The hypothesis is that an average decrement of 22.1% in nickel undergoes a reaction producing at least 18.0% of oxygen together with alpha and neutron emissions. Secondly, another average decrement in Ni of 1% can transform it into silicon (+0.9%) and additional neutrons. Thirdly, an average depletion of 2% in Fe produces about 1.9% of Al accompanied by neutron emissions (+0.1%). The calculated O increment of 18.0% is not far from the experimental value of 19.5%. Considering the average concentrations measured before and after the experiment, the three considerations outlined above can be summarized by the following balances: Ni (−22.1%) = O (+18.0%) + He (+3.0%) + neutrons (+1.1%); Ni (−1.0%) = Si (+0.9%) + neutrons (+0.1%); Fe (−2.0%) = Al (+1.9%) + neutrons (+0.1%).

11.6 Conclusions

Neutron emissions up to one order of magnitude higher than the natural background level are observed during the operating time of the Pd and Ni electrolytic cell. In particular, after a time period of about 7.5 h, neutron emission peaks of about 3 times the background level are measured. After 9 h, it is possible to observe neutron emission peaks of about one order of magnitude greater than the natural background level.

A delay or incubation time appears to be needed for phono-fission reactions to occur, as in previously considered situations, for which damage is a cumulative function of time (see for example the case of fatigue). On the other hand, an alternative hypothesis is that the resonance of the atomic lattice itself needs some specific time to achieve, with its diverging vibrations, the critical nuclear energy necessary for the nucleus disintegration or fission.

By the EDX analysis performed on the two electrodes, significant compositional variations are recorded. In general, the decrements in Pd at the first electrode seem to be almost perfectly matched by the increments in lighter elements like oxygen. As far as the second electrode is concerned, the Ni decrement is almost perfectly matched by the O and Si increments. At the same time, the Fe decrement may be considered to transform into the Al concentration increment after the test.

The results reported in the present chapter give a valid confirmation to the previous results obtained using Ni–Fe and Co–Cr electrodes, see Chap. 10. Also in this case, the chemical changes and the subatomic particle emissions may be accounted for direct evidence of phono-fission reactions. The last are correlated to microcrack formation and/or propagation after hydrogen embrittlement. The most relevant process emerging from the experiment is the primary fission of palladium into iron and calcium. Then, secondary fissions of the products, as well as of nickel in the other electrode, in turn appear producing oxygen atoms, alpha particles, and neutrons.

References

1. Borghi DC, Giori DC, Dall'Olio A (1992) Experimental evidence on the emission of neutrons from cold hydrogen plasma. In: Proceedings of the international workshop on few-body problems in low-energy physics, Alma-Ata, Kazakhstan, pp 147–154; Unpublished Communication (1957); Comunicacao no. 25 do CENUFPE, Recife, Brazil (1971)
2. Diebner K (1962) Fusionsprozesse mit Hilfe konvergenter Stosswellen—einige aeltere und neuere Versuche und Ueberlegungen. Kerntechnik 3:89–93
3. Kaliski S (1978) Bi-conical system of concentric explosive compression of D-T. J Tech Phys 19:283–289
4. Winterberg F (1984) Autocatalytic fusion–fission implosions. Atomenergie-Kerntechnik 44:146
5. Derjaguin BV et al (1989) Titanium fracture yields neutrons? Nature 34:492
6. Fleischmann M, Pons S, Hawkins M (1989) Electrochemically induced nuclear fusion of deuterium. J Electroanal Chem 261:301
7. Bockris JM, Lin GH, Kainthla RC, Packham NJC, Velev O (1990) Does tritium form at electrodes by nuclear reactions? In: The first annual conference on cold fusion. National Cold Fusion Institute, University of Utah, Research Park, Salt Lake City
8. Preparata G (1991) Some theories of cold fusion: a review. Fusion Technol 20:82
9. Preparata G (1991) A new look at solid-state fractures, particle emissions and "cold" nuclear fusion. Il Nuovo Cimento 104(A):1259–1263
10. Mills RL, Kneizys P (1991) Excess heat production by the electrolysis of an aqueous potassium carbonate electrolyte and the implications for cold fusion. Fusion Technol 20:65

11. Notoya R, Enyo M (1992) Excess heat production during electrolysis of H_2O on Ni, Au, Ag and Sn electrodes in alkaline media. In: Proceedings of third international conference on cold fusion, Nagoya Japan. Universal Academy Press, Tokyo
12. Miles MH, Hollins RA, Bush BF, Lagowski JJ, Miles RE (1993) Correlation of excess power and helium production during D_2O and H_2O electrolysis using palladium cathodes. J Electroanal Chem 346:99–117
13. Bush RT, Eagleton RD (1993) Calorimetric studies for several light water electrolytic cells with nickel fibrex cathodes and electrolytes with alkali salts of potassium, rubidium, and cesium. In: Fourth international conference on cold fusion, Lahaina, Maui. Electric Power Research Institute, Palo Alto, p 13
14. Fleischmann M, Pons S, Preparata G (1994) Possible theories of cold fusion. Nuovo Cimento, Soc Ital Fis A 107:143
15. Szpak S, Mosier-Boss PA, Smith JJ (1994) Deuterium uptake during Pd-D codeposition. J Electroanal Chem 379:121
16. Sundaresan R, Bockris JOM (1994) Anomalous reactions during arcing between carbon rods in water. Fusion Technol 26:261
17. Arata Y, Zhang Y (1995) Achievement of solid-state plasma fusion ("cold-fusion"). Proc Jpn Acad 71(B):304–309
18. Ohmori T, Mizuno T, Enyo M (1996) Isotopic distributions of heavy metal elements produced during the light water electrolysis on gold electrodes. J New Energy 1(3):90–99
19. Monti RA (1996) Low energy nuclear reactions: experimental evidence for the alpha extended model of the atom. J New Energy 1(3):131
20. Monti RA (1998) Nuclear transmutation processes of lead, silver, thorium, uranium. In: The seventh international conference on cold fusion. ENECO, Vancouver/Salt Lake City
21. Ohmori T, Mizuno T (1998) Strong excess energy evolution, new element production, and electromagnetic wave and/or neutron emission in light water electrolysis with a Tungsten cathode. Infin Energy 20:14–17
22. Mizuno T (1998) Nuclear transmutation: the reality of cold fusion. Infinite Energy Press
23. Little SR, Puthoff HE, Little ME (1998) Search for excess heat from a Pt electrode discharge in K_2CO_3-H_2O and K_2CO_3-D_2O electrolytes. Infin Energy 5:34
24. Ohmori T, Mizuno T (2000) Nuclear transmutation reaction caused by light water electrolysis on tungsten cathode under incandescent conditions. J New Energy 4(4):66–78
25. Ransford HE (1999) Non-stellar nucleosynthesis: transition metal production by DC plasma-discharge electrolysis using carbon electrodes in a non-metallic cell. Infin Energy 4(23):16–22
26. Storms E (2000) Excess power production from platinum cathodes using the Pons-Fleischmann effect. In: 8th international conference on cold fusion, Lerici (La Spezia). Italian Physical Society, Bologna, Italy, pp 55–61
27. Storms E (2007) Science of low energy nuclear reaction: a comprehensive compilation of evidence and explanations about cold fusion. World Scientific Publishing, Singapore
28. Mizuno T et al (2000) Production of heat during plasma electrolysis. Jpn J Appl Phys A 39:6055
29. Warner J, Dash J, Frantz S (2002) Electrolysis of D_2O with titanium cathodes: enhancement of excess heat and further evidence of possible transmutation. In: The ninth international conference on cold fusion. Tsinghua University, Beijing, China
30. Fujii MF et al (2002) Neutron emission from fracture of piezoelectric materials in deuterium atmosphere. Jpn J Appl Phys 41:2115–2119
31. Mosier-Boss PA et al (2007) Use of CR-39 in Pd/D co-deposition experiments. Eur J Appl Phys 40:293–303
32. Swartz M (2008) Three physical regions of anomalous activity in deuterated palladium. Infin Energy 14:19–31
33. Mosier-Boss PA et al (2010) Comparison of Pd/D co-deposition and DT neutron generated triple tracks observed in CR-39 detectors. Eur J Appl Phys 51(2):20901–20911
34. Kanarev M, Mizuno T (2002) Cold fusion by plasma electrolysis of water. New Energy Technol 1:5–10
35. Cardone F, Cherubini G, Petrucci A (2009) Piezonuclear neutrons. Phys Lett A 373:862–866

36. Carpinteri A, Cardone F, Lacidogna G (2009) Piezonuclear neutrons from brittle fracture: early results of mechanical compression tests. Strain 45:332–339; Atti dell'Accademia delle Scienze di Torino 33:27–42
37. Cardone F, Carpinteri A, Lacidogna G (2009) Piezonuclear neutrons from fracturing of inert solids. Phys Lett A 373:4158–4163
38. Carpinteri A, Cardone F, Lacidogna G (2010) Energy emissions from failure phenomena: mechanical, electromagnetic, nuclear. Exp Mech 50:1235–1243
39. Carpinteri A, Lacidogna G, Manuello A, Borla O (2012) Piezonuclear fission reactions: evidences from microchemical analysis, neutron emission, and geological transformation. Rock Mech Rock Eng 45:445–459
40. Carpinteri A, Lacidogna G, Manuello A, Borla O (2013) Piezonuclear fission reactions from earthquakes and brittle rocks failure: evidence of neutron emission and nonradioactive product elements. Exp Mech 53:345–365
41. Milne I, Ritchie RO, Karihaloo BL (2003) Comprehensive structural integrity: fracture of materials from nano to macro, vol 2, Ch 6. Elsevier, pp 31–33
42. Liebowitz H (1971) Fracture: an advanced treatise. Academic Press, New York/San Francisco/ London
43. Carpinteri A, Borla O, Goi A, Manuello A, Veneziano D (2014) Mechanical conjectures explaining cold nuclear fusion. In: Advancement of optical methods in experimental mechanics. Conference proceedings of the society for experimental mechanics, Lombard, Illinois, 2013, Paper No. 481, vol 3, pp 353–367
44. Veneziano D, Borla O, Goi A, Manuello A, Carpinteri A (2013) Mechanical conjectures based on hydrogen embrittlement explaining cold nuclear fusion. In: Proceedings of the 21st Congresso Nazionale di Meccanica Teorica ed Applicata (AIMETA), Torino, Italy, 2013, CD-ROM
45. Carpinteri A, Borla O, Goi A, Manuello A, Veneziano D (2015) Cold nuclear fusion explained by hydrogen embrittlement and piezonuclear fissions of the metallic electrodes—part I: Ni-Fe and Co-Cr electrodes. In: Carpinteri A et al (eds) Acoustic, electromagnetic, neutron emissions from fracture and earthquakes. Springer

Chapter 12
Electrolysis Experiments with Pd and Ni Electrodes: High Repeatability of Mass Balances in Chemical Composition Changes

Abstract In the last few decades, several scientific papers have reported experimental evidence of anomalous nuclear reactions occurring in condensed matter during electrolytic phenomena or mechanical instabilities such as fracture (in solids) and cavitation (in liquids). Despite the numerous research activities carried out in the field of the so-called "Cold Nuclear Fusion", it remains today not fully understood. As already treated in Chaps. 10 and 11, the formation of cracks on the external surfaces of the electrodes used during electrolysis tests, together with chemical composition changes and anomalous subatomic particle emissions, are detected. A macromechanical interpretation of the experimental evidence can be based on low-energy phono-fission nuclear reactions, which are a consequence of hydrogen embrittlement, microcracking, and THz phonons. In the present chapter, the results of four identical experimental test repetitions with Pd and Ni electrodes are discussed. In particular, the excellent repeatability of nuclear and stoichiometric balances between the chemical compositions of both electrodes, before and after the test, is consistently emphasized.

Keywords Electrolysis · Hydrogen embrittlement · Microcracking · TeraHertz phonons · Phono-fission reactions · Neutron and alpha particle emissions · Nuclear and stoichiometric balances · Cold nuclear fusion

12.1 Preliminary Remarks

Since the 1920s, it has been assumed that nuclear fusion could also occur at temperatures much lower than tens of million degrees by hydrogen adsorption in a metal catalyst [1].

In the present Part IV, a totally different interpretation of the phenomenon usually called "Cold Nuclear Fusion" is proposed. The produced helium atoms or ions are proved to be fragments from fission reactions (alpha particles) and not products from hydrogen or deuterium nuclear fusion.

A. Carpinteri, *Terahertz Phonons and Nanomechanical Instabilities*,
https://doi.org/10.1007/978-3-032-14692-2_12

Among the several experiments conducted in this research field, the most relevant statements about cold fusion were made by Stanley Pons and Martin Fleischmann in 1989 [2]. Their conjecture was that the high compression ratio and the mobility of deuterium (a stable isotope of hydrogen) during electrolysis tests might result in an unexpected nuclear fusion. In particular, their experiments were carried out by means of a palladium cathode and heavy water (D_2O) placed inside a thermally insulated calorimeter, in order to measure the heat generated by the electrolytic process. The electrical power supply was continuously applied and the heavy water was renewed at regular intervals [2]. For most of the time, the power absorbed by the cell was equal to the calculated power within the measurement accuracy, and the cell temperature was stable around 30 °C. However, in some experiments, the temperature suddenly increased to about 50 °C without changes in the input power [2]. In support to the presumed occurrence of nuclear reactions giving rise to the temperature increment, Fleischmann and Pons reported the presence of neutrons and tritium [2].

In 1998, the Japanese researchers Mizuno and Ohmori [3, 4] announced the possibility of obtaining cold fusion reactions without using palladium and heavy water, but only through tungsten electrodes immersed in a solution of common water and potassium carbonate (K_2CO_3). By applying a voltage supply ranging from 160 up to 300 V, the temperature of the solution exceeded 70–80 °C and a plasma bubble formed around the immersed portion of the tungsten electrode. Then, a positive energy balance was estimated with a thermal energy emission 20–100% above the electrical energy used to trigger the reaction. Moreover, several medium-weight elements like calcium, titanium, chromium, manganese, iron, cobalt, copper, and zinc, which were not present before operating the electrolytic cell [5], were observed on the cathode after the experiment.

In 2007, Mosier-Boss et al. [6, 7] obtained important evidence of anomalous heat generation, alpha particle emissions, and chemical composition changes during electrolysis experiments by means of a palladium cathode immersed in a solution of lithium chloride and deuterated water. As in the case of Mizuno et al. [5], elements such as Fe, Cr, Ni, and Al were detected at the end of the tests [8]. To explain the formation of these new elements, Mosier-Boss and co-workers suggested a multi-body deuteron fusion phenomenon and the palladium lattice disintegration [8].

In addition to the experimental results described above, many other examples of unexpected nuclear reactions in condensed matter have been reported by various Authors [9–36]. All these tests are characterized by anomalous heat generation, as well as by neutron and alpha particle emissions. Furthermore, some of the Authors cited significant examples of compositional changes and microcracking of the electrodes [3, 4, 30–32].

Preparata commented as follows: "Despite the great amount of experimental results observed by a large number of scientists, a unified interpretation and theory of these phenomena has not been accepted and their comprehension still remains unsolved" [19, 20]. On the other hand, as discussed in several subsequent papers on the same subject, one of the common features regarding the experiments is the appearance of microcracks on the electrode surfaces after the tests: i.e., the common environment, in which low-energy nuclear reactions occur, is provided by cracks of a critical size accompanied by a resonance process dissipating energy from those sites [30, 31]. Such scenario might be directly correlated to hydrogen embrittlement of the material composing the metal electrodes. This phenomenon, well-known in Metallurgy and Fracture Mechanics, characterizes metal atomic lattices during forming or finishing operations [37].

Recent and innovative experiments provided strong evidence of anomalous fission reactions and nuclear transmutations occurring in condensed matter, not only during electrolysis tests, but also in fracture of solids and in cavitation of liquids [38–46]. Based on these experimental observations, a macromechanical explanation for the so-called cold nuclear fusion is suggested [38, 46].

Hydrogen embrittlement seems to play an essential role in the observed microcracking of the electrode host metals. In particular, the metal matrix is subjected to mechanical damage and cracking due to hydrogen atoms (produced by the electrolysis itself) penetrating into the atomic lattice and forcing it during gas loading. The hydrogen atoms generate an internal compression stress that apparently lowers the fracture toughness of the metal, so that brittle crack growth can occur with a hydrogen partial pressure below 1 atm [38]. Consequently, the hypothesis is that phono-fission reactions may occur in correspondence to nano- and microcrack formation and/or propagation [38]. Terahertz phonons are, in fact, produced at the nano-scale, showing a frequency equivalent to that of thermal neutrons and being able to trigger fission reactions in medium-weight chemical elements. As a matter of fact, thermal neutrons are characterized by a frequency of 6.05 THz (according to the well-known law that links energy and frequency through the Planck's constant), which is very close to the Uranium atomic lattice resonance frequency (Debye frequency) of 6.24 THz, as well as to that of Fe (7.77 THz) and Ca (4.79 THz).

This new kind of reactions has been observed from the laboratory to the Earth's crust or tectonic scale, when particularly high-frequency waves originate from fracture phenomena as well as before or in correspondence to seismic events [38, 40–45, 47–54]. Possible theoretical explanations for this anomalous nuclear phenomenon are described in [38, 55–61]. An alternative explanation based on the resonant interactions between phonons, plasmons, and metal atomic lattices will be provided in Part VII.

In the present chapter, the results of phono-fission reactions during electrolysis tests are presented. In order to confirm the earlier results obtained by Co–Cr and Ni–Fe electrodes [38, 46], as well as by Pd and Ni electrodes (Test 1) [38, 46], three additional electrolysis experiments (Test 2, Test 3, and Test 4) are conducted using a 100% palladium cathode. These new experimental tests intend to emphasize the repeatability of the physical phenomenon and to more deeply understand the different phases that characterize the experiments and the various catalyzing factors. For these reasons, the three further series of tests are arranged in three trials each, with a duration of 2.5, 5, and 10 h. This protocol is adopted to evaluate whether the chemical changes observed in the previous experiment are concentrated during a single trial or, otherwise, are distributed over the entire duration of the test.

The experimental results show significant neutron emissions together with relevant chemical composition changes and the appearance of previously absent elements. Moreover, evident micro- and macrocracks are observed in the electrodes, thus emphasizing the macromechanical origin of the phenomenon. Further experimental evidence of cracking and shredding of the copper and tungsten cathode wires during electrolysis experiments were also described by Widom et al. [60, 61].

12.2 Experimental Set-Up

Over the last fifteen years, specific experiments have been conducted using an electrolytic reactor (Owners: Mr. Alessandro Goi et al.) to investigate whether the anomalous heat generation may be correlated to a new type of nuclear reaction occurring during electrolysis experiments [38, 46].

The experimental device, when filled with a salt solution of water and Potassium Carbonate (K_2CO_3), triggers the electrolysis by using two metal electrodes immersed in the aqueous solution (Fig. 12.1).

The solution container, or reaction chamber in the following, is an AISI 316L steel cylinder, 100 mm in diameter, 150 mm in height, and with a wall thickness of 5 mm.

The base of the chamber consists of a ceramic plate preventing the direct contact between liquid solution and Teflon lid. Two threaded holes secure the electrodes, which are screwed into the bottom of the chamber. A valve at the top of the cell allows the gas to escape from the reactor and to condense in an external collector. Externally, two circular Inox steel flanges, fastened by means of four threaded supports, hold the Teflon layers. The inferior steel flange of the reactor is connected to four supports isolated from the ground by means of a rubber-based material.

The voltage supply is provided by means of a RTS40CE dimmer connected to the 220VDC domestic power line and sets the voltage using a potentiometer. In order to obtain a pulsating direct voltage/current (VDC), a diode bridge rectifier is added

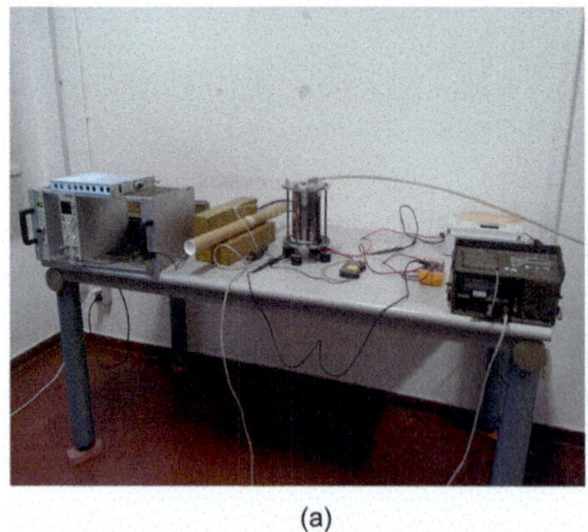

(a)

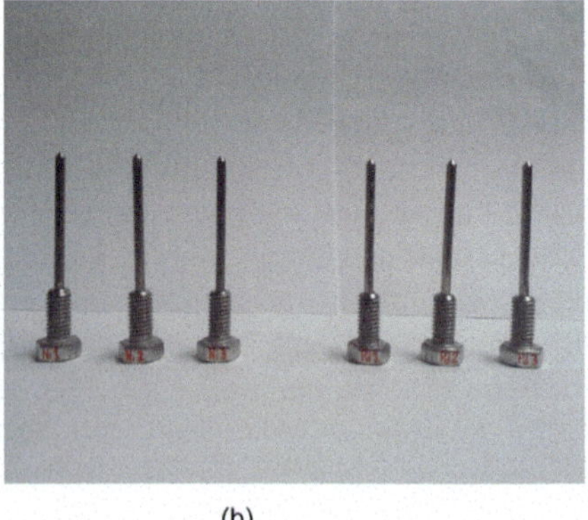

(b)

Fig. 12.1 Experimental set-up (**a**); Pd and Ni electrodes (**b**) to the output of the RTS40CE. All the electrical parameters are monitored during the tests by a virtual multimeter specially programmed for the experiment. The Pd electrode is employed as cathode (negative pole), whereas the Ni electrode as anode (positive pole).

The electrolytic cell is switched from low to high working parameters by letting the current and voltage vary in the ranges of 3–5 A and 20–120 V, respectively.

Neutron emission is monitored using a ^{3}He proportional counter with pre-amplification, amplification, and discrimination electronics directly connected to the detector tube (Fig. 12.1a). The detector is supplied with a high voltage of ~1.3 kV via a Nuclear Instrument Module (NIM). The logic output producing the TTL (transistor–transistor logic) pulses is connected to a NIM counter. The logic output of the detector is enabled for analog signals exceeding 300 mV. This discrimination threshold is a consequence of the sensitivity of the ^{3}He detector to the gamma rays produced during neutron emission in ordinary nuclear processes. This value is determined by measuring the analog signal of the detector using a Co-60 gamma source. The detector is also calibrated at the factory for the measurement of thermal neutrons; its sensitivity is 65 cps/$n_{thermal}$ ($\pm$10% according to the factory). Therefore, the flux of thermal neutrons is one thermal neutron/s cm^2, corresponding to a count rate of 65 cps.

Eventually, before and after the experiments, Energy Dispersive X-ray Spectroscopy is carried out in order to identify possible direct evidence of phono-fission reactions that can take place during the electrolysis experiment. The elemental analyses are performed by a ZEISS Auriga field emission scanning electron microscope (FESEM) equipped with an Oxford INCA energy-dispersive X-ray detector (EDX), with a resolution of 124 eV@MnKa. The energy used for the analyses is 18 keV.

12.3 Neutron Emission Measurements

Neutron emission monitoring performed during the first experimental activity on Pd and Ni electrodes (Test 1) [38] is shown in Fig. 12.2. The measurement performed by the ^{3}He detector is conducted for a total time of about 20 h. The natural background level is equal to $(3.23 \pm 1.49) \times 10^{-2}$ cps.

After about 7.5 h (460 min), a neutron peak of about 3 times the background level is detected [38]. Similarly, after 9 h (545 min) from the beginning of the test, it is possible to observe a neutron emission peak of about one order of magnitude greater than the background. Eventually, after 20 h (1,200 min), another neutron peak up to 7 times the background is measured. More details about this experimental test can be found in [38].

The neutron emissions monitored during the second experimental test (Test 2) are reported in Fig. 12.3. In particular, the neutron levels detected during the three distinct time intervals of 2.5, 5, and 10 h are shown. A neutron background of $(6.00 \pm 2.45) \times 10^{-2}$ cps is measured for the first two intervals (2.5 and 5 h), whereas

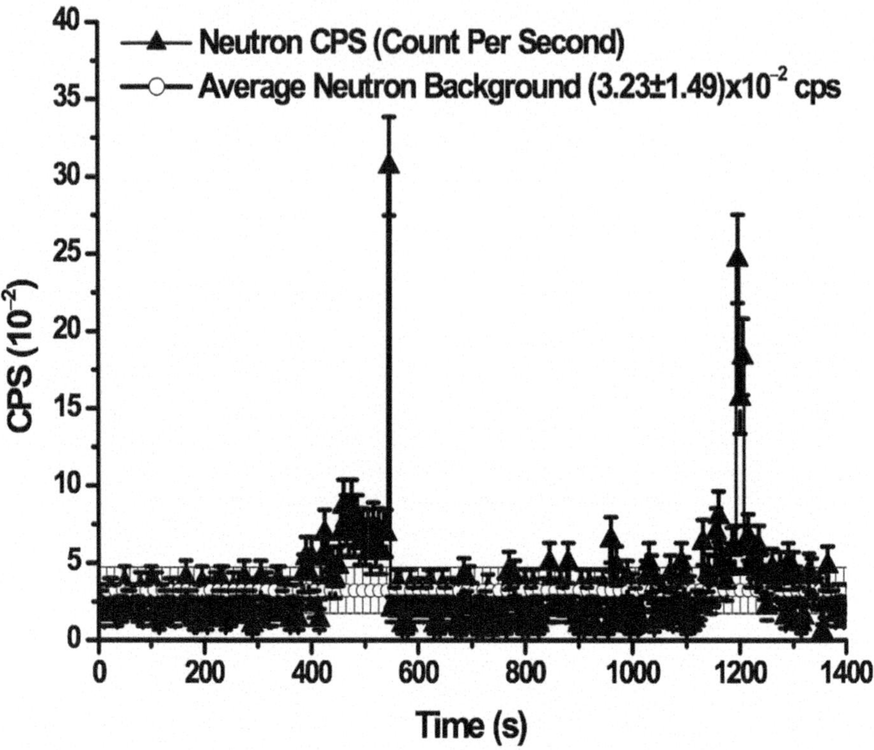

Fig. 12.2 Neutron emission measurements detected during the first experiment (Test 1) with a time duration of 20 h and using Pd and Ni electrodes

neutron peaks about 2 times (Fig. 12.3a) and 5 times (Fig. 12.3c) higher than the background are observed just before switching off the cell.

As additional supporting evidence, the cumulative count curves are compared to the cumulative average background (Fig. 12.3b, d). In this way, it is possible to observe how, in both cases, the neutron emission is comparable to the background level. Conversely, during the third trial, the cumulative curve appears to be significantly higher, suggesting that the anomalous neutron emission takes place after an incubation time and not immediately after the cell switching on. In this case, the environmental background is found to be equal to $(7.85 \pm 1.62) \times 10^{-2}$ cps. After about 80 min from the beginning of the test, neutron emissions greater than the background level are detected (Fig. 12.3e). In particular, after 200 min and up to 400 min, neutron peaks greater than ten times the background are measured. This evidence can also be observed in the cumulative curve shown in Fig. 12.3f.

The observed behavior appears to be consistent with claims made by Fleishmann and Pons and other researchers: anomalous emissions and heat generation are concentrated in some periods of the experiment and not continuously distributed over its duration [2–5].

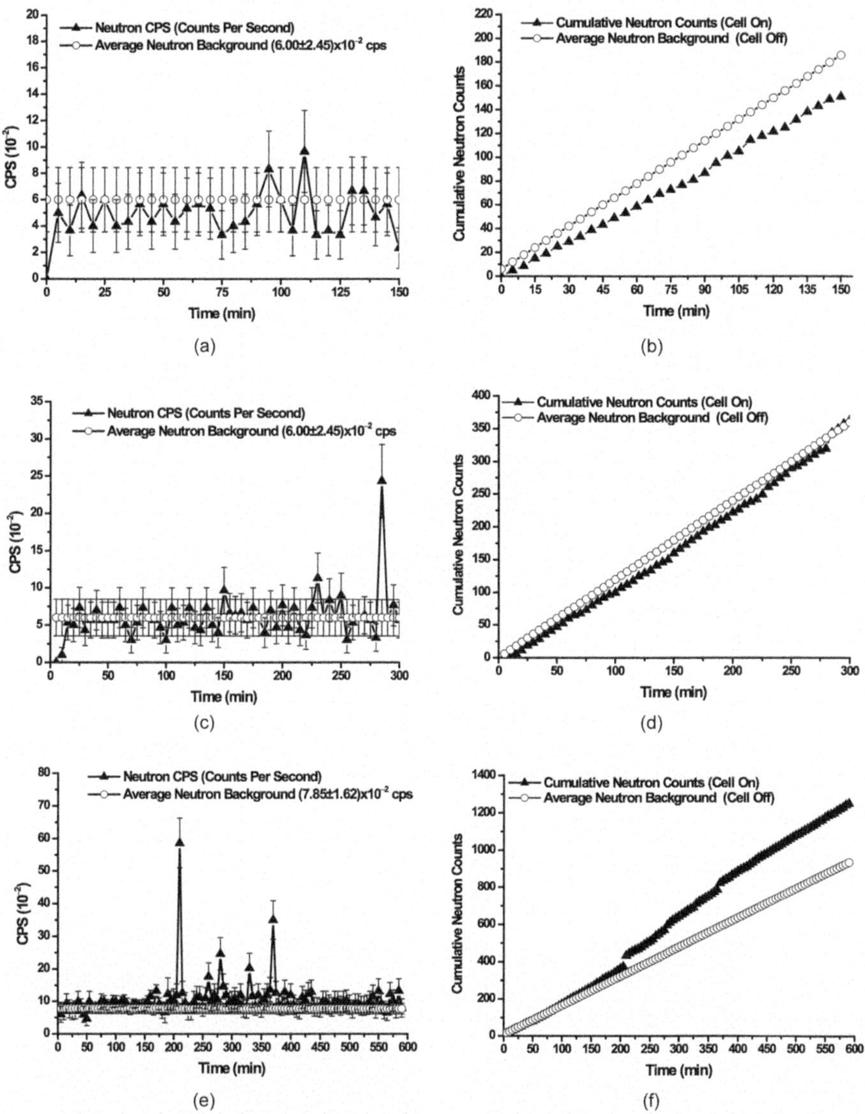

Fig. 12.3 Neutron emission measurements for the second experiment on Pd and Ni electrodes (Test 2) after 2.5 h (**a**); 5 h (**c**); and 10 h (**e**). Cumulated curves compared to the cumulated average background for the same durations (**b, d, f**)

12.4 Chemical Composition Changes on the Electrode Surfaces

In the following, the chemical composition analyses performed before and after all the experiments on Pd and Ni electrodes are summarized. The reported percentages represent the average values based on 15 independent measurements performed in different equidistant locations along each metallic electrode (40 mm in length).

Hydrogen embrittlement hypothesis, based on the presence of micro- and mesocracks on the electrode surface, supports the macromechanical interpretation of the nuclear phenomenon [38, 46]. On the Pd electrode, where a mesoscopic fracture took place during the test, the crack presents an opening displacement of about 40 μm and a length of a few millimeters (Fig. 12.4a, b).

Palladium is widely utilized in the experiments related to the so-called cold nuclear fusion due to its hydrogen adsorption and embrittlement capability. The authors investigate with Pd and Ni electrodes, firstly in a test of 20 h (Test 1, see also Chap. 11) [38, 46]. The most impressive result in this case is the huge Pd depletion, with a final decrement of 28.7% and the appearance of chemical elements before absent (Fe, Ca, O, Si, K) [38, 46]. Additionally, a total decrement in nickel content of 23.1% is measured on the Ni electrode.

In particular, considering the average decline in Pd (−28.7%) after 20 h (Table 12.1) and according to the phono-fission hypothesis, the following reaction can be postulated, as already done in Chap. 11:

$$\mathrm{Pd}_{46}^{106} \rightarrow \mathrm{Ca}_{20}^{40} + \mathrm{Fe}_{26}^{56} + 10 \text{ neutrons} \tag{12.1}$$

Referring to reaction (12.1), the Pd decrement of 28.7% can be balanced by the Ca and Fe increments of 10.8% and 15.2%, respectively (Table 12.1).

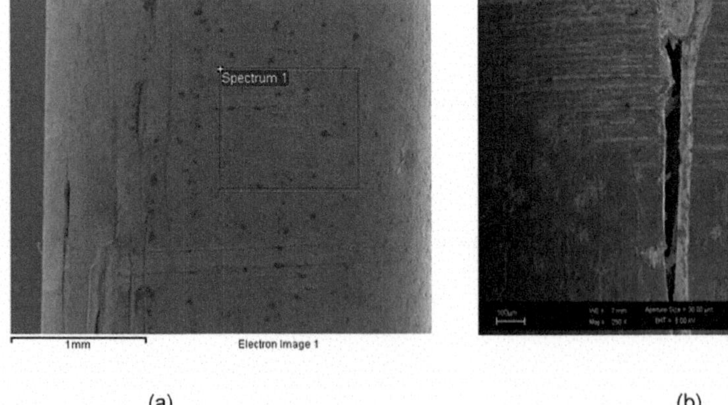

Fig. 12.4 Image of the Pd electrode before the experiment (**a**). The electrode after several operating hours presents a crack of about 40 μm width observable also at naked eyes (**b**)

Table 12.1 Test 1 performed on Pd and Ni electrodes. Element concentrations before the experiment and after 20 h from the beginning (a); nuclear (b) and stoichiometric (c) balances

(a)

Pd electrode element concentrations

	Pd (%)	O (%)	Fe (%)	K (%)	Si (%)	Mg (%)
Before the experiment	100	0.0	0.0	0.0	0.0	0.0
After 20 h	71.3	18.5	2.0	2.0	1.1	1.0

Ni electrode element concentrations

	Ni (%)	O (%)	Fe (%)	Si (%)	Al (%)
Before the experiment	91.6	2.0	2.4	0.3	0.0
After 20 h	68.5	21.5	0.4	1.1	1.8

(b)

Pd electrode nuclear balances

$Pd_{46}^{106} \rightarrow Ca_{20}^{40} + Fe_{26}^{56} + 10n_0^1$

$Fe_{26}^{56} \rightarrow 3O_8^{16} + He_2^4 + 4n_0^1$

$Ca_{20}^{40} \rightarrow 2O_8^{16} + 2He_2^4$

$Ca_{20}^{40} \rightarrow O_8^{16} + Mg_{12}^{24}$

$Ca_{20}^{40} \rightarrow K_{19}^{39} + H_1^1$

$Ca_{20}^{40} \rightarrow C_6^{12} + Si_{14}^{28}$

Ni electrode nuclear balances

$Ni_{28}^{58} \rightarrow 3O_8^{16} + 2He_2^4 + 2n_0^1$

$Ni_{28}^{58} \rightarrow 2Si_{14}^{28} + 2n_0^1$

$Fe_{26}^{56} \rightarrow 2Al_{13}^{27} + 2n_0^1$

(c)

Pd electrode stoichiometric balances

$Pd(-28.7\%) = Ca(+10.8\%) + Fe(+15.2\%) + n(+2.7\%)$

$Fe(-13.2\%) = O(+11.3\%) + He(+0.9\%) + n(+0.9\%)$

$Ca(-6.0\%) = O(+4.8\%) + He(+1.2\%)$

$Ca(-1.7\%) = O(+0.7\%) + Mg(+1.0\%)$

$Ca(-1.5\%) = K(+1.5\%) + H(+0.0\%)$

$Ca(-1.6\%) = C(+0.5\%) + Si(+1.1\%)$

Ni electrode stoichiometric balances

$Ni(-22.3\%) = O(+18.4\%) + He(+3.1\%) + n(+0.8\%)$

$Ni(-0.8\%) = Si(+0.8\%) + n(+0.0\%)$

$Fe(-1.8\%) = Al(+1.8\%) + n(+0.0\%)$

These variations should be accompanied by a neutron emission corresponding to the remaining 2.7% of the mass. The iron increment, according to reaction (12.1), could be considered as the starting element for the production of other lighter elements. Hence, a second hypothesis can be advanced involving Fe as the starting element and O as the end-product, together with alpha and neutron emissions:

$$Fe_{26}^{56} \rightarrow 3O_{8}^{16} + He_{2}^{4} + 4 \text{ neutrons} \tag{12.2}$$

According to reaction (12.2) and considering the residual Fe increment of 2% measured at the end of the test, an iron depletion of 13.2% produces 11.3% of Oxygen with alpha particles (He) and neutron emissions. The total measured increment in Oxygen after the experiment is equal to 18.5 (Table 12.1). This amount is only partially explained by reaction (12.2), although part of the remaining O concentration could be justified by other reactions involving Ca (generated in reaction (12.1)) as the starting element:

$$Ca_{20}^{40} \rightarrow O_{8}^{16} + Mg_{12}^{24} \tag{12.3}$$

$$Ca_{20}^{40} \rightarrow 2O_{8}^{16} + 2He_{2}^{4} \tag{12.4}$$

From reaction (12.3), a decrement in Ca concentration of 1.7% with the formation of 1.0% of Mg and 0.7% of O can be considered. From reaction (12.4), a decrement in Ca of 6.0% corresponding to an increment of about 4.8% in O, with alpha particle emissions, is observed. Considering the O increment coming from reactions (12.2), (12.3) and (12.4), which once combined give 16.8%, and the experimental evidence reporting a total measured O concentration of 18.5%, Oxygen is at least partly implicated in the assumed reaction cascade.

The remaining Ca decrement of 3.1% can be explained by the following reactions:

$$Ca_{20}^{40} \rightarrow K_{19}^{39} + H_{1}^{1} \tag{12.5}$$

$$Ca_{20}^{40} \rightarrow C_{6}^{12} + Si_{14}^{28} \tag{12.6}$$

At the same time, the Mg, K, and Si increments observed after the experiment can be explained by reactions (12.3), (12.5) and (12.6).

In a similar manner, let us consider the Nickel electrode. Table 12.1 summarizes the concentration changes after the electrolysis. The most significant result concerns the nickel depletion equal to 23.1% and the Oxygen increment of 19.5%.

The following phono-fission reactions imply neutron and alpha particle emissions, which would also explain the entire Ni decrement and the corresponding O increment:

$$Ni_{28}^{58} \rightarrow 2Si_{14}^{28} + 2 \text{ neutrons} \tag{12.7}$$

Table 12.2 Test 2 on Pd and Ni electrodes. Element concentrations before the experiment, and after 2.5, 5, 10 h from the beginning (a); nuclear (b) and stoichiometric (c) balances

(a)

Pd electrode element concentrations

	Pd (%)	O (%)	Fe (%)	K (%)	Zn (%)	Si (%)	Mg (%)
Before the experiment	99.8	0.1	0.0	0.0	0.0	0.0	0.0
After 2.5 h	92.7	6.2	0.6	0.4	0.0	0.0	0.0
After 5 h	88.3	9.0	2.7	0.0	0.0	0.0	0.0
After 10 h	70.9	21.2	1.3	3.3	1.4	0.6	0.4

Ni electrode element concentrations

	Ni (%)	O (%)	Fe (%)	Si (%)	Al (%)
Before the experiment	91.4	4.0	2.0	0.7	0.4
After 2.5 h	76.6	18.5	0.0	0.3	0.9
After 5 h	71.4	22.8	0.1	0.5	1.7
After 10 h	72.6	16.9	0.3	0.5	1.6

(b)

Pd electrode nuclear balances

$Pd^{106}_{46} \rightarrow Ca^{40}_{20} + Fe^{56}_{26} + 10n^{1}_{0}$

$Pd^{106}_{46} \rightarrow Zn^{64}_{30} + Si^{28}_{14} + He^{4}_{2} + 10n^{1}_{0}$

$Fe^{56}_{26} \rightarrow 3O^{16}_{8} + He^{4}_{2} + 4n^{1}_{0}$

$Ca^{40}_{20} \rightarrow O^{16}_{8} + Mg^{24}_{12}$

$Ca^{40}_{20} \rightarrow 2O^{16}_{8} + 2He^{4}_{2}$

$Ca^{40}_{20} \rightarrow K^{39}_{19} + H^{1}_{1}$

Ni electrode nuclear balances

$Ni^{58}_{28} \rightarrow 3O^{16}_{8} + 2He^{4}_{2} + 2n^{1}_{0}$

$Fe^{56}_{26} \rightarrow 3O^{16}_{8} + He^{4}_{2} + 4n^{1}_{0}$

$Fe^{56}_{26} \rightarrow 2Al^{27}_{13} + 2n^{1}_{0}$

(c)

Pd electrode stoichiometric balances

Pd(−26.6%) = Ca(+10.0%) + Fe(+14.1%) + n(+2.5%)

Pd(−2.3%) = Zn(+1.4%) + Si(+0.6%) + He(+0.1%) + n(+0.2%)

Fe(−12.8%) = O(+11.0%) + He(+0.9%) + n(+0.9%)

Ca(−0.7%) = O(+0.3%) + Mg(+0.4%)

Ca(−6.0%) = O(+4.8%) + He(+1.2%)

Ca(−3.3%) = K(+3.3%) + H(+0.0%)

Ni electrode stoichiometric balances

Ni(−18.8%) = O(+15.6%) + He(+2.6%) + n(+0.6%)

Fe(−0.5%) = O(+0.4%) + He(+0.05%) + n(+0.05%)

Fe(−1.2%) = Al(+1.2%) + n(+0.1%)

$$Ni_{28}^{58} \rightarrow 3O_8^{16} + 2He_2^4 + 2\text{ neutrons} \tag{12.8}$$

On the other hand, Fe (–2.0%) and Al (+1.8%) content variations can be justified by the following reaction:

$$Fe_{26}^{56} \rightarrow 2AI_{13}^{27} + 2\text{ neutrons} \tag{12.9}$$

Considering the experimental results of Tests 2, 3, and 4, the observed changes in chemical composition result to be highly repeatable.

The results related to Test 2 are reported in Table 12.2. Note that only small chemical composition changes are detected in the first two trials (2.5 and 5 h), whereas a very appreciable Pd depletion (−28.9%) is measured at the end of the 10 h trial. In particular, compositional variations very similar to those observed in the case of the 20 h experiment (Test 1) are found.

The remaining two experiments (Tests 3 and 4) provide totally comparable results, thus confirming the repeatability of the four Pd–Ni experimental tests. As noted above, the most significant chemical changes are found at the end of the experimental tests.

In Tables 12.3 and 12.4, the Pd and Ni electrode compositional concentrations evaluated before and after the electrolytic experiments are reported for Tests 3 and 4. The data satisfy the nuclear and stoichiometric balances. The phono-fission reactions postulated above can explain the chemical concentration changes and the related balances.

It is remarkable that the depletion of Pd and Ni is closely balanced by the increments in lighter elements and fragments with an approximation of 2.2% for the Pd electrode and of 1.3% for the Ni electrode in Test 1. An approximations of 5.0% for the Pd electrode and of 3.3% for the Ni electrode are found in Test 2. Finally, values of 1.1% for the Pd electrode and 0.8% for the Ni electrode can be observed in Test 3, as well as of 3.9% for the Pd electrode and 1.1% for the Ni electrode in Test 4.

12.5 Conclusions

Several examples are reported in the literature of unexpected nuclear reactions occurring in condensed matter during electrolysis experiments, which are characterized by subatomic particle emissions, chemical composition changes, and excess heat generation [9–36]. However, despite the great amount of experimental results, a unified theory explaining all these phenomena has not yet been proposed and accepted [19, 20].

As shown in different papers devoted to the so-called cold nuclear fusion, one of the principal and recurring features is the appearance of microcracks on electrode surfaces after the tests [30, 31]. Such evidence might be directly correlated to hydrogen embrittlement that plays a crucial role in the observed microcracking of the electrode host metal. This is subjected to mechanical damage and microcracking due

Table 12.3 Test 3 on Pd and Ni electrodes. Element concentrations before the experiment, and after 2.5, 5, 10 h from the beginning (a); nuclear (b) and stoichiometric (c) balances

(a)

Pd electrode element concentrations

	Pd (%)	O (%)	Fe (%)	Ca (%)	K (%)	Si (%)	Mg (%)
Before the experiment	99.9	0.1	0.0	0.0	0.0	0.0	0.0
After 2.5 h	93.4	5.2	0.6	0.0	0.7	0.0	0.0
After 5 h	86.3	8.3	1.9	0.0	0.5	0.2	0.2
After 10 h	74.0	15.8	2.7	0.7	1.0	0.5	1.1

Ni electrode element concentrations

	Ni (%)	O (%)	Fe (%)	Si (%)	Al (%)
Before the experiment	94.6	1.9	1.7	0.1	0.0
After 2.5 h	87.6	11.3	0.9	0.3	0.7
After 5 h	83.4	12.8	0.2	0.6	1.1
After 10 h	82.0	11.9	0.0	0.5	0.9

(b)

Pd electrode nuclear balances

$Pd_{46}^{106} \rightarrow Ca_{20}^{40} + Fe_{26}^{56} + 10n_0^1$

$Fe_{26}^{56} \rightarrow 3O_8^{16} + He_2^4 + 4n_0^1$

$Ca_{20}^{40} \rightarrow O_8^{16} + Mg_{12}^{24}$

$Ca_{20}^{40} \rightarrow 2O_8^{16} + 2He_2^4$

$Ca_{20}^{40} \rightarrow K_{19}^{39} + H_1^1$

$Ca_{20}^{40} \rightarrow C_6^{12} + Si_{14}^{28}$

Ni electrode nuclear balances

$Ni_{28}^{58} \rightarrow 3O_8^{16} + 2He_2^4 + 2n_0^1$

$Ni_{28}^{58} \rightarrow 2Si_{14}^{28} + 2n_0^1$

$Fe_{26}^{56} \rightarrow 3O_8^{16} + He_2^4 + 4n_0^1$

$Fe_{26}^{56} \rightarrow 2Al_{13}^{27} + 2n_0^1$

(c)

Pd electrode stoichiometric balances

Pd(−25.9%) = Ca(+9.8%) + Fe(+13.7%) + n(+2.4%)

Fe(−11.0%) = O(+9.4%) + He(+0.8%) + n(+0.8%)

Ca(−1.8%) = O(+0.7%) + Mg(+1.1%)

Ca(−5.6%) = O(+4.5%) + He(+1.1%)

Ca(−1.0%) = K(+1.0%) + H(+0.0%)

Ca(−0.7%) = C(+0.2%) + Si(+0.5%)

Ni electrode stoichiometric balances

Ni(−12.2%) = O(+10.1%) + He(+1.7%) + n(+0.4%)

(continued)

Table 12.3 (continued)

(c)
Ni(−0.4%) = Si(+0.4%) + n(+0.0%)
Fe(−0.8%) = O(+0.7%) + He(+0.05%) + n(+0.05%)
Fe(−0.9%) = Al(+0.9%) + n(+0.0%)

to electrolytically produced hydrogen ions that are adsorbed into the metal atomic lattice. The hydrogen ions reduce the apparent metal fracture toughness, triggering brittle crack formation and/or propagation.

Based on the extensive experiments carried out by the authors, a macromechanical interpretation is proposed of the aforementioned phenomena induced by phono-fission reactions in the TeraHertz frequency range.

The original results reported in the present chapter confirm the repeatability of the data obtained during the early experiments, see Chaps. 10 and 11.

In particular, neutron emission peaks up to one order of magnitude higher than the natural background level are measured. In addition, the EDX analyses performed on the electrode surfaces provide evidence that the depletions of Pd and Ni are almost perfectly balanced by the increments in lighter elements. In particular, the relatively small differences between experimentally measured concentrations and expected theoretical values are mainly due to the Oxygen and Potassium excess, which can be interpreted as a consequence of K_2CO_3 aqueous solution deposition.

It is observed that the most significant nuclear transmutations have usually appeared in the later time periods of the experiments, and not continuously over their total duration, suggesting an incubation time period for the phenomenon to occur.

Evidence of diffused cracking emerges on the electrode external surface after the experiments, emphasizing the mechanical explanation based on hydrogen embrittlement, microcracking, and THz phonons emitted by the microcrack formation and/or propagation. These results suggest that, during the gas loading with hydrogen or deuterium, the host atomic lattice is subjected to mechanical damage and cracking due to hydrogen ion adsorption on the external surface and in the bulk of the electrode. Helium atoms or, better to say, alpha particles are in this case fragments of phono-fission reactions occurring at the electrodes rather than products of hypothetical fusion reactions between hydrogen or deuterium atoms.

The content of this chapter was preliminarily published in [62].

Table 12.4 Test 4 on Pd and Ni electrodes. Element concentrations before the experiment, and after 2.5, 5, 10 h from the beginning (a); nuclear (b) and stoichiometric (c) balances

(a)

Pd electrode element concentrations

	Pd (%)	O (%)	Fe (%)	Ca (%)	K (%)	Si (%)	Mg (%)
Before the experiment	100.0	0.0	0.0	0.0	0.0	0.0	0.0
After 2.5 h	93.1	4.2	1.1	0.3	0.8	0.0	0.0
After 5 h	85.7	10.2	1.2	0.7	1.0	0.3	0.6
After 10 h	77.4	14.5	0.9	1.5	2.0	1.6	1.5

Ni electrode element concentrations

	Ni (%)	O (%)	Fe (%)	Si (%)	Al (%)
Before the experiment	90.9	3.2	1.5	0.4	0.0
After 2.5 h	77.1	17.3	1.0	0.7	1.3
After 5 h	76.4	19.8	0.4	0.4	1.1
After 10 h	74.5	13.6	0.6	2.3	1.5

(b)

Pd electrode nuclear balances

$Pd_{46}^{106} \rightarrow Ca_{20}^{40} + Fe_{26}^{56} + 10n_{0}^{1}$

$Fe_{26}^{56} \rightarrow 3O_{8}^{16} + He_{2}^{4} + 4n_{0}^{1}$

$Ca_{20}^{40} \rightarrow O_{8}^{16} + Mg_{12}^{24}$

$Ca_{20}^{40} \rightarrow 2O_{8}^{16} + 2He_{2}^{4}$

$Ca_{20}^{40} \rightarrow K_{19}^{39} + H_{1}^{1}$

$Ca_{20}^{40} \rightarrow C_{6}^{12} + Si_{14}^{28}$

Ni electrode nuclear balances

$Ni_{28}^{58} \rightarrow 3O_{8}^{16} + 2He_{2}^{4} + 2n_{0}^{1}$

$Ni_{28}^{58} \rightarrow 2Si_{14}^{28} + 2n_{0}^{1}$

$Fe_{26}^{56} \rightarrow 2Al_{13}^{27} + 2n_{0}^{1}$

$Si_{14}^{28} \rightarrow Al_{13}^{27} + H_{1}^{1}$

(continued)

Table 12.4 (continued)

(c)
Pd electrode stoichiometric balances
Pd(−22.6%) = Ca(+8.5%) + Fe(+11.9%) + n(+2.1%)
Fe(−11.0%) = O(+9.4%) + He(+0.8%) + n(+0.8%)
Ca(−2.5%) = O(+1.0%) + Mg(+1.5%)
Ca(−0.2%) = O(+0.2%) + He(+0.0%)
Ca(−2.0%) = K(+2.0%) + H(+0.0%)
Ca(−2.3%) = C(+0.7%) + Si(+1.6%)
Ni electrode stoichiometric balances
Ni(−13.9%) = O(+11.5%) + He(+1.9%) + n(+0.5%)
Ni(−2.5%) = Si(+2.5%) + n(+0.0%)
Fe(−0.9%) = Al(+0.9%) + n(+0.0%)
Si(−0.6%) = Al(+0.6%) + H(+0.0%)

Acknowledgements Special thanks for their collaboration in the chemical analyses are due to Dr. Angelica Chiodoni and Dr. Salvatore Guastella (Politecnico di Torino). In addition, the Owner of the electrolytic cell, Mr. Alessandro Goi, is gratefully acknowledged.

References

1. Paneth F, Peters K (1926) The reported conversion of hydrogen into helium. Nature 118:526
2. Fleischmann M, Pons S, Hawkins M (1989) Electrochemically induced nuclear fusion of deuterium. J Electroanal Chem 261:301
3. Mizuno T (1998) Nuclear transmutation: the reality of cold fusion. Infinite Energy Press
4. Ohmori T, Mizuno T (1998) Strong excess energy evolution, new element production, and electromagnetic wave and/or neutron emission in light water electrolysis with a tungsten cathode. Infin Energy 20:14–17
5. Mizuno T, Akimoto T, Ohmori T (2003) Confirmation of anomalous hydrogen generation by plasma electrolysis. In: 4th meeting of Japan CF research society, Iwate, Japan
6. Mosier-Boss PA et al (2007) Use of CR-39 in Pd/D co-deposition experiments. Eur J Appl Phys 40:293–303
7. Mosier-Boss PA et al (2010) Comparison of Pd/D co-deposition and DT neutron generated triple tracks observed in CR-39 detectors. Eur J Appl Phys 51(2):20901–20911
8. Szpak S, Mosier-Boss PA, Young C, Gordon FE (2005) The effect of an external electric field on surface morphology of co-deposited Pd/D films. J Electroanal Chem 580(2):284–290
9. Bridgman PW (1927) The breakdown of atoms at high pressures. Phys Rev 29:188–191
10. Batzel RE, Seaborg GT (1951) Fission of medium weight elements. Phys Rev 82:607–615
11. Borghi DC, Giori DC, Dall'Olio A (1992) Experimental evidence on the emission of neutrons from cold hydrogen plasma. In: Proceedings of the international workshop on few-body problems in low-energy physics, Alma-Ata, Kazakhstan, pp 147–154. Unpublished Communication (1957). Comunicacao no. 25 do CENUFPE, Recife, Brazil (1971)
12. Diebner K (1962) Fusionsprozesse mit Hilfe konvergenter Stosswellen—einige aeltere und neuere Versuche und Ueberlegungen. Kerntechnik 3:89–93

13. Fulmer CB et al (1967) Evidence for photofission of iron. Phys Rev Lett 19:522–523
14. Goradovskii TY (1967) Hard radiation from solids failing in shear. JETP Lett 5:64–67
15. Kaliski S (1978) Bi-conical system of concentric explosive compression of D-T. J Tech Phys 19:283–289
16. Winterberg F (1984) Autocatalytic fusion–fission implosions. Atomenergie-Kerntechnik 44:146
17. Derjaguin BV et al (1989) Titanium fracture yields neutrons? Nature 34:492
18. Bockris JOM, Lin GH, Kainthla RC, Packham NJC, Velev O (1990) Does tritium form at electrodes by nuclear reactions? In: The first annual conference on cold fusion. National Cold Fusion Institute, University of Utah Research Park, Salt Lake City
19. Preparata G (1991) Some theories of cold fusion: a review. Fusion Technol 20:82
20. Preparata G (1991) A new look at solid-state fractures, particle emissions and "cold" nuclear fusion. Il Nuovo Cimento 104(A):1259–1263
21. Mills RL, Kneizys P (1991) Excess heat production by the electrolysis of an aqueous potassium carbonate electrolyte and the implications for cold fusion. Fusion Technol 20:65
22. Notoya R, Enyo M (1992) Excess heat production during electrolysis of H_2O on Ni, Au, Ag and Sn electrodes in alkaline media. In: Proceedings of the third international conference on cold fusion, Nagoya, Japan. Universal Academy Press, Tokyo
23. Miles MH, Hollins RA, Bush BF, Lagowski JJ, Miles RE (1993) Correlation of excess power and helium production during D_2O and H_2O electrolysis using palladium cathodes. J Electroanal Chem 346:99–117
24. Bush RT, Eagleton RD (1993) Calorimetric studies for several light water electrolytic cells with nickel fibrex cathodes and electrolytes with alkali salts of potassium, rubidium, and cesium. In: Fourth international conference on cold fusion, Lahaina, Maui. Electric Power Research Institute, Palo Alto, California
25. Fleischmann M, Pons S, Preparata G (1994) Possible theories of cold fusion. Nuovo Cimento, Soc Ital Fis A 107:143
26. Sundaresan R, Bockris JOM (1994) Anomalous reactions during arcing between carbon rods in water. Fusion Technol 26:261
27. Arata Y, Zhang Y (1995) Achievement of solid-state plasma fusion ("cold-fusion"). Proc Jpn Acad 71:304–309
28. Little SR, Puthoff HE, Little ME (1998) Search for excess heat from a Pt electrode discharge in K_2CO_3-H_2O and K_2CO_3-D_2O electrolytes. Infin Energy 5:34
29. Ransford HE (1999) Non-stellar nucleosynthesis: transition metal production by DC plasma-discharge electrolysis using carbon electrodes in a non-metallic cell. Infin Energy 4(23):16
30. Storms E (2000) Excess power production from platinum cathodes using the Pons-Fleischmann effect. In: 8th international conference on cold fusion, Lerici (La Spezia). Italian Physical Society, Bologna, Italy, pp 55–61
31. Storms E (2007) Science of low energy nuclear reaction: a comprehensive compilation of evidence and explanations about cold fusion. World Scientific Publishing, Singapore
32. Warner J, Dash J, Frantz S (2002) Electrolysis of D_2O with titanium cathodes: enhancement of excess heat and further evidence of possible transmutation. In: The Ninth international conference on cold fusion. Tsinghua University, Beijing, China, p 404
33. Fujii MF et al (2002) Neutron emission from fracture of piezoelectric materials in deuterium atmosphere. Jpn J Appl Phys 41:2115–2119
34. Lipson AG et al (2005) DD reaction enhancement and X-ray generation in a high-current pulsed glow discharge in deuterium with titanium cathode at 0.8–2.45 kV. J Exp Theor Phys 100(6)
35. Swartz M (2008) Three physical regions of anomalous activity in deuterated palladium. Infin Energy 14:19–31
36. Kanarev M, Mizuno T (2002) Cold fusion by plasma electrolysis of water. New Energy Technol 1:5–10
37. Gangloff RP (2003) Hydrogen-assisted cracking. In: Milne I, Ritchie RO, Karihaloo BL (eds) Comprehensive structural integrity, Pergamon, pp 31–101

38. Carpinteri A, Lacidogna G, Manuello A (eds) (2015) Acoustic, electromagnetic, neutron emissions from fracture and earthquakes. Springer
39. Cardone F, Mignani R (2004) Energy and geometry, Chapter 10. World Scientific Publishers, Singapore
40. Cardone F, Mignani R (2007) Deformed spacetime, Chapters 16 and 17. Springer, Dordrecht
41. Carpinteri A, Cardone F, Lacidogna G (2009) Piezonuclear neutrons from brittle fracture: early results of mechanical compression tests. Strain 45:332–339
42. Cardone F, Carpinteri A, Lacidogna G (2009) Piezonuclear neutrons from fracturing of inert solids. Phys Lett A 373:4158–4163
43. Carpinteri A, Cardone F, Lacidogna G (2010) Energy emissions from failure phenomena: mechanical, electromagnetic, nuclear. Exp Mech 50:1235–1243
44. Carpinteri A, Lacidogna G, Manuello A, Borla O (2012) Piezonuclear fission reactions: evidences from microchemical analysis, neutron emission, and geological transformation. Rock Mech Rock Eng 45:445–459
45. Carpinteri A, Lacidogna G, Manuello A, Borla O (2013) Piezonuclear fission reactions from earthquakes and brittle rocks failure: evidence of neutron emission and nonradioactive product elements. Exp Mech 53:345–365
46. Carpinteri A, Borla O, Manuello A, Veneziano D, Goi A (2015) Hydrogen embrittlement and piezonuclear reactions in electrolysis experiments. J Condens Matter Nucl Sci 15:162–182
47. Carpinteri A, Borla O, Lacidogna G, Manuello A (2010) Neutron emissions in brittle rocks during compression tests: monotonic vs. cyclic loading. Phys Mesomech 13:264–274
48. Carpinteri A, Lacidogna G, Manuello A, Borla O (2011) Energy emissions from brittle fracture: neutron measurements and geological evidences of piezonuclear reactions. Strength, Fract Complex 7:13–31
49. Carpinteri A, Chiodoni A, Manuello A, Sandrone R (2011) Compositional and microchemical evidence of piezonuclear fission reactions in rock specimens subjected to compression tests. Strain 47(2):267–281
50. Carpinteri A, Manuello A (2011) Geomechanical and geochemical evidence of piezonuclear fission reactions in the earth's crust. Strain 47(2):282–292
51. Carpinteri A, Lacidogna G, Borla O, Manuello A, Niccolini G (2012) Electromagnetic and neutron emissions from brittle rocks failure: experimental evidence and geological implications. Sadhana 37:59–78
52. Carpinteri A, Borla O (2018) Nano-scale fracture phenomena and TeraHertz pressure waves as the fundamental reasons for geochemical evolution. Strength, Fract Complex 11:149–168
53. Carpinteri A, Borla O (2017) Fracto-emissions as seismic precursors. Eng Fract Mech 177:239–250
54. Carpinteri A, Borla O (2019) Acoustic, electromagnetic, and neutron emissions as seismic precursors: the lunar periodicity of low-magnitude seismic swarms. Eng Fract Mech 210:29–41
55. Lucia U, Carpinteri A (2015) GeV plasmons and spalling neutrons from crushing of iron-rich natural rocks. Chem Phys Lett 640:112–114
56. Hagelstein PL, Letts D, Cravens D (2010) TeraHertz difference frequency response of PdD in two-laser experiments. J Condens Matter Nucl Sci 3:59–76
57. Hagelstein PL, Chaudhary IU (2015) Anomalies in fracture experiments and energy exchange between vibrations and nuclei. Meccanica 50:1189–1203
58. Widom A, Swain J, Srivastava YN (2013) Neutron production from the fracture of piezoelectric rocks. J Phys G: Nucl Part Phys 40(15006):1–8
59. Widom A, Swain J, Srivastava YN (2015) Photo-disintegration of the iron nucleus in fractured magnetite rocks with magnetostriction. Meccanica 50:1205–1216
60. Widom A, Srivastava YN, Swain J, de Montmollin G, Rosselli L (2017) Reaction products from electrode fracture and Coulomb explosions in batteries. Eng Fract Mech 184:88–100
61. Widom A, Srivastava YN, Swain J, de Montmollin G (2018) Tensile and explosive properties of current carrying wires. Eng Fract Mech 197:114–127
62. Carpinteri A, Borla O, Manuello A (2023) Hydrogen embrittlement, microcracking, and THz vibrations in the metal electrodes of cold-fusion electrolysis experiments: repeatability of nuclear and stoichiometric balances. J Condens Matter Nucl Sci 37:68–83

BY

Chapter 13
Electrolysis Experiments with Pd and Ni Electrodes: Evidence of Heat Generation During the Neutron Emission-Peak Time-Periods

Abstract In the last few decades, several scientific investigations have described the experimental evidence of non-conventional nuclear phenomena happening in condensed matter during electrolysis, fracture, or cavitation experiments. In spite of the several studies carried out in the area of the so-called "Cold Nuclear Fusion", the class of these phenomena remains not completely understood. As reported in different papers, the development of cracks on the external surface of the electrodes employed in electrolysis, the chemical composition changes, as well as the subatomic particle emissions during the tests, are evident (Carpinteri et al. in J Condens Matter Nucl Sci 37:68–83, 2023). A mechanical interpretation of the experimental evidence can be based on low-energy phono-fission nuclear reactions, which are a consequence of hydrogen embrittlement, microcracking, and emitted THz phonons (Carpinteri et al. in J Condens Matter Nucl Sci 37:68–83, 2023). In the present chapter, the results of different experiments on Pd and Ni electrodes are discussed under the light of calorimetric and heat generation measurements in correspondence to the neutron emission peaks. The evidence of a positive energy balance is repeatedly confirmed in correspondence to limited time intervals containing the neutron bursts.

Keywords Electrolysis · Hydrogen embrittlement · Microcracking · THz phonons · Phono-fission reactions · Neutron and alpha particle emissions · Energy balance · Calorimetric measurements · Heat generation · Cold nuclear fusion

13.1 Preliminary Remarks

Already at the beginning of the twentieth century, it was conjectured that nuclear fusion could occur also at temperatures well below tens of million degrees, through hydrogen adsorption by a catalyst [1]. In more recent studies, a totally different interpretation of the phenomenon called "cold nuclear fusion" has been proposed [2]. The Helium atoms produced are fragments of fission reactions (alpha particles) and not nuclear fusion products of Hydrogen or Deuterium [2]. Among several experiments conducted in this field, the most well-known indications were made by Stanley Pons

A. Carpinteri, *Terahertz Phonons and Nanomechanical Instabilities*,
https://doi.org/10.1007/978-3-032-14692-2_13

and Martin Fleischmann in 1989 [3]. Their idea was that the high compression ratio and mobility of deuterium (a stable isotope of hydrogen) during electrolysis tests could lead to unexpected nuclear fusion reactions. The power was applied continuously and heavy water was renewed at regular intervals [3]. Most of the time, the power absorbed by the cell was the same as that calculated, within the accuracy of the measurement, and the cell temperature remained stable around 30 °C. However, in some experiments, the temperature suddenly increased to around 50 °C with no change in the input power [3]. In support to the presumed presence of nuclear reactions at the origin of the temperature increment, Fleischmann and Pons reported the presence of neutrons and tritium [3].

In 1998, the Japanese researchers Mizuno and Ohmori [4, 5] claimed the possibility of achieving cold fusion reactions without palladium and heavy water, but using only tungsten electrodes immersed in a solution of common water and potassium carbonate (K_2CO_3). Putting on a voltage of 160–300 V, the temperature of the solution exceeded 70–80 °C and a plasma bubble formed around the immersed tungsten electrodes. A positive energy balance was estimated with thermal energy excess 20–100% higher than the electrical energy input used to trigger the reaction. In addition, several medium-weight elements such as calcium, titanium, chromium, manganese, iron, cobalt, copper, and zinc were observed on the cathode, which were not present before the operation of the electrolytic cell [6].

About ten years later, Mosier-Boss et al. [7, 8] showed important evidence of anomalous heat generation, alpha particle emission, and chemical composition changes during electrolysis experiments with a palladium cathode immersed in a solution of lithium chloride and deuterated water. In these last experiments and in those conducted by Mizuno et al. [6], elements such as Fe, Cr, Ni, and Al were obtained and detected at the end [9]. To explain the formation of these new elements, Mosier-Boss and coauthors suggested a multi-body deuteron fusion phenomenon and palladium lattice disintegration [9].

Preparata observed that: "*Despite the large number of experimental results observed by a large number of scientists, a unified interpretation and theory of these phenomena has not been accepted, and their understanding remains unresolved*" [10, 11]. Numerous examples of unexpected nuclear reactions in condensed matter were reported by several authors [9–36]. Most of these tests are accompanied by heat generation and emission of neutrons and/or alpha particles. In addition, some Authors mention significant chemical composition changes and microcracking of the electrodes [4, 5, 31–33].

As reported in different subsequent articles on cold fusion, one of the common features of the tests is the appearance of microcracks on electrode surfaces after the tests. It was assumed that the common environment in which low-energy nuclear reactions take place is provided by microcracks of a critical size accompanied by a resonance process that dissipates energy from those sites [31, 32]. This interpretation could be directly related to hydrogen embrittlement of the metal. This phenomenon, well-known in Metallurgy and Fracture Mechanics, characterizes metals during forming or finishing processes [37–39].

The first attempts by the research team at Politecnico di Torino to exploit the electrolysis to produce anomalous nuclear reactions started in 2012 [2]. In this first series of experimental tests, the electrolytic reactor was equipped by Ni–Fe and Co–Cr electrodes immersed in a water solution of Potassium Carbonate (K_2CO_3). During these tests, neutron emission peaks were revealed up to ten times higher than the natural background, as well as time intervals of anomalous alpha particle radiation activity [2]. On the other hand, the compositional X-ray spectroscopy analyses of the surfaces of the electrodes, which resulted to be extensively microcracked at the end of the tests due to hydrogen embrittlement, allowed to reveal significant compositional changes with respect to the conditions before the tests. The decrement in heavier elements was found to be almost perfectly balanced in mass by the increment in lighter elements [2], see Chap. 12.

Subsequently, in the context of the activities performed within the European project CleanHME, a second experimental campaign on the specific thermal effects of LENR is carried out. The electrolytic reactor is equipped, like in the case reported in the previous chapter, with a 100% Pd cathode and a 91% Ni anode (Figs. 13.1 and 13.2). Although the first series of experiments presented different trial times of 2.5, 5, and 10 h [2], the present experimental campaign is characterized by longer operating times (10 and 20 h) in order to have more accurate calorimetric measurements.

As in the previous case (see Chap. 12), a water solution of Potassium Carbonate is adopted. In addition to the alpha particle and neutron proportional counters, the electrolytic reactor is equipped with a FLIR A300 thermographic camera, a specific virtual multimeter, and a potentiometer in order to estimate the energy balance during the electrolytic reactor activity. The new tests allow to detect relevant neutron peaks, even several times higher than the natural background, and significant chemical composition changes on the microcracked surfaces of the electrodes, which always fulfill the stoichiometric mass balances and show the decrement in heavier elements (−30% for Pd) and the increment in lighter ones [2]. All the information related to the devices used for the neutron measurements are reported in [2] and in the previous chapter. Furthermore, the thermal monitoring highlights sensible variations in the temperature of the electrolytic cell during the tests. As a consequence, the energy balance assessment, which is performed during the intervals of intense neutron emission, allows to detect energy excesses. The energy balance equation considers an input term, i.e., the energy provided to the cell by the electric network, and two output contributions given by the vaporization energy of the water volume in the collector and the energy externally transmitted by heat convection. The energy balances, carried out on the different intervals of intense neutron emission, show values of output energy between 2 and 3 times higher than the electrical input energy. Therefore, these anomalous heat generations may be explained as a direct consequence of the low-energy nuclear reactions revealed during the electrolytic activity.

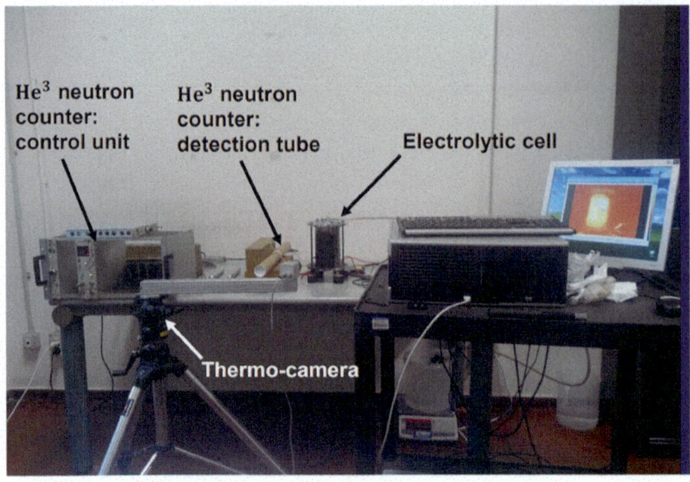

Fig. 13.1 Image of the experimental set-up together with the thermal measurement instrument (FLIR A300 thermographic camera)

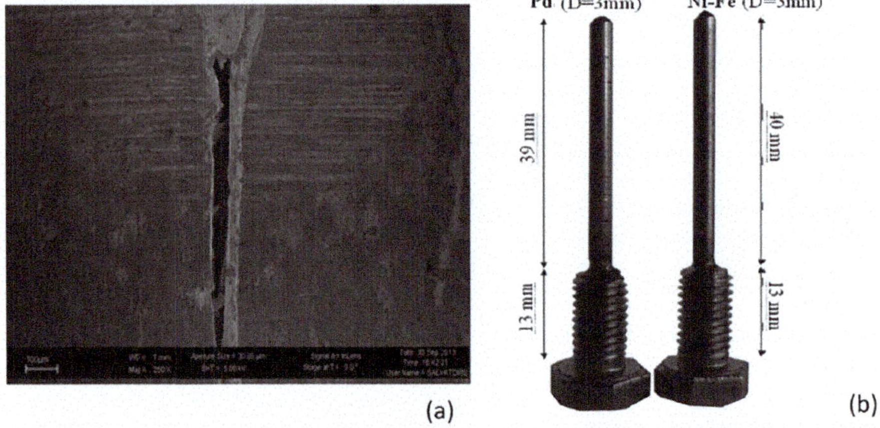

Fig. 13.2 The Pd electrode, after several operating hours, presents cracks of 40 μm width observable also at naked eyes (**a**). Electrode dimensions (**b**)

13.2 Calorimetric Measurements

In the calorimetric tests, the input energy E_{in} is related to the electric power P_{in} exchanged between the two electrodes and can be quantified by means of electric power consumption. The corresponding electric power $P_{in,a}$ can be calculated as the average power absorbed by the system and is given by direct electric measurements:

$$P_{\text{in,a}} = \frac{1}{t_{\text{f}} - t_{\text{i}}} \int_{t_{\text{i}}}^{t_{\text{f}}} P_{\text{in}}(t)\text{d}t \tag{13.1}$$

where t_{f} and t_{i} are the final and initial time instants, respectively, of the testing session. The instant input power $\text{P}_{\text{in}}(t)$ is:

$$P_{\text{in}}(t) = V_{\text{in}} I(t) \tag{13.2}$$

where V_{in} is the measured voltage between the electrodes and $I(t)$ is the electric current intensity measured by means of a virtual oscilloscope placed before the cell, neglecting the circuit dissipations. The main terms of energy transformation during testing are: (i) vaporization, and (ii) heat convection exchange. The energy balance equation, involving the main energy terms, has the following formulation in steady state conditions:

$$E_{\text{in}} + E_{\text{x}} = E_{\text{v}} + E_{\text{H}} \tag{13.3}$$

where E_{in} is the power of the electric circuit, at the connection between the circuit and the electrodes just before the cell; E_{x} represents the unknown energy term correlated to the LENR; E_{v} and E_{H} represent the terms due to vaporization and convection/irradiation, respectively. Electrolytic transformations and turbulent flow are considered quantitatively negligible at a first approximation.

For the first neutron peak of Test 1, the energy from the electric network is 889 kJ (E_{in}), the vaporization energy E_{v} is equal to 1,578 kJ, the measured energy transmitted by heat convection and irradiation is 584 kJ (E_{H}), and, finally, the energy from the "fuel" can be obtained as equal to 1,273 kJ ($E_{\text{v}} + E_{\text{H}} - E_{\text{in}}$) (Fig. 13.3). For the second neutron peak of Test 1, the energy from the electric network is measured in the amount of 763 kJ (E_{in}), the vaporization energy E_{v} is equal to 1,325 kJ, the measured energy transmitted by heat convection and irradiation is 491 kJ (E_{H}), and finally the energy from the "fuel" can be obtained as in the previous case and is equal to 1,053 kJ ($E_{\text{v}} + E_{\text{H}} - \text{E}_{\text{in}}$). The ratio of E_{out} to E_{in} for the first two peaks, considering that $\text{E}_{\text{out}} = (E_{\text{v}} + E_{\text{H}})$, is equal to 2.43 and 2.38, respectively (Table 13.1).

For the third neutron peak of Test 1, the energy from the electric network is 835 kJ (E_{in}), the vaporization energy E_{v} is equal to 1,333 kJ, the measured energy transmitted by heat convection and irradiation is 494 kJ (E_{H}), and finally the energy from the "fuel" can be obtained as equal to 992 kJ ($E_{\text{v}} + E_{\text{H}} - \text{E}_{\text{in}}$). For the fourth neutron peak of Test 1, the energy from the electric network is 749 kJ (E_{in}), the vaporization energy E_{v} is equal to 1,521 kJ, the measured energy transmitted by heat convection and irradiation is 563 kJ (E_{H}), and finally the energy from the "fuel" can be obtained as 1,335 kJ ($E_{\text{v}} + E_{H} - E_{\text{in}}$). The ratio $E_{\text{out}}/E_{\text{in}}$ amounts, for the last two peaks of the first experiment, considering $E_{\text{out}} = (E_{\text{v}} + E_{\text{H}})$, to 2.18 and 2.78, respectively (Table 13.1).

In a very similar way, for the first neutron peak of Test 2 the energy from the electric network is 587 kJ (E_{in}), the vaporization energy E_{v} is equal to 1,006 kJ, the measured

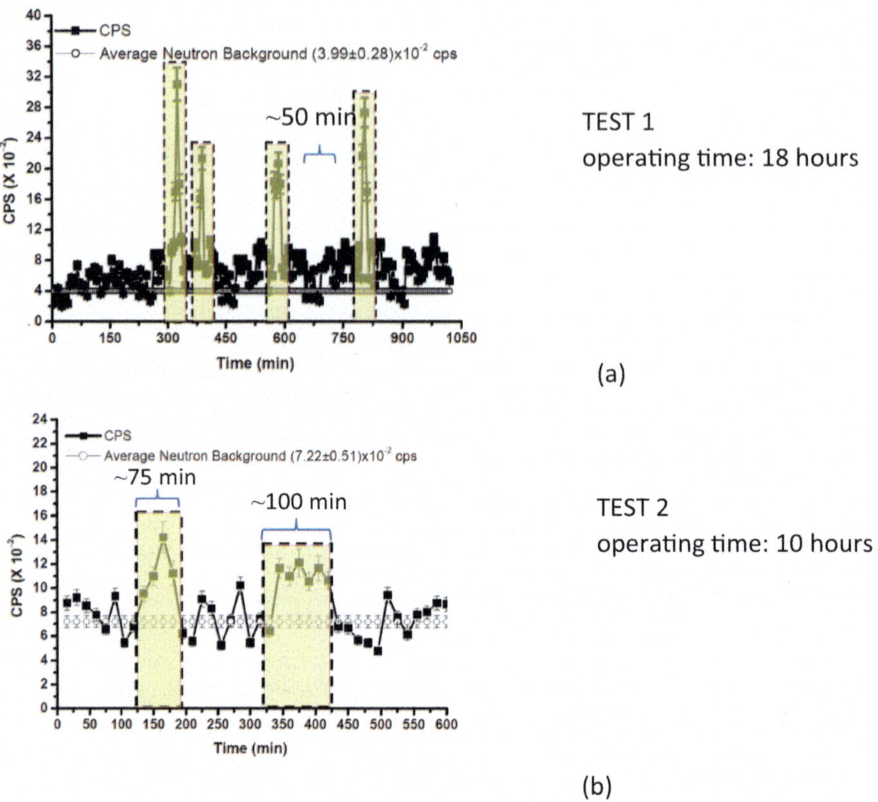

Fig. 13.3 Tests 1 and 2. Neutron emissions during the electrolysis experiments. Time intervals, when neutron bursts are detected and energy balances considered, range from 30 to 100 min

Table 13.1 E_{out}/E_{in} ratios during the time intervals around the main neutron peaks (Tests 1–4)

Test 1	E_{out}/E_{in}	**Test 3**	E_{out}/E_{in}
Neutron peak 1	2.43	No measured neutron peaks	–
Neutron peak 2	2.38		
Neutron peak 3	2.18	**Test 4**	E_{out}/E_{in}
Neutron peak 4	2.78	Neutron peak 1	2.17
Mean value (test 1)	**2.44**	Neutron peak 2	1.62
		Mean value (test 4)	**1.89**
Test 2	E_{out}/E_{in}		
Neutron peak 1	2.30		
Neutron peak 2	2.97		
Mean value (test 2)	**2.63**		

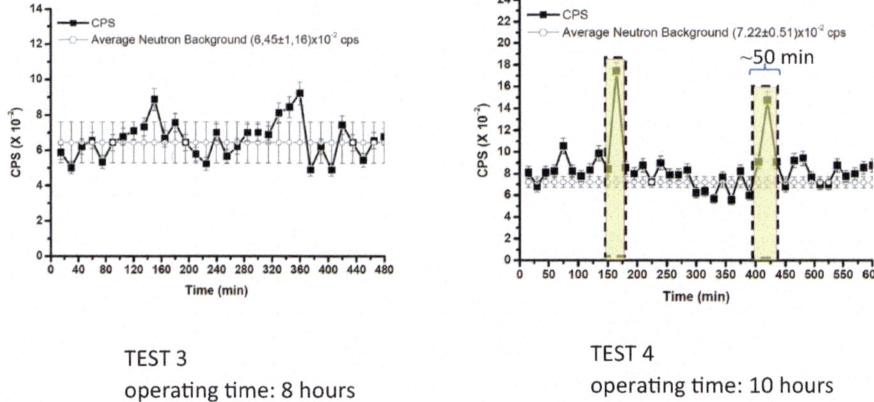

Fig. 13.4 Tests 3 and 4. Neutron emissions during the electrolysis experiments. The time intervals, when neutron bursts are detected and energy balances considered, range from 30 to 100 min

energy transmitted by heat convection and irradiation is 347 kJ (E_H), and finally the energy from the "fuel" can be obtained as 766 kJ ($E_v + E_H - E_{in}$) (Fig. 13.3). For the second neutron peak of Test 2, the energy from the electric network is 605 kJ (E_{in}), the vaporization energy E_v is equal to 1,227 kJ, the measured energy transmitted by heat convection and irradiation is 571 kJ (E_H), and finally the energy from the "fuel" can be obtained as 1,193 kJ ($E_v + E_H - E_{in}$). The ratio E_{out}/E_{in} for the two peaks of the second test, considering $E_{out} = (E_v + E_H)$, is equal to 2.30 and 2.97, respectively (Table 13.1).

As far as Test 3 is concerned, no neutron peaks were detected (Fig. 13.4).

In the case of Test 4, for the first neutron peak the energy from the electric network is 411 kJ (E_{in}), the vaporization energy E_v is equal to 595 kJ, the measured energy transmitted by heat convection and irradiation is 298 kJ (E_H), and finally the energy from the "fuel" can be computed as 482 kJ ($E_v + E_H - E_{in}$). For the second neutron peak of test 4, the energy from the electric network is 490 kJ (E_{in}), the vaporization energy E_v is equal to 555 kJ, the measured energy transmitted by heat convection and irradiation is 239 kJ (E_H), and finally the energy from the "fuel" can be computed as 304 kJ ($E_v + E_H - E_{in}$). The ratio E_{out}/E_{in}, for the last two neutron peaks, considering $E_{out} = (E_v + E_H)$, is equal to 2.17 and 1.62, respectively (Table 13.1).

Figure 13.5 shows the neutrons cumulated for Tests 1, 2, and 4. In a comparison to the cumulated natural background, the increments are 60.4%, 20.7%, and 15.1%, respectively.

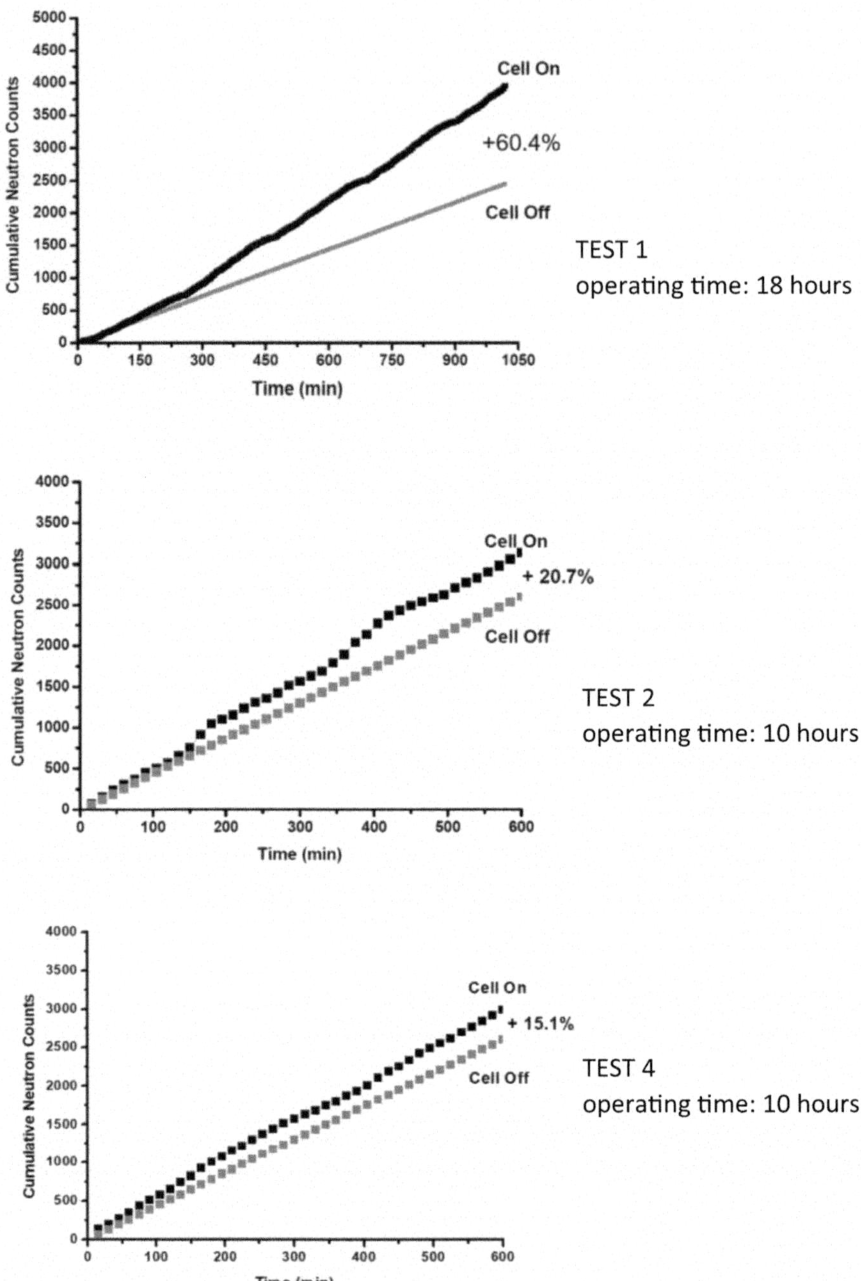

Fig. 13.5 Cumulative neutron emissions for tests 1, 2, and 4

13.3 Conclusions

Important calorimetric results are found in correlation to neutron bursts well above the natural background level, as well as to chemical composition changes detectable on the external surface of the electrodes subjected to several operating hours [2]. The present chapter, focusing on thermal measurements performed during experimental testing and concomitant with neutron bursts, shows the occurrence of positive energy balances (coefficients of performance, COP, defined as the ratio of the output to the input energy, above the unity). This evidence, together with those mentioned above, aims at further supporting the presence of LERNs in electrolysis experiments, which may occur or not depending on different parameters during the experiment: (i) the catalyst temperature, (ii) the number of hours of cell operation, (iii) the chemical composition of the electrodes at the beginning of the test, (iv) the power supply, and (v) the geometric dimensions and proportions of the electrodes.

Acknowledgements Special thanks for her collaboration in the realization of the thermal analyses are due to Prof. Raffaella Sesana. In addition, the Owner of the electrolytic cell, Mr. Alessandro Goi, is gratefully acknowledged.

References

1. Paneth F, Peters K (1926) The reported conversion of hydrogen into helium. Nature 118:526
2. Carpinteri A, Borla O, Manuello A (2023) Hydrogen embrittlement, microcracking, and THz vibrations in the metal electrodes of cold-fusion electrolysis experiments: repeatability of nuclear and stoichiometric balances. J Condens Matter Nucl Sci 37:68–83
3. Fleischmann M, Pons S, Hawkins M (1989) Electrochemically induced nuclear fusion of deuterium. J Electroanal Chem 261:301
4. Mizuno T (1998) Nuclear transmutation: the reality of cold fusion. Infinite Energy Press
5. Ohmori T, Mizuno T (1998) Strong excess energy evolution, new element production, and electromagnetic wave and/or neutron emission in light water electrolysis with a tungsten cathode. Infin Energy 20:14–17
6. Mizuno T, Akimoto T, Ohmori T (2003) Confirmation of anomalous hydrogen generation by plasma electrolysis. In: 4th meeting of Japan CF research society, Iwate, Japan
7. Mosier-Boss PA et al (2007) Use of CR-39 in Pd/D co-deposition experiments. Eur J Appl Phys 40:293–303
8. Mosier-Boss PA et al (2010) Comparison of Pd/D co-deposition and DT neutron generated triple tracks observed in CR-39 detectors. Eur J Appl Phys 51(2):20901–20911
9. Szpak S, Mosier-Boss PA, Young C, Gordon FE (2005) The effect of an external electric field on surface morphology of co-deposited Pd/D films. J Electroanal Chem 580(2):284–290
10. Preparata G (1991) Some theories of cold fusion: a review. Fusion Tech 20:82
11. Preparata G (1991) A new look at solid-state fractures, particle emissions and "cold" nuclear fusion. Il Nuovo Cimento 104(A):1259–1263
12. Bridgman PW (1927) The breakdown of atoms at high pressures. Phys Rev 29:188–191
13. Batzel RE, Seaborg GT (1951) Fission of medium weight elements. Phys Rev 82:607–615
14. Borghi DC, Giori DC, Dall'Olio A (1992) Experimental evidence on the emission of neutrons from cold hydrogen plasma. In: Proceedings of the international workshop on few-body problems in low-energy physics, Alma-Ata, Kazakhstan: 147–154. Unpublished Communication (1957). Comunicacao no. 25 do CENUFPE, Recife, Brazil (1971)

15. Diebner K (1962) Fusionsprozesse mit Hilfe konvergenter Stosswellen—einige aeltere und neuere Versuche und Ueberlegungen. Kerntechnik 3:89–93
16. Fulmer CB et al (1967) Evidence for photofission of iron. Phys Rev Lett 19:522–523
17. Goradovskii TY (1967) Hard radiation from solids failing in shear. JETP Lett 5:64–67
18. Kaliski S (1978) Bi-conical system of concentric explosive compression of D-T. J Tech Phys 19:283–289
19. Winterberg F (1984) Autocatalytic fusion–fission implosions. Atomenergie-Kerntechnik 44:146
20. Derjaguin BV et al (1989) Titanium fracture yields neutrons? Nature 34:492
21. Bockris JOM, Lin GH, Kainthla RC, Packham NJC, Velev O (1990) Does tritium form at electrodes by nuclear reactions? In: The first annual conference on cold fusion. National Cold Fusion Institute, University of Utah Research Park, Salt Lake City
22. Mills RL, Kneizys P (1991) Excess heat production by the electrolysis of an aqueous potassium carbonate electrolyte and the implications for cold fusion. Fusion Technol 20:65
23. Notoya R, Enyo M (1992) Excess heat production during electrolysis of H_2O on Ni, Au, Ag and Sn electrodes in alkaline media. In: Proceedings of third international conference on cold fusion, Nagoya, Japan. Universal Academy Press, Tokyo
24. Miles MH, Hollins RA, Bush BE, Lagowski JJ, Miles RE (1993) Correlation of excess power and helium production during D_2O and H_2O electrolysis using palladium cathodes. J Electroanal Chem 346:99–117
25. Bush RT, Eagleton RD (1993) Calorimetric studies for several light water electrolytic cells with nickel fibrex cathodes and electrolytes with alkali salts of potassium, rubidium, and cesium. In: Fourth international conference on cold fusion, Lahaina, Maui. Electric Power Research Institute, Palo Alto, California
26. Fleischmann M, Pons S, Preparata G (1994) Possible theories of cold fusion. Il Nuovo Cimento, Soc Ital Fis A 107:143
27. Sundaresan R, Bockris JOM (1994) Anomalous reactions during arcing between carbon rods in water. Fusion Technol 26:261
28. Arata Y, Zhang Y (1995) Achievement of solid-state plasma fusion ("cold-fusion"). Proc Jpn Acad 71:304–309
29. Little SR, Puthoff HE, Little ME (1998) Search for excess heat from a Pt electrode discharge in K_2CO_3-H_2O and K_2CO_3-D_2O electrolytes. Infin Energy 5:34
30. Ransford HE (1999) Non-stellar nucleosynthesis: transition metal production by DC plasma-discharge electrolysis using carbon electrodes in a non-metallic cell. Infin Energy 4(23):16
31. Storms E (2000) Excess power production from platinum cathodes using the Pons-Fleischmann effect. In: 8th international conference on cold fusion, Lerici (La Spezia). Italian Physical Society, Bologna, Italy, pp 55–61
32. Storms E (2007) Science of low energy nuclear reaction: a comprehensive compilation of evidence and explanations about cold fusion. World Scientific Publishers, Singapore
33. Warner J, Dash J, Frantz S (2002) Electrolysis of D_2O with titanium cathodes: enhancement of excess heat and further evidence of possible transmutation. In: The ninth international conference on cold fusion. Tsinghua University, Beijing, China, p 404
34. Fujii MF et al (2002) Neutron emission from fracture of piezoelectric materials in deuterium atmosphere. Jpn J Appl Phys 41:2115–2119
35. Lipson AG et al (2005) DD reaction enhancement and X-ray generation in a high-current pulsed glow discharge in deuterium with titanium cathode at 0.8–2.45 kV. J Exp Theor Phys 100(6)
36. Swartz M (2008) Three physical regions of anomalous activity in deuterated palladium. Infin Energy 14:19–31
37. Kanarev M, Mizuno T (2002) Cold fusion by plasma electrolysis of water. New Energy Technol 1:5–10
38. Gangloff RP (2003) Hydrogen-assisted cracking. In: Milne I, Ritchie RO, Karihaloo BL (eds) Comprehensive structural integrity. Pergamon, pp 31–101

39. Carpinteri A, Borla O, Goi A, Manuello A, Veneziano D (2015) Cold nuclear fusion explained by hydrogen embrittlement and piezonuclear fissions in metallic electrodes: part I: Ni-Fe and Co-Cr electrodes. In: Acoustic, electromagnetic, neutron emissions from fracture and earthquakes. Springer

Chapter 14
Hydrodynamic Cavitation Experiments: Neutron Emissions, Chemical Composition Changes, and Correlated Energy Aspects

Abstract Analogously to the formation of nano-cracks in solid matter, the implosion of nano-bubbles in liquid media (water solutions of iron salts) due to cavitation phenomena produces TeraHertz phonons that cause, also in this case, anomalous nuclear reactions. In the present chapter, some calorimetric tests performed at Politecnico di Torino by a hydrodynamic cavitation pilot-plant are described. The diameter of the bubbles produced by hydrodynamic cavitation and the temperature increments are monitored by means of the Phase Doppler Anemometry technique and by thermocouples, respectively. These experiments suggest that the energy emitted per each imploded nano-bubble is approximately constant despite the different bubble geometries, populations, and temperatures. The effect of the different initial concentrations of iron salts on the temperature versus time evolution is thoroughly evaluated. The performed tests show an increment in the final temperature by increasing the initial concentration of iron salts until the optimal concentration of 20 ppm is reached. Beyond that limit, the water solution saturation occurs and a decrement in the final temperature is detected.

Keywords Hydrodynamic cavitation · Nano-bubble implosion · TeraHertz phonons · Phono-fission nuclear reactions · Neutron emissions · Iron depletion · Energy aspects

14.1 Preliminary Remarks

In the last two decades, a clear evidence has been achieved by the research group at Politecnico di Torino about chemical composition changes and subatomic particle emissions, both revealed during fracture tests on rock stones [1–9] and metal specimens [10], as well as during electrolysis experiments [11, 12] on cells filled with water solutions. As the microcracking observed in rocks and embrittled electrodes [11], so the implosion of micro- and nano-bubbles produced by hydrodynamic cavitation [13] in a solution made by ultra-pure water and iron salts is supposed to be the

A. Carpinteri, *Terahertz Phonons and Nanomechanical Instabilities*,
https://doi.org/10.1007/978-3-032-14692-2_14

cause of the chemical composition changes and of the anomalous neutron emissions revealed during the cavitation tests [12, 14].

The chemical composition changes in the metal salts dissolved in the water solution and the anomalous neutron emissions, which suggest the nuclear nature of the phenomenon, are proposed to be a consequence of the nano-scale mechanical instabilities produced in the liquid medium by the implosion of nano-bubbles [13]. Indeed, the development of the nano-scale instabilities produces TeraHertz pressure waves (phonons) that propagate through the solution and provoke resonance with the iron salt molecules, consequently triggering phono-fission reactions. In order to demonstrate the supposed effects of cavitation on the aqueous solution of iron chloride ($FeCl_3$), in terms of neutron emissions, compositional changes, and possible heat generation, a pilot-plant is adopted.

14.2 Pilot-Plant for the Study of Heat Generation from Hydrodynamic Cavitation

The pilot-plant used for the investigation on possible heat generation from phono-fission reactions produced by hydrodynamic cavitation was realized and installed by the company MetalWork at the Laboratory of Fracture Mechanics of Politecnico di Torino in 2016. This pilot-plant, in the following also defined as "cavitation reactor," is composed of a three-phase volumetric pump, a main duct, and a storage tank aimed at the collection of the water solution (Fig. 14.1). The three main components of the reactor are connected by PVC pipes, which are characterized by an internal diameter of 11.6 mm and a total length of about 9.70 m. The three-phase volumetric pump is able to reach a maximum nominal pressure of 20 atmospheres and is assembled with polymeric components directly in contact with the water solution in order to exclude any metal contamination to the solution itself (Model AR 45 bp EM, CC31222 three-phase). The pump is provided by a safety valve (bypass), which is aimed at avoiding possible failures of parts of the plant due to excessive pressure levels. The bypass automatically activates by exercise pressures exceeding the value of 15 atmospheres.

The main duct holds the brass nozzle called "cavitator" and the greater part of the measurement devices. It is composed of a series of modular elements made of a polymeric material (PMMA), which allows the transmission of the neutrons produced by cavitation (Fig. 14.2). The modular elements are connected to each other by means of PTFE ring joints and also by four steel passing screws, which ensure the maintenance of a constant total length of the duct and the adequate contrast to the high testing pressures. The internal diameter of the main duct is uniformly equal to 11.6 mm, whereas the external diameter is varying. The central segment of the main duct, defined as "cavitation room," contains the cavitator, a brass nozzle with an internal geometry conveniently designed in order to guarantee a relevant pressure drop in correspondence to the throat section and, in particular, the formation of micro- and nano-bubbles in the water solution.

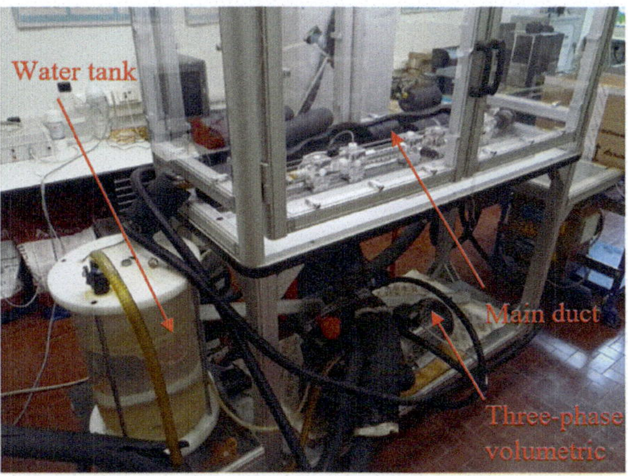

Fig. 14.1 Pilot-plant of hydrodynamic cavitation with its principal components

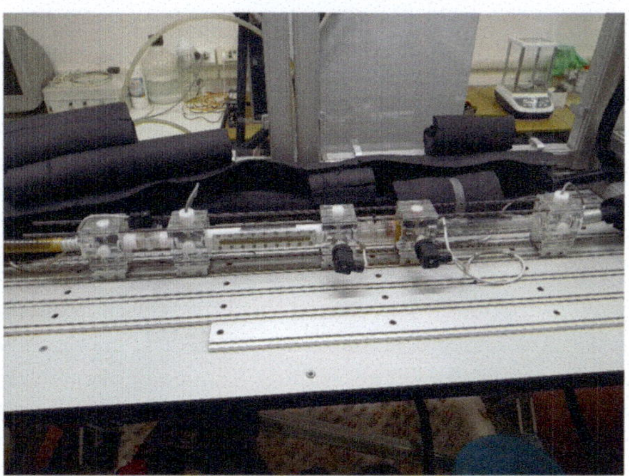

Fig. 14.2 Main duct of the cavitation reactor

The storage tank, which is made of the same polymeric material of the main duct (PMMA), is able to contain approximately 25 L of water solution, and it has an internal diameter of 290 mm and a total height of 500 mm.

As regards the monitoring systems, the cavitation reactor is equipped with several different sensors. More in detail, three temperature sensors are installed in different positions along the cavitation reactor. Two thermocouples are placed 30 cm before and after the cavitator, whereas the third thermocouple is located inside the tank. During the design of the cavitation reactor, it was decided to adopt a sampling frequency for the temperature measurement equal to six acquisition data per minute.

The pressure of the fluid is measured by two pressure gauges that are placed just before and after the cavitator in order to evaluate the pressure drop caused by the nozzle (Fig. 14.3). Furthermore, the cavitation plant is equipped with a flow monitoring device that measures the fluid velocity in a cross-section of the main duct

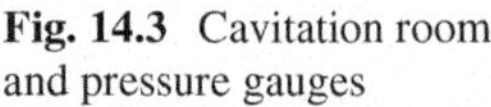
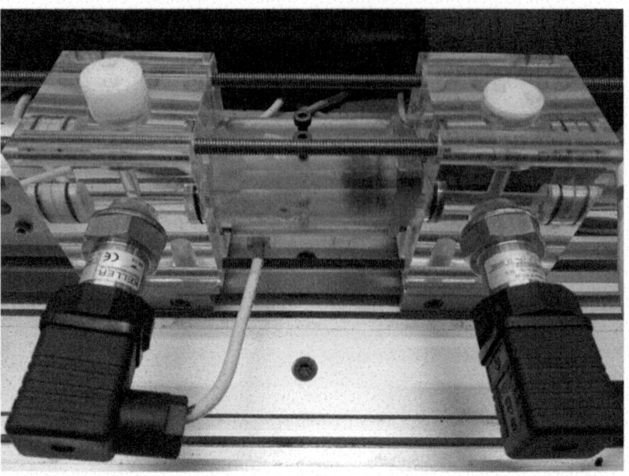

Fig. 14.3 Cavitation room and pressure gauges

located after the cavitator. The flow is then automatically evaluated by the product of the measured speed times the known area of the monitoring section.

The hydraulic scheme of the cavitation reactor is reported in Fig. 14.4a, whereas the related legenda of the symbols appears in Fig. 14.4b, according to the present UNI Italian Standards.

All the above-described measurement devices are connected to a control unit, which allows to save the acquired experimental data with a predetermined time interval. It is worth to highlight that the same control unit is able to acquire the electric power supplied by each one of the three phases of the pump (by means of a SIEMENS PAC3200 device), so that the total electric energy absorbed by the cavitation plant can be recorded with a sampling frequency of one value per minute. Eventually, the cavitation plant is equipped with a neutron emission detector, a ^{3}He neutron proportional counter produced by the industrial company Canberra (France). It is composed by a ^{3}He detector tube powered with 1.25 kV by means of a NIM (Nuclear Instrumentation Module) provided by a NIM counter [5] for the TTL (Transistor–Transistor Logic) pulses, and systems of preamplification, amplification, and discrimination of the acquired signals. The device is directly calibrated by the company for the detection of thermal neutrons with the aim at excluding false signals due to gamma-ray acquisitions. Its sensitivity is equal to a flux of one $\frac{n_{\text{thermal}}}{\text{s cm}^2}$, with a declared variability of $\pm 10\%$ [4].

The geometrical characteristics of the pilot-plant are reported in Table 14.1.

14.3 Energy Emitted per Each Imploded Nano-bubble

The implosion of nano-bubbles in a water solution is supposed to be able to cause anomalous nuclear reactions in the dissolved iron salts by means of the production of TeraHertz phonons in the liquid medium. In order to validate that assumption,

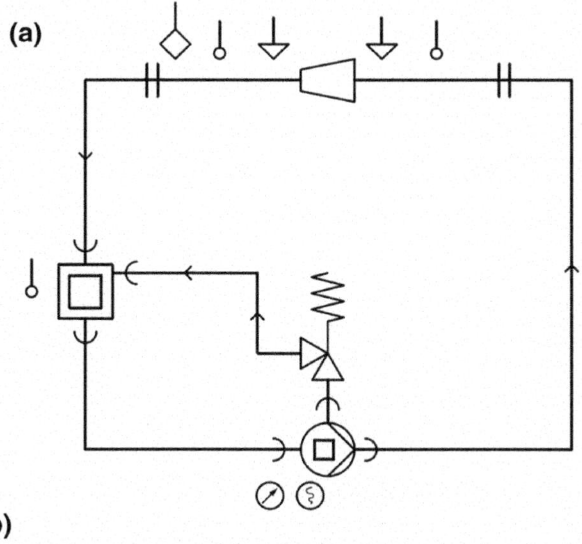

Nomenclature according to UNI 9511 regulation (effectively adopted nomenclature)	Simbolo
Spigot joint	
Flange joint	
Flux way	
Concentric reduction (Cavitator)	
Storage tank (Tank)	
Safety valve (Bypass)	
Volumetric pump	
Temperature probe (Thermocouple)	
Pressure detector	
Flow rate detector	
Direct-lecture indicator device	
Recording device	

Fig. 14.4 **a** Hydraulic scheme of the hydrodynamic pilot-plant. **b** Legenda of the symbols adopted in the hydrodynamic pilot-plant scheme (see above)

Table 14.1 Geometrical characteristics of the pilot-plant

Main duct	
L (m)	1.10
$Ø_{int}$ (m)	0.0116
$Ø_{ext}$ (m)	0.0209
t (m)	0.00465
PVC tubes and pump	
L (m)	9.70
$Ø_{int}$ (m)	0.0116
$Ø_{ext}$ (m)	0.0265
t (m)	0.00745
Tank	
L (m)	0.40
$Ø_{int}$ (m)	0.290
$Ø_{ext}$ (m)	0.300
t (m)	0.005
Insulating material (Polyflex)	
t (m)	0.019

a preliminary phase of the experimental investigation on the cavitation pilot-plant is focused on the evaluation of the energy emitted per each imploded nano-bubble. With the scope of measuring the diameter of the bubbles produced by each one of the three brass cavitators to be tested (indicated as M2, M4, and M5), the Phase Doppler Anemometry (PDA) technique is adopted, which allows to determine also size-distribution and velocity of spherical particles or bubbles in the solution on the basis of laser interferometry. The measurement set-up, which can be seen in Fig. 14.5, is composed of two laser emitters, a lens that forces the laser beams to converge to a monitoring section of 0.125 mm^2, where they are refracted by the spherical inclusions, and two photo-multipliers, which are detectors receiving the refracted beams.

Due to Doppler effect, the difference in frequency between emitted and refracted signals allows to evaluate the velocity of the bubbles, whereas the phase difference allows to estimate the bubble diameter. The device is set for the acquisition of a fixed number of 20,000 bubbles during a maximum time interval of 100 s for each measurement, and, as a consequence of that, a statistics on the basis of 20,000 bubbles is adopted. It is relevant to note that PDA is characterized by a range of measurable bubble diameters between nanometers and one millimeter, and therefore a statistical extrapolation by means of a Gaussian size-distribution is applied for the assessment of the nano-bubble population.

The cavitation tests are carried out for the different brass cavitators at their optimal pressure levels, which are found to maximize the heat generation. The total emitted energy is evaluated in terms of the temperature increment in the water solution, which

Fig. 14.5 Phase Doppler Anemometry (PDA) set-up for the measurements of bubble diameter and velocity

is monitored by means of the thermocouples. Considering the whole thermal energy emitted as just a consequence of the nano-bubble implosions, the energy emitted per each imploded nano-bubble is evaluated by the simple ratio of the total thermal energy to the total number of nano-bubbles. The results of these evaluations are reported in Table 14.2.

As is possible to observe (Fig. 14.6), the values of the emitted energy per each imploded nano-bubble are found to be almost constant, being they comprised within a very narrow error band of approximately ±5%. This result is supporting the phono-fission conjecture as a consequence of cavitation. As a matter of fact, although the main parameters of the tests change (i.e., the geometry of the cavitators, the imposed pressures, the number of bubbles, the total emitted energy, as well as the final temperatures), the constancy of the emitted energy per each imploded nano-bubble confirms the relevant effect exerted by nanometric instabilities and by the emitted TeraHertz phonons, which are promoting anomalous nuclear reactions in the iron-rich water solution.

Table 14.2 Evaluation of the energy emitted per each imploded nano-bubble

	M2	M4	M5
Bubbles produced in 1 h	183,200,000	219,840,000	412,200,000
Nano-bubbles produced in 1 h	3,874,200	7,254,720	4,946,400
Thermal energy produced at the optimal pressure in 1 h (kJ)	139	237	171
Energy emitted per imploded nano-bubble (mJ)	36.13	32.67	34.57

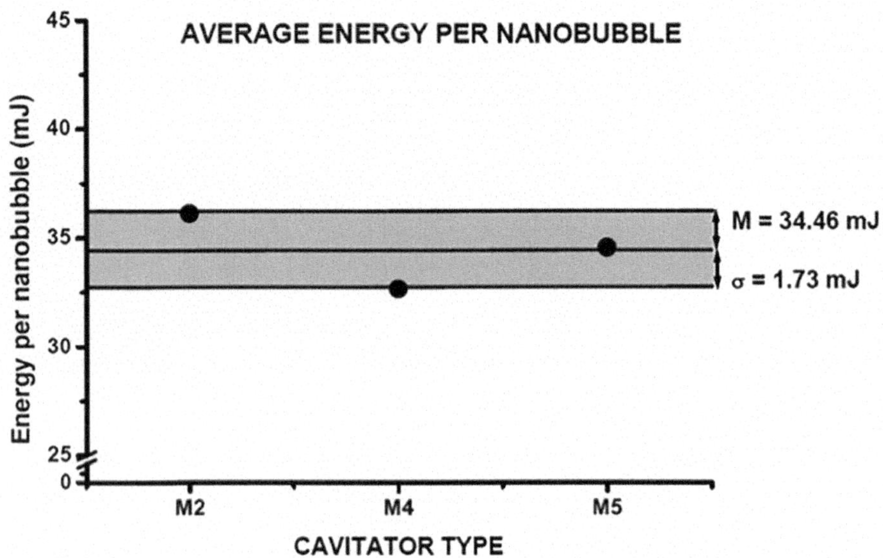

Fig. 14.6 Energy emitted per each imploded nano-bubble in the three different brass cavitators (M2, M4, M5)

14.4 Neutron Emission Monitoring

In the early stages of the experimental activity on hydrodynamic cavitation, a great importance is bestowed to the monitoring of neutron emissions produced by the assumed nuclear reactions. The continuous acquisition of neutron emission signals during the cavitation tests is performed by means of the He^3 proportional counter. Figure 14.7 represents the results obtained in terms of neutron count rate for a test lasted six hours.

In conjunction to the cavitation activity in the water solution with iron salts, neutron emission peaks up to 100% higher than the natural background are observed. Such a trend is confirmed by the comparison between the two cumulative curves corresponding to cavitation and to natural background (Fig. 14.8). The diagram shows a final positive deviation from the cumulative natural background of 34.4%.

14.5 Chemical Composition Changes After Cavitation

In addition to calorimetric and neutron monitoring, it was established to perform also a chemical composition analysis on metal ions dissolved in the water solution before, during, and after the cavitation test. A sampling rate of one analysis every sixty minutes is adopted, and the results of the analyses, in terms of iron and aluminum contents, are shown in Fig. 14.9. A sensible decrement in the concentration of iron

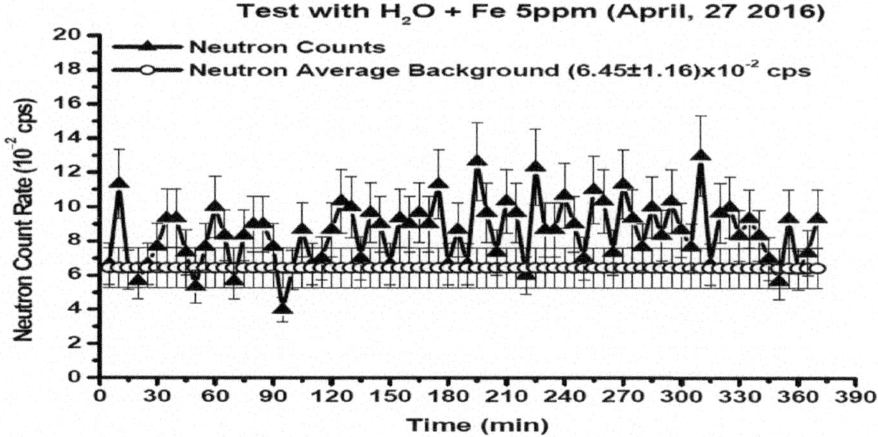

Fig. 14.7 Neutron emission rate during a cavitation test

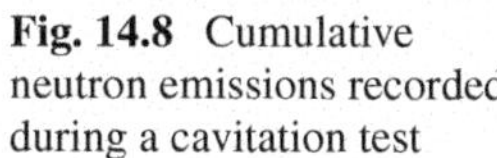

Fig. 14.8 Cumulative neutron emissions recorded during a cavitation test

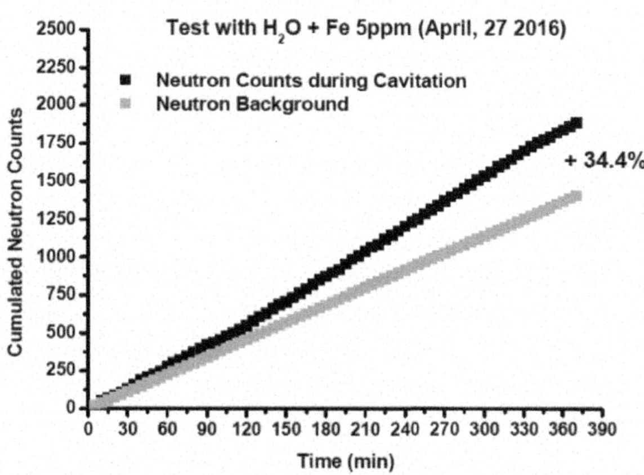

ions in the water solution is observed, from the initial 5 ppm to the final 4 ppm. At the same time, an even more important relative increment is observed in the concentration of aluminum ions, which passes from the starting value of 5 to 10 ppb at the end of the test. The reasons for the lack of linearity in the obtained trends and for the unbalanced chemical changes are still unclear.

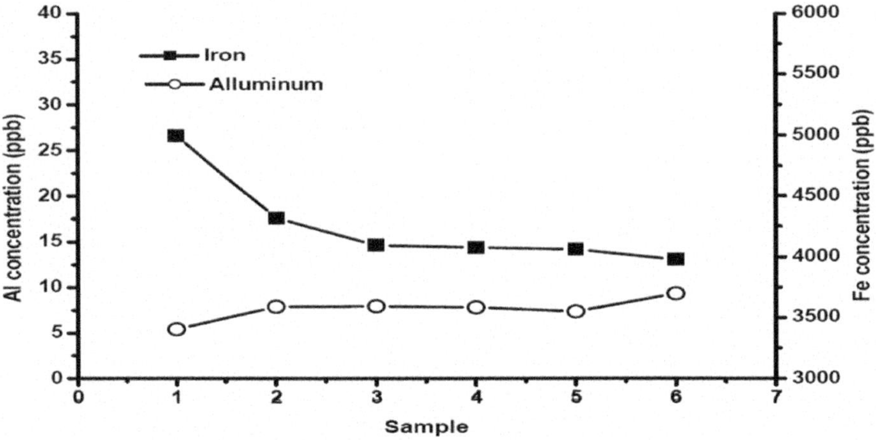

Fig. 14.9 Chemical composition changes for iron and aluminum in samples collected during a cavitation test

14.6 Dependence of Heat Generation on the Initial Concentration of Iron Chloride in the Water Solution

A series of tests is performed by the pilot-plant, in two consecutive stages, with different initial concentrations of Iron Chloride dissolved in an ultra-pure water solution. At the first stage, tests with the initial concentrations of 0, 5, 10, 15, and 20 ppm are carried out, followed by a washing with ultra-pure water (zero ppm at the beginning), and then by a second stage of tests with 25 and 30 ppm of Iron Chloride at the beginning.

The temperature versus time curves recorded by the thermocouple in the tank during all the tests are reported in Fig. 14.10, for all the different initial Iron-salt concentrations.

These calorimetric curves show rather small relative variations, with final temperatures comprised between a maximum value of 39.6 °C, for 20 ppm, and a minimum value of 37.1 °C, for a starting concentration of zero ppm. From the diagram it is possible to observe that the tests performed after the washing, with the concentrations of 25 and 30 ppm, are characterized by final temperature values lower than those expected on the basis of the previous tests. In other terms, no monotonic trends emerge. Such remark leads us to separate the calorimetric results obtained before the washing from those obtained after the washing.

In Fig. 14.11, the temperature versus time curves related to the initial concentrations between 0 and 20 ppm are reported. The diagram shows that the temperature tends to increase by increasing the initial salt concentration, with the relevant exception of the test for 5 ppm. This anomaly can not be explained by any excess in the input electric energy as shown in Table 14.3. Therefore, such a curve is removed from the set of Fig. 14.11. In this way, the diagram shows a monotonic growth of

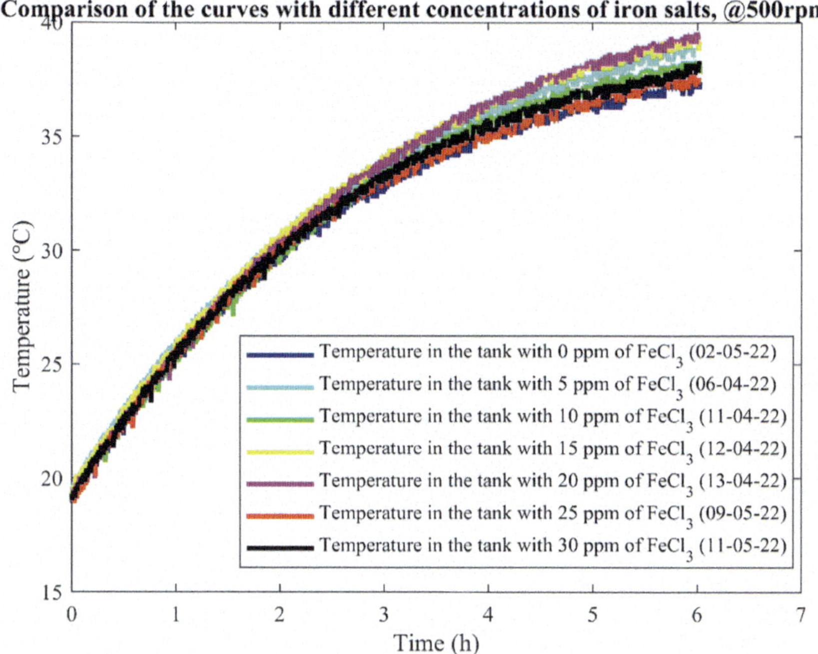

Fig. 14.10 Temperature versus time curves for different initial salt concentrations (0–30 ppm)

the final temperature as a function of the initial salt concentration. In particular, between the curves for 10 and 15 ppm, an average difference during the whole test of 0.69 °C is revealed, whereas an average difference of 1.03 °C is observed during the last 120 s. On the other hand, the average difference between the curves for 15 and 20 ppm presents the slightly negative value of −0.10 °C, whereas a positive difference during the last 120 s equal to 0.19 °C is observed.

A confirmation to the saturation occurring for initial concentrations higher than 20 ppm is given by the tests that are performed after the washing. As is possible to see in Fig. 14.12, the average difference between 25 and 30 ppm during the whole test presents the value of 0.18 °C, and only during the last 120 s it increases up to 0.69 °C.

Table 14.4 summarizes the average differences for 0–20 ppm, during the whole test and during the last 120 s. For a comparison, the temperature values at the time of the plant switching off are reported in the last column for each test.

Analogously, Table 14.5 reports the synthesis of the average temperature differences of the tests following the washing (25–30 ppm), and the related final temperatures.

On the basis of what has been reported, it is possible to conclude that exceeding initial concentration values of 20 ppm is not convenient, since reduced temperature

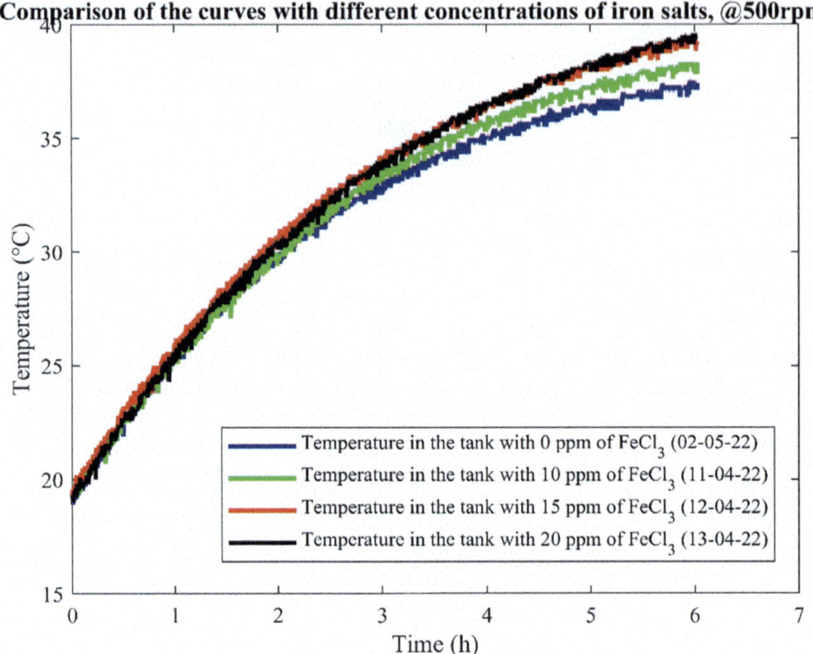

Fig. 14.11 Temperature versus time curves of the tests that precede the washing (0–20 ppm), excluding the first test (5 ppm)

Table 14.3 Average electric power provided to the pilot-plant during the single tests

Test date	$\overline{W}_{el}$ (W)
06/04/22	456.52
11/04/22	458.14
12/04/22	456.68
13/04/22	457.68
02/05/22	456.34
09/05/22	457.65
11/05/22	455.07

increments are found for higher amounts of dissolved salts, due to the saturation and precipitation of the same salts.

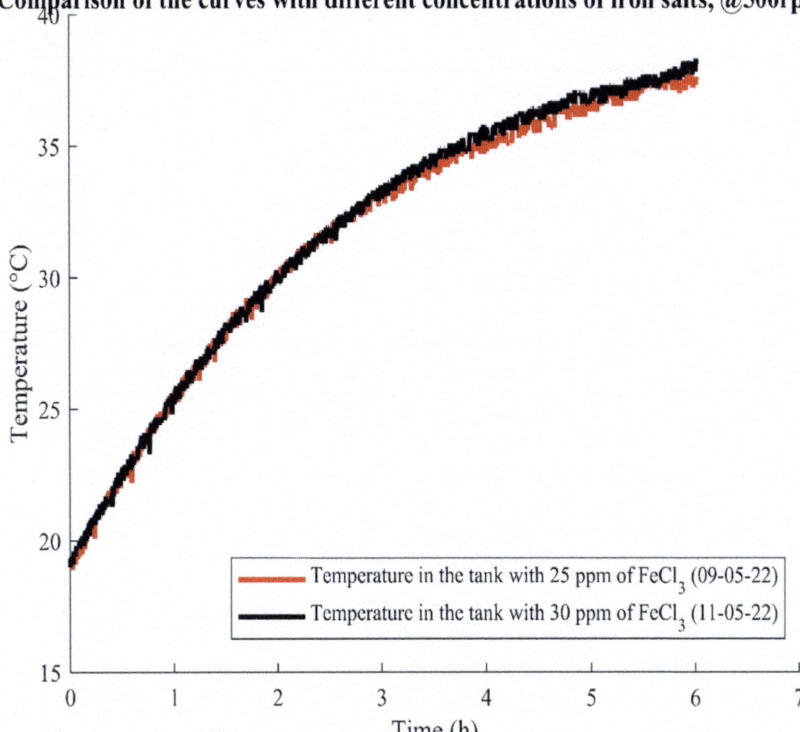

Fig. 14.12 Temperature versus time curves of the tests that follow the washing (25–30 ppm)

Table 14.4 Final temperatures in the tests preceding the washing (0–20 ppm)

Test date	Concentration (ppm)	Average difference with respect to the previous test on the whole duration (°C)	Average difference with respect to the previous test on the last 120 s (°C)	Final temperature (°C)
	0	–		37.1
11/04/22	10	–	–	37.8
12/04/22	15	0.69	1.03	39.1
13/04/22	20	−0.10	0.19	39.6

Table 14.5 Final temperatures in the tests following the washing (25–30 ppm)

Test date	Concentration (ppm)	Average difference with respect to the previous test on the whole duration (°C)	Average difference with respect to the previous test on the last 120 s (°C)	Final temperature (°C)
09/05/22	25	–	–	37.3
11/05/22	30	0.18	0.69	38.1

Acknowledgements MetalWork SpA is thankfully acknowledged for designing and providing the pilot-plant of hydrodynamic cavitation. In particular, we would like to recall the enthusiastic support by the President Erminio Bonatti and the Manager Fausto Rodella.

References

1. Cardone F, Carpinteri A, Lacidogna G (2009) Piezonuclear neutrons from fracturing of inert solids. Phys Lett A 373:4158–4163
2. Carpinteri A, Cardone F, Lacidogna G (2009) Piezonuclear neutrons from brittle fracture: early results of mechanical compression tests. Strain 45:332–339
3. Carpinteri A, Chiodoni A, Manuello A, Sandrone R (2011) Compositional and microchemical evidence of piezonuclear fission reactions in rock specimens subjected to compression tests. Strain 47(Suppl 2):282–292
4. Carpinteri A, Lacidogna G, Manuello A, Borla O (2011) Energy emissions from brittle fracture: neutron measurements and geological evidences of piezonuclear reactions. Strength, Fract Complex 7:13–31
5. Carpinteri A, Lacidogna G, Manuello A, Borla O (2012) Piezonuclear fission reactions in rocks: evidences from microchemical analysis, neutron emission, and geological transformation. Rock Mech Rock Eng 45:445–459
6. Lucia U, Carpinteri A (2015) GeV plasmons and spalling neutrons from crushing of iron-rich natural rocks. Chem Phys Lett 640:112–114
7. Carpinteri A, Manuello A (2011) Geomechanical and geochemical evidence of piezonuclear fission reactions in the earth's crust. Strain 47(Suppl 2):267–281
8. Carpinteri A, Lacidogna G, Borla O (2012) Piezonuclear neutrons from earthquakes as a hypothesis for the image formation and the radiocarbon dating of the Turin Shroud. Sci Res Essays 7:2603–2612
9. Carpinteri A, Lacidogna G, Manuello A, Borla O (2013) Piezonuclear fission reactions from earthquakes and brittle rocks failure: evidence of neutron emission and nonradioactive product elements. Exp Mech 53:345–365
10. Cardone F, Manuello A, Mignani R, Petrucci A, Santoro E, Sepielli M, Carpinteri A (2016) Ultrasonic piezonuclear reactions in steel and sintered ferrite bars. J Adv Phys 5:69–75
11. Carpinteri A, Borla O, Manuello A, Veneziano D, Goi A (2015) Hydrogen embrittlement and piezonuclear reactions in electrolysis experiments. J Condens Matter Nucl Sci 15:162–182
12. Carpinteri A, Borla O, Manuello A, Niccolini G (2018) Energy balance during electrolysis and cavitation experiments. Soc Exp Mech, Micro Nanomech 5:37–40
13. Brennen CE (1995) Cavitation and bubble dynamics. Oxford University Press

14. Manuello A, Malvano R, Borla O, Palumbo A, Carpinteri A (2016) Neutron emission from hydrodynamic cavitation. In: Fracture, fatigue, failure and damage evolution, Conference proceedings of the society for experimental mechanics series, vol 8. The Society for Experimental Mechanics, pp 175–182

Part V
Fracto-Emissions from Earthquakes and Volcanic Eruptions

Chapter 15
Neutron Emissions from Seismic Events

Abstract Neutron emission detections from cracking and fracture in solids logically lead to consider also the Earth's Crust, in addition to cosmic rays, as a relevant source of neutron flux variations. Neutron emissions measured in seismic areas of the Pamir region (4,200 m a.s.l.) exceeded the natural background level up to three orders of magnitude in correspondence to seismic activity, more precisely an earthquake of magnitude equal to 4 in the Richter scale. An additional analysis, with respect to those already carried out by Russian research groups, is here presented. Reference is made to data acquired at the "Testa Grigia" Laboratory of Plateau Rosa, Cervinia (Italy), during an experimental campaign on the evaluation of neutron radiation from cosmic rays. Further data refer to dedicated experimental trials carried out at the seismic district of "Val Trebbia", Bettola (Piacenza, Italy). The assessment of neutron radiation at the environmental level can help to make a clear distinction between the component from Cosmic Rays and the component from Earth's Crust (seismic events).

Keywords Earthquakes · Neutron emissions · Phono-fission nuclear reactions · Cosmic rays

15.1 Preliminary Remarks

Monitoring the three different forms of fracto-emission (Acoustic Emission AE, Electromagnetic Emission EME, and Neutron Emission NE), during the failure of natural and artificial brittle materials, deserves an accurate interpretation in terms of mechanical damage and fracture. The mechanical energy emissions are usually measured through the signals captured by acoustic sensors [1–5] or electromagnetic detectors [6–13]. Nowadays, the AE technique is well-known in the scientific community and applied for structural health monitoring purposes. In addition, based on the scale-invariant analogy between microcracking and seismic activity, AE associated to microcracking presents a statistical law of frequency (of occurrence) versus magnitude that is formally identical to the seismic power-law due to Richter. Also

A. Carpinteri, *Terahertz Phonons and Nanomechanical Instabilities*,
https://doi.org/10.1007/978-3-032-14692-2_15

the EME signals are related to brittle materials, in which fracture propagation occurs suddenly and is accompanied by abrupt stress drops in the stress–strain diagram. Several experimental studies revealed the existence of EME signals during fracture experiments carried out on a wide range of materials [6]. It was also observed that the EME signals detected during failure of materials in the laboratory are analogous to the anomalous radiation of geoelectromagnetic waves observed before major earthquakes [7], reinforcing the idea that also the EME effect can be applied as a forecasting tool to seismic events. In fact, magnetic phenomena associated to earthquakes and volcanic eruptions have been studied worldwide utilizing very sensitive instruments.

As regards the neutron emissions, original experimental tests were performed by Carpinteri et al. [14–18] on natural rock specimens. Different kinds of compression test under monotonic, cyclic, or ultrasonic mechanical loading were carried out, fully confirming the hypothesis of phono-fission reactions accompanied by neutron emissions up to three orders of magnitude higher than the natural background level at the time of catastrophic failure of the specimens. A theoretical explanation to these phenomena was provided by Widom et al. [19, 20]. An alternative explanation will be given in Chap. 22.

Phono-fission reactions have important implications also at the Earth's crust or tectonic scale. Neutron emission detections by Volodichev et al. [21, 22], Kuzhevskij et al. [23, 24], and Antonova et al. [25] led to consider also the Earth's crust, in addition to cosmic rays, as a relevant source of neutron flux variations. Citing Volodichev et al., neutron emissions measured in seismic areas of the Pamir region (4,200 m a.s.l.) exceeded the natural background "up to two orders of magnitude in correspondence to seismic activity and rather appreciable earthquakes, greater than or equal to the 4th degree in the Richter scale magnitude" [22]. Considering the altitude dependence of neutron radiation (Pfotzer profile [26]), values approximately ten times higher than natural background at sea level are generally detected at 5,000 m altitude. Therefore, the same earthquake occurring at sea level should produce a neutron flux up to 1,000 times higher than the local natural background. More recent neutron emission observations were performed before the Sumatra earthquake of December 2004 [27]. Variations in thermal neutron measurements were observed in different areas (Crimea, Kamchatka) a few days before that catastrophic event.

Neutron components exceeding the natural background level in correspondence to seismic activity were detected in the "Testa Grigia" Laboratory of Plateau Rosa, Cervinia (Italy), during an experimental campaign on neutron radiation from cosmic rays [28–30]. In particular, the assessment of the neutron radiation at the environmental level could help to make a clear distinction between the component of cosmic origin and the component from the Earth's crust (phono-fission reactions due to seismic activity).

By integrating all these signals (AE, EME, NE)—and also considering the gaseous radon emission that appears to be one of the most reliable seismic precursors—it should be possible to set up a sort of alarm system based on a regional warning network. This could combine the signals from different alarm stations to prevent the effects of seismic events and to identify the epicenter of the earthquake. Similar networks, just based on seismic accelerations, are presently utilized worldwide (Mexico, Taiwan, Turkey, Romania, and Japan [31]).

Phono-fission reactions related to neutron emissions from active faults may be considered as the principal cause of magnesium depletion and the consequent carbon formation during seismic activity. In this way, the CO_2 atmospheric level may be considered as an appreciable precursor, together with acoustic, electromagnetic, and neutron emissions, before relevant earthquakes. Significant changes in the diffuse emission of carbon dioxide were recorded in a geochemical station located at El Hierro, in the Canary Islands [32], before the occurrence of different seismic events during the year 2004. Appreciable CO_2 emissions were observed some days before such seismic events.

15.2 Atmospheric Neutron Measurements at High Altitude Observatories

Galactic cosmic radiation generates secondary ionizing particles in the atmosphere (Fig. 15.1a), such as neutrons, electrons, positrons, protons, muons, and photons. Usually, the dose is varying in a complicated way with altitude and geomagnetic coordinates (longitude and latitude), being larger toward the Polar Regions and lower in the vicinity of the Equator. It also depends on the solar activity, which varies according to a cycle approximately 11 years long. Besides radiation components originating from the galactic cosmic radiation, the sun may occasionally contribute an additional component of solar particle events (SPE) (Fig. 15.1b).

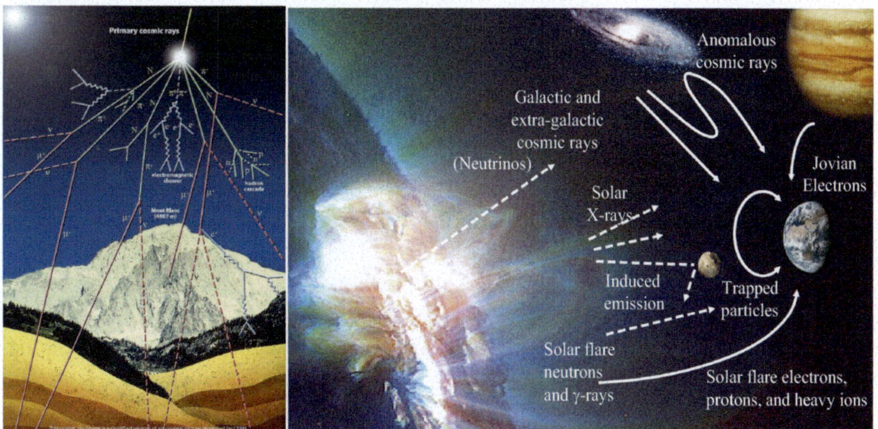

Fig. 15.1 Radiation field generated by the hadronic shower in the atmosphere (left) (available at http://cosmicrays.le.infn.it). Cosmic rays origin and composition (right)

Usually, the neutron energy distribution is influenced by the atmospheric composition: as a matter of fact, neutrons are mainly produced by the reaction of primary protons with atmospheric nuclei of N (78%) and O (21%).

Since 1997, dedicated experimental campaigns [28–30] have been performed at High Altitude Observatories (HMOs), in both Northern and Southern Hemispheres, to obtain information on the variability of atmospheric neutron spectra with solar activity. During the research activity in these laboratories, specific techniques for neutron spectrometry and dosimetry were set up, suitable for being applied to different fields, such as aircrew exposure to Cosmic Rays (CRs) in high-altitude flights or in space missions. This confirms the relevance of the research activity at HMOs for environment, space, and health studies.

The experimental evaluation of neutron spectra in the wide energy range of interest and in the complex radiation field generated by the hadronic shower in the atmosphere requires a special technique. This experimental technique, based on passive neutron detectors with different thresholds and energy responses, allows the reconstruction of the neutron spectra in the energy range of interest. The results of neutron spectra measurement were obtained with passive instruments coupled with the unfolding code BUNTO, whereas the monitoring of the integral neutron dose was performed by a REM (Roentgen Equivalent Man) counter.

The short range spectrometric system (from 10 keV to 20 MeV) is based on the passive Bubble Detector Spectrometer (BDS, BTI, Ontario, Canada) [33]. It is constituted of polycarbonate vials filled with a tissue equivalent gel, in which tiny superheated liquid (Freon) droplets are dispersed. Neutrons interact with the gel and produce recoil charged particles, which give rise to the boiling of droplets. This leads to the formation of visible bubbles that are trapped within the gel. The number of bubbles is related to the neutron dose. Six different types of detector (with different chemical composition) are available; each of them corresponds to a different energy threshold (10, 100, 600, 1,000, 2,500, 10,000 keV).

The unfolding package BUNTO [34] was especially developed to process the responses of the wide and short range spectrometers. In order to get an appropriate solution from the system of Fredholm's equations, which are obtained from measurements affected by large experimental uncertainties, a special method is introduced: it is based on the random sampling of unfolding data from a normal distribution, whose parameters (mean value and standard deviation) are the average experimental reading and the associated statistic uncertainty. The BUNTO final spectrum is the calculated mean of possible solutions of the unfolding procedure, weighted on the mean standard deviation. BUNTO fixes the maximum variation between possible solution and mean value within 20%: this is assumed as "percent error" on the experimental spectrum points.

As regards the monitoring of neutron dose, the ALNOR REM counter was used. This device is able to detect the contribution of the equivalent ambient neutron dose in a very wide energy range (neutron sensitivity from thermal energy to 17 MeV).

In Fig. 15.2, typical neutron spectra are shown as an example in terms of neutron fluence rate obtained at the "Testa Grigia" laboratory (geographical position: 3,480 m a.s.l., 45°56′ N, 7°42′ E) during the experimental campaigns of November 1997,

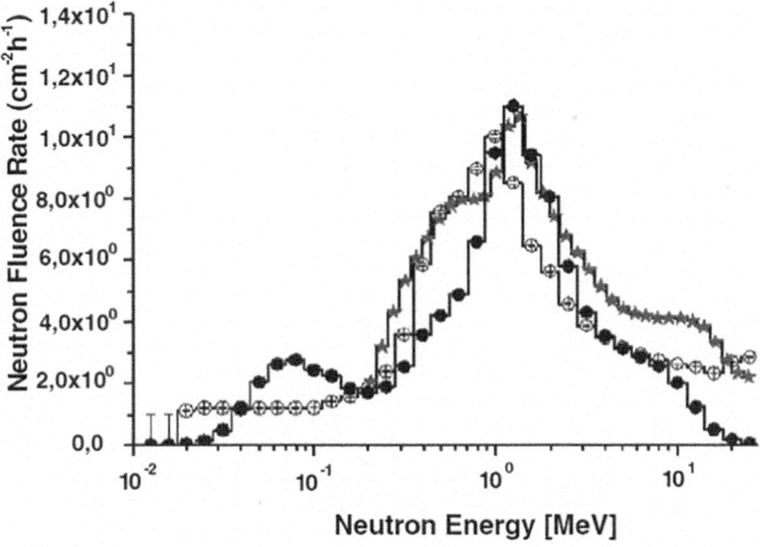

Fig. 15.2 Neutron spectra measured at the "Testa Grigia" laboratory during different solar activity periods (stars: November 1997; circles: March 2003; filled circles: December 2007), by using the BDS spectrometer (energy range: 10 keV–20 MeV) [28]

March 2003, and December 2007, by using the BDS Spectrometer. The neutron spectra were measured during different solar activity periods (mean sunspot number: 10–20/11/1997: 34; 24–31/03/2003: 88; 05–10/12/2007: 19; from the website of the National Geophysical Data Center [35]). Due to the different values of solar activity, the energy spectra show different fluence intensities but similar shapes, as expected, with evidence of a main peak at about 1 MeV, the so-called evaporative contribution.

15.3 Phono-Fission Reactions: From the Laboratory to the Earth's Crust Scale

The confirmation that the environmental neutron rate is linked to neutrons coming from galactic events but also from phono-fission reactions occurring in the Earth's Crust was assessed during experimental tests conducted at Politecnico di Torino on different natural rocks [14–18]. In particular, neutron emission measurements, by means of ^{3}He devices and neutron bubble detectors, were performed during three different kinds of compression tests: (i) under displacement control, (ii) under cyclic loading, and (iii) by ultrasonic vibration. The tested materials were Luserna granite, basalt, magnetite, Carrara marble, mortar enriched with iron oxides, gypsum, and quartz (see Part II).

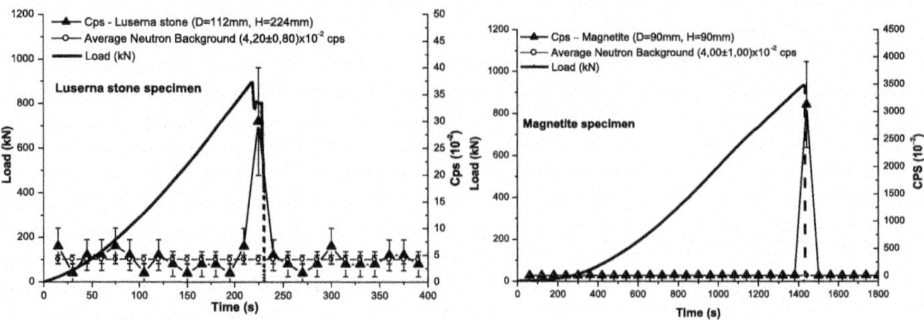

Fig. 15.3 Luserna stone (left) and magnetite (right) specimens. Load versus time diagrams and neutron emission count rates

During compression tests on specimens characterized by a brittle material and a sufficiently large size, the neutron flux was found to be up to three orders of magnitude higher than the natural background level at the time of catastrophic failure. For specimens with more ductile behavior (metallic materials, see Part III), neutron emissions significantly higher than the background level were however found. As an example, in Fig. 15.3a, b, the load versus time diagram and the neutron count rate evolution for Luserna stone and magnetite specimens are shown.

Since the analyzed materials contain different amounts of iron, the conjecture by Carpinteri et al. [14–18] is that phono-fission reactions of iron into aluminum, or into magnesium and silicon, have occurred during compression damage and crushing failure. This hypothesis is systematically and repeatedly confirmed by energy dispersive X-ray spectroscopy (EDS) analysis.

From the results and the experimental evidence reported in [14–18], it can be clearly seen that phono-fission reactions are possible in inert non-radioactive solids. From the EDS results on crushed samples, the evidence for Fe (phengite mineral) and Al content variations leads to the conclusion that the phono-fission reaction:

$$\mathrm{Fe}_{26}^{56} \rightarrow 2\mathrm{Al}_{13}^{27} + 2 \text{ neutrons} \tag{15.1}$$

has occurred [14–18, 36, 37]. Moreover, considering the evidence for the biotite content variations in Fe, Al, Si, and Mg, it is possible to conjecture that an additional phono-fission reaction has occurred during the crushing tests [14–18]:

$$\mathrm{Fe}_{26}^{56} \rightarrow \mathrm{Mg}_{12}^{24} + \mathrm{Si}_{14}^{28} + 4 \text{ neutrons} \tag{15.2}$$

Taking into account that granite is a common and widely present type of intrusive, sialic, igneous rock, and that it is characterized by a high concentration in the rocks that make up the Earth's crust ($\approx$60% of the Earth's crust), the phono-fission reactions expressed above can be generalized from the laboratory to the Earth's crust scale, where mechanical phenomena of brittle fracture, due to plate collision and fault subduction, take place continuously in the most seismic areas.

15.4 Neutron Emissions from Earthquakes

As regards the observations described in [28–30], in the period from July 30 to August 3, 2008, an additional experimental campaign was conducted at the "Testa Grigia" laboratory. These measurements were performed to integrate those of December 2007. Neutron monitoring was carried out by means of the short range bubble detector spectrometer (BDS) and the REM ALNOR counter. During the data acquisition, an evident increment in neutron radiation was monitored between July 31 and August 1st. This variation was detected in real time by the REM counter and later confirmed by the analysis of bubble dosimeters unfolded by the BUNTO code. An increment of about six times in the neutron dose with respect to the natural background dose was observed (Fig. 15.4a). This phenomenon was monitored for a period of about two hours. Then the values decreased to the usual background level.

The subsequent estimation of the neutron energy spectrum (Fig. 15.4b) showed the anomalous event. In addition to the usual evaporative peak, at about 700 keV–1 MeV, a considerable high-energy neutron component of about 8 MeV was monitored. The fact that two different instruments, with different data acquiring methods, simultaneously monitored the same anomaly excludes any type of misfunctioning of the instruments. As usual, the assumptions made for the explanation of this event have firstly focused on possible effects of cosmic origin. However, from the analysis of data relating to solar and galactic events, no event of such great intensity was apparently found. As a matter of fact, no significant sunspot activity was recorded during the data acquisition time window. Analogously, during the same period, no anomalies in the cosmic ray flux were detected (Fig. 15.5) [38]. This is also demonstrated by the data acquired at the laboratory of Jungfraujoch (geographical position: 3,450 m a.s.l., 46°32′ N, 7°59′ E), a few hundred kilometers away from the "Testa Grigia" laboratory. In particular, in Fig. 15.5, the cosmic ray variations acquired at the laboratory of Jungfraujoch [38], in the period from July 30 to August 3, 2008, is reported. The fluctuation of few percentages of cosmic rays flux is absolutely normal and it can not

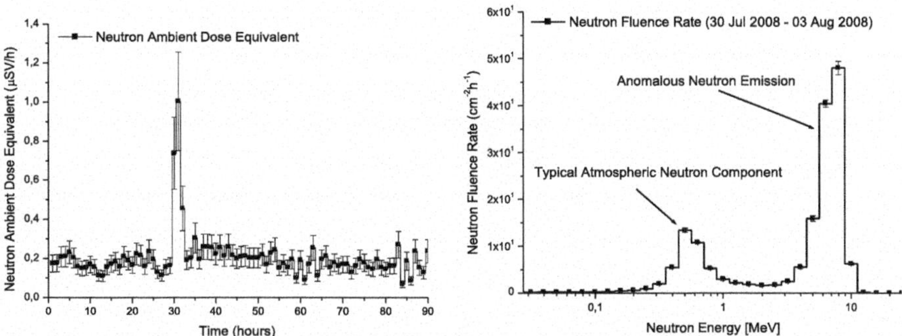

Fig. 15.4 Neutron ambient dose equivalent (left) measured by REM ALNOR counter at the "Testa Grigia" laboratory during the experimental campaign of July–August 2008. Neutron spectrum (right) measured by using the BDS spectrometer (energy range: 10 keV–20 MeV)

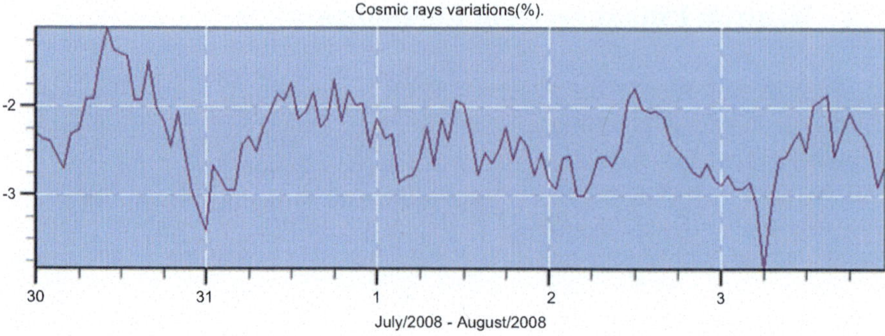

Fig. 15.5 Cosmic rays variation acquired at the laboratory of Jungfraujoch [38] in the period from July 30 to August 3, 2008

explain the so marked variations in the observed environmental neutron background level shown in Fig. 15.4a, b.

On the other hand, considering the phenomenon of neutron emission before earthquakes, an investigation for earthquakes occurred in the immediate vicinity of the laboratory was carried out later. A significant seismic activity [39] was observed during the period July–August 2008 in a region a few hundred kilometers away from the laboratory (Table 15.1). In particular, about 20 days after the anomalous increment in neutron radiation, a seismic event of the 3rd degree in the Richter magnitude scale occurred approximately 130 km away. This interpretation is consistent with the observations of Kuzhevskij et al. [23, 24] and it provides further experimental evidence of the correlation between neutron emission and seismic events of appreciable intensity.

Some researchers impute these environmental neutron fluctuations to solar eclipses or full moon periods, which for gravitational attraction reasons can be the cause of earthquakes on the Earth's surface [21]. Nevertheless, the monitored anomalous neutron emissions and the related seismic activity occurred in a granitic geographical area, therefore strengthening the phono-fission hypothesis, experimentally demonstrated at the laboratory scale.

Table 15.1 Seismic activity in the surrounding area of the "Testa Grigia" laboratory in August 2008 [39]. The distance between the earthquake epicenter and the laboratory is also reported

"Testa Grigia" laboratory—geographical posit.: 3,480 m a.s.l., 45°56′ N, 7°42′ E						
Year	Month	Day	Latitude	Longitude	Magnitude	Distance (km)
2008	08	10	44°18′ N	7°15′ E	2.7	199
2008	08	14	44°43′ N	7°19′ E	2.8	171
2008	08	20	44°78′ N	7°30′ E	3.0	131
2008	08	21	46°65′ N	8°47′ E	2.5	99
2008	08	21	44°86′ N	6°62′ E	2.9	145

15.5 Early Experiments in "Val Trebbia" and "Murisengo" Seismic Stations

From December 28, 2012, to January 6, 2013, a dedicated experimental campaign was conducted in Bettola, Piacenza, located in northern Italy, at the "Val Trebbia" seismic district (Fig. 15.6a) (geographical position: 410 m a.s.l., 44°46′ N, 9°30′ E). The seismic risk level of this geographical area is changed after the disastrous earthquakes that have stricken the Emilia-Romagna region in Spring 2012. At the present time, the area is considered as a medium–high seismic zone. In Fig. 15.6b, all the seismic events observed in the year 2012, within a circular area of 100 km radius centered in Bettola, are reported. A total of 218 earthquakes were detected [40], of which 194 have recorded a magnitude lower than or equal to 2.5.

During the experimental trial, the neutron monitoring was carried out in "continuous modality" by means of a ^{3}He neutron radiation monitor. The AT1117M (ATOMTEX, Minsk, Republic of Belarus) neutron detector is a multifunctional portable instrument with a digital readout consisting of a processing unit (PU) with an internal Geiger-Müller tube and external smart probes (BDKN-03 type). This type of device provides high sensitivity and wide measuring ranges (neutron energy range: 0.025 eV–14 MeV), with a fast response to radiation field change, ideal for environmental monitoring purposes.

Three evident peaks in neutron radiation were monitored between December 30, 2012, and January 2, 2013. An increment of about six times in the neutron dose with respect to the natural background dose was observed in two cases (Fig. 15.7). These phenomena were monitored for a period of at least three hours. Then the values decreased up to the usual background level.

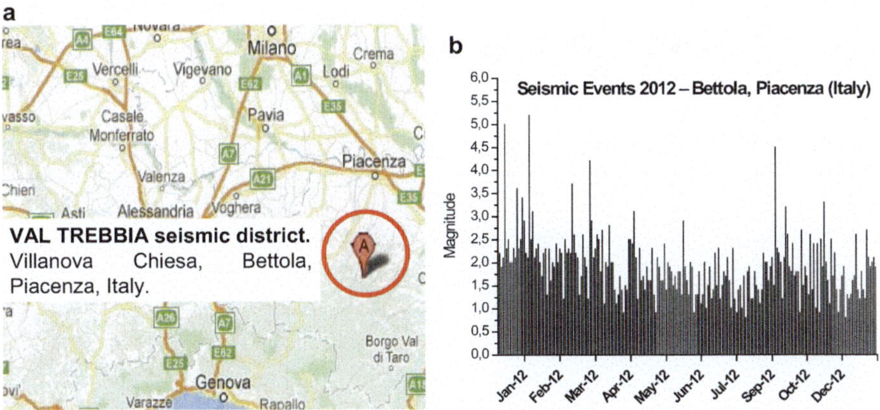

Fig. 15.6 **a** Val Trebbia seismic district, Villanova Chiesa, near Bettola, Piacenza (Italy). **b** Seismic events recorded in 2012 within a circular area of 100 km radius centered in Bettola

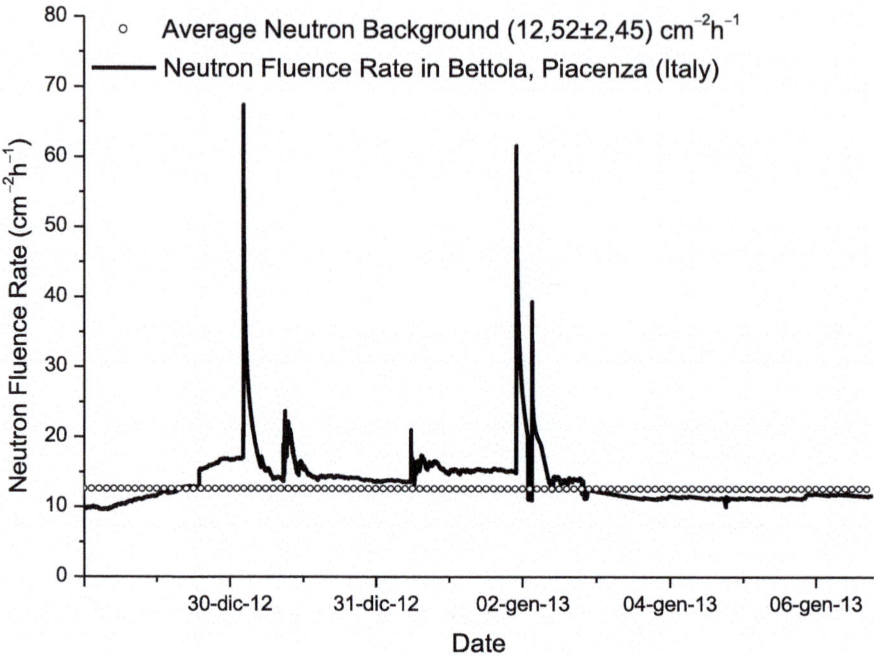

Fig. 15.7 Neutron ambient dose equivalent measured by ATOMTEX radiation monitor at Villanova Chiesa, near Bettola, Piacenza (Italy), during the experimental campaign of December 2012–January 2013

As in the case of "Testa Grigia" laboratory, no plausible explanation of a cosmic or galactic origin was found. As a matter of fact, only small fluctuations of few percent points in the cosmic ray flux were detected (Fig. 15.8) [38].

An investigation for earthquakes occurred in the immediate vicinity of the monitored area was carried out later. The usual seismic activity (average magnitude of

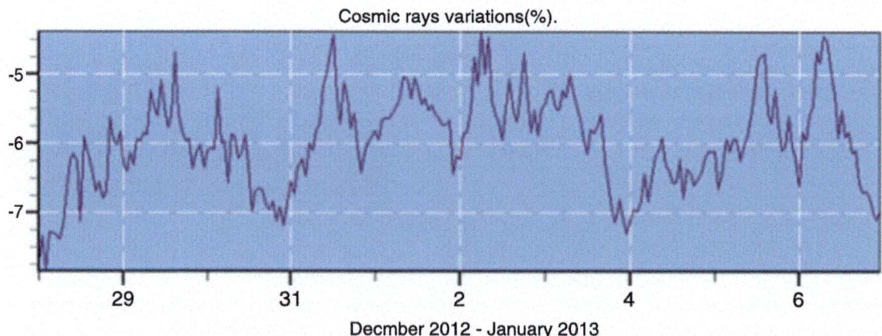

Fig. 15.8 Cosmic ray variations acquired at the laboratory of Jungfraujoch [38] in the period from December 28, 2012, to January 6, 2013

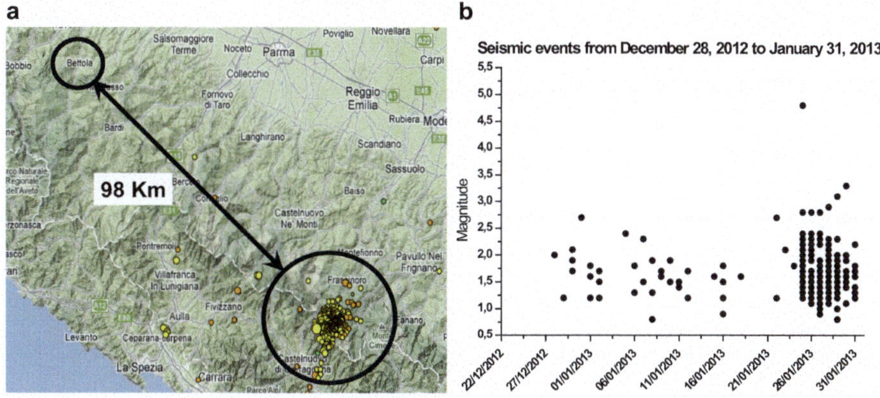

Fig. 15.9 **a** Seismic swarm occurred about 100 km away from the monitored area. **b** Seismic events from December 28, 2012, to January 31, 2013

2.5) [40] was observed during the period from December 28, 2012, to January 6, 2013.

Most importantly, 25 days after the anomalous increment in neutron radiation, an earthquake of the 5th degree in the Richter scale of magnitude was recorded (geographical position: 44°16′ N, 10°31′ E). This event belongs to a seismic swarm occurred in the "Garfagnana" district less than 100 km from Bettola (Fig. 15.9a). In particular, in Fig. 15.9b, the seismic events from December 28, 2012, to January 31, 2013, are reported.

From July 1st, 2013, a dedicated in-situ monitoring station started at the San Pietro-Prato Nuovo gypsum mine, near Murisengo (Alessandria, Italy). The monitoring activities went on for over six years, up to the beginning of the year 2020, when the mine was closed due to the COVID-19 pandemic. In the next chapter, the very relevant and absolutely repeatable results of that six-year monitoring will be reported and deeply discussed.

Initially, some rock pillars of the mine, located at about 100 m below the ground level, were subjected to a multi-parameter monitoring in order to assess their structural health. The structural monitoring was principally conducted by the AE technique, although, in the following, the seismic risk of the surrounding area was also evaluated with the detection of EME fluctuations and environmental neutron fields.

Thanks to the position of the monitoring station (100 m under the ground level), the acoustic and electromagnetic noise of human origin is greatly reduced, as well as the neutron background is more than one order of magnitude lower than that on the ground.

These aspects make the mine a convenient place for the monitoring of the three fracto-emissions produced by seismic events. The experimental activity has extensively and undoubtedly demonstrated that a strict correlation exists between underground seismic activity and acoustic/electromagnetic/neutron emissions.

15.6 Conclusions

Based on early experimental data unexpectedly acquired at the "Testa Grigia" laboratory of Plateau Rosa, Cervinia (Italy), further investigations—besides those already known from the literature [21–25]—are presented to confirm the hypothesis that the Earth's crust, in addition to cosmic rays, is a relevant source of neutron flux variations. This phenomenon seems to take place several days before a significant seismic event occurs in the monitored area, or even rather far from it.

References

1. Mogi K (1962) Study of elastic shocks caused by the fracture of heterogeneous materials and its relation to earthquake phenomena. Bull Earthq Res Inst 40:125–173
2. Lockner DA, Byerlee JD, Kuksenko V, Ponomarev A, Sidorin A (1991) Quasi static fault growth and shear fracture energy in granite. Nature 350:39–42
3. Shcherbakov R, Turcotte DL (2003) Damage and self-similarity in fracture. Theoret Appl Fract Mech 39:245–258
4. Ohtsu M (1996) The history and development of acoustic emission in concrete engineering. Mag Concr Res 48:321–330
5. Carpinteri A, Lacidogna G, Pugno N (2006) Richter's laws at the laboratory scale interpreted by acoustic emission. Mag Concr Res 58:619–625
6. Miroshnichenko M, Kuksenko V (1980) Study of electromagnetic pulses in initiation of cracks in solid dielectrics. Sov Phys-Solid State 22:895–896
7. Warwick JW, Stoker C, Meyer TR (1982) Radio emission associated with rock fracture: possible application to the great Chilean earthquake of May 22, 1960. J Geophys Res 87:2851–2859
8. O'Keefe SG, Thiel DV (1995) A mechanism for the production of electromagnetic radiation during fracture of brittle materials. Phys Earth Planet Inter 89:127–135
9. Scott DF, Williams TJ, Knoll SJ (2004) Investigation of electromagnetic emissions in a deep underground mine. In: Proceedings of the 23rd international conference on ground control in mining, Morgantown, 3–5 August 2004, pp 125–132
10. Frid V, Rabinovitch A, Bahat D (2003) Fracture induced electromagnetic radiation. J Phys D 36:1620–1628
11. Rabinovitch A, Frid V, Bahat D (2007) Surface oscillations. A possible source of fracture induced electromagnetic oscillations. Tectonophysics 431:15–21
12. Lacidogna G, Carpinteri A, Manuello A, Durin G, Schiavi A, Niccolini G, Agosto A (2010) Acoustic and electromagnetic emissions as precursor phenomena in failure processes. Strain 47(2):144–152
13. Carpinteri A, Lacidogna G, Manuello A, Niccolini G, Schiavi A, Agosto A (2010) Mechanical and electromagnetic emissions related to stress-induced cracks. Exp Tech 36(3):53–64
14. Carpinteri A, Cardone F, Lacidogna G (2009) Piezonuclear neutrons from brittle fracture: early results of mechanical compression tests. Strain 45:332–339
15. Cardone F, Carpinteri A, Lacidogna G (2009) Piezonuclear neutrons from fracturing of inert solids. Phys Lett A 373:4158–4163
16. Carpinteri A, Cardone F, Lacidogna G (2010) Energy emissions from failure phenomena: mechanical, electromagnetic, nuclear. Exp Mech 50:1235–1243
17. Carpinteri A, Borla O, Lacidogna G, Manuello A (2010) Neutron emissions in brittle rocks during compression tests: monotonic vs cyclic loading. Phys Mesomech 13:268–274
18. Carpinteri A, Lacidogna G, Manuello A, Borla O (2011) Energy emissions from brittle fracture: neutron measurements and geological evidences of piezonuclear reactions. Strength, Fract Complex 7:13–31

19. Widom A, Swain J, Srivastava YN (2013) Neutron production from the fracture of piezoelectric rocks. J Phys G: Nucl Part Phys 40(015006):1–8
20. Widom A, Swain J, Srivastava YN (2014) Photo-disintegration of the iron nucleus in fractured magnetite rocks with magnetostriction. Meccanica 50:1205–1216
21. Volodichev NN, Kuzhevskij BM, Nechaev OYu, Panasyuk MI, Shavrin PI (1997) Phenomenon of neutron intensity bursts during new and full moons. Cosm Res 31(2):135–143
22. Volodichev NN, Kuzhevskij BM, Nechaev OYu, Panasyuk MI, Podorolsky AN, Shavrin PI (2000) Sun-moon-earth connections: the neutron intensity splashes and seismic activity. Astron Vestnik 34:188–190
23. Kuzhevskij M, Nechaev OYu, Sigaeva EA, Zakharov VA (2003) Neutron flux variations near the earth's crust. A possible tectonic activity detection. Nat Hazards Earth Syst Sci 3:637–645
24. Kuzhevskij M, Nechaev OYu, Sigaeva EA (2003) Distribution of neutrons near the Earth's surface. Nat Hazards Earth Syst Sci 3:255–262
25. Antonova VP, Volodichev NN, Kryukov SV, Chubenko AP, Shchepetov AL (2009) Results of detecting thermal neutrons at Tien Shan high altitude station. Geomag Aeron 49(6):761–767
26. Pfotzer G, Regener E (1935) Vertical intensity of cosmic rays by threefold coincidence in the stratosphere. Nature 136:718–719
27. Sigaeva EA, Nechaev O, Panasyuk M, Bruns A, Vladimirsky B, Kuzmin Yu (2006) Thermal neutrons' observations before the Sumatra earthquake. Geophys Res Abstr 8:00435
28. Zanini A, Storini M, Visca L, Durisi EAM, Fasolo F, Perosino M, Borla O, Saavedra O (2005) Neutron spectrometry at high mountain observatories. J Atmos Solar Terr Phys 67:755–762
29. Mishev A, Bouklijski A, Visca L, Borla O, Stamenov J, Zanini A (2008) Recent cosmic ray studies with lead free neutron monitor at basic environmental observatory Moussala. Sun Geosph 3(1):26–28
30. Zanini A, Storini M, Saavedra O (2009) Cosmic rays at high mountain observatories. Adv Space Res 44(10):1160–1165
31. Allen R (2011) Seconds before the big one. Seismology, ScientificAmerican.com, pp 54–59
32. Padron E, Melina G, Marrero R, Nolasco D, Barrancos J, Padilla G, Hernandez PA, Perez NM (2008) Changes on diffuse CO_2 emission and relation to seismic activity in and around El Hierro, Canary Islands. Pure Appl Geophys Special Issue on "Terrestrial Fluids, Earthquakes and Volcanoes: The Hiroshi Wakita" 165(3):95–114
33. BTI (2003) Instruction manual for the bubble detector spectrometer (BDS). Bubble Technology Industries, Chalk River, Ontario, Canada
34. Ongaro C, Zanini A, Tommasino L (2001) Unfolding technique with passive detectors in neutron dosimetry. In: Proceedings of the workshop "Neutron Spectrometry and Dosimetry: Experimental Techniques and MC Calculations", Stockholm, October 18–20, 2001, Otto Editor, pp 117–128
35. Information on National Geophysical Data Center (2012) Sunspot Numbers. http://www.ngdc.noaa.gov/stp/solar/ssndata.html. Accessed April 2012
36. Carpinteri A, Chiodoni A, Manuello A, Sandrone R (2010) Compositional and microchemical evidence of piezonuclear fission reactions in rock specimens subjected to compression tests. Strain 47(Suppl 2):282–292
37. Carpinteri A, Manuello A (2010) Geomechanical and geochemical evidence of piezonuclear fission reactions in the earth's crust. Strain 47(Suppl 2):267–281
38. Information on Jungfraujoch Neutron Monitor (18igy) (2013). http://cr0.izmiran.rssi.ru/jun1/main.htm. Accessed March 2013
39. Information on National Geophysical Data Center/World Data Center (NGDC/WDC) (2012) Significant earthquake database, Boulder, CO, USA. http://www.ngdc.noaa.gov/nndc/struts/form?t=101650&s=1&d=1. Accessed April 2012
40. ISIDe Working Group (INGV, 2010). Italian Seismological instrumental and parametric database. http://iside.rm.ingv.it. Accessed March 2013

BY

Chapter 16
Acoustic, Electromagnetic, and Neutron Emissions as Seismic Precursors

Abstract Three different forms of fracto-emission can be used as earthquake precursors. At the macro-scale, Acoustic Emission (AE) prevails, as well as Electromagnetic Emission (EME) at the meso-scale, and Neutron Emission (NE) at the micro- and nano-scale. TeraHertz phonons are in fact produced at the last extremely small scale by mechanical instability phenomena, and fracture experiments on natural rocks have recently revealed that these high-frequency waves are able to induce nuclear fission reactions with neutron and/or alpha particle emissions. Relevant applications to earthquake precursors are proposed. The results obtained at a gypsum mine located in Northern Italy are presented. To avoid interference with human activities and cosmic rays, the measurement instruments are located at one hundred meters underground. The experimental results obtained from July 1st, 2013, to June 30, 2019 (six years) are analyzed by means of a suitable multi-modal statistics procedure. The experimental observations reveal a high correlation between the three fracto-emission peaks and the major earthquakes occurring in the areas closest to the seismic station. The three fracto-emissions result to be regularly, repeatably, and systematically shifted with respect to the next seismic event (approximately one day, 3–4 days, and one week before, respectively for AE, EME, and NE). In this context, an innovative dynamic extension of the concept of earthquake preparation zone is also proposed. Eventually, the evident correlation between small-magnitude seismic swarms occurring in the surroundings of the mine and Moon phases is emphasized.

Keywords Fracture · Microcracking · TeraHertz phonons · Phono-fission reactions · Neutron emissions · Electromagnetic emissions · Acoustic emissions · Seismic precursors · Multi-modal statistical analysis · Earthquake preparation zone · Lunar periodicity

A. Carpinteri, *Terahertz Phonons and Nanomechanical Instabilities*,
https://doi.org/10.1007/978-3-032-14692-2_16

16.1 Preliminary Remarks

Solids that break in a brittle way are subjected to a rapid emission of energy involving the generation of pressure waves that travel at a characteristic speed with the order of magnitude of 10^3 m/s. Considering the very important case of earthquakes, it is possible to observe that, as fracture at the nanoscale (10^{-9} m) emits pressure waves (phonons) at the frequency of TeraHertz (10^{12} Hz), so fracture at the microscale (10^{-6} m) emits pressure waves at the frequency of GigaHertz (10^9 Hz), at the scale of millimeter (meso-scale) emits pressure waves at the frequency of MegaHertz (10^6 Hz), at the scale of meter (nacro-scale) emits pressure waves at the frequency of kiloHertz (10^3 Hz), and eventually faults at the kilometer (tectonic) scale emit pressure waves at the Hertz frequency, which is the typical and most likely frequency of seismic oscillations (Fig. 16.1 up) [1].

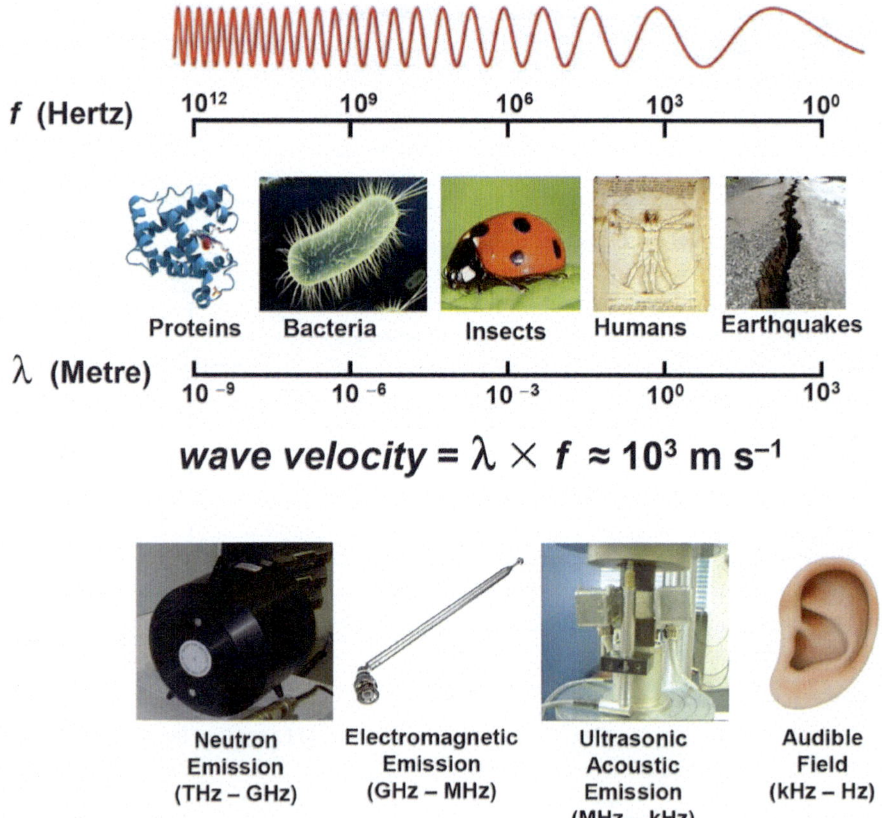

Fig. 16.1 Correlation between scale of the wavelength (length of the forming crack) and scale of the frequency of emitted pressure wave, by assuming a constant pressure wave velocity (up). Fracto-emission measurement instruments (down)

The animals with sensitive hearing in the ultrasonic range (frequency >20 kHz) "feel" the earthquake up to one day in advance, when the active cracks are still below the macroscopic scale (meter). Ultrasounds are in fact a well-known seismic precursor [2, 3]. With frequencies between Mega- and GigaHertz, and therefore cracks between micron and millimeter scale, pressure waves generate electromagnetic waves of the same frequency, which turn out to be even a more powerful seismic precursor (up to a few days before the event) [4, 5]. When pressure waves show frequencies between Giga- and TeraHertz, and then with cracks below the micron scale, we can witness a phenomenon partially unexpected: phonons resonate with the atomic lattice and, through a complex cascade of events (acceleration of electrons, bremsstrahlung gamma radiation, photo-fission, etc.), produce nuclear fission reactions [6–16]. It can be shown experimentally how such fission reactions can emit neutrons [17–19] like in the well-known case of uranium-235, but without gamma radiation and radioactive wastes. Note that the Debye frequency, i.e., the fundamental frequency of free vibration in atomic lattices, is in the order of magnitude of TeraHertz, and this is not a coincidence, since it is simply due to the fact that the inter-atomic distance is just around the nanometer, as indeed the minimum size of the lattice defects. As the chain reactions are sustained by thermal neutrons in a nuclear power plant, so the phono-fission reactions are triggered by phonons that have a frequency close to the resonance frequency of the atomic lattice and an energy close to that of thermal neutrons [8]. Emitted neutrons therefore appear to be as the most powerful seismic precursor (up to three weeks before) [20–27].

If all the precursory phenomena are simultaneously analyzed in suitable monitoring sites, they can provide the basis for the prediction of the three main data of an earthquake: place and time of occurrence, as well as magnitude of the seismic event. In this framework, the use of the fracto-emission precursors (Acoustic Emission AE, Electromagnetic Emission EME, and Neutron Emission NE) represents a huge step forward, not only for their monitoring capabilities during the earthquake, but also for their forecasting potentialities before the event [28, 29].

The prediction strategy should take into account an integral approach including the evaluation of several physical quantities and discriminating true signals from the environmental background or noise. Hence, ad-hoc measuring devices are used for the correct acquisition of the three fracto-emissions in their respective ranges of frequency: proportional counter for environmental neutrons, telescopic antenna and oscilloscope for electromagnetic waves, and piezoelectric transducers for ultrasonic acoustic emissions. Finally, during the seismic event, low-frequencies prevail, typically in the audible range (Fig. 16.1 down).

In addition, it is important to consider that, rather in advance before the earthquake occurrence, a very wide area of cracking rocks is active under the influence of tectonic stresses and in a critical condition around the future (still unknown) earthquake epicenter. In particular, Dobrovolsky et al. [30] tried to evaluate the dimension of this *earthquake preparation zone* as a function of the magnitude of the incoming earthquake, considering average heterogeneity and anisotropy values of the Earth's surface. Assuming that the zone of effective deformation is a circle centered in the epicenter of the incoming earthquake, the radius R of this strained zone is up to 100 km for earthquakes with a magnitude M equal to 3 in the Richter scale, 1,000 km for M = 6, and 10,000 km for M = 9, i.e., the preparation zone tends to the whole Earth surface for M = 9 (extreme seismic events like, for example, Sumatra 2004, Chile 2010, Japan 2011). The comparison between theoretical and field results showed a satisfactory agreement [30]. It was also observed that all the precursors tend to emerge within the circle.

An innovative and dynamic extension of the earthquake preparation zone is that, in addition to being proportional to the magnitude of the incoming earthquake [30], it also depends on time, i.e., on the average size of the cracks forming and growing in the Earth's crust before the seismic event. Approaching the earthquake occurrence, this area shrinks because of the closure of the pre-existing smaller cracks outside it, resulting in a smaller preparation zone where the remaining open cracks coalesce to form larger cracks. Eventually, on the day of the earthquake, the area will coincide with the quake epicenter. A typical strain localization develops in the time period before the seismic event. It can be supposed that, in the early stages of evolution of a seismic event, the preparation zone presents its maximum size. The nano- and micro-cracks, as well as the THz and GHz phonons, will prevail and, as a consequence, the neutron emission is more likely. This means that the neutron component is detected and monitored even at a considerable distance from the epicenter of the future earthquake. As an example, this experimental evidence was observed for the Sumatra earthquake of 2004 when, some days before the earthquake occurrence, significant anomalies in the neutron environmental flux were measured in Crimea and Kamchatka [23].

In the following stage, tectonic stresses tend to concentrate closer to the earthquake epicenter. As a result, the dimension of the preparation zone will shrink, the crack size will increase (from the micro up to the millimeter scale), and electromagnetic emissions in the GHz-MHz frequency range will occur [31–35].

Approaching the seismic event, a further size reduction of the preparation zone is expected, which is characterized by larger cracks (from the millimeter up to the meter scale), which are able to generate ultrasonic acoustic waves.

Finally, at the last stage, the preparation zone collapses to the earthquake epicenter, macro-cracks along the seismic faults coalesce and the earthquake takes place. In Fig. 16.2, a graphical representation of the conjectured model of preparation zone localization is illustrated. Each circle represents the border of the dynamic preparation zone, inside which the different fracto-emissions are subsequentially generated: NE (violet), EME (blue), and AE (red), whereas the black dot identifies the epicenter of the earthquake.

In the present chapter, after recalling the results obtained during two different experimental investigations at the "Testa Grigia" laboratory (Plateau Rosa, Cervinia, Italy) and at the seismic district of "Val Trebbia" (Bettola, Piacenza, Italy) [36, 37], the results are described that were acquired at a gypsum mine situated in Northern Italy (Murisengo, Alessandria) for the evaluation of acoustic, electromagnetic, and nuclear

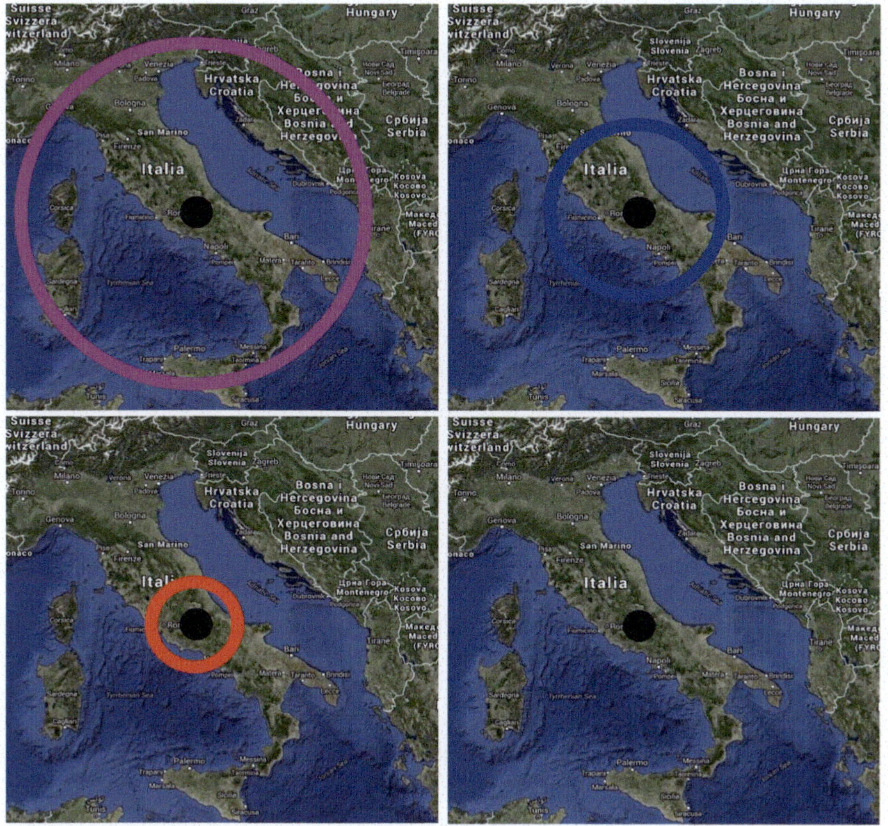

Fig. 16.2 Evolution of the earthquake preparation zone. Each circle represents the border of the evolving preparation zone, inside which the three different fracto-emissions are sequentially generated: NE (violet), EME (blue), AE (red). The black dot identifies the earthquake epicenter

phenomena. The novelty of this experimental investigation consists in the simultaneous acquisition of the three forms of fracto-emission and in their temporal correlation with the incoming seismic event that is obtained by means of an appropriate multi-modal statistical analysis [28, 29].

The monitoring system, based on the simultaneous acquisition of the various physical quantities, can assess the seismic risk. The results obtained during the in-situ monitoring reveal a high correlation between AE/EME/NE peaks and the major earthquakes occurring in the surrounding area. In particular, it appears very clear how the three fracto-emissions tend to anticipate the following seismic swarm peak with an evident and chronologically ordered sequence [28, 29].

Eventually, the correlation between seismic activity and lunar periodicity is emphasized. As a matter of fact, some studies have found that a large number of earthquakes take place during the periods of full or new Moon, when the gravitational forces acting on the Earth's crust are at their maximum level [38–43]. The gravitational forces, working as an additional stress field applied to the fault system and close to the critical condition, can trigger the cracking process that brings to fault evolution and propagation. In this context, the correlation between Moon cycle and seismicity strongly emerges.

16.2 Fracto-Emissions as Seismic Precursors

The seismic precursors are phenomena taking place well in advance with respect to the earthquake occurrence. Their parameters are of various kinds, such as ground deformation, changes in tilt and Earth tidal strain, in geoacoustic and geomagnetic fields, in radon and carbon dioxide contents, in environmental temperature and radioactivity, etc.

In the last few decades, a great number of laboratory tests and experimental observations have evidenced that mechanical, electromagnetic, and subatomic particle emissions, together with radon levels, carbon dioxide emanations, and temperature variations, are the most reliable natural phenomena that can be linked to earthquake preparation [44–46].

In particular, the experimental tests carried out at Politecnico di Torino since 2008 have demonstrated how the monitoring of the different forms of energy (AE, EME, NE), which are emitted during the failure of natural and artificial rocks, can enable a rational interpretation of the precursory phenomena, not only at the scale of the laboratory, but also at the Earth's crust tectonic scale [47–55].

Nowadays, the AE technique is well-known in the scientific community and applied for structural health monitoring purposes [3, 56–60]. In addition, the relation between AE and geological structure has been investigated for several years in order to clarify the AE implications as a precursor of earthquakes and volcanic eruptions [2, 61–63]. For example, an AE paroxysm, i.e. a large and almost abrupt increment in the AE signals, was observed at about 400 km of distance from the epicentral area before the occurrence of the Assisi earthquake [61, 62].

Considering that earthquakes always affect structural stability, an interpretation, in which AE and seismic events are linked both in space and time, seems to be possible. According to this approach, the correlation between AE activity in a masonry building and regional seismicity was investigated [2, 56, 57, 64].

In another recent work [65], a new method for evaluating seismic risk in regional areas based on the acoustic emission (AE) technique is proposed. Two important buildings of the Italian cultural heritage were considered: a chapel of the "Sacred Mountain of Varallo" and the "Asinelli Tower" in Bologna. The structures were monitored during earthquake sequences and, by using the Grassberger-Procaccia algorithm, a statistical method for space–time correlation between AE and seismic events was developed. That study emphasized how, under certain conditions, AE precedes the earthquake. In particular, a uniform AE activity in the 24 h prior to the earthquake was monitored.

On the other hand, EME signals are emitted by brittle materials, in which the fracture propagation occurs suddenly and is accompanied by abrupt stress drops in the stress–strain diagram. A number of laboratory studies revealed the existence of EME signals during fracture experiments carried out on a wide range of materials [5, 66–71]. It was also observed that the EME signals detected during material failure are analogous to the anomalous radiation of geo-electromagnetic waves observed before major earthquakes [71], reinforcing the idea that the EME effect can be applied as a forecasting tool for seismic events.

Several scientific reports emphasize how an objective correlation between anomalous electromagnetic signals and earthquakes is more and more plausible [72–77]. As an example, the Kobe earthquake occurred in Japan in 1995 was considered as a "big impact" in terms of pre-seismic electromagnetic changes monitored by several scientists working independently worldwide [75]. Moreover, kHz and MHz electromagnetic anomalies were also recorded before the L'Aquila (Italy) earthquake occurred on April 6, 2009 [78, 79]. Clear EME anomalies were revealed between 2 and 8 days before the occurrence of the earthquake.

As regards the neutron emissions (NE), their detections by Russian scientists [20–27] have led to consider also the Earth's crust, in addition to cosmic rays, as a relevant source of neutron flux variations. In particular, quoting from Volodichev: Neutron emissions measured in seismic areas of the Pamir region (4,200 m a.s.l.) exceeded the usual neutron background "up to two orders of magnitude in correspondence to seismic activity and rather appreciable earthquakes, greater than or equal to the 4th degree in the Richter scale magnitude" [27]. Considering the altitude dependence of neutron radiation (Pfotzer profile [80]), values approximately ten times higher than natural background at the sea level are usually detected at 5,000 m of altitude. Therefore, the same earthquakes occurring at the sea level should produce a neutron flux up to 1,000 times higher than the natural background at the same level.

Anomalous neutron measurements were carried out by Sigaeva et al. [23] before the Sumatra earthquake of December 26, 2004. Variations in the neutron flux were observed in different distant regions (Crimea and Kamchatka) some days before the earthquake. The peaks reached hundreds of background percentage.

As anticipated in Chap. 15, preliminary in-situ investigations were carried out following those by the Russian research groups. These data were acquired at the "Testa Grigia" laboratory of Plateau Rosa, Cervinia, during an experimental campaign on the evaluation of neutron radiation from cosmic rays [36, 37, 81–85]. Additional data refer to a dedicated experimental trial carried out in Northern Italy, at the seismic district of "Val Trebbia", Bettola, Piacenza [36, 37, 84].

16.3 Multi-modal Statistical Analysis of Seismic Data

Several seismic monitoring networks, just based on seismic acceleration, are being utilized worldwide presently, in California, Mexico, Taiwan, Turkey, Romania, and Japan [86]. However, any sort of multi-parameter monitoring system, which takes into account the simultaneous observation of different precursory phenomena, does not exist yet. To this purpose, from July 1st, 2013, to June 30, 2019, a dedicated in-situ monitoring seismic station at the San Pietro-Prato Nuovo gypsum mine, located in Murisengo (Alessandria, Northern Italy), has performed its activity. For six years, continuously, a rock pillar of the mine, located at about 100 m below the ground level, was subjected to a multi-parameter monitoring in order to evaluate the seismic risk of the surrounding area. The structural health monitoring was principally conducted through the AE technique, whereas the seismic risk of the surrounding area was monitored by the detection of AE/EME fluctuations, as well as by the environmental neutron flux fluctuations (NE).

Thanks to the position of the monitoring station (100 m underground), the acoustic and electromagnetic noise of human origin was greatly reduced, as well as the neutron background was between one and two orders of magnitude lower than that at the ground level, due to the remarkable reduction in neutrons of cosmic origin. These aspects make the mine an ideal place for fracto-emission monitoring.

The AE equipment consists of six USAM® units that can be synchronized for multi-channel data processing. Each unit contains a pre-amplified wideband piezo-electric sensor (PZT) sensitive to the frequency range between 50 and 800 kHz. The AE signals are pre-amplified and filtered through a bandpass filter, in order to have a high signal-to-noise ratio and a flat frequency response over a broad range. In addition, before the beginning of the acoustic emission monitoring, specific tests were carried out in order to exclude any sort of interference induced by the excavation activity (i.e., drilling and blasting) [87]. In particular, any type of noise was excluded, since the shock-wave frequency spectrum produced by such activities was observed to be outside the sensitivity range of the instrument used.

On the other hand, the EME device consists of a telescopic antenna having a maximum length of 125 cm. By pulling to the convenient length, the antenna can be tuned to operate at different frequencies. For this reason, it is a "wide band" device in the sense that it is possible to adjust its length according to the frequency/wavelength that the operator wishes to receive. If these frequencies are conveniently selected, the EME peak will be at mid-time between the AE and NE peaks. The antenna is then coupled with an Agilent Microwave300 oscilloscope for the acquisition of EME signals with frequencies up to 300 MHz.

As regards neutrons, it is well-known that they are electrically neutral particles, so that they cannot directly produce ionization in a detector, and therefore cannot be directly detected. This means that neutron detectors must rely upon a conversion process where an incident neutron interacts with a nucleus to produce a secondary charged particle. These charged particles are then detected, and from them the neutrons' presence is deduced. In particular, during the experimental trials at the gypsum mine, the environmental neutron flux monitoring was carried out by means of the AT1117M (ATOMTEX, Minsk, Republic of Belarus) neutron device. This type of detector provides a high sensitivity and wide measuring ranges (neutron energy range: 0.025 eV–14 MeV), with a fast response to radiation field change, ideal for any environmental monitoring purpose.

Fig. 16.3 The monitoring seismic station equipped with AE piezoelectric sensors, EME telescopic antenna and oscilloscope, and NE proportional counter

In Fig. 16.3, the monitoring set-up of the underground station is illustrated. From July 1st, 2013, the acquisition of acoustic and neutron environmental radiation parameters was performed, whereas only from February 15, 2015, the measuring platform was integrated with the acquisition of electromagnetic emissions by means of the telescopic antenna coupled with the oscilloscope.

The acquisition of the experimental data of acoustic emission, electromagnetic emission, neutron environmental flux, and seismic activity was carried out daily and collected on a monthly basis. Moreover, the period of monitoring was conveniently and graphically reported over twelve distinct semesters: from July 1st, 2013, to June 30, 2019.

The statistical analysis of the seismic distribution and of the three fracto-emissions was performed by means of a multi-modal (multi-peak) and Gaussian statistical approach. The software used for the statistical analysis is Microcal Origin. Given a specific discrete distribution of experimental points (one per each day), and applying suitable computational algorithms, the software determines the relative maxima of the distribution and, for each peak, evaluates the best Gaussian fitting by symmetrical or non-symmetrical bell-shaped curves.

Regarding the seismic activity, during the twelve semesters of measurement (from July 1st, 2013, to June 30, 2019), 572 earthquakes were observed of local magnitude greater than or equal to 1.8 in the Richter scale, within a circular geographical area of radius 100 km [87–89]. The threshold magnitude of 1.8 is selected because, considering the experimental evidence, this was found to be a sort of seismic off-set, below which no significant changes in the neutron flux are observed with respect to the natural background. By performing the multi-modal statistical analysis of temporal distributions of the 572 earthquakes detected during the twelve semesters of monitoring, 63 distinct seismic swarms (Gaussian peaks) of maximum epicenter magnitude between 2.5 and 4.7 in the Richter scale are identified. These earthquakes tend to generate a preparation zone estimated by the criteria proposed by Dobrovolsky et al. [30] and characterized by a radius between some tens and some hundreds kilometers. It has to be emphasized the difference between magnitude at the epicenter (M = 2.5–4.7) and local magnitude in Murisengo (M = 1.8–3.5).

Similar multi-modal statistical evaluations are also performed for acoustic, electromagnetic, and neutron emissions. In the case of AE events, the discrete distribution of the daily total number of observed acoustic events is considered, and twelve AE distributions (one per each semester) are identified with a total of 63 seismic swarms (Gaussian peaks).

For what concerns the electromagnetic emissions, the same multi-modal statistical analysis is performed, considering the daily total number of anomalous electromagnetic events. In this case, the data processing is limited to only nine semesters, since the acquisition of electromagnetic signals started on February 15, 2015. Compared to the case of AE distributions, only 46 EME Gaussian peaks are found.

Also for the neutron emissions, the multi-modal analysis is performed. As for the previous fracto-emissions, the daily integrated neutron flux (dose) is considered and 63 Gaussian peaks over the twelve semesters of monitoring are found.

As an example of application of multi-modal statistical analysis, the discrete distributions of the earthquakes and of the three fracto-emissions (AE, EME, NE) during the first half of the year 2015 are shown in Fig. 16.4a–d.

A temporal correlation between the seismic distributions and the corresponding fracto-emission distributions is considered. The comparison between the temporal distribution of earthquakes and those of fracto-emissions is provided in the following for the semesters of monitoring (Figs. 16.5a–n, 16.6a–i and 16.7a–n).

From the comparison between the different diagrams, it is evident the high correlation between acoustic, electromagnetic, neutron signals and the seismic swarms occurring in the surrounding areas. The three fracto-emission peaks tend to anticipate the next seismic peak (swarm) with an evident and chronologically ordered sequence.

As a matter of fact, this very peculiar behavior is observed in an extremely repeatable and systematic way for all the 63 detected seismic swarms. In particular, it is really impressive to observe how the acoustic emission peak anticipates the seismic peak (swarm) by approximately one day, the electromagnetic emission peak by 3–4 days, whereas the neutron peak by approximately one week. Therefore, they can

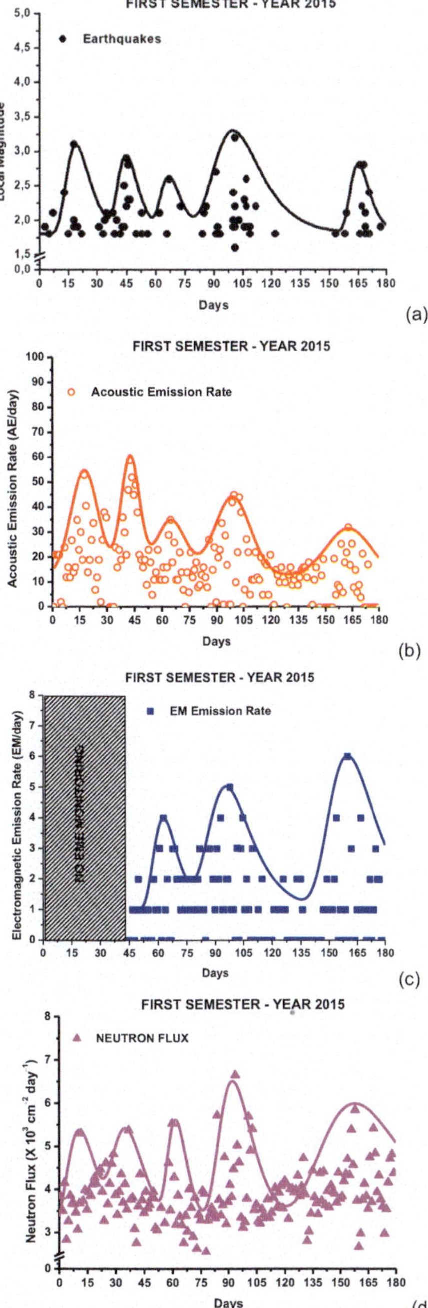

Fig. 16.4 Multi-modal Gaussian statistical analysis during the first semester 2015, for earthquakes (**a**), acoustic emissions (**b**), electromagnetic emissions (**c**), and neutron emissions (**d**)

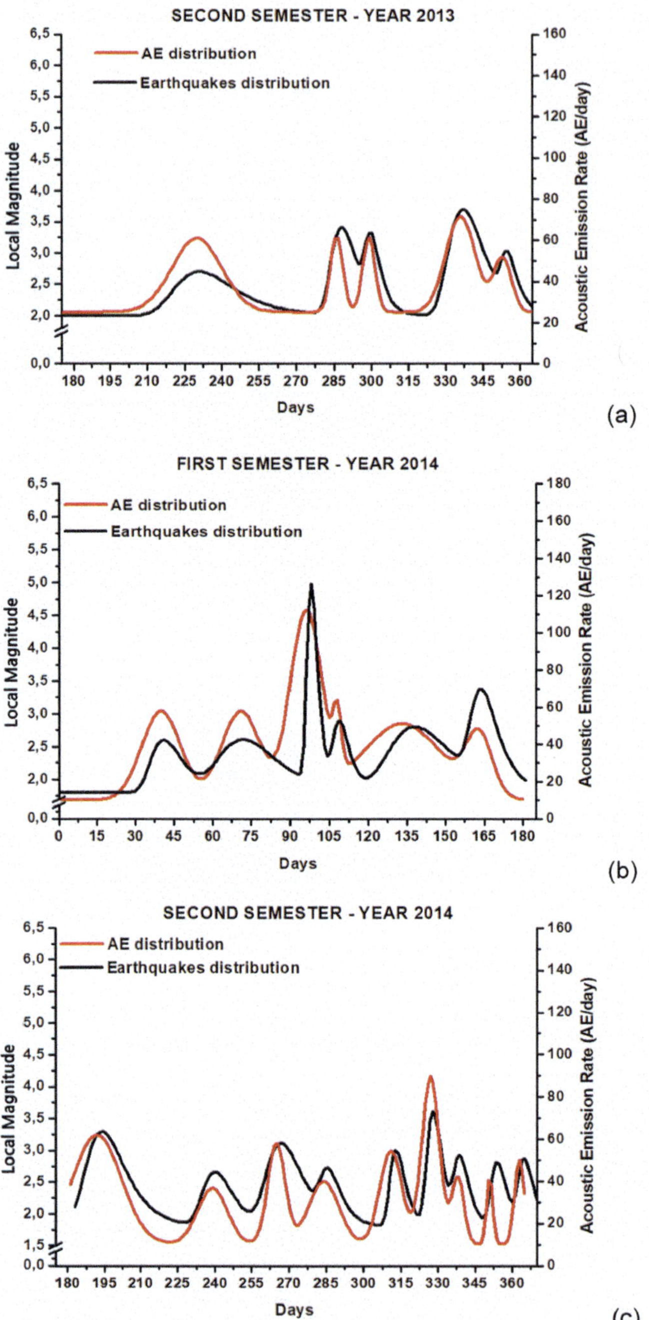

Fig. 16.5 a–n Earthquake versus acoustic emission temporal distributions for the twelve semesters of monitoring

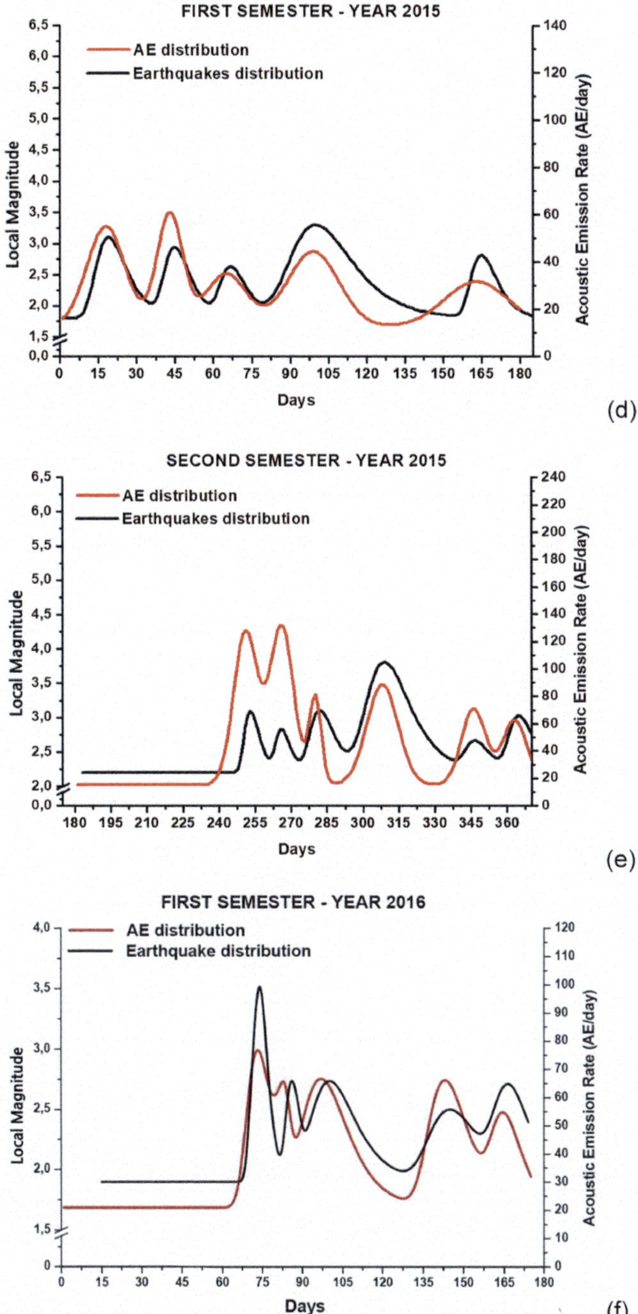

Fig. 16.5 (continued)

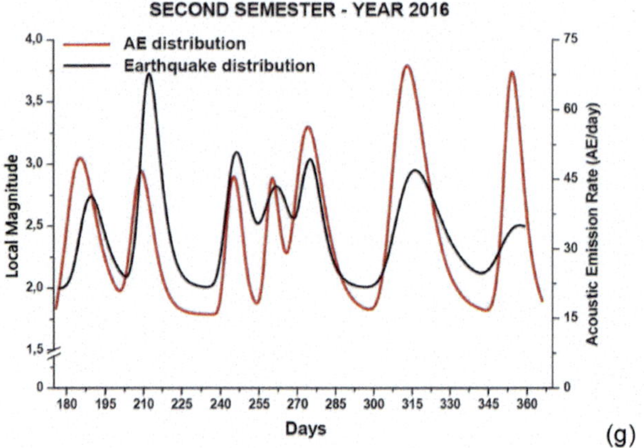

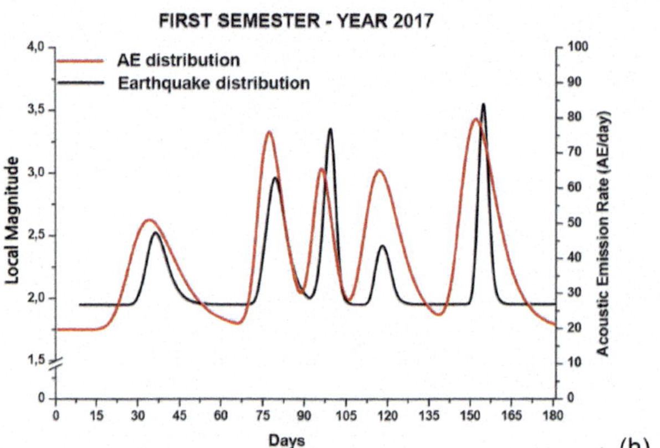

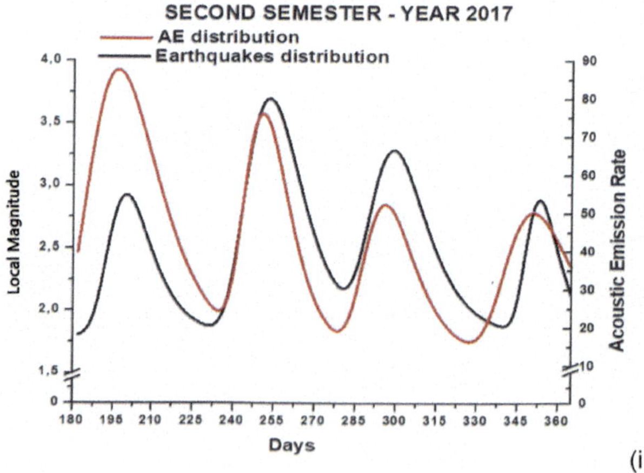

Fig. 16.5 (continued)

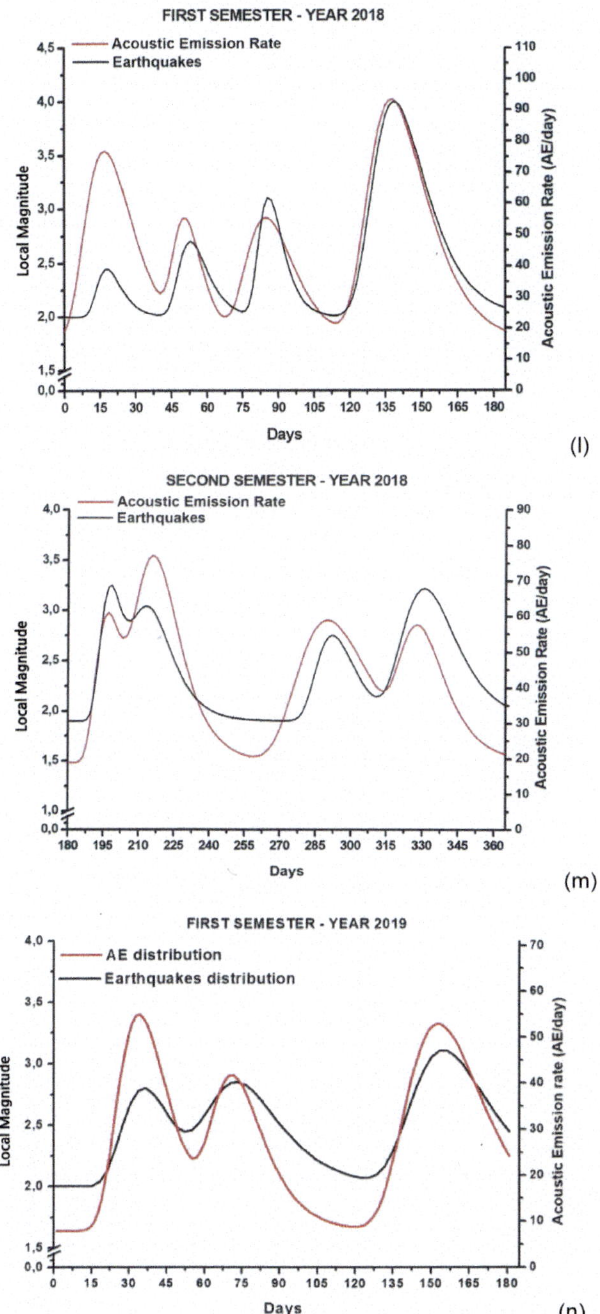

Fig. 16.5 (continued)

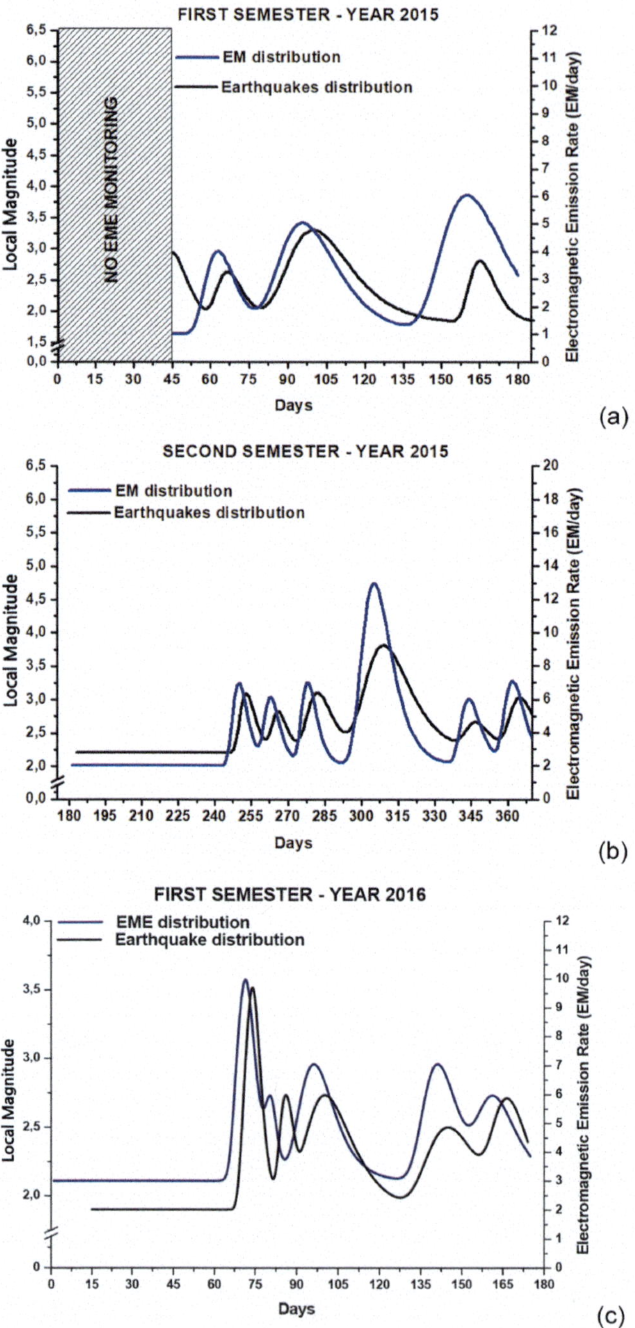

Fig. 16.6 a–i Earthquake versus electromagnetic emission temporal distributions for the nine semesters of monitoring

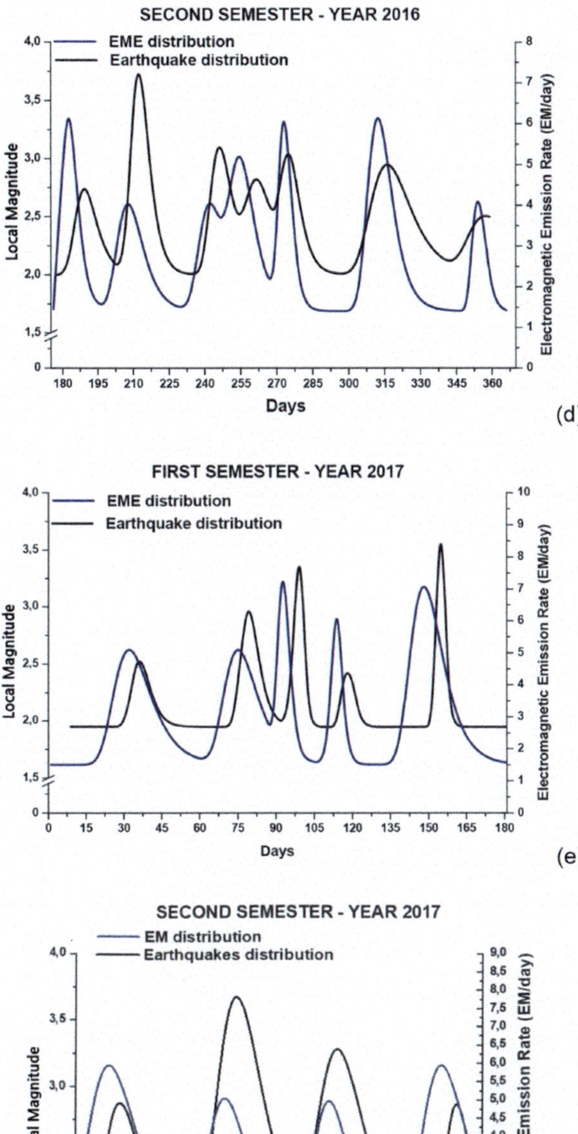

Fig. 16.6 (continued)

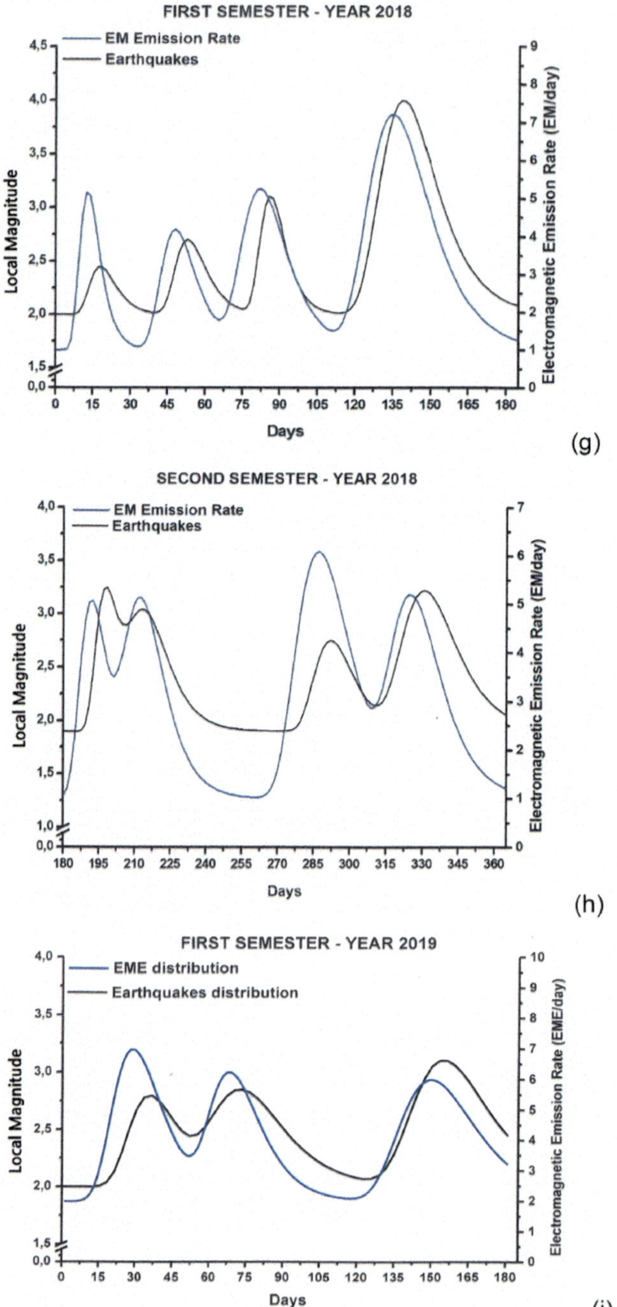

Fig. 16.6 (continued)

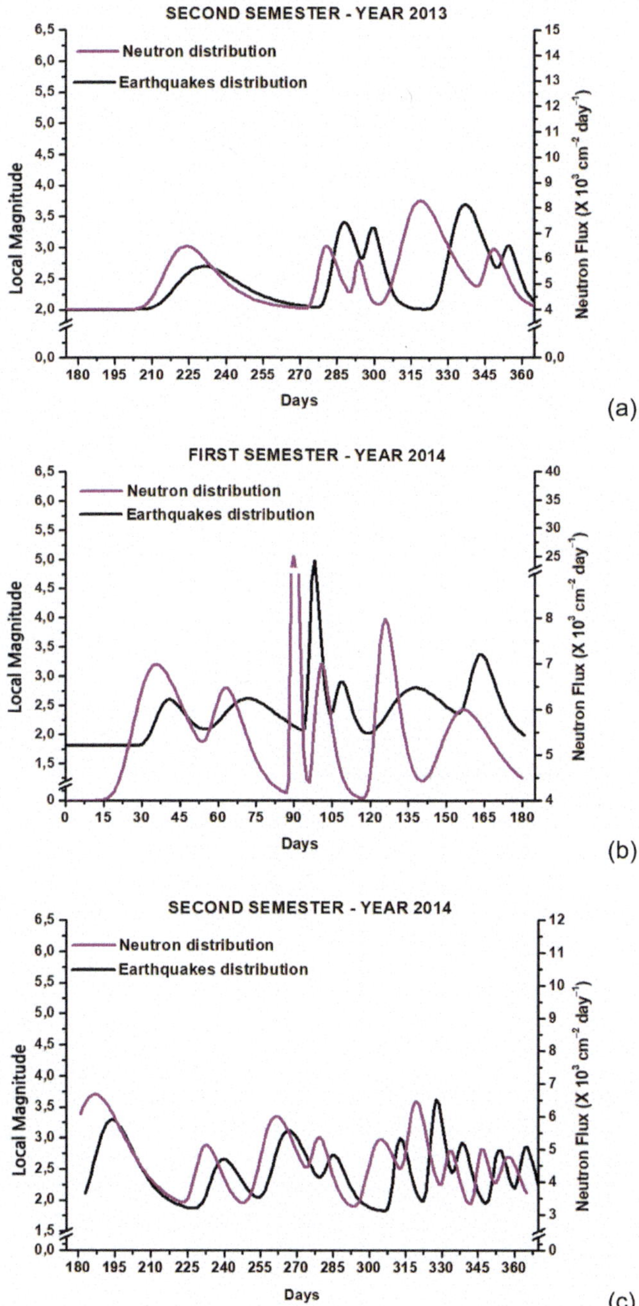

Fig. 16.7 **a–n** Earthquake versus neutron emission temporal distributions for the twelve semesters of monitoring

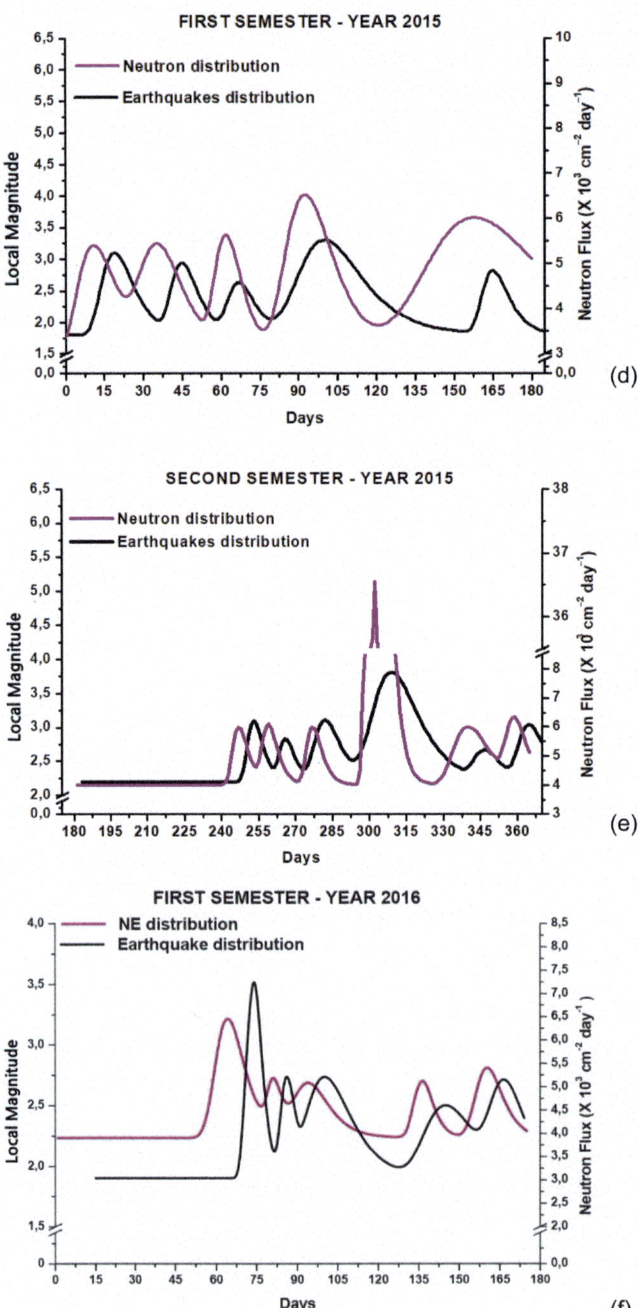

Fig. 16.7 (continued)

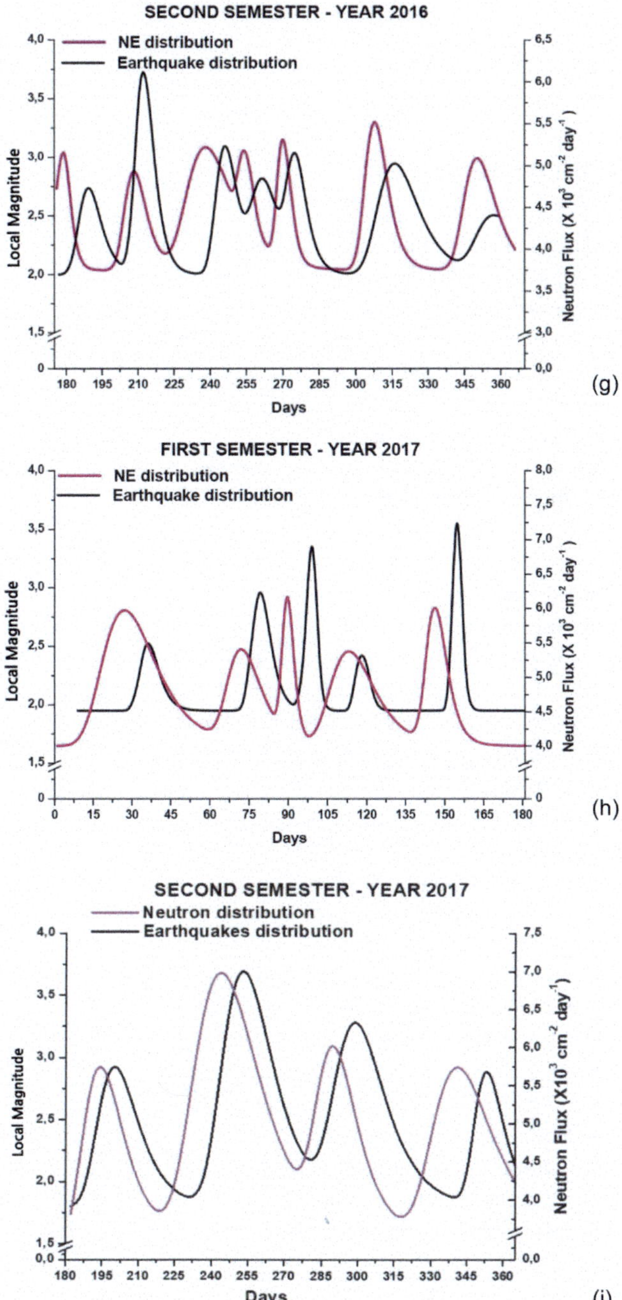

Fig. 16.7 (continued)

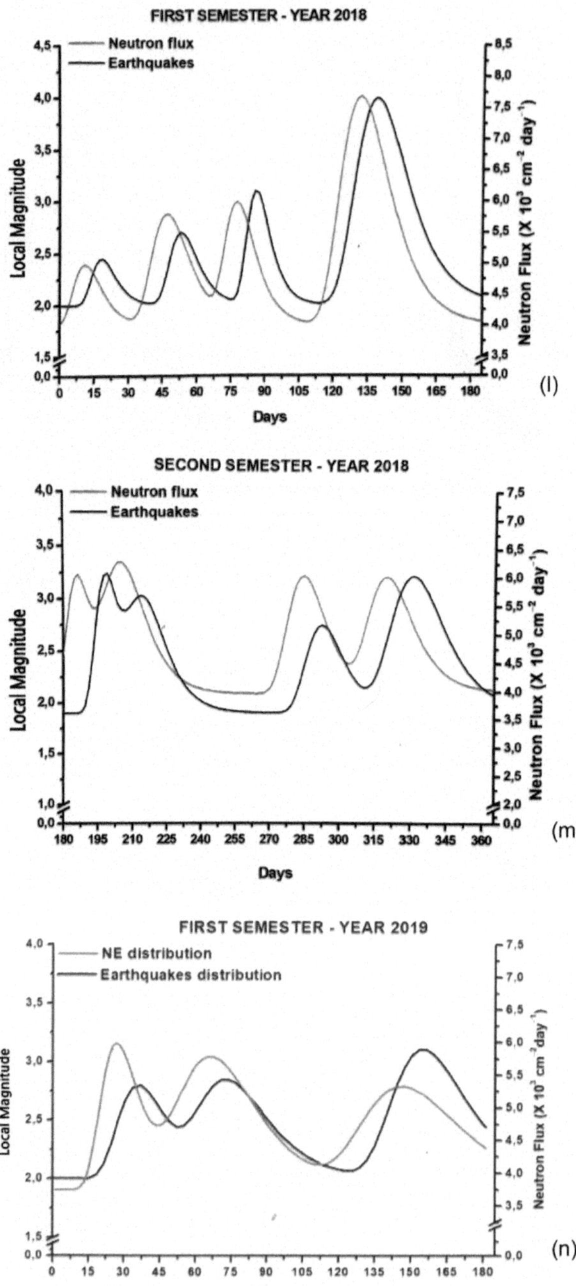

Fig. 16.7 (continued)

be considered as precursors of the following major earthquake, rather than aftershocks of the preceding one, on the basis of the statistical signal processing and of the different temporal distances.

In Figs. 16.8a–c and 16.9a–c, two of the 63 seismic swarms and the correlated fracto- emissions are considered. The seismic events refer to the swarms of April 2015 and November 2016, whose main events occurred on April 11, 2015, (Local $M = 3.2$) and on November 11, 2016, (Local $M = 3.0$), respectively. The fracto-emission peaks anticipate the seismic activity peak very clearly and with a regular and repeatable temporal sequence [89].

The lead-time standard deviation for each fracto-emission, equal to 31% for AE, 29% for EME, and 17% for NE, indicates a rather small statistical dispersion of the experimental data (Fig. 16.10a–c).

On the other hand, as regards the correlation between amplitude of the single fracto-emission peaks and the corresponding seismic swarm peak local magnitude, some preliminary considerations can be made. In Fig. 16.11a–c, the correlation diagrams of the fracto-emission peak intensity versus the local magnitude of the seismic peak are reported. We have enough data only up to $M = 4$, whereas it would be necessary to know additional data for the more intense seismic events, so to be in the condition to extend the correlation diagrams and to establish procedures and protocols for a possible evacuation of the population.

16.4 Correlation Between Seismic Activity and Lunar Periodicity

Since the end of the nineteenth century, several investigations have been carried out concerning the question whether gravitational forces induced by Moon and Sun can trigger earthquakes and volcanic activities [38–43]. In particular, these studies emphasize that the relationship between earthquake occurrence and lunar phases seems to be stronger during the periods of full or new Moon, that are when Sun, Moon, and Earth are aligned. As a matter of fact, in that case the gravitational attractions and the tidal forces acting on the Earth's surface are more intense. On the contrary, at first and third quarter Moon phases, lunar and solar tides are in opposition and the tidal forces are at their minimum.

Two different types of tide can induce elastic deformation on the Earth's crust: the solid Earth's tide and the oceanic tide [90]. The former is due to the gravitational forces exerted by Sun and Moon onto the Earth crust; it generates strains and, consequently, leads the surface of the Earth to yield radially up to tens of centimeters. This phenomenon happens inadvertently, since the wavelength of the ground oscillation is of the order of thousands km. However, if it occurs in a critical area where a great amount of strain energy (or stress) is accumulated and the pressure gradient is close to the breakpoint, the tidal effect can trigger the nucleation of failure processes. Then, the crack growth along the seismic faults will accelerate until the

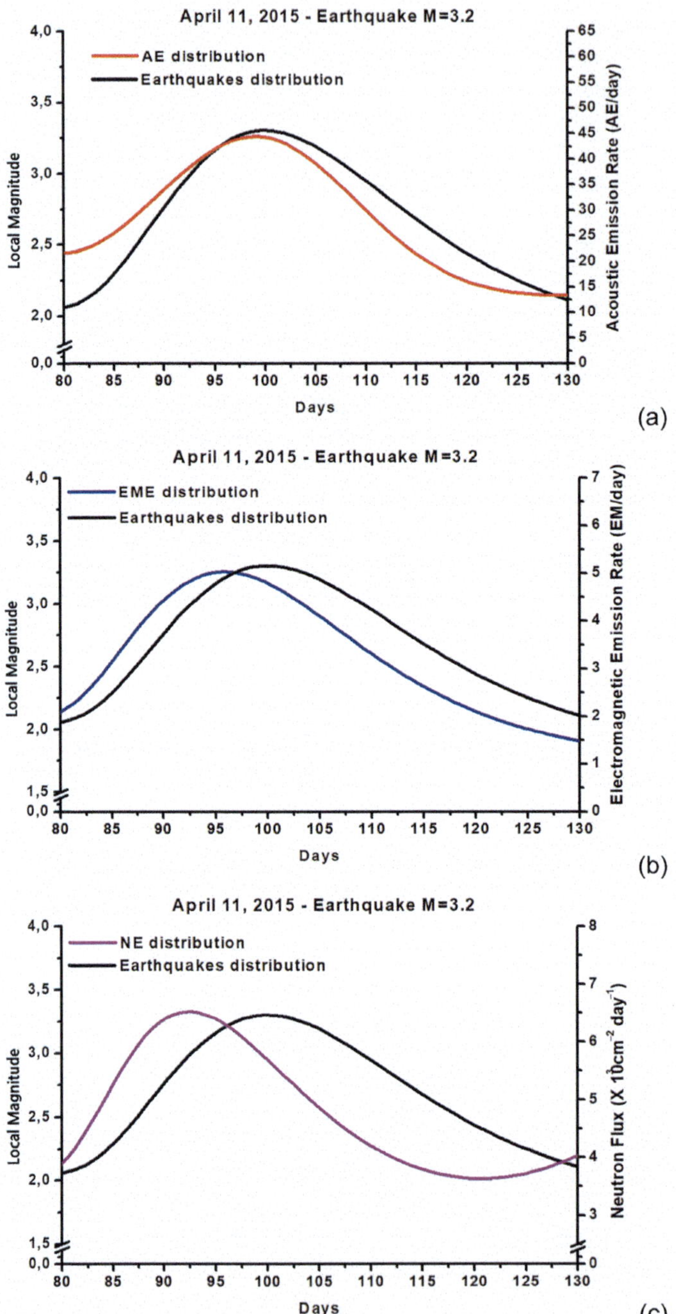

Fig. 16.8 a–c Anticipated and differently shifted Gaussian distribution peaks of AE/EME/NE emissions for the earthquake of April 11, 2015

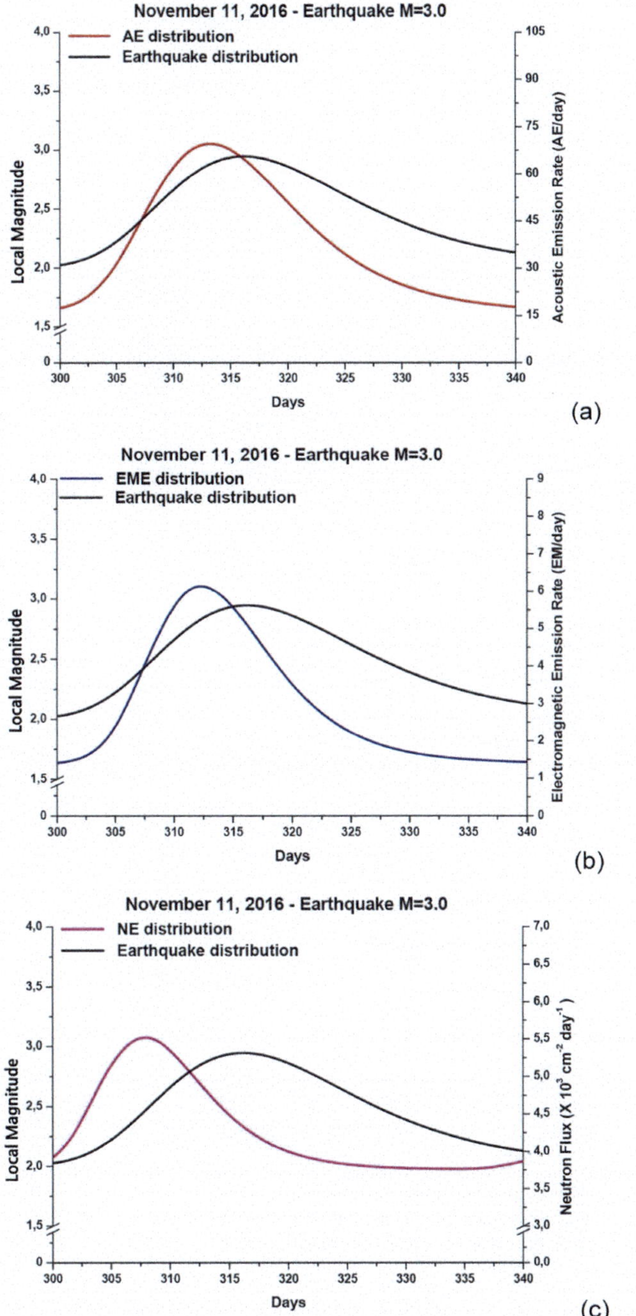

Fig. 16.9 a–c Anticipated and differently shifted Gaussian distribution peaks of AE/EME/NE emissions for the earthquake of November 11, 2016

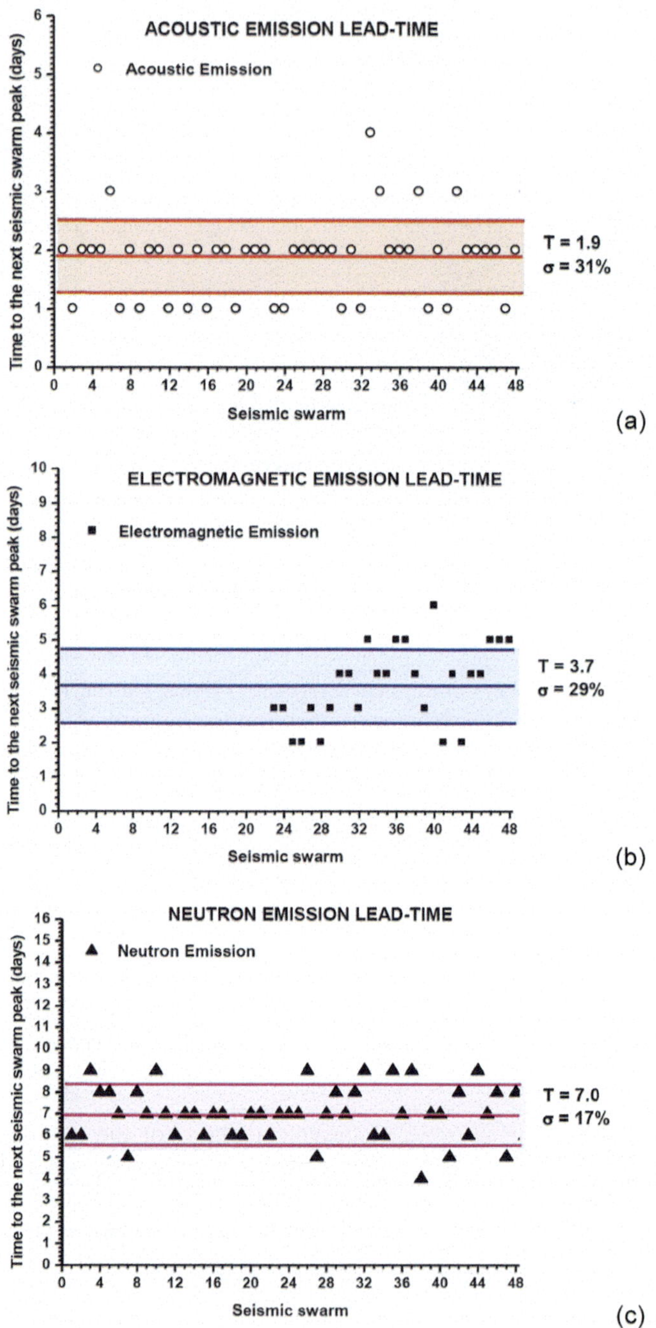

Fig. 16.10 a–c Lead-time statistical dispersion of the three fracto-emissions (first four years of monitoring)

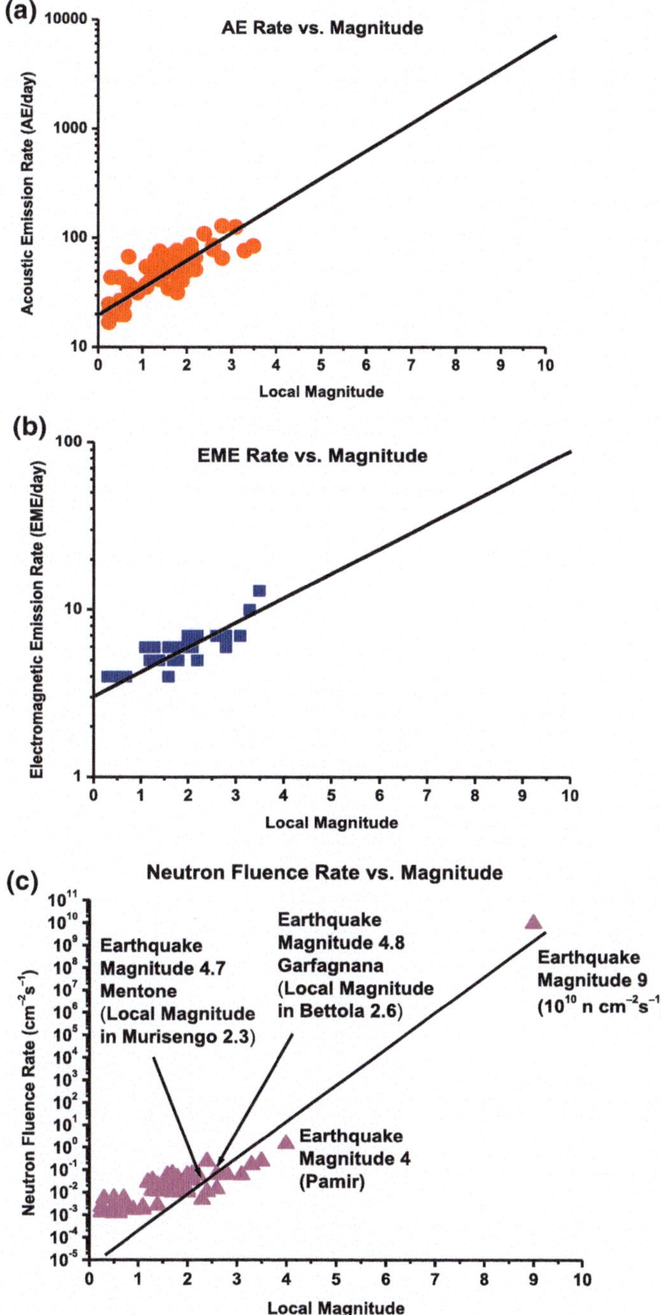

Fig. 16.11 **a–c** Correlation between fracto-emission peak intensity and local magnitude of the incoming earthquake

earthquake occurrence. The latter term, instead, is caused by the effect of the oceanic tides on the Earth's crust. Considering then the individual effects of Sun and Moon, the latter presents a larger gravitational pull force on the Earth than the Sun because, even if the Sun has a greater mass, the Earth-Sun distance is much larger than that Earth-Moon.

Although the assumption that tidal forces can trigger earthquakes is still debated, it is becoming more plausible thanks to some research studies conducted along the San Andrea fault in California [43, 91, 92]. By detecting the several seismic events occurring in the surrounding area of the fault, the researchers observed that, when the gravitational forces caused by Sun and Moon are added to the tectonic force that pushes the Pacific Ocean plaque against the American continent, more earthquakes occur.

A further experimental evidence was published in the Nature Geoscience journal [93], where the Authors assess that, when tides are large enough, small earthquakes tend to grow because the strong variation in the tide intensity accelerates the tectonic movement along a fault and increases the probability of seismic events. Then, the gravitational forces acting on the Earth's crust change the stress field along the faults leading to seismic tremors up to significant or even very strong earthquakes. As a matter of fact, the research carried out by Ide et al. [93] reports that the Sumatra earthquake of 2004, with a magnitude of 9.1, and the seismic event of Maule (Chile) in 2010, with a magnitude of 8.8, took place during the full Moon phase, when the tidal stress level was the maximum [93]. Hence, although these stress variations due to gravitational forces (stress jumps of about 10^3 Pa) are much smaller than typical stress drops during a seismic event (up to 10^7 Pa) [94], nevertheless their combined effect, applied to a fault system close to failure, can achieve the critical condition.

Also the sunspots, and in turn the solar activity, could influence the earthquake occurrence [95]. As a consequence, the solar-terrestrial interaction could play some role on seismicity, involving a sort of coupling among Sun, solar wind, magnetosphere, and lithosphere. In addition, geomagnetic storms, caused by Coronal Mass Ejections (CMEs) might induce eddy electric currents in the rocks along faults, reducing their shear resistance [96, 97], and thus possibly increasing the seismic activity.

16.4.1 Experimental Evidence at the "San Pietro-Prato Nuovo" Gypsum Mine (Murisengo)

The correlation between lunar phases and seismic activity in the surrounding area of Murisengo gypsum mine is explored considering the 48 seismic swarms occurred during the first eight semesters of monitoring, from July 1st, 2013, to June 30, 2017.

It is possible to observe that the mean waiting time between two consecutive seismic swarms is of about 30 days (48 seismic swarms in 48 months), i.e., the average duration of a synodic month, which is the period of Moon's revolution.

Therefore, this experimental evidence appears to be not a simple coincidence. As a matter of fact, comparing the seismic swarm occurrence times with the times of full and new Moon, an appreciable temporal correlation emerges.

In Fig. 16.12a–h, the temporal distributions of the 48 seismic swarms together with the lunar phases are reported. A sinusoidal curve is used to represent the lunar cycle. In particular, the maximum and minimum values of the sinusoidal function represent the full and new Moon, respectively. On the other hand, the occurrence of the seismic swarm peak is identified by a black dot. Considering the various diagrams, a rather close correlation appears between lunar periodicity and seismic activity. For the majority of cases, the seismic swarm takes place on the same day, or less than 2–3 days before/after the occurrence of full or new Moon. Therefore, the experimental evidence at Murisengo seems to show that, not only high magnitude earthquakes can be affected by the gravitational forces exerted by the Sun and the Moon [93], but also seismic swarms and low-magnitude earthquakes tend to happen under the influence of our Star and our Satellite.

16.5 Conclusions

The experimental results obtained at the San Pietro-Prato Nuovo gypsum mine in Murisengo (Alessandria, Italy) emphasize the very close correlation between acoustic, electromagnetic, neutron emissions and seismic activity. In particular, it was detected that the three different fracto-emissions regularly, repeatably, and systematically, with a temporally ordered sequence, anticipate the seismic event by approximately one day, 3–4 days, and one week, respectively.

This experimental investigation can be considered as a starting point for the design and installation of additional monitoring stations in other geographical areas where the seismic activity is significantly greater than that of Murisengo, such as in Sicily (Italy), Greece, Guangdong (China), California (USA), or Japan. Applying the methodology and the experimental approach of the gypsum mine, it would be possible to create a multi-parameter monitoring system to identify epicenter, time, and magnitude of the incoming earthquake, and to prevent its effects well in advance.

The installation of experimental devices for the acquisition of fracto-emission data in geographical regions characterized by high-magnitude earthquakes could be likely performed at the ground level and not necessarily underground as in the case of the Murisengo gypsum mine. More intense fracto-emission peaks, better distinguishable from the natural environmental background levels, should in fact be expected. In particularly seismic areas, it should also be possible to define more accurately the seismic off-set based on neutron emission (currently evaluated as equal to 1.8 in the Richter scale), and to better evaluate the time correlation between the three fracto-emission peaks and the incoming earthquake. A longer temporal shift of the three fracto-emission peaks with respect to the seismic event is in fact expected.

An innovative extension of the concept of earthquake preparation zone is proposed with reference to the temporal evolution of it. The preparation zone depends not only

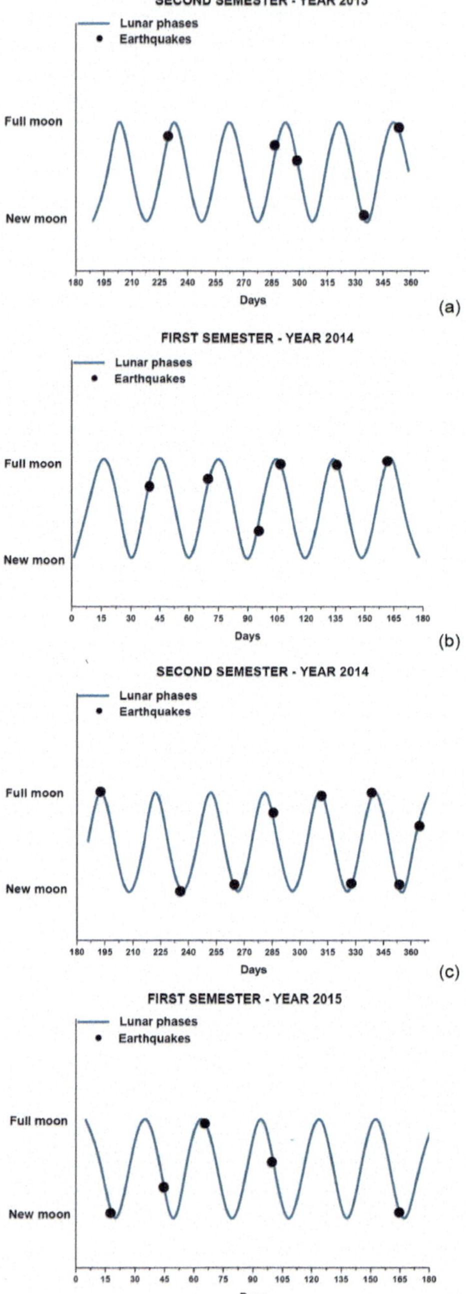

Fig. 16.12 a–h Correlation between lunar periodicity and earthquake occurrence for the first eight semesters of monitoring (July 2013–June 2017)

Fig. 16.12 (continued)

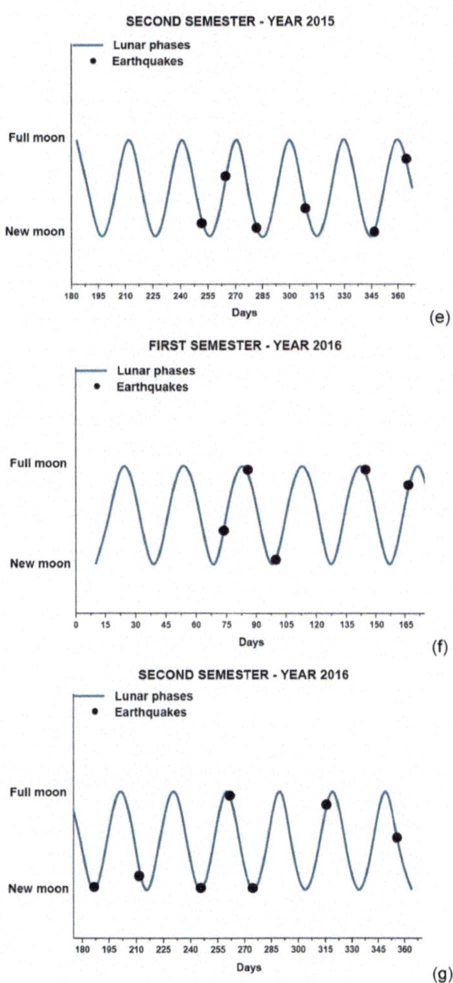

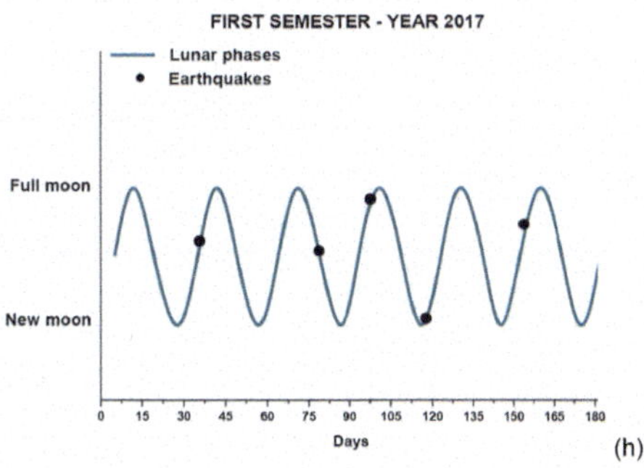

on the magnitude of the incoming earthquake, but also on time through the average crack size development in the Earth's crust. Approaching the earthquake occurrence, the preparation zone shrinks because of the closure of the pre-existing smaller cracks, resulting in a new area where the remaining cracks grow and coalesce to form longer cracks. Accordingly, the preparation zone continues to reduce in size until it coincides with the epicenter of the earthquake.

Eventually, a correlation between lunar phases and seismic activity monitored in the surroundings of Murisengo is verified. As a matter of fact, during the first eight semesters of monitoring, 48 major seismic swarms were observed, approximately one every 30 days, which is the typical duration of a synodic month. The experimental observations have evidenced that the seismic swarm peaks take place on the same day, or less than 2–3 days before/after, as the occurrence of full or new Moon, when the gravitational forces acting on the Earth's surface are at their maximum level. Therefore, similarly to the assumption that high-magnitude earthquakes are linked to lunar cycles, so also seismic swarms or low-magnitude earthquakes are influenced by the tidal forces exerted by the Sun and the Moon.

References

1. Ashcroft NW, Mermin DN (2013) Solid state physics, cengage learning, Delhi
2. Carpinteri A, Lacidogna G, Niccolini G (2007) Acoustic emission monitoring of medieval towers considered as sensitive earthquake receptors. Nat Hazard 7:251–261
3. Lockner DA et al (1991) Quasi-static fault growth and shear fracture energy in granite. Nature 350:39–42
4. Carpinteri A et al (2012) Mechanical and electromagnetic emissions related to stress-induced cracks. Exp Tech 36:53–64
5. Rabinovitch A, Frid V, Bahat D (2007) Surface oscillations. A possible source of fracture induced electromagnetic oscillations. Tectonophysics 431:15–21
6. Batzel RE, Seaborg GT (1951) Fission of medium weight elements. Phys Rev 82:607–615
7. Bridgman PW (1927) The breakdown of atoms at high pressures. Phys Rev 29:188–191
8. Carpinteri A (2015) TeraHertz phonons and piezonuclear reactions from nano-scale mechanical instabilities. In: Carpinteri A, Lacidogna G, Manuello A (eds) Acoustic, electromagnetic, neutron emissions from fracture and earthquakes. Springer International Publishing, Switzerland, pp 1–10
9. Cook ND (2010) Models of the atomic nucleus, 2nd edn. Springer, Dordrecht
10. Cook ND, Manuello A, Veneziano D, Carpinteri A (2015) Piezonuclear fission reactions simulated by the lattice model of the atomic nucleus. In: Carpinteri A, Lacidogna G, Manuello A (eds) Acoustic, electromagnetic, neutron emissions from fracture and earthquakes. Springer International Publishing, Switzerland, pp 219–235
11. Fulmer CB et al (1967) Evidence for photofission of iron. Phys Rev Lett 19:522–523
12. Hagelstein PL, Letts D, Cravens D (2010) Terahertz difference frequency response of PdD in two-laser experiments. J Cond Mat Nucl Sci 3:59
13. Hagelstein PL, Chaudhary IU (2015) Anomalies in fracture experiments and energy exchange between vibrations and nuclei. Meccanica 50:1189–1203
14. Lucia U, Carpinteri A (2015) GeV plasmons and spalling neutrons from crushing of iron-rich natural rocks. Chem Phys Lett 640:112–114
15. Widom A, Swain J, Srivastava YN (2013) Neutron production from the fracture of piezoelectric rocks. J Phys G: Nucl Part Phys 40(15006):1–8

16. Widom A, Swain J, Srivastava YN (2015) Photo-disintegration of the iron nucleus in fractured magnetite rocks with magnetostriction. Meccanica 50:1205–1216
17. Derjaguin BV et al (1989) Titanium fracture yields neutrons? Nature 34:492
18. Diebner K (1962) Fusionsprozesse mit hilfe konvergenter stosswellen—Einige aeltere und neuere versuche und ueberlegungen. Kerntechnik 3:89–93
19. Fujii MF et al (2002) Neutron emission from fracture of piezoelectric materials in deuterium atmosphere. Jpn J Appl Phys Pt.1 41:2115–2119
20. Antonova VP, Volodichev NN, Kryukov SV, Chubenko AP, Shchepetov AL (2009) Results of detecting thermal neutrons at Tien Shan high altitude station. Geomag Aeron 49(6):761–767
21. Kuzhevskij M et al (2003) Neutron flux variations near the Earth's crust. A possible tectonic activity detection. Nat Hazards Earth Syst Sci 3:637–645
22. Kuzhevskij M, Nechaev OY, Sigaeva EA (2003) Distribution of neutrons near the Earth's surface. Nat Hazard 3:255–262
23. Sigaeva EA et al (2006) Thermal neutrons' observations before the Sumatra earthquake. Geophys Res Abstr 8:00435
24. Sobolev GA, Shestopalov IP, Kharin EP (1998) Implications of solar flares for the seismic activity of the Earth. Izvestiya. Phys Solid Earth 34:603–607
25. Volodichev NN, Kuzhevskij BM, Nechaev OY, Panasyuk MI, Shavrin PI (1997) Phenomenon of neutron intensity bursts during new and full Moons. Cosm Res 31(2):135–143
26. Volodichev NN et al (1999) Lunar periodicity of the neutron radiation burst and seismic activity on the Earth. In: Proceedings of the 26th international cosmic ray conference, Salt Lake City
27. Volodichev NN, Kuzhevskij BM, Nechaev OY, Panasyuk MI, Podorolsky AN, Shavrin PI (2000) Sun-moon-earth connections: the neutron intensity splashes and seismic activity. Astron Vestnik 34:188–190
28. Carpinteri A, Borla O (2017) Fracto-emissions as seismic precursors. Eng Fract Mech 177:239–250
29. Carpinteri A, Borla O (2019) Acoustic, electromagnetic, and neutron emissions as seismic precursors: the lunar periodicity of low-magnitude seismic swarms. Eng Fract Mech 210:29–41
30. Dobrovolsky IP, Zubkov SI, Miachkin VI (1979) Estimation of the size of earthquake preparation zones. Pure Appl Geophys 117(5):1025–1044
31. Eftaxias K, Potirakis SM (2013) Current challenges for pre-earthquake electromagnetic emissions: shedding light from micro-scale plastic flow, granular packings, phase transitions and self-affinity notion of fracture process. Nonlinear Processes Geophys 20:771–792
32. Potirakis SM, Contoyiannis Y, Melis NS, Kopanas J, Antonopoulos G, Balasis G, Kontoes C, Nomicos C, Eftaxias K (2016) Recent seismic activity at Cephalonia (Greece): a study through candidate electromagnetic precursors in terms of non-linear dynamics. Nonlinear Processes Geophys 23:223–240
33. Potirakis SM, Contoyiannis Y, Eftaxias K, Koulouras G, Nomicos C (2015) Recent field observations indicating an Earth system in critical condition before the occurrence of a significant earthquake. IEEE Geosci Remote Sens Lett 12(3):631–635
34. Donner RV, Potirakis SM, Balasis G, Eftaxias K, Kurths J (2015) Temporal correlation patterns in pre-seismic electromagnetic emissions reveal distinct complexity profiles prior to major earthquakes. Phys Chem Earth 85–86:44–55
35. Kalimeris A, Potirakis SM, Eftaxias K, Antonopoulos G, Kopanas J, Nomikos C (2016) Multi-spectral detection of statistically significant components in pre-seismic electromagnetic emissions related with Athens 1999, $M = 5.9$ earthquake. J Appl Geophys 128:41–57
36. Borla O, Lacidogna G, Zanini A, Carpinteri A (2012) The phenomenon of neutron emission from earthquakes. In: Fracture mechanics for durability, reliability and safety (Proceedings of the 19th European conference on fracture, Kazan, Russia, 2012), CD-ROM, Paper N. 620
37. Borla O, Lacidogna G, Carpinteri A (2016) Piezonuclear neutron emissions from earthquakes and volcanic eruptions. In: Carpinteri A, Lacidogna G, Manuello A (eds) Acoustic, electromagnetic, neutron emissions from fracture and earthquakes. Springer International Publishing, Switzerland, pp 135–151

38. Schuster A (1897) On lunar and solar periodicities of earthquakes. Proc R Soc London 61:455–465
39. Cotton LA (1922) Earthquake frequency with spatial reference to tidal stresses in the lithosphere. Bull Seism Soc Am 12:47–198
40. Mcnutt SR, Beavan RJ (1981) Volcanic earthquakes at Pavlof volcano correlated with the solid Earth tide. Nature 294:615–618
41. Rydelek PA et al (1988) Tidal triggering of earthquake swarms at Kilauea Volcano, Hawaii. J Geophys Res 93:4401–4411
42. Emter D (1997) Tidal triggering of earthquakes and volcanic events. Lect Notes Earth Sci 66:293–309
43. Witze A (2016) Moon's pull can trigger big earthquakes. Nat News. https://doi.org/10.1038/nature.2016.20551
44. Potirakis S, Minadakis G, Eftaxias K (2012) Analysis of electromagnetic preseismic emissions using Fisher information and Tsallis entropy. Phys A 391:300–306
45. Immè G, Morelli D (2012) Radon as earthquake precursor. In: D'Amico S (ed) Earthquake research and analysis—Statistical studies, observations and planning. InTech. https://doi.org/10.5772/29917
46. Padron E et al (2008) Changes on diffuse CO_2 emission and relation to seismic activity in and around El Hierro, Canary Islands. Pure Appl Geophys 165:95–114
47. Carpinteri A, Lacidogna G, Manuello A (eds) (2015) Acoustic, electromagnetic, neutron emissions from fracture and earthquakes. Springer
48. Carpinteri A, Cardone F, Lacidogna G (2009) Piezonuclear neutrons from brittle fracture: early results of mechanical compression tests. Strain 45:332–339
49. Carpinteri A, Borla O, Lacidogna G, Manuello A (2010) Neutron emissions in brittle rocks during compression tests: monotonic versus cyclic loading. Phys Mesomech 13:264–274
50. Carpinteri A, Cardone F, Lacidogna G (2010) Energy emissions from failure phenomena: mechanical, electromagnetic, nuclear. Exp Mech 50:1235–1243
51. Carpinteri A, Lacidogna G, Manuello A, Borla O (2011) Energy emissions from brittle fracture: neutron measurements and geological evidences of piezonuclear reactions. Strength Fracture Complex 7:13–31
52. Carpinteri A, Chiodoni A, Manuello A, Sandrone R (2011) Compositional and microchemical evidence of piezonuclear fission reactions in rock specimens subjected to compression tests. Strain 47:267–281
53. Carpinteri A, Lacidogna G, Manuello A, Borla O (2012) Piezonuclear fission reactions in rocks: evidences from microchemical analysis, neutron emission, and geological transformation. Rock Mech Rock Eng 45:445–459
54. Carpinteri A, Lacidogna G, Borla O, Manuello A, Niccolini G (2012) Electromagnetic and neutron emissions from brittle rocks failure: experimental evidence and geological implications. Sadhana 37:59–78
55. Carpinteri A, Lacidogna G, Manuello A, Borla O (2013) Piezonuclear fission reactions from earthquakes and brittle rocks failure: evidence of neutron emission and nonradioactive product elements. Exp Mech 53:345–365
56. Carpinteri A, Lacidogna G (2006) Damage monitoring of an historical masonry building by the acoustic emission technique. Mater Struct 39:161–167
57. Carpinteri A, Lacidogna G (2006) Structural monitoring and integrity assessment of medieval towers. J Struct Eng 132:1681–1690
58. Mogi K (1962) Study of elastic shocks caused by the fracture of heterogeneous materials and its relation to earthquake phenomena. Bull Earthq Res Inst 40:125–173
59. Ohtsu M (1996) The history and development of acoustic emission in concrete engineering. Mag Concr Res 48:321–330
60. Shcherbakov R, Turcotte DL (2003) Damage and self-similarity in fracture. Theoret Appl Fract Mech 39:245–258
61. Gregori GP, Paparo G (2004) Acoustic emission (AE). A diagnostic tool for environmental sciences and for non destructive tests (with a potential application to gravitational antennas).

In: Schroeder W (ed) Meteorological and geophysical fluid dynamics, Science edn., Bremen, pp 166–204
62. Gregori GP, Paparo G, Poscolieri M, Zanini A (2005) Acoustic emission and released seismic energy. Nat Hazards Earth Syst Sci 5:777–782
63. Gregori G, Poscolieri M, Paparo G, De Simone S, Rafanelli C, Ventrice G (2010) Storms of crustal stress and AE earthquake precursors. Nat Hazard 10:319–337
64. Carpinteri A, Lacidogna G (2003) Damage diagnosis in concrete and masonry structures by acoustic emission technique. J Facta Universitatis Ser: Mech Autom Control Robot 3(13):755–764
65. Lacidogna G, Cutugno P, Niccolini G, Invernizzi S, Carpinteri A (2015) Correlation between earthquakes and AE monitoring of historical buildings in seismic areas. Appl Sci 5:1683–1698
66. Carpinteri A, Lacidogna G, Manuello A, Niccolini G, Schiavi A, Agosto A (2010) Mechanical and electromagnetic emissions related to stress-induced cracks. Exp Tech 36(3):53–64
67. Frid V, Rabinovitch A, Bahat D (2003) Fracture induced electromagnetic radiation. J Phys D 36:1620–1628
68. Lacidogna G, Carpinteri A, Manuello A, Durin G, Schiavi A, Niccolini G, Agosto A (2010) Acoustic and electromagnetic emissions as precursor phenomena in failure processes. Strain 47(2):144–152
69. Miroshnichenko M, Kuksenko V (1980) Study of electromagnetic pulses in initiation of cracks in solid dielectrics. Soviet Phys-Solid State 22:895–896
70. O'Keefe SG, Thiel DV (1995) A mechanism for the production of electromagnetic radiation during fracture of brittle materials. Phys Earth Planet Int 89:127–135
71. Warwick JW, Stoker C, Meyer TR (1982) Radio emission associated with rock fracture: possible application to the great Chilean earthquake of May 22, 1960. J Geophys Res 87:2851–2859
72. Biagi P (1999) Seismic effects on LF radiowaves. In: Hayakawa M (ed) Atmospheric and ionospheric electromagnetic phenomena associated with earthquakes. Terrapub, Tokyo, pp 535–542
73. Hayakawa M, Fujinawa Y (eds) (1994) Electromagnetic phenomena related to earthquake prediction, Terrapub, Tokyo, p 677
74. Lighthill S (ed) (1996) A critical review of VAN. In: Earthquake prediction from seismic electric signals. World Scientific Publishing, Singapore, p 276
75. Nagao T, Enomoto Y, Fujinawa Y, Hata M, Hayakawa M, Huang Q, Izutsu J, Kushida Y, Maeda K, Oike K, Uyeda S, Yoshino T (2002) Electromagnetic anomalies associated with 1995 Kobe earthquake. J Geodyn 33(4–5):349–359
76. Telesca L, Cuomo V, Lapenna V (2001) A new approach to investigate the correlation between geoelectrical time fluctuations and earthquakes in a seismic area of southern Italy. Geophys Res Lett 28:4375–4378
77. Varotsos P, Lazaridou M, Eftaxias K, Antonopoulos G, Makris J, Kopanas J (1996) Short-term earthquake prediction in Greece by seismic electric signals. In: Ligthhill Sir J (ed) A critical review of VAN: earthquake prediction from seismic electric signals. World Scientific Publishing, Singapore, pp 29–76
78. Eftaxias K, Athanasopoulou L, Balasis G, Kalimeri M, Nikolopoulos S, Contoyiannis Y, Kopanas J, Antonopoulos G, Nomicos C (2009) Unfolding the procedure of characterizing recorded ultra low frequency, kHz and MHz electromagnetic anomalies prior to the L'Aquila earthquake as pre-seismic ones—Part 1. Nat Hazards Earth Syst Sci 9:1953–1971
79. Eftaxias K (2009) Footprints of nonextensive Tsallis statistics, self-affinity and universality in the preparation of the L'Aquila earthquake hidden in a pre-seismic EM emission. Phys A 389:133–140
80. Pfotzer G, Regener E (1935) Vertical intensity of cosmic rays by threefold coincidence in the stratosphere. Nature 136:718–719
81. Mishev A, Bouklijski A, Visca L, Borla O, Stamenov J, Zanini A (2008) Recent cosmic ray studies with lead free neutron monitor at basic environmental observatory Moussala. Sun Geosphere 3(1):26–28

82. Zanini A, Storini M, Visca L, Durisi EAM, Fasolo F, Perosino M, Borla O, Saavedra O (2005) Neutron spectrometry at high mountain observatories. J Atmos Solar Terr Phys 67:755–762
83. Zanini A, Storini M, Saavedra O (2009) Cosmic rays at high mountain observatories. Adv Space Res 44(10):1160–1165
84. Information on Jungfraujoch Neutron Monitor (18igy) (2013). http://cr0.izmiran.rssi.ru/jun1/main.htm. Accessed Mar 2013
85. Information on National Geophysical Data Center/World Data Center (NGDC/WDC) Significant Earthquake Database (2012) Boulder, CO, USA. http://www.ngdc.noaa.gov/nndc/struts/form?t=101650&s=1&d=1. Accessed Apr 2012
86. Allen R (2011) Seconds before the big one. Seismology, ScientificAmerican.com, pp 54–59
87. Borla O, Lacidogna G, Di Battista E, Niccolini G, Carpinteri A (2014) Electromagnetic emission as failure precursor phenomenon for seismic activity monitoring. In: Conference & exposition on experimental and applied mechanics (SEM), vol 5, Greenville, South Carolina, USA, pp 221–229
88. ISIDe Working Group, INGV (2010) Italian seismological instrumental and parametric database. http://iside.rm.ingv.it
89. Borla O, Lacidogna G, Carpinteri A (2015) Piezonuclear neutron emissions from earthquakes and volcanic eruptions. In Carpinteri A, Lacidogna G, Manuello A (eds) Acoustic, electromagnetic, neutron emissions from fracture and earthquakes. Springer International Publishing, Switzerland
90. Jentzsch G (1997) Earth tides and ocean tidal loading. In: Wilhelm H, Zürn W, Wenzel HG (eds) Tidal phenomena. Lecture notes in earth sciences, p 66
91. van der Elst NJ et al (2016) Fortnightly modulation of San Andrea tremor and low-frequency earthquakes. Proc Natl Acad Sci 113:8601–8605
92. Thomas AM, Bürgmann R, Shelly DR, Beeler NM, Rudolph ML (2012) Tidal triggering of low frequency earthquakes near Parkfield, California: implications for fault mechanics within the brittle-ductile transition. J Geophys Res 117:B05301
93. Ide S, Yabe S, Tanaka Y (2016) Earthquake potential revealed by tidal influence on earthquake size-frequency statistics. Nat Geosci 9:834–837
94. Kikuchi M (1992) Strain drop and apparent strain for large earthquakes. Tectonophysics 211:107–113
95. Wolf R (1853) On the periodic return of the minimum of sun-spots: the agreement between those periods and the variations of magnetic declination. Philos Mag 5:67
96. Sobolev GA, Demin VM (1980) Electromechanical phenomena in the earth. Nauka, Moscow, pp 1–163
97. Han Y, Guo Z, Wu J, Ma L (2004) Possible triggering of solar activity to big earthquakes (Ms $\geq$ 8) in faults with near west–east strike in China. Sci China Ser B Phys Mech Astron 47:173–181

Chapter 17
Radiocarbon Dating of Organic Materials Subjected to Very-High Magnitude Earthquakes: The Case-Study of Turin Shroud

Abstract Phillips and Hedges, in the scientific magazine Nature (1989), suggest that neutron radiation could be liable of wrong radiocarbon dating, whereas proton radiation could be responsible of the image formation of a human body on the Turin Shroud. On the other hand, no plausible physical reason has been proposed so far to explain the radiation source origin, and its effects on the linen fibers. Some recent studies, carried out at the Laboratory of Fracture Mechanics of Politecnico di Torino, found that it is possible to generate neutron emissions from brittle failure of natural rocks in compression through phono-fission reactions. Analogously, neutron flux increments, in correspondence to seismic activity, are a result of the same reactions. Russian scientists measured a neutron flux exceeding the natural background level by three orders of magnitude in correspondence to a not so strong earthquake (Magnitude 4 in Richter scale). The possibility is herein considered that neutron emissions from high-magnitude earthquakes have induced the image formation on the Shroud's linen fibers through their capture by nitrogen nuclei, so providing also a wrong radiocarbon dating due to the unexpected increment in the C_6^{14} content. Let us observe that, although the calculated integral flux of 10^{13} neutrons per square centimeter is 10 times greater than the cancer therapy dose, nevertheless it is 100 times smaller than the lethal dose.

Keywords Radiocarbon dating · Shroud of Turin · Neutron emissions · High-magnitude earthquakes

17.1 Preliminary Remarks

After the first photographs of the Shroud, taken by Secondo Pia during the Exposition of 1898 in Turin [1], a widespread interest has been generated among scientists and curious observers to explain the image formation and to evaluate its dating. The first results of a radiocarbon analysis were published in 1988. They showed that the Shroud is at most 728 years old [2]. Later, some researchers suggested that neutron radiation is liable of a wrong radiocarbon dating of the linen tissue [3–5]. However, no

A. Carpinteri, *Terahertz Phonons and Nanomechanical Instabilities*,
https://doi.org/10.1007/978-3-032-14692-2_17

plausible physical reasons have been proposed so far to explain the neutron radiation source origin.

Different documents in the literature attest the occurrence of disastrous earthquakes in the "Old Jerusalem" of 33 A.D., during the Christ's death [6–11]. On the other hand, recent neutron emission detections have led to consider the Earth's crust as a relevant source of neutron flux variations. Russian scientists measured neutron fluxes exceeding the natural background by three orders of magnitude in correspondence to seismic activity and rather appreciable earthquakes [12–17]. In this chapter, we consider that neutron emissions from earthquakes—as for the conventional gadolinium-like neutron imaging technique—could have induced the image formation on Shroud linen fibers through thermal neutron capture by nitrogen nuclei, so providing also a wrong radiocarbon dating due to an unexpected increment in C_6^{14} content (carbon-14 isotope). Some recent studies, carried out at the Laboratory of Fracture Mechanics of Politecnico di Torino, found that it is possible to generate neutron emissions from brittle failure of rock specimens in compression through phono-fission reactions. The neutron flux increment, in correspondence to seismic activity, is a consequence of the same mechanical phenomena [18–20].

Based on the first photographs of the Shroud, which highlighted an image of a human body undraped with hands crossed (Fig. 17.1), a wide debate on the cause that may have produced such an image was conducted in the scientific community. The image appears with reversed lights and shades as in a sort of negative photograph. Vignon [1] asserts that the image is produced by a radiographic action from the body, which, according to ancient texts, was wrapped in a shroud impregnated by a mixture of oil and aloes. Other authors, instead, disapprove the remarks by Vignon. In particular, Waterhouse [21] affirms that, if a body were wrapped in a linen cloth, under the conditions stated in the Gospels, it would be impossible for such a detailed impression to be produced in the manner suggested by Vignon.

Further studies have focused on the Shroud dating, especially since 1986 [22], when the Roman Catholic Church declared that pieces of the Shroud of Turin had been sent to seven laboratories around the world, later reduced to only three [23], for radiocarbon dating. In 1988, Dickman [2], after weeks of rumors and speculations, declared that the official carbon dating results for the Turin Shroud were released in Zurich. The results, also published in [24], provided evidence that the linen tissue of the Shroud of Turin is medieval, dated between 1260 and 1390. More recently, an exhaustive study on the statistical aspects of radiocarbon dating, due to the heterogeneity caused by the division of the samples into subsamples, was published [25].

Phillips, in the paper "Shroud irradiated with neutrons?" [3], supposes that the Shroud may have been irradiated with neutrons, which would have transformed some of the carbon nuclei into different isotopes by neutron capture. In particular, Phillips assumes that part of the C_6^{14} nuclei could have generated from C_6^{13} nuclei, and that an integrated flux of 2×10^{16} thermal neutrons cm^{-2} could have produced an apparent carbon-dated age of just 670 years. In the reply to the same paper, Hedges [4] asserts that the integrated flux assumed by Phillips [3] is excessively high and that «including

Fig. 17.1 Front image of the Turin Shroud

the neutron capture by nitrogen in the cloth, an integrated thermal neutron flux of 2 $\times 10^{13}$ would be appropriate» for the wrong radiocarbon dating of the Shroud.

Also Rinaudo [26] argues that simultaneous fluxes of protons and neutrons could explain at the same time the imprint on the cloth (by protons) and the shifting to the thirteenth century of the C_6^{14} carbon dating (by neutrons). According to our hypothesis, the increment in C-14 was not due to the additional time with respect to the historical date, but to the additional neutrons given by a disastrous earthquake.

Fanti [27] confirms the hypothesis of Rinaudo stating that, at the same time of the neutron and proton emissions, also an electron emission could have generated the body image formation. Other authors have recently introduced the hypothesis of radon emissions as a possible trigger for surface electrostatic discharges (ESD) and then for the image impression [28].

17.2 Historical Earthquakes and Neutron Emissions from the Earth's Crust

Scientific data of the historical earthquake occurred during the year 33 A.D. in the Jerusalem area are mentioned in the "Significant Earthquake Database" of the American Scientific Agency NOAA (National Oceanic and Atmospheric Administration) [29]. This database contains information on destructive earthquakes from 2150 B.C. to present days. The "Old Jerusalem" earthquake is classified as an average devastating seismic event that destroyed also the city of Nisaea, the port of Megara located at the West of the Isthmus of Corinth [6]. It would have implied a total reconstruction cost that, currently, would be between 1.0 and 5.0 million US Dollars.

In addition, if we assume the image imprinted on the Shroud to belong to the Man who died during the Passover of 33 A.D., there are at least three documents in the non-scientific literature attesting the occurrence of disastrous earthquakes during and after that event.

Within his chronicle in Greek language, a historian named Thallos, probably living in Rome in the middle of the first century, left mention of events occurred on the Christ's death day: the darkening of the sky and the happening of an earthquake [7, 8]. The work by Thallos was lost, but the quotation of his passage about Jesus was inserted in the Chronographia by Sextus Julius Africanus, a Christian Palestinian Author who died in Nicopolis around the year 240 A.D.: «The most dreadful darkness fell over the whole world, the rocks were torn apart by an earthquake and much of Judaea and the rest of the land were torn down. Thallos calls this darkness an eclipse of the Sun in the third book of his Histories, without reason it seems to me... How can we believe that an eclipse happened when the Moon was diametrically opposite to the Sun?».

Thallos, due to the quotation by Julius Africanus, is generally considered by the historians as a witness to the Gospel story of the "darkness" during the death of Christ, see Mark 15:33, Luke 23:44, and Mattew 27:45 [9]. However, the most interesting fact is that Julius Africanus criticizes Thallos, judging as impossible the eclipse on the day of the Passover, which occurs in a full Moon period, but he does not dispute that on the same day there was an earthquake.

On the other hand, Matthew wrote that there was a strong earthquake at the moment of Christ's death: «When the centurion and those who were with him, keeping watch over Jesus, saw the earthquake and what took place, they were filled with awe and said: "Truly this was the Son of God!"» (Matthew 27:54) [9]. He wrote that there was another even stronger earthquake at the time of the resurrection: «And behold, there was a great earthquake. An angel of the Lord descended from heaven and rolled back the stone and sat on it. His appearance was like lightning and his clothing white as snow. And for fear of him the guards trembled and became like dead men» (Matthew 28:2–4) [9].

A further document is provided by the narrative of Joseph of Arimathea: «And behold, after He had said this, Jesus gave up the ghost, on the day of the preparation, at the ninth hour. And there was darkness over all the earth; and from a great earthquake that happened, the sanctuary fell down, and the wing of the temple» (The Narrative of Joseph, Chap. 3, The Good Robber, 5) [10].

That event is also mentioned by Dante Alighieri, XXI Canto, Inferno, as the most violent earthquake that had ever shaken the Earth: « Poi disse a noi: "Più oltre andar per questo/iscoglio non si può, però che giace/tutto spezzato al fondo l'arco sesto./E se l'andare avante pur vi piace,/andatevene su per questa grotta;/presso è un altro scoglio che via face./Ier, più oltre cinqu'ore che quest'otta,/mille dugento con sessanta sei/anni compié che qui la via fu rotta"» (Inferno, XXI Canto:106–114) [11]. Since most scholars believe that the journey of Dante began on the anniversary of the Christ's death, during the Jubilee of 1,300 A.D., the chronology goes back to 33 A.D., on the Friday when, according to the tradition, Christ was put to death. Therefore, it was the earthquake after the Christ's death to cause disasters and crashes, including the Sanctuary of Jerusalem and the wing of the Solomon's Temple [10].

Nevertheless, the results from historical studies have a value for Earth scientists only when the information is converted into data representing epicentral location and magnitude of the events. Modern scholars affirm that Jerusalem is situated relatively close to the active Dead Sea fault. They accept the occurrence of the Resurrection earthquake, to which they assign the severity of a catastrophic event, characterized by a local magnitude $ML = 8.2$, as well as of another earthquake that took place in Bithynia, during the same period, with an even greater magnitude [12]. From a geophysical and stratigraphical point of view, it can be understood that the "Old Jerusalem" earthquake should have occurred between 26 and 36 A.D. [32].

Based on a detailed analysis of paleo-earthquakes along the major active faults in the Earth's crust, some studies give evidence of their spatial and temporal distributions, as well as of their regional recurrent behaviour [30]. From these studies, it can be argued that a hypothetical earthquake of magnitude 11 in the Richter scale may have a recurrence time of about 1,000 years worldwide, as well as of about 100 years one of magnitude 10, and of about 10 years one of magnitude 9.

Analogously, in the active faults of the Mediterranean basin and of the Middle East region, about one hundredth of the earthquakes recorded during long periods over the entire Earth surface take place [12]. In this case, an earthquake of magnitude 9 may have a recurrence time of about 1,000 years, as well as of about 100 years one of magnitude 8, and so on. The last statistical remark would give further scientific value, as well as historical and archaeological importance, to the hypothesis that, in the "Old Jerusalem", a strong earthquake occurred, very close to magnitude 9 in the Richter scale.

Pioneering neutron emission detections by Volodichev et al. [13], Kuzhevskij et al. [14, 15], and Antonova et al. [16] have led to consider the Earth's crust as a relevant source of neutron flux variations. Neutron emissions measured in seismic areas of the Pamir region (4,200 m a.s.l.) exceeded the usual neutron natural background "up to two orders of magnitude in correspondence to seismic activity and rather appreciable earthquakes, greater than or equal to the 4th degree in Richter scale magnitude" [13].

On the other hand, it is important to note that the flux of atmospheric neutrons increases linearly starting from the top of the atmosphere up to a maximum value corresponding to an altitude of about 20 km. From this altitude, it decreases exponentially up to the sea level, where it is relatively negligible (Pfotzer profile [31]). Considering the altitude dependence of neutron radiation, values about 10 times higher than the natural background at sea level are generally detected at 5,000 m of altitude. Therefore, the same earthquake occurring at sea level should produce a neutron flux up to 1,000 times higher than the natural background. More recent neutron emission observations have been performed before the Sumatra earthquake of December 2004 [17]. Variations in thermal neutron measures were observed in different areas (Crimea, Kamchatka) a few days before that earthquake.

17.3 Neutron Radiography and Imaging on Linen Fibers

Considering the possibility that a neutron flux could have induced appreciable effects on the linen fibers of Shroud, it is possible to develop some hypotheses based on phono-fission reactions [33]. In the following, it is briefly described the process of image formation induced by neutron and proton radiation, likely occurred during the earthquake in the "Old Jerusalem" of 33 A.D., and the thirteen-century apparent time shifting that should have been produced on the linen cloth.

Neutron radiography is an imaging technique that utilizes the transmission of neutron radiation to obtain a static picture of given objects [34]. The object under examination is placed in the path of the incident radiation. The transmitted and scattered neutrons "bring the visual information" of the object that is recorded by an appropriate imaging system.

Thanks to the high thermal neutron cross-section of the gadolinium nucleus (~254,000 barn), the most important detection reaction used in neutron imaging is:

$$\mathrm{Gd}_{64}^{157} + \mathrm{n}_0^1 \rightarrow \mathrm{Gd}_{64}^{158} + \text{gamma} + \text{conversion electrons (8.5MeV)}, \qquad (17.1)$$

in which the converter material (gadolinium) captures neutrons and emits secondary charged particles that reproduce the irradiated object on neutron imaging plates (NIP). Usually, a thermal neutron flux of 10^5 neutrons cm^{-2} s^{-1} is employed, with an irradiation time of few minutes for a total integrated flux of about 10^8 neutrons cm^{-2}, with typical NIP enriched by more than 20% in weight of Gd_2O_3 [35].

Neutron imaging is a technique different, although complementary, to X-ray radiography. While the X-rays are more sensitive to materials with rather high density, one of the advantages of neutron radiation is its ability to affect preferentially elements with low atomic numbers such as hydrogen and nitrogen [33]. For this reason, neutron rays give sharp images of biological soft tissue samples, whereas X-rays are more sensitive to tissues rich in calcium like bones.

The most important nuclear reaction of thermal neutrons on nitrogen nuclei is represented by:

$$N_7^{14} + n_0^1 \rightarrow C_6^{14} + H_1^1 \quad (17.2)$$

which is liable of radiocarbon (C-14) formation also in the Earth Atmosphere.

As regards the chemical composition of linen fibers, a typical nitrogen concentration of 1,000 ppm could be supposed [4]. The thermal neutron capture cross-section of nitrogen is of about 1.83 barn. By comparison, the Gd_{64}^{15} cross-section is 10^5 times greater than the nitrogen one. Neglecting the different concentrations in gadolinium and nitrogen, which mainly affect the image resolution, and taking into account only their cross-sections, the thermal neutron flux necessary to nitrogen imaging should be approximately equal to or higher than 10^{10} neutrons cm^{-2} s^{-1}.

The hypothetical reaction induced by neutrons on nitrogen nuclei might have contributed to the image formation by means of proton radiation (as assumed by Rinaudo [26]), triggering chemical combustion reactions or oxidation processes on linen fibers. Usually, image formation from a neutron beam can be accomplished in a variety of ways by using a suitable conversion screen. Thus, through an etching process with a chemical reagent (like KOH or NaOH) and under appropriate lighting, the image will become visible. Similarly, in the case of linen, neutrons could have interacted with nitrogen nuclei, and the protons—produced as secondary particles—may have assumed the function of the reagent, triggering oxidation or combustion phenomena and making the image visible. Hypotheses and experimental confirmations that oxidative phenomena generated by earthquakes can provide 3D images on the linen clothes have also been proposed by de Liso [36].

17.4 Earthquake and Neutron Effects on the Shroud Radiocarbon Dating

Taking into account the historical sources attesting the occurrence of a disastrous earthquake in 33 A.D., and assuming a hypothetical magnitude 9 in Richter scale [12], it is possible to provide an evaluation of the consequent neutron flux.

Richter scale is logarithmic (with base 10). This means that, for each increasing degree in Richter scale, amplitude and acceleration of the ground motion both increase by 10 times. From the displacement or acceleration viewpoint, the seismic event occurred in 33 A.D. was 10^5 times more intense than the reference event of magnitude 4 in Pamir. On the other hand, from the energy viewpoint, it was 10^{10} times more intense than the same reference event [13, 37].

Assuming a typical thermal neutron natural background of 10^{-3} cm^{-2} s^{-1} at the sea level, in correspondence of earthquakes of magnitude $M = 4$, a thermal neutron flux of 10^0 cm^{-2} s^{-1} is expected, as previously calculated following the Pfotzer profile [16, 31].

Thus, an earthquake of magnitude 9 in the Richter scale would provide a thermal neutron flux ranging around 10^{10} cm^{-2} s^{-1}, if proportionality between released energy and neutron flux holds (see Fig. 16.11c). A so powerful event could produce chemical and/or nuclear reactions, contributing to both the image formation and the C_6^{14} increment in the linen fibers of the Shroud, if it had totally lasted for at least 15 min. In this way, a sufficient integrated thermal neutron flux of 10^{13} neutrons cm^{-2} is obtained, exactly as assumed by Hedges [4]. Let us consider that, although the calculated integral flux of 10^{13} neutrons per square centimeter is 10 times greater than the cancer therapy dose, nevertheless it is 100 times smaller than the lethal dose.

As a confirmation to the previous assumptions, one of the most powerful earthquakes, the so-called "Greatest Chile Earthquake", occurred in Valdivia on May 22, 1960, had a complicated seismogram and lasted for more than 15 min [38]. Further information about the intensity and duration of this earthquake are reported in [39].

Neutron emissions were detected not only at the Earth's crust scale, but also in laboratory compression experiments, as is well-known from the previous parts of the book [18–20]. The tests were carried out by using suitable He^3 and bubble type BD thermodynamic neutron detectors. One of the materials employed for the tests was a non-radioactive Luserna stone, a metamorphic rock deriving from a granitoid protolith. Neutron emissions from this material were found to be of about one order of magnitude higher than the natural background level at the time of the catastrophic failure. For basaltic rocks, the neutron flux achieved a level even two orders of magnitude higher than the background. A first theoretical explanation was provided by Widom et al. [40, 41]. More recently, a different theoretical explanation was proposed by Carpinteri, Lucia, Zucchetti, and Borla, see Chap. 22 of this book.

17.5 Conclusions

Recent neutron emission detections have led to consider the Earth's crust as a relevant source of neutron flux variations. According to this experimental evidence, the hypothesis is considered that neutron emissions from a very destructive historical earthquake led to appreciable effects on Shroud linen fibers. Recalling the historical documents attesting the occurrence in the "Old Jerusalem" of a disastrous earthquake in the year 33 A.D., it is assumed that a seismic event of magnitude ranging between 8 and 9 in Richter scale produced a thermal neutron flux of 10^{10} cm^{-2} s^{-1}. Through thermal neutron capture by nitrogen nuclei, this phenomenon may have contributed to both the image formation (protons) and the increment in C_6^{14} on linen fibers of the Shroud. Let us observe that, such unexpected increment seems to be the cause of the wrong radiocarbon dating in the 1980s.

References

1. Vignon M (1902) M. Vignon's researches and the "Holy Shroud." Nature 66:13–14
2. Dickman S (1988) Shroud a good forgery. Nature 335:663
3. Phillips TJ (1989) Shroud irradiated with neutrons? Nature 337:594

4. Hedges REM (1989) Replies to: Shroud irradiated with neutrons? Nature 337:594
5. Fanti G (2012) Open issues regarding the Turin Shroud. Sci. Res. Essays 7(29):2504–2512
6. Mallet R (1853) Catalogue of recorded earthquakes from 1606 B.C. to 1850 A.D., Part I, 1606 B.C. to 1755 A.D. Report of the 22nd Meeting of the British association for the advancement of science held at Hull, Sept. 1853, John Murray, London, pp 1–176
7. Rigg H (1941) Thallus: the samaritan? Harv Theol Rev 34:111–119
8. Prigent P (1978) Thallos, Phlégon et le Testimonium Flavianum témoins de Jésus? Paganisme, Judaïsme, Christianisme. Influences et Affrontements dans le Monde Antique, F. Bruce Paris, pp 329–334
9. Gospels Acts, Matthew, Mark, Luke and John. http://bible.cc. Last accessed May 2011
10. The Narrative of Joseph of Arimathea. http://christianbookshelf.org/. Last accessed on May 2011
11. Alighieri D La Divina Commedia, Inferno, Canto XXI
12. Ambraseys N (2005) Historical earthquakes in Jerusalem—a methodological discussion. J Seismolog 9:329–340
13. Volodichev NN, Kuzhevskij BM, Nechaev OY, Panasyuk M, Podorolsky MI (1999) Lunar periodicity of the neutron radiation burst and seismic activity on the Earth. In: Proceedings of the 26th International cosmic ray conference, Salt Lake City
14. Kuzhevskij M, Nechaev OY, Sigaeva EA (2003) Distribution of neutrons near the Earth's surface. Nat Hazard 3:255–262
15. Kuzhevskij M, Nechaev OY, Sigaeva EA, Zakharov VA (2003) Neutron flux variations near the Earth's crust. a possible tectonic activity detection. Nat Hazard 3:637–645
16. Antonova VP, Volodichev NN, Kryukov SV, Chubenko AP, Shchepetov AL (2009) Results of detecting thermal neutrons at Tien Shan high altitude station. Geomag Aeron 49:761–767
17. Sigaeva EA, Nechaev O, Panasyuk M, Bruns A, Vladimirsky B, Kuzmin Y (2006) Thermal neutrons' observations before the Sumatra earthquake. Geophys Res Abstr 8:00435
18. Carpinteri A, Cardone F, Lacidogna G (2009) Piezonuclear neutrons from brittle fracture: early results of mechanical compression tests. Strain 45:332–339
19. Cardone F, Carpinteri A, Lacidogna G (2009) Piezonuclear neutrons from fracturing of inert solids. Phys Lett A 373:4158–4163
20. Carpinteri A, Cardone F, Lacidogna G (2010) Energy emissions from failure phenomena: mechanical, electromagnetic, nuclear. Exp Mech 50:1235–1243
21. Waterhouse J (1903) The holy Shroud of Turin. Nature 67:317
22. Campbell P (1986) Shroud to be dated. Nature 323:482
23. Dutton D (1988) The Shroud of Turin. Nature 332:300
24. Damon PE (1989) Radiocarbon dating of the Shroud of Turin. Nature 337:611–615
25. Riani M, Atkinson AC, Fanti G, Crosilla F (2013) Regression analysis with partially labelled regressors: carbon dating of the Shroud of Turin. J Stat Comput 23:551–561
26. Rinaudo JB (1998) Image formation on the Shroud of Turin explained by a protonic model affecting radiocarbon dating. III Congresso Internazionale di Studi sulla Sindone, Torino
27. Fanti G (2010) Can a corona discharge explain the body image of the Turin Shroud? J Imag Sci Technol 54(2):020508–1/10
28. Amoruso V, Lattarulo F (2012) A physicochemical interpretation of the Turin Shroud imaging. Specl Issue Sci Res Essays 7(29):2554–2569
29. National Geophysical Data Center/World Data Center (NGDC/WDC) (2011) Significant earthquake database, Boulder, CO, USA. http://www.ngdc.noaa.gov/nndc/struts/form?t=101650&s=1&d=1. Last accessed on May 2011
30. Min W, Zhang PZ, Deng QD (2000) Primary study on regional paleoearthquake recurrence behaviour. Acta Seismol Sin 13:180–188
31. Pfotzer G, Regener E (1935) Vertical intensity of cosmic rays by threefold coincidence in the stratosphere. Nature 136:718–719
32. Williams JB et al (2012) An early first-century earthquake in the Dead Sea. Int Geol Rev 54:1219–1228

33. Carpinteri A, Lacidogna G, Manuello A, Borla O (2012) Piezonuclear neutrons from earthquakes as a hypothesis for the image formation and the radiocarbon dating of the Turin Shroud. Sci Res Essays 7(29):2603–2612
34. Anderson IS, McGreevy R, Bilheux HZ (eds) (2009) Neutron imaging and application. In: A reference for the imaging community series: neutron scattering applications and techniques XVI. Springer, pp 1–341
35. Cipriani F, Castagna JC, Lehmann MS, Wilkinson C (1995) A large image-plate detector for neutrons. Physica B 213–214:975–977
36. de Liso G (2010) Shroud-like experimental image formation during seismic activity. In: Proceedings of the International workshop on the scientific approach to the acheiropoietos images, ENEA, Frascati, Italy
37. Richter CF (1958) Elementary seismology. W. H. Freeman, San Francisco and London
38. Barrientos SE, Ward SN (1990) The 1960 Chile earthquake: Inversion for slip distribution from surface deformation. Geophys J Int 103:589–598
39. Kanamori H, Cipar JJ (1974) Focal process of the great Chilean earthquake May 22, 1960. Phys Earth Planet Inter 9:128–136
40. Widom A, Swain J, Srivastava YN (2013) Neutron production from the fracture of piezoelectric rocks. J Phys G Nucl Part Phys 40(015006):1–8
41. Widom A, Swain J, Srivastava YN (2014) Photo-disintegration of the iron nucleus in fractured magnetite rocks with magnetostriction. Meccanica 50:1205–1216

Part VI
Chemical Evolution of Our Planet and the Planets of Solar System

Chapter 18
Earth's Crust, Ocean, and Atmosphere: From Iron and Calcium Depletion to Carbon and Water Appearance

Abstract The Earth's chemical composition and evolution are topics that give rise to unanswered questions. However, some evident data involving Geology, Geophysics, and the climate change of our planet seem to imply a possible common explanation. Recently, several data coming from Geochemistry and Geomechanics have emphasized how tectonic activity is strictly correlated to the most important chemical composition changes in the Earth's Crust over the last 4.5 Billion years (life time of our planet). At the same time, significant measurements of neutron emissions are observed at the Earth's Crust scale during and before seismic events. On the other hand, at the laboratory scale, original experiments, performed on non-radioactive natural rocks under mechanical compression, have shown repeatable neutron emissions in correspondence to microcracking and macroscopic fracture. After these experiments, a considerable reduction in the Iron content appears to be consistently balanced by an equivalent increment in Al, Si, and Mg contents, as occurred at the planetary scale. In the same context, also the Calcium depletion can surprisingly explain the sudden appearance of water on our planet. The almost perfect ponderal balances between chemical compositions, before and after the major tectonic events, permit to exclude matter migration in favor of matter transformation of a nuclear origin. Summarizing, an overall decrement in ferrous elements (Fe, Ni) of 12% is totally balanced by consistent increments in silicon and aluminum (SI-AL) in the crust, and in carbon in the atmosphere. Analogously, an overall decrement in alcalyne-earth (or alcalyne-terrous) elements (Mg, Ca) of 8.7% is totally balanced by consistent increments in alcalyne elements (Na, K), and in oxygen.

Keywords Earth's crust · Ocean formation · Great oxidation event (GOE) · Primordial atmosphere carbon increment · Neutron emissions · Geochemical evolution · Phono-fission reactions · Depletion of iron and calcium

A. Carpinteri, *Terahertz Phonons and Nanomechanical Instabilities*,
https://doi.org/10.1007/978-3-032-14692-2_18

18.1 Preliminary Remarks

Over the last century, most of the recently established scientific disciplines, such as Cosmology, Astrophysics, and Geology, have tried to answer questions concerning the origin of the Earth, the planets, and the Universe [1]. Such questions have now given place to even more precise and advanced interrogatives concerning the substance composing the Universe, the heterogeneous distribution of the main chemical elements on the Earth, and their evolution in time [1–7]. Significant and evident phenomena, such as the relatively abrupt changes in element abundances [2, 3, 6], the Great Oxidation Event (GOE) between 2.7 and 2.4 Gyrs ago [8, 9], the strong iron and nickel depletions in Earth's crust and oceans [10–12], the transition from a basaltic to a sialic condition in the Continental Crust [6, 13, 14], the present level of CO_2 and N_2 concentrations in the Earth's atmosphere [8, 9], the alkaline-hearth element (Ca, Mg) depletions and the concomitant alkaline element (Na, K) increments, the salinity increment in the oceans, as well as the appreciable precursory role of CO_2 and neutron emissions before relevant earthquakes [15–20], are just some of the major events pertaining to dynamics and evolution of the chemical element abundances in Earth's crust, oceans, and atmosphere. They still remain unsolved. Recent investigations and new instruments for data analysis led the specialists to study these unexplained phenomena more deeply.

Another question that still remains unanswered concerns the non-homogeneous composition of Oceanic and Continental Crusts. The 85% of Earth's volcanic eruptions take place at the sea bottom, in correspondence to the mid ocean-ridges [21]. These submarine volcanoes generate the solid underpinnings of all the Earth's oceans (Oceanic Crust) [2, 3, 7, 22, 23]. Comparing the data presented in the literature concerning the composition of the two different types of terrestrial crust, it can be noted that the iron concentration changes from ~8% in the Oceanic Crust, to ~4% in the Continental one. Analogously, nickel changes from ~0.03% in the Oceanic Crust, to ~0.01% in the Continental one (about a three-fold decrement). Vice versa, Si, Al, and Na vary from ~24, ~7, and ~ 1% in the Oceanic Crust, to ~28, ~ 8, and ~ 2.9% in the Continental Crust, respectively [2, 3, 7]. Considering that approximately 50% of the Continental Crust has originated, over the last 3.8 Gyrs, from the Oceanic Crust subduction [2, 3, 21–23], the considerable variations in the composition of the Continental with respect to the Oceanic Crust would remain a mystery [21].

18.2 Chemical Evolution of the Earth's Crust

The locations of Al and Fe mineral reservoirs appear to be inexplicably related to the geological periods when the different continental plaques were formed [2, 23–26]. This fact seems to suggest that our planet has undergone a continuous evolution, from the most ancient geological regions, which currently are in the continental cores rich in Fe, to more recent areas where the concentrations of Si and Al oxides present very high mass percentages [2, 23–26]. The main iron reservoir locations (Magnetite and Hematite mines) are reported in Fig. 18.1a [24, 26]. At the same time, the concentrations of Al oxides and andesitic formations (the Rocky Mountains and the Andes) are shown, together with the most important subduction lines, tectonic plate trenches, and rifts, in Fig. 18.1b [2, 25]. The bauxite mine locations show that the largest concentrations of Al reservoirs can be found in correspondence to the most seismic areas, without any exception (Fig. 18.1b). The largest iron mines are instead exclusively located in the oldest and interior parts of continents, i.e., in geographic areas with a reduced seismic risk and always far from the main fault lines. What is the relationship between areas rich in aluminum and poor in iron and the fault system? Why does the concentration of certain chemical elements seem to be connected to the geological periods in which tectonic plates were formed? These questions are still unanswered by the scientific community.

Several additional questions concern the fact that the Earth, differently from the other rocky planets of Solar System, shows a particularly strong chemical composition evolution over the last 4.57 Billion years. From 4.0 to 2.0 Gyrs ago, in fact, Fe can be considered as one of the most common bio-essential elements required for the metabolic activity of all living organisms [5–7, 10, 27]. Today, the deficiency of this nutrient suggests it to be a limiting factor for the development of marine phytoplankton and, therefore, of life on Earth [6, 10, 27]. Elements such as Fe and Ni in the Earth's protocrust had higher concentrations during the Hadean (4.5–3.8 Gyr ago) and the Archean (3.8–2.5 Gyr ago) periods, if compared to the present values [2–7, 10–14]. On the other hand, Si and Al concentrations were lower than they are today [2–7, 14]. The estimated concentrations of Fe, Ni, Al, and Si in the Hadean and Archean Earth's protocrust and in the present Earth's Continental Crust are reported in Fig. 18.2. The data for the Hadean period (4.5–3.8 Gyrs ago) are referred to the composition of Earth's protocrust, considering the assumptions by Foing [28] and by Taylor and Mc Lennan [3]. The most abrupt changes in element concentrations, shown in Fig. 18.2, appear to be objectively related to the tectonic activity of the Earth. The vertical drops in the concentrations of Fe and Ni, as well as the vertical jumps in the concentrations of Si and Al, occurred 3.8 and 2.5 Gyrs ago, coincide with the times of the tectonic plate formation and of the most intense tectonic activity, respectively [2, 3].

From the data reported in Fig. 18.2, a decrement of ~7% in Fe and ~0.2% in Ni concentrations are observed between the Hadean period (4.5–3.8 Gyrs ago) and the Archean period (3.8–2.5 Gyrs ago) [2–7, 14, 29, 30]. At the same time, Al and Si concentrations increase of ~3.0 and ~2.4%, respectively. Similarly, a decrement in the concentrations of Fe (~4.0%) and Ni (~0.8%), as well as an increment in the concentrations of Si (~2.4%) and Al (~1.0%) are shown between the Archean period (3.8–2.5 Billion years ago) and more recent times. The balances between heavier/ferrous (Fe, Ni) and lighter/medium-weight (Al, Si) elements can be considered as perfectly satisfied, if we take into account also two stepwise increments in C, occurred in the primordial atmosphere (not in the protocrust).

In addition to Fe, Ni, Al, Si, and C, it is possible to consider also the remaining most abundant elements such as Mg, Ca,, Na, K, and O, which are also dramatically

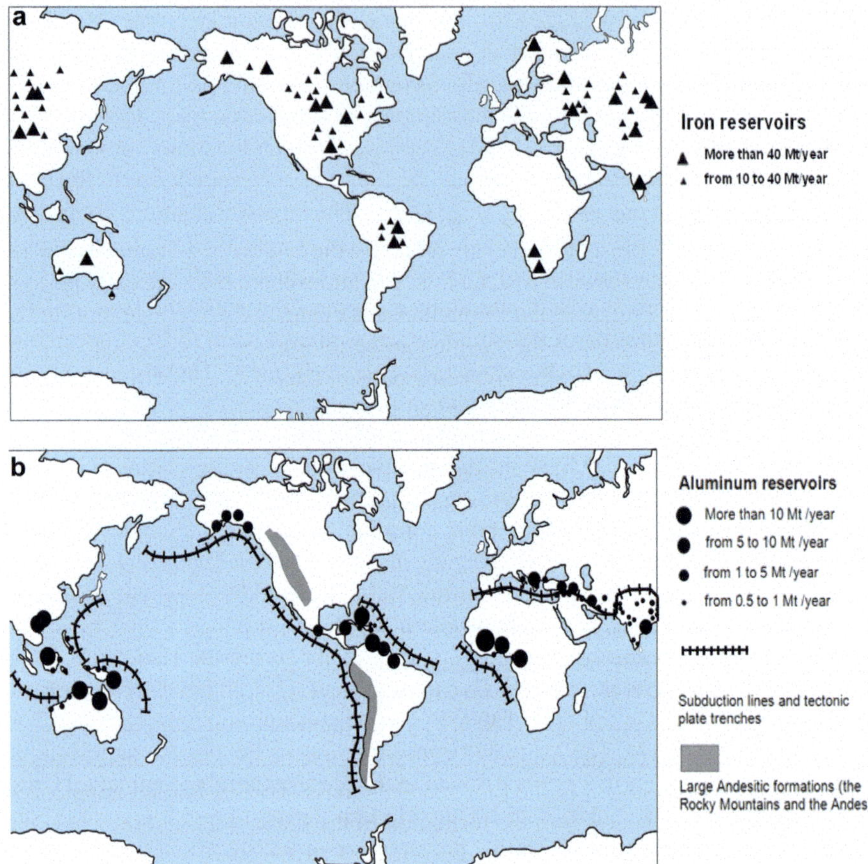

Fig. 18.1 **a** Locations of the largest iron mines in the world [2, 24–26]. Iron ore reservoirs (Magnetite and Hematite mines) are located in geographic areas with reduced seismic risks and always far from fault lines. The most abundant iron reservoirs are located in North-Central USA, Eastern Canada, North-Central Brazil, Central Australia, Ukraine, Russia, Mongolia, and North-Central China. **b** Locations of the largest aluminum (bauxite) reservoirs. The largest bauxite and alumina mines are located in Jamaica, Mexico, the North-Eastern littorals of Brazil, Guyana, the Gulf of Guinea, India, the Chinese littorals along the East China Sea, Greece, the South of Italy, the Philippines, New Guinea, and the Australian coast [2, 25]. The largest concentrations of Al reservoirs and the largest Andesitic formations are in the Rocky Mountains and the Andes, and, in general, in correspondence to the most seismic areas of the Earth: subduction lines, tectonic plate trenches and rifts

involved in the Earth's crust chemical evolution. The variations in mass percentage of Mg, Ca, Na, K, and O, between Hadean and Archean periods, and in the Earth's Continental crust, are reported in Fig. 18.3, analogously to Fig. 18.2 [2, 3, 14]. The decrement in the mass concentrations of Mg and Ca is balanced by an increment in Na, K, and O, during the Earth's lifetime. In particular, between the Hadean and the Archean eras, and between the latter and more recent times, it is possible to observe

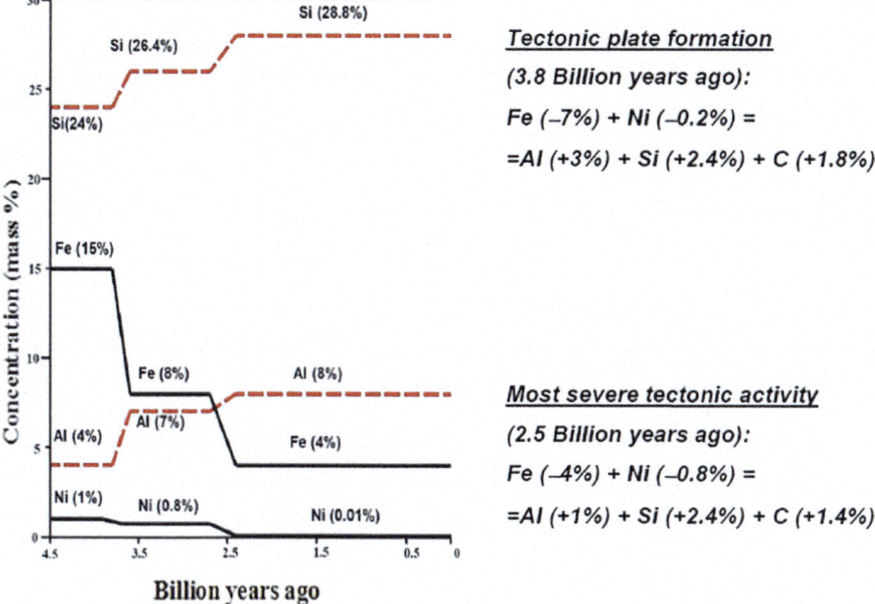

Fig. 18.2 The estimated concentrations of Fe, Ni, Al, and Si in the Hadean and Archean Earth's protocrust, and in the Earth's Continental Crust, are reported. The Archean Earth's protocrust (3.8–2.5 Gyrs ago) had a less basaltic composition (Fe ~8%, Ni ~0.8%, Al ~ 7%, Si ~ 26.4%) [2, 3, 10, 11, 13, 28, 30] compared to the previous period (Hadean Era, 4.5–3.8 Gyrs ago) [3, 11, 12], and a less sialic composition compared to the Continental Crust today: Fe ~ 4%, Ni ~ 0.01%, Al ~ 8%, Si ~ 28.8% [3, 7, 13, 14, 22, 23, 28, 30]. It can be observed, in particular, that the overall 12% decrement in ferrous elements (Fe, Ni) is nearly perfectly balanced by the Al, Si, and C increments (C increment in the atmosphere)

an overall decrement of ~4.7% for Mg and of ~4.0% for Ca. This decrement in the two alkaline-earth (or -terrous) elements (Mg, Ca) is nearly perfectly balanced by the increment in the concentrations of the two alkaline elements (Na, K), which increase by 2.7 and 2.8%, respectively, and by a total increment of ~3.3% in O, which changes from ~44 to ~47.3% (the latter being the present Oxygen concentration in the Earth's Crust), see Fig. 18.3. Also in this case, the greatest chemical changes in the Earth's crust occur during the most intense tectonic activity.

Plate tectonics and the related subduction phenomena involve the Earth's Crust chemical composition evolution and the changes in the most abundant chemical elements. These strong correlations have often been ignored by orthodox views. At the same time, the objective evidence emerging from Figs. 18.1, 18.2, and 18.3 cannot be explained just by simple chemical element migrations [2, 3]. According to the traditional explanation, the changes are caused by the heavy metals (iron, nickel, and other ferrous elements) migrating toward the core of the planet, whereas the light elements (oxygen, silicon, aluminum, magnesium, potassium, sodium, etc.) migrate to the upper layers, i.e., the mantle and, in particular, the crust [2, 3]. Matter

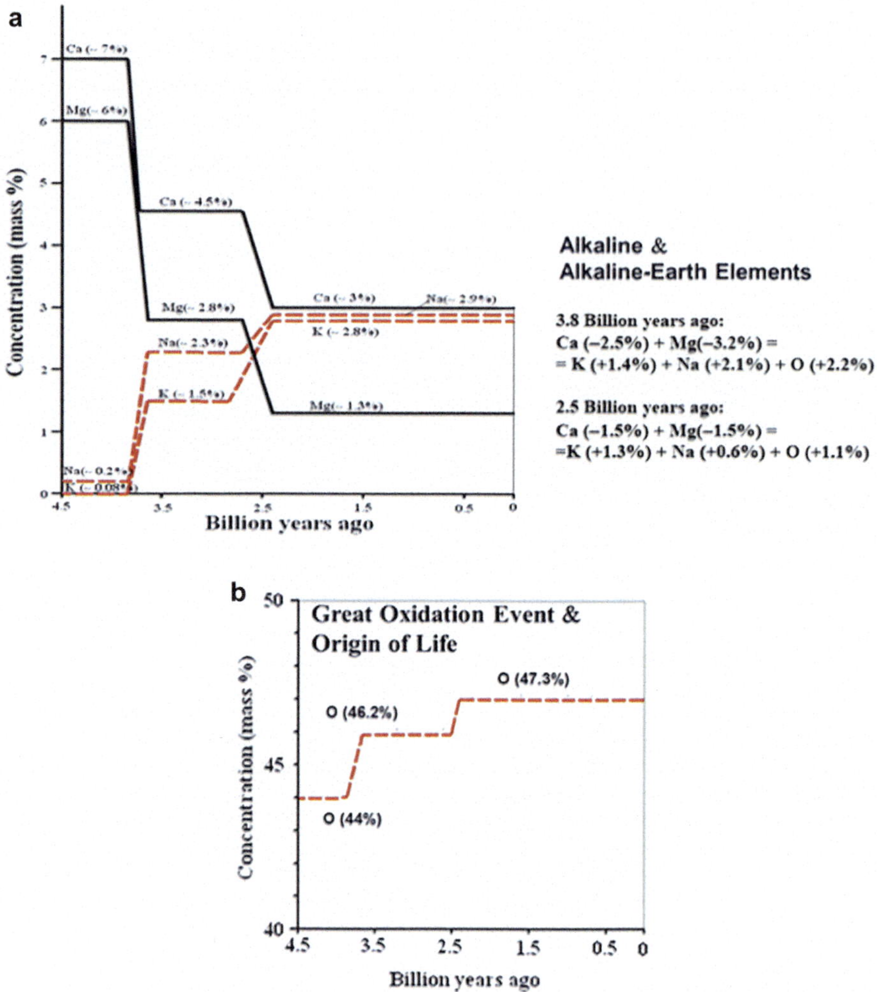

Fig. 18.3 The variations in mass percentage for Mg, Ca, Na, K **a**, and Oxigen **b** in the Hadean and Archean Earth's protocrust. The overall 8.7% decrement in alkaline-earth elements (Mg, Ca) is nearly perfectly balanced by the increment in alkaline elements (Na, K) and Oxygen (also the last in the crust)

migrations would occur simply according to atomic weights and a sort of floating principle. On the other hand, the traditional approach can not explain how the oceanic crust, characterized by a more basaltic composition, generated the margins of the continents, which are composed by the lighter sialic elements. In other terms, if the mechanism were only that traditionally proposed, the oceanic crust would be sialic as the continental one. Moreover, the traditional approach cannot explain another contradictory evidence such as that regarding uranium and thorium concentrations.

These two very heavy elements should be expected to be abundant in the core; on the contrary, they are concentrated in the crust and the mantle of our planet.

The temporal correspondence between most important tectonic activities and most evident chemical changes in the crust, together with the nearly perfect chemical concentration balances, suggest that alternative interpretations of the Earth's crust evolution should be taken into consideration. Plate tectonics, TeraHertz phonons, and chemical evolution of our planet imply the phono-fission reactions.

This last interpretation is even more convincing if we consider recent measurements, performed by Kuzhevskij et al. [15, 16], Antonova, Volhodichev et al., and Sigaeva et al. [17–19], in correspondence to seismic activity. The results presented in these papers lead to consider also the Earth's Crust, in addition to cosmic rays, as a relevant source of neutron flux variations. Neutron emissions exceeded the natural background up to 1,000 times in correspondence to earthquakes with a Richter magnitude equal to 4 [18]. These measurements can be directly related to the chemical evolution in the Earth's Crust and to mechanical phenomena of fracture, crushing, fragmentation, comminution, erosion, friction, occurring during seismic events [31–40].

18.3 From the Earth's Crust to the Laboratory: Experimental Confirmation of Chemical Changes Due to Fracture

The evidence of the Earth's crust chemical evolution and the recent measurements of anomalous neutron emissions from tectonic activity have been both confirmed by short-time experiments performed at the laboratory scale. Original studies have shown a surprising although clear and repeatable evidence of anomalous neutron emissions from the failure of non-radioactive and inert iron-bearing rocks [31–40].

It was shown that pressure, exerted indifferently on radioactive or inert solids, can generate reproducible neutron emissions. In particular, anomalous nuclear emissions and heat generation were detected during fracture or fragmentation of fissile [41–43] or deuterated [44–46] materials, in pressurized deuterium gases [47], or in liquids containing radioactive deuterium and solicited by ultrasounds and cavitation [48].

The experiments performed by Carpinteri et al. and by Cardone et al. [31–42] follow a different path, and represent the first evidence of neutron emissions from inert, stable, and non-radioactive solids under compression, as well as from non-radioactive liquids subjected to ultrasonic cavitation [48]. Successively, Carpinteri et al. obtained similar results varying the size-scale of the specimens and considering the brittleness scale effects [49, 50], as well as they performed cyclic and ultrasonic loading tests (see Part II of this book). The maximum neutron emissions were obtained from specimens exceeding a certain critical volume [39, 40].

Neutron emission results, obtained from iron chloride liquid solutions subjected to cavitation and from non-radioactive rocks subjected to fracture, fatigue, and ultrasounds, are summarized in Table 18.1 [31–40, 48]. Liquids subjected to cavitation were characterized by neutron emissions up to 2.5 times the natural background level. The same neutron emission level was obtained in the case of steel specimens subjected to compression or tension up to the final failure [39]. For granitic rocks, presenting a small amount of iron content (Fe ~ 1.5%), a neutron emission up to one order of magnitude greater than the natural background level was detected during several repeatable tests [31–38]. For basalt and magnetite, where the iron concentration is much higher (Fe ~ 15% and ~ 72.5%, respectively), neutron emissions, respectively, up to 10^2 and 10^3 times the background level were measured during compression failure experiments [42]. On the other hand, in the case of marble no neutron emissions greater than the natural background level were observed during compression failure experiments. This fact was firstly and erroneously explained considering the lack of iron in the marble chemical composition, as well as the softening behavior of this material in comparison to the much more brittle behavior of iron-rich natural rocks [31, 32].

After these tests, the Energy Dispersive X-ray Spectroscopy (EDS) analysis was performed on different spots of external or fracture surfaces belonging to the granite specimens, before and after the experiments, in order to get an average information about possible changes in the chemical composition. The results for phengite and biotite (minerals with the highest Fe concentration in granite) show that the relative reduction in iron abundance (~25%) is balanced by the increment in lighter elements such as Al, Si, and Mg (Figs. 18.4 and 18.5) [37–39]. In particular, the distribution of Fe concentrations shows two different values equal to 6.2% and 4.0% for external and fracture surfaces, respectively (Fig. 18.4a). Similarly, the Al concentration shows an average value changing from 12.5% (external surface) to 14.5% (fracture surface). The evidence that the values of iron decrement (−2.2%) and of Al increment (+2.0%) are approximately equal is really impressive (Fig. 18.4b).

In Fig. 18.5a–d, the results for Fe, Al, Si, and Mg concentrations in biotite crystalline phase are shown. In this case, the iron decrement is about 3.0%, from 21.2% (external surface) to 18.2% (fracture surface), see Fig. 18.5a. At the same time, Al content variations show an average increment of about 1.5% (Fig. 18.5b). In

Table 18.1 Neutron emission (compared to the natural background) produced by fracture and cavitation in solids and liquids, respectively

Liquids (Cavitation)	Neutron emission
Iron chloride solution	Up to 2.5 times the background
Solids (Fracture)	
Steel	Up to 2.5 times the background
Granite	Up to 10^1 times the background
Basalt	Up to 10^2 times the background
Magnetite	Up to 10^3 times the background
Marble	Background

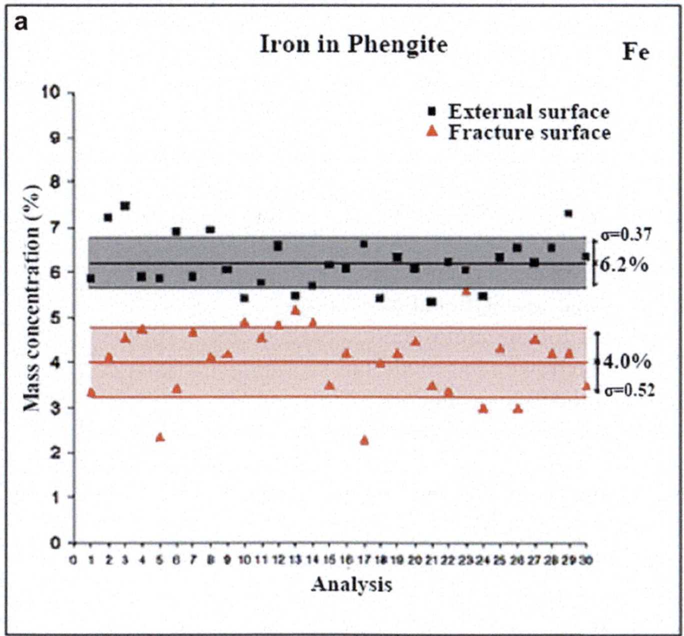

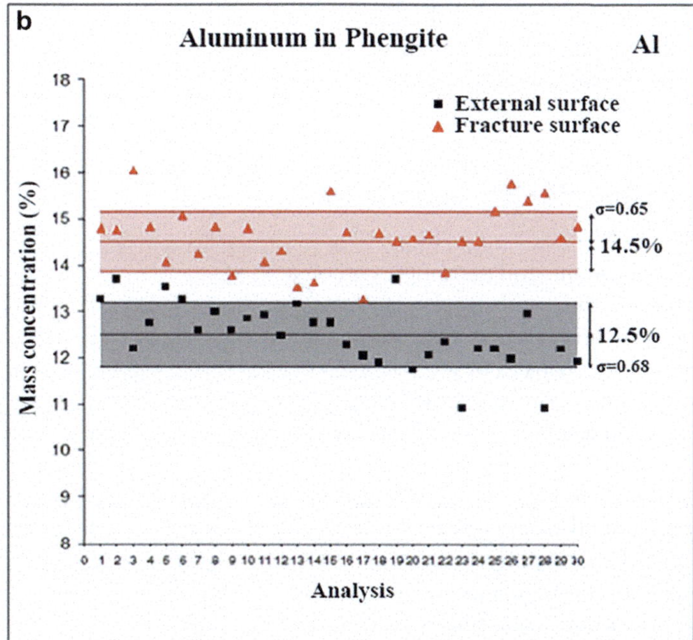

Fig. 18.4 2.2%. **b** Al concentrations on external surface (black squares) and on fracture surface (red triangles). The Al increment, considering the two mean values of the two distributions, is equal to 2.0% [36]. The evidence that the iron decrement and the Al increment plus the excess neutrons are approximately equal seems to imply that reaction 18.1, reported in Table 18.2, occurs on the fracture surface

Table 18.2 Phono-fission reactions from the direct evidence of EDS analysis of crushed rock specimens, and also emerging from the evolution of the continental crust

	Earth's crust evolution
18.1	$Fe_{26}^{56} \rightarrow 2Al_{13}^{27} + 2$ neutrons
18.2	$Fe_{26}^{56} \rightarrow Mg_{12}^{24} + Si_{14}^{28} + 4$ neutrons
18.3	$Fe_{26}^{56} \rightarrow Ca_{20}^{40} + C_{6}^{12} + 4$ neutrons
18.4	$Co_{27}^{59} \rightarrow Al_{13}^{27} + Si_{14}^{28} + 4$ neutrons
18.5	$Ni_{28}^{59} \rightarrow 2\ Si_{14}^{28} + 3$ neutrons
18.6	$Ni_{28}^{59} \rightarrow Na_{11}^{23} + Cl_{17}^{35} +1$ neutron
	Atmosphere evolution, ocean formation, and origin of life
18.7	$Mg_{12}^{24} \rightarrow 2C_{6}^{12}$
18.8	$Mg_{12}^{24} \rightarrow Na_{11}^{23} + H_{1}^{1}$
18.9	$Mg_{12}^{24} \rightarrow O_{8}^{16} + 4H_{1}^{1} + 4$ neutrons
18.10	$Ca_{20}^{40} \rightarrow 3C_{6}^{12} + He_{2}^{4}$
18.11	$Ca_{20}^{40} \rightarrow K_{19}^{39} + H_{1}^{1}$
18.12	$Ca_{20}^{40} \rightarrow 2O_{8}^{16} + 4H_{1}^{1} + 4$ neutrons
	Greenhouse gas formation
18.13	$O_{8}^{16} \rightarrow C_{6}^{12} + He_{2}^{4}$
18.14	$Al_{13}^{27} \rightarrow C_{6}^{12} + N_{7}^{14} + 1$ neutron
18.15	$Si_{14}^{28} \rightarrow 2N_{7}^{14}$
18.16	$Si_{14}^{28} \rightarrow C_{6}^{12} + O_{8}^{16}$
18.17	$Si_{14}^{28} \rightarrow 2C_{6}^{12} + He_{2}^{4}$
18.18	$Si_{14}^{28} \rightarrow O_{8}^{16} + 2He_{2}^{4} + 2H_{1}^{1} + 2$ neutrons

Fig. 18.5c and d, it is shown that, in the case of biotite, Si and Mg contents present considerable variations. The mass percentage of Si changes from a mean value of 18.4% (external surface) to a mean value of 19.6% (fracture surface), with an increment of 1.2% (Fig. 18.5c). Similarly, the mean value of Mg concentration changes from 1.5% (external surface) to 2.2% (fracture surface), with an increment of 0.7% (Fig. 18.5d). Therefore, the absolute iron decrement (−3.0%) in biotite is balanced by increments in aluminum (+1.5%), silicon (+1.2%), and magnesium (+0.7%) [36–39] (Table 18.1).

Similar results were observed in the case of basalt subjected to fatigue cycles up to final failure. The basalt tested in the experiments [51] is dark gray with few mm-sized crystals of plagioclase, clinopyroxene, and olivine. Its rupture was characterized by an equivalent neutron dose of about 20 times the background dose. In Fig. 18.6a–c, the distributions of Fe, Si, and Mg concentrations are reported from sixty spots on the external surface, as well as sixty from the fracture surface. All these measurements were carried out on olivine, the only iron-rich mineral in basalt [51]. It can be observed that the distribution of Fe content on the external surface shows an average value equal

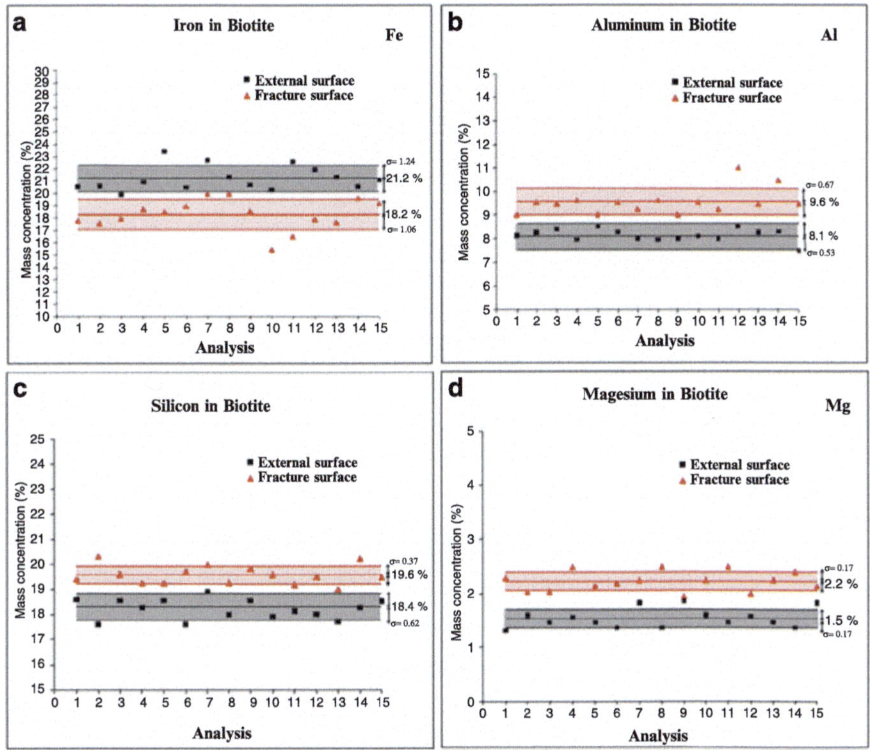

Fig. 18.5 3.0%) in biotite is balanced by increments in aluminum (+1.5%), silicon (+1.2%), and magnesium (+0.7%) [36]. These results suggest that reaction 18.2, reported in Table 18.2, occurs on the fracture surface

to 18.4% (Fig. 18.6a). In the same graph, the distribution of Fe concentrations on the fracture surface shows a significant variation. The mean value is equal to 14.4%, considerably lower than the mean value on the external surface. In Fig. 18.6b, the Si mass percentages are considered: the increment is approximately equal to 2.2%. The average value of Si concentrations changes from 18.3% on the external surface to 20.5% on the fracture surface. In Fig. 18.6c, it is shown that, in the case of olivine, also the Mg content presents a considerable variation. The mass percentage concentration of Mg changes from a mean value of 21.2% (external surface) to a mean value of 22.8% (fracture surface), with an increment of 1.6%. Therefore, the iron decrement (−4.0%) in olivine is balanced by increments in silicon (+2.2%) and magnesium (+1.6%).

Additional results are reported for Carrara marble specimens [52]. In this case, XPS quantitative composition analyses were carried out in order to detect any variation in chemical composition after the brittle crushing failure of cylindrical specimens [52]. XPS survey scan (pass energy of 187.85 eV) of Carrara marble surfaces was performed. Thirty measurements on the external surface and twenty on the fracture

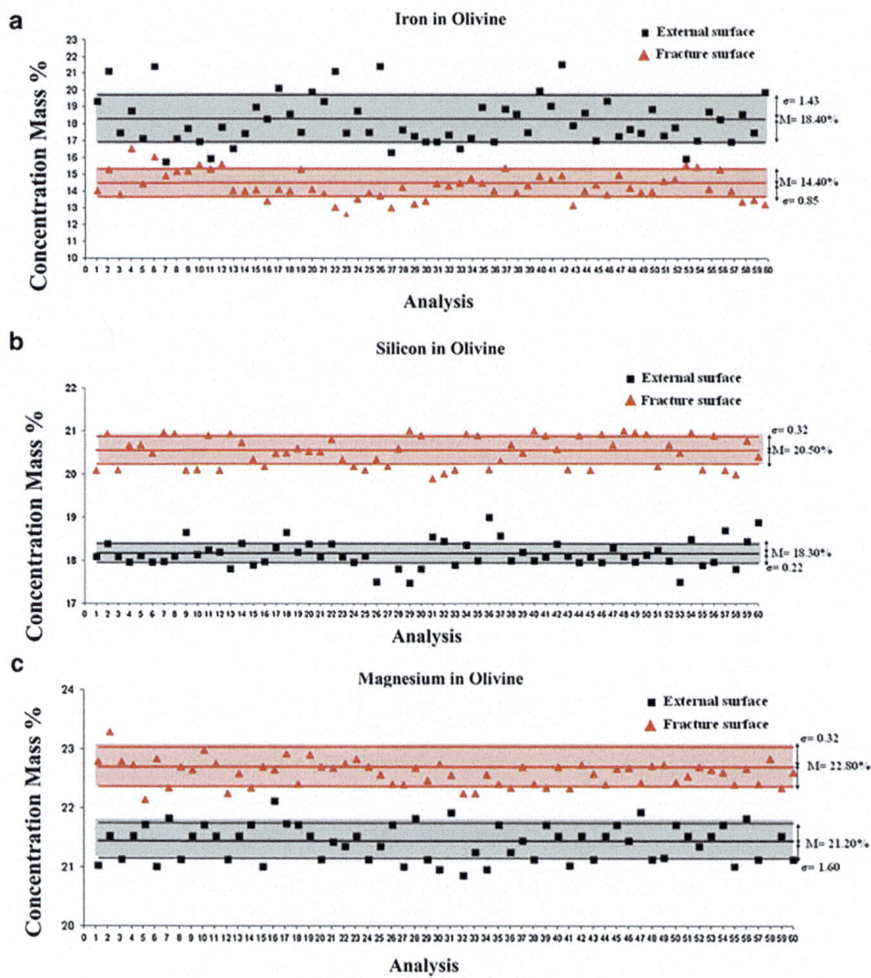

Fig. 18.6 Olivine chemical changes after crushing failure of basalt specimens: **a** The Fe decrement (−4.0%) is balanced by the increments in Si **b** (+2.2%) and Mg **c** (+1.6%) [51]

surface were performed and analyzed. In Figs. 18.7a-d, the results for the O, Ca, Mg, and C concentrations are shown. It can be observed that the distributions of O, Ca, and Mg concentrations on the external surface present average values, respectively, equal to 45.8, 13.4, and 0.7%. In the same diagrams, the distributions of O, Ca, and Mg concentrations on the fracture surface present significant decrements. The mean values of the measurements performed on the fracture surface are, respectively, equal to 36.8, 9.8, and 0.3%. The carbon mass percentage increment of 13.0% is equal to the total decrement in O, Ca, and Mg. The average value of C concentration changes from 40.1% on the external surface to 53.1% on the fracture surface.

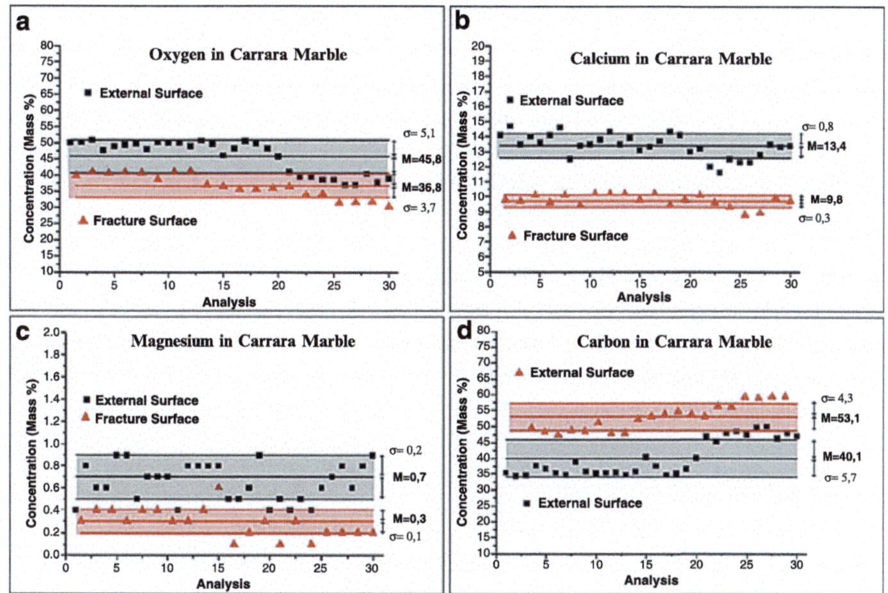

Fig. 18.7 O **a**, Ca **b**, Mg **c**, and C **d** concentrations in Carrara marble are reported on external and fracture surfaces of samples taken from crushing failure experiments [52]

18.4 Primordial Atmosphere Carbon Increment and Ocean Formation: A New and Unified Explanation

Considering the data reported in Figs. 18.2 and 18.3, and also that granite, basalt, and calcareous rocks (like marble) are common and widely diffused natural rocks constituting the Earth's Crust, the evidence from EDS and XPS chemical analysis techniques, reported in Figs. 18.4, 18.5, 18.6 and 18.7, can be extrapolated from the laboratory to the Earth's Crust scale, where mechanical phenomena of brittle fracture, due to plate collision and subduction, take place continuously in the most seismic areas.

The common explanation is that, at the scale of the Earth's crust as well as at the laboratory scale, phono-fission reactions take place where the mechanical conditions are most severe. Therefore, based on the evidence in both Earth's crust (Figs. 18.1, 18.2 and 18.3) and experimental results (Figs. 18.4, 18.5 and 18.6), the phono-fission reactions 18.1 and 18.2 reported in Table 18.2 have occurred during fracture phenomena [31–34], without gamma ray emission or production of radioactive wastes.

Considering the present natural abundances of the major chemical elements, such as Fe, Ni, Al, Si, Mg, Ca, Na, K, C, O in the Continental Crust and in the Atmosphere [2–7, 14, 29, 30], it is possible to conjecture that the additional nuclear reactions 18.3

to 18.6, reported in Table 18.2, take place in correspondence to tectonic plate collision and subduction.

The large concentrations of granite minerals, such as fengite, biotite, quartz, and feldspar in the Continental Crust, and, to a lesser extent, of magnesite, halite, and zeolite (MgO, Na_2O, Cl_2O_3), as well as the low concentrations of magnetite, hematite, bunsenite, and cobaltite minerals (predominantly composed of Fe, Co, and Ni), could be ascribed to the phono-fission reactions 18.1 to 18.6. The transition between basalt composition of the Oceanic Crust to sialic composition of the Continental Crust may be explained by the same phono-fission reactions 18.1 to 18.6.

In particular, phono-fission reactions 18.1, 18.2, 18.5, and 18.7 seem to be the cause of the abrupt changes shown in Fig. 18.2. The overall 12% decrement in ferrous elements (Fe, Ni) is perfectly balanced by the Al, Si, and C global increment (+12%) over the last 3.8 Billion years. For the transformations of 3.8 Billion years ago, in fact, we can consider the following balance: Fe (−7.0%) + Ni (−0.2%) = Al (+3.0%) + Si (+2.4%) + C (+1.8%). Analogously, at 2.5 Billion years ago we have: Fe (−4.0%) + Ni (−0.8%) = Al (+1.0%) + Si (+2.4%) + C (+1.4%). The C increment is due to the phono-fission reactions 18.2 and 18.7. The latter reaction provides important explanations concerning the composition of the atmosphere in the past geological eras and the present natural CO_2 emissions. The global increment of + 3.2% in C, involved, at the same time, in reactions 18.2 and 18.7, could explain the high level of carbon in the primordial atmosphere and, consequently, the higher atmospheric pressure with respect to the present one. Taking into account an average density of 3.6×10^3 kg m^{-3} and an average thickness of 60 km for the Hadean and Archean Crusts, the mass of the protocrust involved in reaction 18.7 is equal to 3.4×10^{21} kg and implies a pressure on the Earth's surface of about 660 atm (considering the Earth's surface equal to 5.1×10^{14} m^2, like today). This pressure is very high if compared to that of the present Earth's atmosphere, but it can be considered a plausible value for the primordial atmosphere (between 3.8 and 2.5 Billion years ago). Several Authors, in fact, describe a primordial Earth's atmosphere saturated in CO_2 and CH_4, with a pressure of several hundreds atmospheres. In particular, Liu [53] suggests a value of approximately 650 atm for the proto-atmosphere, composed principally by carbon gasses and H_2O, and similar to that of Mars and Venus [53, 54]. After the intense tectonic activity that involved our planet during the passage from Hadean to Archean periods, the high CO_2 concentration, principally due to phono-fission reaction 18.7, started to decrease. This decrement can be explained considering the planetary air leak of gaseous elements and molecules, such as H, He, and CO_2, which has affected the atmosphere during more recent times [54].

Phono-fission reaction 18.7 can also be put into relation to the increment in seismic activity that has occurred over the last century [55]. Recent evidence has shown CO_2 emissions in correspondence to seismic activity: significant increments in the emission of carbon dioxide were recorded in a geochemical station at El Hierro, in the Canary Islands, before the occurrence of seismic events in the year 2004. Appreciable precursory CO_2 emissions were observed to start before seismic events of relevant magnitude and to reach their maximum values some days before the earthquakes [20]. The evidence, emerging from the XPS analyses at the laboratory scale on marble, that the percentage of oxygen, calcium, and magnesium globally decreases (−13.0%), whereas that of carbon increases (+13.0%), emphasizes how phono-fission reactions 18.7, 18.10, and 18.13 may take place in calcareous rocks. On the other hand, absence of neutron emissions does not imply absence of phono-fission reactions (see the case of marble).

Reaction 18.7 is not the only one that involves Mg as a starting element. From a close examination of the data reported in Fig. 18.3, it is possible to conjecture a series of phono-fission reactions involving Mg, Ca, Na, K, O, C, and H, which represent the real origin of the sharp fluctuations of these chemical elements in the evolution of Earth's Crust (see reactions 18.7 to 18.12 in Table 18.2). Considering phono-fission reactions 18.8, 18.9, 18.11, and 18.12, a general decrement in alkaline-earth elements (Mg, Ca) of 8.7% is balanced by an equal increment in Na, K, and O (see Fig. 18.3). At 3.8 Billion years ago, we have the following balance (see Fig. 18.3 and Table 18.2): Ca (−2.5%) + Mg(−3.2%) = K (+1.4%) + Na (+2.1%) + O (+2.2%). At 2.5 Billion years ago, on the other hand, we have: Ca (−1.5%) + Mg(−1.5%) = K (+1.3%) + Na (+0.6%) + O (+1.1%). Also in this case, the mass percentage changes in the major elements can be perfectly explained considering the new kind of nuclear fission reactions (see Fig. 18.3).

In particular, the global decrement in Ca (-4.0%) can be balanced by an increment in K (+2.7%), reaction 18.11, and by two molecules of H_2O (+1.3%), reaction 18.12. Considering the mass of the protocrust equal to 1.08×10^{23} kg, the decrement of 1.3% in Ca concentration corresponds to 1.40×10^{21} kg. This value is very close to the present mass of water in the oceans (1.35×10^{21} kg, considering a global ocean surface of 3.60×10^{14} m^2 and an average depth of 3,950 m). In this way, reaction 18.12 can be considered as responsible for the formation of oceans during the Earth's lifetime. These last considerations seem to be even more important, if we take into account the recent results from European Union Rosetta Mission about the origin of water on the planet Earth. The evidence coming from the exploration of the Comet 67P led to exclude the extraterrestrial origin of the oceans [56].

Taking into account reaction 18.6, a small portion of the overall Ni decrement reported in Fig. 18.2 could be balanced by the salinity level in the oceans. Considering the NaCl average concentration (3.8%) in the sea water, the mass of dissolved sodium chloride can be quantified in 5.14×10^{19} kg. This value corresponds to about 0.05% of the Earth's protocrust and could be ascribed to a small part of the total Ni depletion between Archean period and present times (−0.8%). This conjecture can be justified considering: (i) the increment in NaCl concentration in the Earth's oceans, about twice, between 2.7 and 2.4 Gyrs ago [22, 57]; (ii) the very high level of salinity in the Ionian and Aegean seas (Eastern Mediterranean Basin) [55], and the highly correlated seismicity in the same area (Fig. 18.8).

Phono-fission reactions 18.13 to 18.18, involving Si, Al, and O as starting elements, and C, N, O, H, and He as resultants, can be considered (see Table 18.2) for greenhouse gas formation. On the other hand, phono-fission reactions 18.1 to 18.12 have affected the evolution of the Earth's crust over the last 4.5 Billion years to the greatest extent.

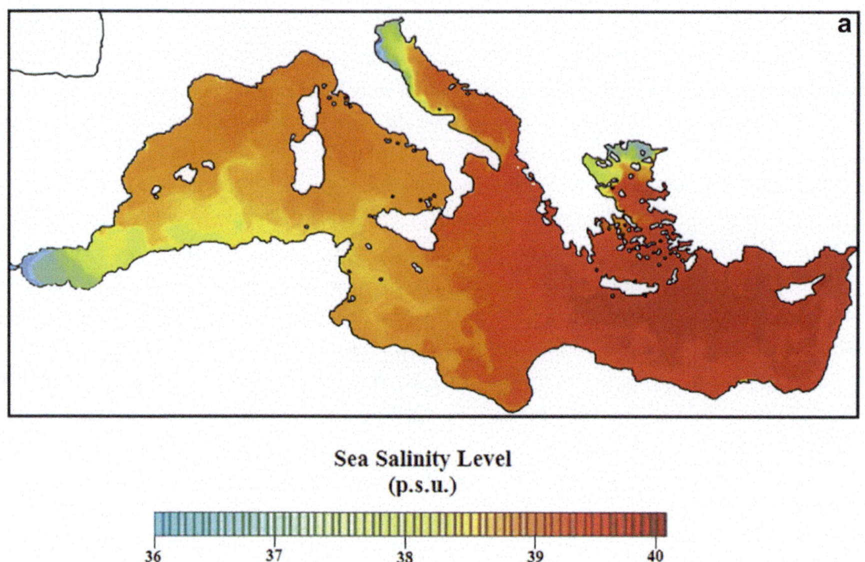

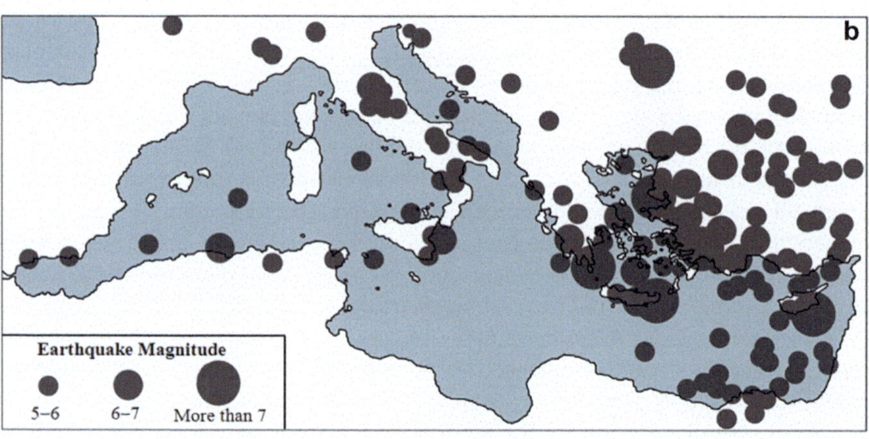

Fig. 18.8 Correlation between Salinity and Seismicity in the Mediterranean Basin (1995–2010 period)

Considering the data shown in Figs. 18.2 and 18.3, it can be noted that reactions 18.1, 18.2, 18.5, 18.8, 18.11, 18.12 should have been particularly recurrent and responsible for a change of approximately 20.8% in the chemical composition of the Earth's crust. Phono-fission reactions 18.7, 18.9, 18.10, 18.13 to 18.18 have instead played a negligible role in the compositional variations of the Earth's crust, but are of a great relevance as far as the increments in H, He, C, N, and O concentrations are concerned in the evolution of the Earth's atmosphere [53]. Reactions 18.9, 18.12,

18.16, and 18.18 played a significant role in phenomena such as the Great Oxidation Event (2.7–2.4 Billion years ago), with a sharp increment in atmospheric oxygen (10^5-fold increment) and the subsequent formation of oceans and origin of life, as well as a high level of N_2 in the primordial atmosphere.

18.5 Conclusions

Phono-fission reactions can be subdivided into two different sets that are typical of our planet:

$$Fe_{26}, Co_{27}, Ni_{28} \rightarrow Mg_{12}, Al_{13}, Si_{14} \rightarrow C_6, N_7, O_8$$

These two jumps represent not only the stepwise temporal variations in the Earth's chemical composition over the last four Billion years, but they also reflect the stepwise spatial variations in the Earth composition from internal core to mantle to crust to atmosphere: Core (NI–FE), Mantle (SI-MA), Crust (SI-AL), and Atmosphere (C, N, O gases).

In addition, as shown by reactions 18.1 and 18.2, the fundamental starting element in the first jump is iron. This element plays also a crucial role in the stellar thermonuclear fusion process in small and medium-sized stars. Iron, in fact, is the heaviest element that is synthesized by thermonuclear fusion reactions before the star collapse. This combustion product is the so-called stellar ash and is the material of which the proto-planets in the Solar System were formed and from which they may evolve, through phono-fission reactions, to a sialic condition (see the case of the Earth). Similarly, the atmospheric elements (C, N, O, H, and He) can be considered as the planetary ashes. These gases are the results of the second jump and they have gradually escaped from Earth toward space by the planetary air leak [54].

The great evidence reported herein cannot be considered as a mere coincidence. It should be considered as a multi-disciplinary signature of the same phenomenon, the phono-fission reactions taking place everywhere, from the laboratory to the planetary scale, and everytime, from the origin of Earth to the present times.

Acknowledgements Special thanks are due to Prof. Riccardo Sandrone of the Environmental Engineering, Land and Infrastructures Department (Politecnico di Torino) for his geological and petrological expertise, and to Dr Angelica Chiodoni of the Italian Institute of Technology (IIT) for her extensive collaboration in the EDS analysis.

References

1. Kolb E (2000) Blind watchers of the sky: the people and ideas that shaped our view of the universe. Oxford University Press, Oxford

2. Favero G, Jobstraibizer P (1996) The distribution of aluminum in the Earth: from cosmogenesis to Sial evolution. Coord Chem Rev 149:367–400
3. Taylor SR, McLennan SM (2009) Planetary crusts: their composition, origin and evolution. Cambridge University Press, Cambridge
4. Hawkesworth CJ, Kemp AIS (2006) Evolution of the continental crust. Nature 443:811–817
5. Canfiled DE (1998) A new model for proterozoic ocean chemistry. Nature 396:450–453
6. Anbar AD (2008) Elements and evolution. Science 322:1481–1482
7. Doglioni C (2007) Interno della Terra, Treccani, Enciclopedia Scienza e Tecnica, pp 595–605
8. Holland HD (2006) The oxygenation of the atmosphere and oceans. Philos Trans R Soc London Ser B 361:903–915
9. Kump LR, Barley ME (2007) Increased subaerial volcanism and the rise of atmospheric oxygen 2.5 Billion years ago. Nature 448:1033–1036
10. Buesseler KO et al (2008) Ocean iron fertilization moving forward in a sea of uncertainty. Science 319:162
11. Saito MA (2009) Less nickel for more oxygen. Nature 458:714–715
12. Konhauser KO et al (2009) Oceanic nickel depletion and a methanogen famine before the great oxidation event. Nature 458:750–754
13. Rudnick RL, Fountain DM (1995) Nature and composition of the continental crust: a lower crustal perspective. Rev Geophys 33(3):267–309
14. Yaroshevsky AA (2006) Abundances of chemical elements in the Earth's crust. Geochem Int 44(1):54–62
15. Kuzhevskij BM, Nechaev OY, Sigaeva EA, Zakharov VA (2003) Neutron flux variations near the earth's crust. a possible tectonic activity detection. Nat Haz Earth Syst Sci 3:637–645
16. Kuzhevskij BM, Nechaev OY, Sigaeva EA (2003) Distribution of neutrons near the Earth's surface. Nat Haz Earth Syst Sci 3:255–262
17. Antonova VP, Volodichev NN, Kryukov SV, Chubenko AP, Shchepetov AL (2009) Results of detecting thermal neutrons at Tien Shan High Altitude Station. Geomag Aeron 49:761–767
18. Volodichev NN, Kuzhevskij BM, Nechaev OY, Panasyuk MI, Podorolsky AN, Shavrin PI (2000) Sun-Moon-Earth connections: the neutron intensity splashes and seismic activity. Astron Vestnik 34(2):188–190
19. Sigaeva EA et al (2006) Thermal neutrons'observations before the Sumatra earthquake. Geophys Res Abs 8:00435
20. Padron E et al (2008) Changes in the diffuse CO_2 emission and relation to seismic activity in and around El Hierro, Canary Islands. Pure Appl Geophys 165:95–114
21. Kelemen PB (2009) The origin of the land under the sea. Sci Am 300(2):42–47
22. Hazen RM et al (2008) Mineral evolution. Am Miner 93:1693–1720
23. Lunine EJI (1998) Earth: evolution of a habitable world. Cambridge University Press, Cambridge/New York/Melbourne
24. Roy I, Sarkar BC, Chattopadhyay A (2001) MINFO-a prototype mineral information database for iron ore resources of India. Comp Geosci 27:357–361
25. World Mineral Resources Map. http://www.mapsofworld.com/world-mineral-map.htm. Last accessed on Oct 2009
26. Key Iron Deposits of the World. http://www.portergeo.com.au/tours/iron2002/-iron2002depm2b.asp. Last accessed on Oct 2009
27. Sigman D, Jaccard S, Haug F (2004) Polar ocean stratification in a cold climate. Nature 428:59–63
28. National Academy of Sciences (1975) Medical and biological effects of environmental pollutants: nickel. Proc Natl Acad Sci Washington, DC
29. Foing B (2005) Earth's childhood attic. Astrobiol Mag (on-line) February 23, 2005
30. Egami F (1975) Minor elements and evolution. J Mol Evol 4(2):113–120
31. Carpinteri A, Cardone F, Lacidogna G (2009) Piezonuclear neutrons from brittle fracture: early results of mechanical compression tests. Strain 45: 332–339. Presented at the Turin Academy of Sciences on December 10, 2008. Proc Turin Acad Sci 33:27–42

32. Cardone F, Carpinteri A, Lacidogna G (2009) Piezonuclear neutrons from fracturing of inert solids. Phys Lett A 373:4158–4163
33. Carpinteri A, Cardone F, Lacidogna G (2009) Energy emissions from failure phenomena: mechanical, electromagnetic, nuclear. Exp Mech 50:1235–1243
34. Carpinteri A, Lacidogna G, Manuello A, Borla O (2010) Piezonuclear transmutations in brittle rocks under mechanical loading: microchemical analysis and geological confirmations. In: Kounadis AN, Gdoutos EE (Eds) Recent advances in mechanics. Springer, pp 361–382
35. Carpinteri A, Borla O, Lacidogna G, Manuello A (2010) Neutron emissions in brittle rocks during compression tests: monotic vs. cyclic loading. Phys Mesomech 13:264–274
36. Carpinteri A, Chiodoni A, Manuello A, Sandrone R (2011) Compositional and microchemical evidence of piezonuclear fission reactions in rock specimens subjected to compression tests. Strain 47:267–281
37. Carpinteri A, Lacidogna G, Manuello A, Borla O (2011) Energy emissions from brittle fracture: neutron measurements and geological evidences of piezonuclear reactions. Strength Fract Complex 7:13–31
38. Carpinteri A, Lacidogna G, Borla O, Manuello A, Niccolini G (2012) Electromagnetic and neutron emissions from brittle rocks failure: experimental evidence and geological implications. Sadhana 37:59–78
39. Carpinteri A, Lacidogna G, Manuello A, Borla O (2012) Piezonuclear fission reactions: evidences from microchemical analysis, neutron emission, and geological transformation. Rock Mech Rock Eng 45:445–459
40. Carpinteri A, Lacidogna G, Manuello A, Borla O (2013) Piezonuclear fission reactions from earthquakes and brittle rocks failure: evidence of neutron emission and nonradioactive product elements. Exp Mech 53(3):345–365
41. Carpinteri A, Manuello A (2012) An indirect evidence of piezonuclear fission reactions: geomechanical and geochemical evolution in the earth's crust. Phys Mesomech 15:14–23
42. Carpinteri A, Borla O, Lacidogna G, Manuello A (2012) Piezonuclear reactions produced by brittle fracture: from laboratory to planetary scale. In: Proceedings of the 19th European conference of fracture, Kazan, Russia
43. Diebner K (1962) Fusionsprozesse mit Hilfe konvergenter Stosswellen—einige aeltere und neuere Versuche und Ueberlegungen. Kerntechnik 3:89–93
44. Kaliski S (1976) Critical masses of mini-explosion in fission–fusion hybrid systems. J Tech Phys 17:99–108
45. Winterberg F (1984) Autocatalytic fusion–fission implosions. Atom Kerntech 44:146
46. Derjaguin BV et al (1989) Titanium fracture yields neutrons? Nature 34:492
47. Arata Y, Zhang Y (1995) Achievement of solid-state plasma fusion ("cold-fusion"). Proc Jpn Acad Ser B 71:304–309
48. Cardone F, Cherubini G, Petrucci A (2009) Piezonuclear neutrons. Phys Lett A 373(8–9):862–866
49. Carpinteri A (1989) Cusp catastrophe interpretation of fracture instability. J Mech Phys Solids 37:567–582
50. Carpinteri A (1990) A catastrophe theory approach to fracture mechanics. Int J Fract 44:57–69
51. Manuello A, Sandrone R, Guastella S, Borla O, Lacidogna G, Carpinteri A (2012) Piezonuclear reactions during mechanical tests of basalt and magnetite. In: Proceedings of the 19th European conference of fracture, Kazan, Russia
52. Lacidogna G, Borla O, Carpinteri A (2012) X-ray Photoelectron spectroscopy on fracture surfaces of Carrara marble specimens crushed in compression. In: Proceedings of the 19th European conference of fracture, Kazan, Russia
53. Liu L (2004) The inception of the oceans and CO_2-athmosphere in the early history of the Earth. Earth Planet Sci Lett 227:179–184
54. Catling CD, Zahnle KJ (2009) The planetary air leak. Sci Am 300(5):24–31
55. Aki K (1983) Strong motion seismology. In: Kanamori H, Boschi E (eds) Earthquakes: observation, theory and interpretation. North Holland Pub, Amsterdam, pp 223–250

56. Altwegg K et al (2014) 67P/Churyumov-Gerasimenko, a Jupiter family comet with a high D/H ratio. Science. https://doi.org/10.1126/science.1261952
57. Knauth P (1998) Salinity history of the Earth's early ocean. Nature 395:554–555

Chapter 19
Earth's Mantle and Outer Core: Analogies to the Crustal Chemical Evolution

Abstract In the preceding chapter, we have seen how the surprising chemical balances at the times of the most intense tectonic plate activities are an indirect evidence of phono-fission reactions. Recent results, observed at the scale of the Earth's crust and reproduced at the scale of the laboratory, may be extended to the different layers of the planet like the atmosphere and the bulk Earth (mantle and outer core). The mantle of our planet is characterized by very high pressures and temperatures (~150 GPa and ~ 4,000 °C) that can favor this kind of reactions. The most important chemical changes in the Earth's crust evolution are recognized also at the internal Earth's layers. Such an internal evolution may be interpreted consistently with and analogously to that of the crust. Summarizing, an overall decrement in iron of 21% is only partially balanced by consistent increments in silicon and magnesium (SI-MA). The remaining unbalanced part of the iron decrement (about 5%) should have contributed (by migration) to the outer core formation, where the iron content is about equal to 80% of the total mass.

Keywords Earth's mantle · Chemical evolution · Iron depletion · Phono-fission reactions

19.1 Preliminary Remarks

Recent geophysical and geochemical studies of our planet present a rich array of large-scale processes and phenomena that are not fully understood [1]. These range from the subducted slabs to the nature of core–mantle boundary; from the mechanisms forming the present-day crust, mantle, and core to the distribution of Earth's crust elements, and the uptake and recycling of volatiles over the Earth's history [1]. It is only recently that pressures similar to those prevailing inside the Earth can be reproduced in the laboratory, as well as the materials can be tested by the necessary instruments, in order to study the interior Earth's composition. The experiments have demonstrated that, under such extreme conditions, the physical behavior of materials can be profoundly altered, causing new and unforeseen reactions and giving rise to

A. Carpinteri, *Terahertz Phonons and Nanomechanical Instabilities*,
https://doi.org/10.1007/978-3-032-14692-2_19

electromagnetic and nuclear transitions in rocks and minerals [1]. Similar results were observed also in rocks of near-surface environment. During the last few years, in fact, surprising results have been reported considering the chemical changes that characterized the Earth's crust and atmosphere evolution, together with the ocean formation, during the last 4.57 billion years [2–9]. These data are represented by the depletions of elements such as iron and calcium, as well as by the increment in lighter elements, just in correspondence to particularly severe tectonic activities in our planet [3–5]. The chemical changes observed at the scale of the Earth's crust can be interpreted by innovative experiments conducted on non-radioactive natural rocks subjected to different mechanical loading conditions [10–13]. These rocks are natural materials that represent the composition of the Earth's near-surface environment, such as granite, basalt, magnetite, and calcareous rocks. All these results provide evidence concerning a new kind of nuclear reactions that take place also during quasi-static or cyclic-fatigue tests [10]. The phenomena observed at the laboratory scale suggest that pressure waves of very high frequency (TeraHertz phonons), suitably exerted on inert and stable nuclides, generate phono-fission reactions accompanied by evident and reproducible chemical changes localized on the fracture surfaces of rock fragments [10–16]. The catalyzing factor producing these new kind of nuclear reactions, with so important geological and geophysical implications, is pressure. In particular, these reactions would be more easily activated where the environment conditions (pressure and temperature) are particularly severe, and mechanical phenomena of fracture, crushing, fragmentation, comminution, erosion, friction, etc., more frequent [8–14]. It is evident that the physical conditions characterizing the microseismic and seismic activity of an impending earthquake may be sufficient to produce these nuclear reactions at the scale of the Earth's crust. Even more severe conditions regard the bulk of our planet [1]. The mantle, presenting huge pressures and temperatures, reaching ~150 GPa and ~4,000 °C, respectively, is characterized by significant chemical composition changes, similar to those observed in the Earth' s crust [1, 3, 17–20]. As will be reported in the following, the chemical evolution in the Earth's mantle may be interpreted in the light of the same low energy nuclear reactions recently discovered and firstly proposed to explain the chemical evolution of the Earth's lithosphere.

19.2 Earth's Crust and Atmosphere Chemical Evolution

The transition from a mafic (basalt) to a sialic (granite) composition of the Earth's crust was discussed by different Authors [2, 3, 8, 9]. In the previous chapter, the hypothesis that the changes in Fe, Ni, Si, and Al are explained by phono-fission reactions is advanced [8–13, 15, 16]. During the first geological period, when the planet's tectonic activity began (3.8 Gyrs ago), the decrement in Fe amounted to 7% of the total mass of the Hadean protocrust, and, through the symmetric fission reaction [8–13]:

$$Fe_{26}^{56} \rightarrow 2Al_{13}^{27} + 2\text{neutrons} \tag{19.1}$$

this was partially balanced by a 3% increment in Al. The remaining 4% can be explained considering the asymmetric fission reaction:

$$Fe_{26}^{56} \rightarrow Mg_{12}^{24} + Si_{14}^{28} + 4 \text{ neutrons} \tag{19.2}$$

In evaluating the effect of this reaction, we can consider a 2.2% increment in Si and an increment in Mg of 1.8% (Table 19.1). In the same period, the decrement in Ni appears to be 0.2%. The latter reduction can be interpreted in the light of the following reaction:

$$Ni_{28}^{59} \rightarrow 2\ Si_{14}^{28} + 3 \text{ neutrons} \tag{19.3}$$

which is balanced by an identical 0.2% increment in Si. From these reactions, the overall increment in Si for this period is thus 2.4% (Table 19.1).

In the second geological period of the Earth's life-time with the most intense tectonic activity (2.5 Gyrs ago), Fe decreased by 4%, and this can be balanced by a 1% increment in Al through reaction (19.1). With regards to reaction (19.2), the remaining 3% can be balanced by increments in Si (+1.6%) and Mg (+1.4%). In the same period, 0.8% decrement in Ni can, by taking reaction (19.3) into account, be balanced by an increment in Si of the same percentage. Nickel can also be involved in reaction 18.6 of Table 18.2, in addition to the reaction (19.3).

The recent studies cited above [8–15] also illustrate the chemical composition changes in the Earth's crust that have taken place for Ca, Mg, K, Na, and O (Table 19.2). In the first critical geological period (3.8 Gyrs ago), the decrement in Ca amounts to 2.5%, through the reaction:

$$Ca_{20}^{40} \rightarrow K_{19}^{39} + H_{1}^{1} \tag{19.4}$$

This decrement is balanced by a 1.4% increment in K. The remaining 1.1% may be balanced by the following reaction:

Table 19.1 Quantitative data of the ferrous element chemical changes in the Earth's crust

	Decrement in Fe and increments in Al, Si, and C				Decrement in Ni and increments in Si and NaCl		
Geological period	Fe (%)	Al (%)	Si (%)	C (%)	Ni (%)	Si (%)	NaCl (%)
Tectonic plate formation 3.8 Gyrs ago	**−7**	**+3**	**+2.2**	**+1.8**	**−0.2**	**+0.2**	**0**
Most severe tectonic activity 2.5 Gyrs ago	**−4**	**+1**	**+1.6**	**+1.4**	**−0.8**	**+0.8**	**+0.05**
Total	**−11**	**+4**	**+3.8**	**+3.2**	**−1**	**+1.0**	**+0.05**

Table 19.2 Quantitative data of the alkaline-terrous (alkaline-earth) element chemical changes in the Earth's crust

	Decrement in Ca and increments in K and H_2O			Decrement in Mg and increments in Na and O		
Geological period	Ca (%)	K (%)	H_2O (%)	Mg (%)	Na (%)	O (%)
Tectonic plate formation 3.8 Gyrs ago	**−2.5%**	**+1.4%**	**+1.1%**	**−3.2%**	**+2.1%**	**+1.1%**
Most severe tectonic activity 2.5 Gyrs ago	**−1.5%**	**+1.3%**	**+0.2%**	**−1.5%**	**+0.6%**	**+0.9%**
Total	**−4.0%**	**+2.7%**	**+1.3%**	**−4.7%**	**+2.7%**	**+2.0%**

$$Ca_{20}^{40} \rightarrow 2O_8^{16} + 4H_1^1 + 4\text{ neutrons} \tag{19.5}$$

This reaction can explain the formation of H_2O, which was firstly in the primordial atmosphere and then condensed to form the oceans (Table 19.2). In the same critical period, there was a 3.2% reduction in Mg:

$$Mg_{12}^{24} \rightarrow Na_{11}^{23} + H_1^1 \tag{19.6}$$

which is balanced by a 2.1% increment in Na and by a total increment in O and H of 1.1% (Table 19.2). During the second critical geological period (2.5 Gyrs ago), the decrement in Ca is 1.5%, which, considering reaction (19.4), is balanced by a 1.3% increment in K. Taking reaction (19.5) into account, the remaining 0.2% is balanced by the formation of H_2O (Table 19.2). The 1.5% decrement in Mg for the same period, see reaction (19.6), can be balanced by a 0.6% increment in Na and by a total increment in O and H of 0.9% (Table 19.2). In both the critical periods of compositional stepwise transition (3.8 and 2.5 Gyrs ago), the reduction in Ca that is not balanced by an increment in K can be balanced by the formation of H_2O molecules, see reaction (19.5) (Table 19.2).

19.3 Greenhouse Gas Increment and Climate Change

The Earth's atmosphere, as the crust, has undergone significant changes in its major constituent elements over the last 4.5 billion years [17–26]. The findings presented here stem from studies on the evolution of individual gases in primordial atmosphere and on the variations in atmospheric pressure.

In the early stages of Earth's formation, our planet underwent complex phenomena of differentiation, that slowly led to the stratigraphic division into crust (continental and oceanic), mantle (upper and lower), and core (outer and inner) [27–31].

There is not any doubt that one of the basic characteristics of the early stages of the proto-Earth was the presence of an atmosphere very different from today's one. The origin of atmosphere and the formation of oceans are still being intensively investigated, with studies that continue to fuel the scientific debate [17–19].

As can be read in the specific literature [32–34], the most common atmospheric elements are today nitrogen (N_2 ~ 78%) and oxygen (O_2 ~ 20%). During the second primordial time, the main constituents were H_2O and CO_2, whereas the secondary constituents, as now on Mars and Venus, were N, O, and CH_4. What is the reason for a so dramatic change? How is it possible to pass from an atmosphere that is highly toxic for present-day forms of life to one compatible with mankind's appearance on the Earth? Although these are still unanswered questions, the literature can provide us with data that suggest a number of intriguing assumptions.

First, it is interesting to note that, having a primordial atmosphere saturated with CO_2 and H_2O, means that the atmospheric pressure reaches some hundreds bar [23–25]. In particular, the hypotheses regarding the concentrations of the various elements make it possible to evaluate the effect exerted by each single gas in terms of atmospheric pressure on the surface of the protoplanet.

On the basis of the data presented in [8, 9] and those given in Table 19.2, reaction (19.2) would result in an overall increment in Mg of 3.2%. However, it is likely involved in a further phono-fission reaction, for which Mg is the starting element and C the resultant:

$$Mg_{12}^{24} \rightarrow 2C_6^{12} \tag{19.7}$$

Today, the Earth's atmosphere consists of a 40 km layer, its mass is 5×10^{18} kg with a density γ of around 1.250 kg/m^3. Given this mass and a terrestrial surface area of 5.10×10^{14} m^2, we thus have an atmospheric pressure of 96,060 Pa or 0.96 atm (~1 atm).

A similar calculation can be made for the primordial atmosphere, with a high concentration of carbon. In this case, the mass to be considered is that resulting from the phono-fission reaction (19.7) and corresponding to 3.2% of the total protocrust mass (3.4×10^{21} kg). Considering the same gravitational acceleration for the proto-Earth, we obtain an atmospheric pressure of 666.70 atm. Although this value is very high in comparison to the current atmosphere, it can be regarded as plausible and consistent with the conditions prevailing between 3.8 and 2.5 billion years ago. In addition, it is corroborated by the studies of a number of authors who present models of a CO_2 and CH_4-rich early atmosphere, with a pressure of some hundreds atm, as was the case of the atmospheric pressure of 660 atm indicated by Liu [17].

The dramatic increment in light molecular components like N_2, O_2, CO_2, and CH_4 can be explained by the phono-fission reactions [8–16]. In particular, the reactions involving Mg, Si, and Al as starting elements and C, N, O, and H as resultants [8, 9]:

$$Mg_{12}^{24} \rightarrow O_8^{16} + 4H_1^1 + 4\text{neutrons} \tag{19.8}$$

$$Si_{14}^{28} \rightarrow 2N_7^{14} \tag{19.9}$$

$$Si_{14}^{28} \rightarrow C_6^{12} + O_8^{16} \tag{19.10}$$

$$Al_{13}^{27} \rightarrow C_6^{12} + N_7^{14} + 1\text{neutron} \tag{19.11}$$

They represent a new explanation to the chemical composition of the primordial atmosphere. Recent data, in fact, have shown that CO_2 and H_2O concentrations increased drastically between 3.8 and 2.5 Gyrs ago (Fig. 19.1), in correspondence to the tectonic activity and to the subduction of the primordial plates. Successively, between 2.5 and 2.0 Gyrs ago, the fast increment in N and O concentrations may be considered to be closely related to the formation of the Earth's continental crust (Fig. 19.1). The oxygen level, in fact, was very low during the Hadean and Archean eras (4.5–2.7 Gyr ago), whereas it increased sharply (Great Oxidation Event) between 2.7 and 2.4 Billion years ago, until the present concentration of ~21% in the Earth's atmosphere was reached, a 10^5-fold higher value than that of the earlier atmospheres (Fig. 19.1).

During the same period (from 2.7 to 2.4 Gyrs ago), which represents 6% of the entire Earth's life-time (4.57 Gyrs), 50% of the continental crustal volume formed in concomitance to the most intense tectonic and continental subduction activities. Considering the scenario of approximately 2.0 Gyr ago, it is also possible to justify the origin of the first aerobic bacteria and multicellular eukaryotic organisms, ancestors of animals and human beings. Based on the compositional time-variations in the Earth's atmosphere shown in Fig. 19.1, and the phono-fission reactions (19.5, 19.7–19.11), it can be observed that the intense tectonic activity caused a sudden increment in CO_2 and, later, in the O and N levels. While the O and N levels remained constant, the high CO_2 concentration, principally due to reaction (19.7), started to decrease after 2.5 Gyr ago (Fig. 19.1). This decrement can be explained considering the planetary air leakage of gaseous elements and molecules, such as H, He, and CO_2, that has affected the atmosphere during the entire Earth's life-time [35]. The continuous decrement in CO_2 over the last 3.5 Gyrs has recently been contrasted by the well-known carbon pollution.

The carbon dioxide (CO_2) variations over the last 500 million years (Phanerozoic period) are reported in Fig. 19.2a. The trend shown in the plot is determined considering Royer's database [36] together with the results of several specific geochemical models [37–39]. The present CO_2 concentration in the Earth's atmosphere is about 400 ppm, whereas 500 million years ago it was about 6,000 ppm. Observing the phenomenon at that time scale, the increment that has occurred over the last 100 years, which has been ascribed to carbon emissions from the industrial revolution, seems to be negligible. In Fig. 19.2b, it is possible to observe the carbon dioxide (CO_2) variations over the last 4×10^5 years [31, 40, 41, 44, 46]. Today, several scientists sustain that, throughout the twentieth century, a new form of carbon pollution and the reactive nitrogen released into terrestrial environment by human activities (synthetic fertilizers, industrial use of ammonia, etc.) have been the only responsible factors for the dramatic increment in CO_2 and N_2 of the contemporary atmosphere [42–44]. However, a doubt remains whether much of these CO_2 (about 3/4 of the total) and

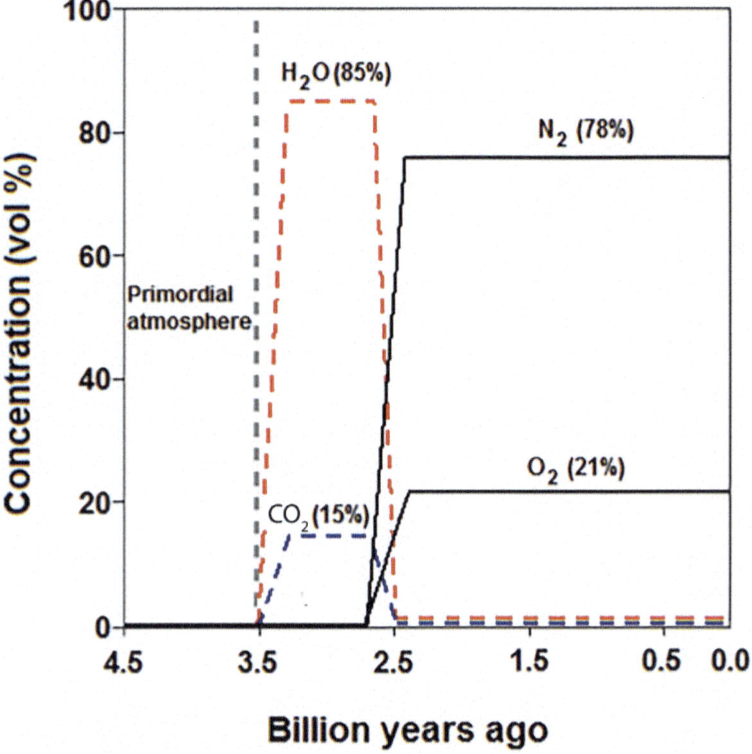

Fig. 19.1 Variations in the atmospheric chemical composition during the Earth's life-time. In the Archean period, 15% of the atmosphere was constituted by CO_2, whereas the remaining part was mainly composed of H_2O. The present Earth's atmosphere composition is dominated by nitrogen (N ~ 78%) and oxygen (O ~ 21%) [17]

N_2 (about 2/3 of the total) are provided by the same causes that have produced the previous cycles of carbon dioxide (Fig. 19.2b) and the nitrogen increment in the Archean atmosphere (Fig. 19.1) [18, 32].

This hypothesis is supported by the evidence presented recently for the Earth's crust [19, 20], together with the usually accepted idea that the early atmosphere's composition derived from mantle degassing [31, 32, 45–47].

19.4 Chemical Evolution in the Earth's Mantle

Studies on the evolution of mantle chemical composition [32, 34, 48, 49] have made it possible to analyze the changes in mass percentage of its main chemical elements and to compare them to those of the Earth's crust [8–11].

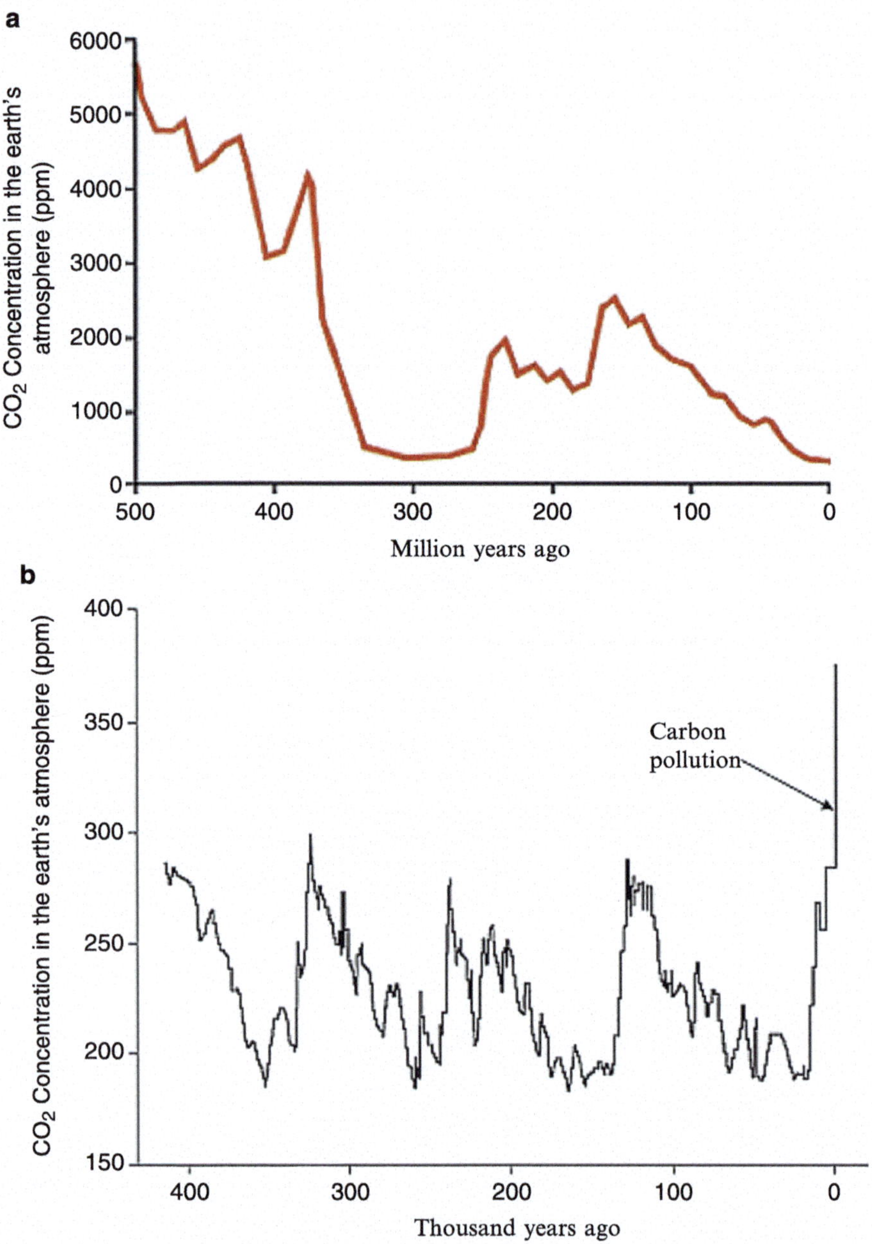

Fig. 19.2 **a** CO_2 concentration in the Earth's atmosphere over the last 500 million years. The trend shown in the plot is determined considering the database reported by Royer [36] together with the results of some geochemical models developed between 2001 and 2004 [38–40]. **b** CO_2 concentration in the Earth's atmosphere over the last 4×10^5 years [38, 39]

These analyses are the basis for the graph reported in Fig. 19.3 and dealing with the changes in mass concentration of Fe, Mg, Al, Si, and Ca during the Earth's mantle evolution. They lead to different considerations concerning the tectonic and convective events that occurred in the late Archean period and their contributions to the chemical evolution of the terrestrial mantle.

As already observed for the Earth's crust [8, 9], major changes in mass concentration took place at the times of greatest tectonic events, transforming heavier (Fe) into lighter (Mg, Al, Si, Ca) elements. The data presented in the literature indicate that the Earth's mantle, like the continental crust [8, 9], saw two abrupt transitions in

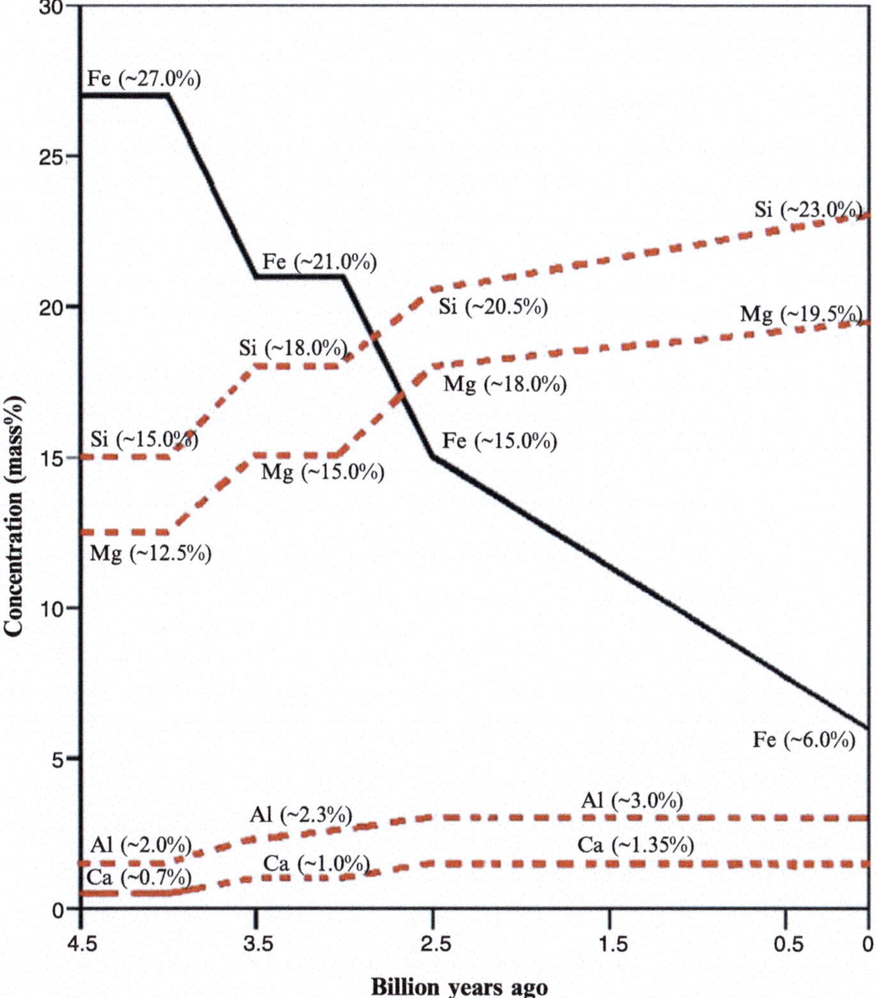

Fig. 19.3 Evolution of the concentrations of Fe, Si, Mg, Al, and Ca in the terrestrial mantle over the Earth's life-time [32]

the concentrations of its constituent elements [32]. Specifically, these periods date between 4.0 and 3.8 billion years ago, and between 2.8 and 2.5 billion years ago. As shown in Fig. 19.3, the mantle's decrement in Fe is balanced by the increment in other elements such as Si, Al, Mg, and Ca, which were involved in phono-fission reactions like those observed for the Earth's crust.

The data for FeO and Fe_2O_3 given in [32] indicate that iron dropped down by about 6% at the first phono-fission transformation, and that the reactions (19.1) and (19.2), and the further reaction:

$$Fe_{26}^{56} \rightarrow Ca_{20}^{40} + C_{6}^{12} + \text{ 4 neutrons} \tag{19.12}$$

can explain the increments of ~3.0% in Si, ~2.5% in Mg, ~0.3% in Al, and ~0.3% in Ca. These contributions are surprisingly precise in balancing the mass reduction in Fe in the mantle. A similar situation occurs in the second critical period. Between 2.8 and 2.5 billion years ago, the decrement in Fe is again around 6%, while the increments of ~2.5% in Si, ~3.0% in Mg, ~0.7% in Al, and ~0.35% in Ca appear to balance the mass reduction in Fe very satisfactorily (Fig. 19.3). The reaction involving Fe as the starting element and Ca and C as resultants not only accounts for the increment in Ca in the mantle, but also offers an alternative for the explanation to the formation of C in the mantle [32]. The current theories sustaining that carbon was transported into the mantle through the subduction of the Earth's crust, in fact, do not seem to be entirely convincing [32].

The nearly perfectly matched balance taking place between iron and the elements increasing their concentration in both critical geological periods may be explained as already described above. However, it is necessary to consider that the Earth's mantle, unlike the crust, went through a further change that involved the same elements analyzed above, from 2.5 billion years ago to the present times (Fig. 19.3). In this last change, we can see that the element concentrations have varied continuously and not stepwise, and that these variations have been not confined to a very narrow span of time as in the previous cases. From 2.5 billion years ago up to the present times, there has been a further decrement in the concentration of Fe equal approximately to 9.0%, and additional increments in the other lighter elements (in particular, Si and Mg), which have only partially balanced the iron reduction [32]. Significant increments can be observed in Si (+2.5%) and Mg (+1.5%), with an overall increment of just 4%. While the remaining not-balanced decrement in Fe is approximately equal to 5.0% of the mantle's mass. This decrement could be explained considering the formation of the Earth's outer core, where the concentration of iron is approximately equal to 80% [8, 9, 28, 32].

19.5 Conclusions

The main results of the present chapter are related to the evidence of chemical composition changes affecting Earth's crust, mantle, and atmosphere from the early periods of their evolution until today. Some anomalies and changes in the concentrations of the most abundant elements in the mantle can be interpreted by the phono-fission reactions already introduced in the preceding parts of the book. In addition, a further reaction involving Fe, Ca, and C is assumed to explain the mantle evolution. This reaction not only accounts for the increment in Ca in the mantle, but also offers an alternative and convincing explanation to the formation of carbon and to its anomalous concentration in the mantle [32]. The current theories sustaining that carbon was transported to the mantle through crustal subduction, in fact, do not seem to be entirely convincing [32].

From 2.5 billion years ago up to the present times, there has been a further progressive decrement in the concentration of iron equal approximately to 9.0% (after the previous stepwise 12.0%), as well as related increments in other elements (in particular, Si and Mg), which have only partially balanced the last iron reduction [32]. Significant increments are observed in Si (+2.5%) and Mg (+1.5%), with an overall increment of just 4%. The decrement in Fe that is not balanced by the increments in Si and Mg amounts to 5.0% of the total mantle's mass. This additional unbalanced decrement can be explained with the formation of the Earth's outer core, where the concentration of iron is approximately equal to 80% [8, 9, 28, 32].

References

1. Mao HR, Hemley J (2007) The high-pressure dimension in earth and planetary science. PNAS 104:914–915
2. Anbar AD (2008) Elements and evolution. Science 322:1481–1482
3. Taylor SR, McLennan SM (2009) Planetary crusts: their composition, origin and evolution. Cambridge University Press, Cambridge
4. Saito MA (2009) Less nickel for more oxygen. Nature 458:714–715
5. Konhauser KO et al (2009) Oceanic nickel depletion and a methanogen famine before the great oxidation event. Nature 458:750–754
6. Favero G, Jobstraibizer P (1996) The distribution of aluminum in the Earth: from cosmogenesis to Sial evolution. Coord Chem Rev 149:367–400
7. Taylor SR, McLennan SM (1995) The geochemical evolution of the continental crust. Rev Geophys 33:241–265
8. Carpinteri A, Manuello A (2011) Geomechanical and geochemical evidence of piezonuclear fission reactions in the Earth's Crust. Strain 47:282–292
9. Carpinteri A, Manuello A (2012) An indirect evidence of piezonuclear fission reactions: geomechanical and geochemical evolution in the Earth's Crust. Phys Mesomech 15:14–23
10. Carpinteri A, Borla O, Lacidogna G, Manuello A (2010) Neutron emissions in brittle rocks during compression tests: monotic versus cyclic loading. Phys Mesomech 13:264–274
11. Carpinteri A, Lacidogna G, Manuello A, Borla O (2011) Energy emissions from brittle fracture: neutron measurements and geological evidences of piezonuclear reactions. Strength Fract Compl 7:13–31

12. Carpinteri A, Lacidogna G, Manuello A, Borla O (2012) Piezonuclear fission reactions: evidences from microchemical analysis, neutron emission, and geological transformation. Rock Mech Rock Eng 45:445–459
13. Carpinteri A, Lacidogna G, Manuello A, Borla O (2013) Piezonuclear fission reactions from earthquakes and brittle rocks failure: evidence of neutron emission and nonradioactive product elements. Exp Mech 53(3):345–365
14. Cardone F, Carpinteri A, Lacidogna G (2009) Piezonuclear neutrons from fracturing of inert solids. Physics Lett A 373:4158–4163
15. Carpinteri A, Cardone F, Lacidogna G (2009) Energy emissions from failure phenomena: mechanical, electromagnetic, nuclear. Exp Mech 50:1235–1243
16. Carpinteri A, Cardone F, Lacidogna G (2009) Piezonuclear neutrons from brittle fracture: early results of mechanical compression tests. Strain 45: 332–339. Atti dell'Accademia delle Scienze di Torino 33:27–42
17. Liu L (2004) The inception of the oceans and CO_2-atmosphere in the early history of the Earth. Earth Planet Sci Lett 227:179–184
18. Weast RC (ed) (1980) CRC handbook of chemistry and physics. CRC Press, New York, 199
19. Garrison TS (2005) Oceanography: an invitation to marine science. Thompson Brooks Cole, Belmont
20. Schopf J (1983) Earth's earliest biosphere: its origin and evolution. Princeton University Press, Princeton
21. Kolb E (2000) Blind watchers of the sky: the people and ideas that shaped our view of the universe. Oxford University Press, Oxford
22. Kolb E, Matarrese S, Notari S, Riotto A (2005) Primordial inflation explains why the Universe is accelerating today. arXiv:hep-th/0503117v1:1–4
23. Williams RP, Da Silva FJR (2003) Evolution was chemically constrained. J Theor Biol 220:323–343
24. Buesseler KO, Doney SC, Karl DM et al (2008) Ocean iron fertilization moving forward in a sea of uncertainty. Science 319:162
25. Holland HD (2006) The oxygenation of the atmosphere and oceans. Philos Trans R Soc Lond Ser B 361:903–915
26. Abbott DH, Burgess L, Longhi J, Smith WHF (1994) An empirical thermal history of the Earth's upper mantle. J Geophys Res 99(13):835–850
27. Vovna GM, Mishkin MA, Sakhno VG, Zarubina NV (2009) Early Archean sialic crust of the Siberian craton: its composition and origin of magmatic protoliths. Dokl Earth Sci 429(2):1439–1442
28. The World Ocean (2007) The Columbia encyclopedia. CD-ROM, 6th edn. Columbia University Press, New York
29. Van Nostrands Scientific Encyclopedia (2008) Ocean volume and depth, 10th edn. Van Nostrands Scientific Encyclopedia, New York
30. The Concise Columbia Electronic Encyclopedia (1994) 3rd edn. Columbia University Press, New York
31. Kasting JF, Ackerman TP (1986) Climatic consequences of very high carbon dioxide levels in the Earth's early atmosphere. Science 234:1383–1385
32. Yung YL, De More WB (1999) Photochemistry of planetary atmospheres. Oxford University Press, New York
33. Ronov AB, Yaroshevsky AA (1978) The chemical composition of Earth's crust and its shells. Tectonosphere of Earth. Nedra, Moscow, pp 376–402
34. Catling CD, Zahnle KJ (2009) The planetary air leak. Sci Am 300:24–31
35. Royer DL, Berner RA, Montanez IP, Tabor NJ, Beerling DJ (2004) CO2 as a primary driver of phanerozoic climate. GSA Today 14:4–10
36. Berner RA (1990) Atmospheric carbon dioxide levels over phanerozoic time. Science 249:1382–1386
37. Berner RA, Kothavala Z (2001) GEOCARB III: a revised model of atmospheric CO_2 over phanerozoic time. Am J Sci 301:182–204

38. Bergman RA, Noam M, Timothy ML, Watson AJ (2004) COPSE: a new model of biogeochemical cycling over Phanerozoic time. Am J Sci 301:182–204
39. Rothman DH (2001) Atmospheric carbon dioxide levels for the last 500 million years. Proc Natl Acad Sci USA 99:4167–4171
40. Fischer H, Wahlen M, Smith J, Mastroianni D, Deck B (1999) Ice core records of atmospheric CO_2 around the last three glacial terminations. Science 283:1712–1714
41. Monnin E, Steig EJ, Siegenthaler U et al (2004) Evidence for substantial accumulation rate variability in Antarctica during the Holocene, through synchronization of CO_2 in the Taylor Dome, Dome C and DML ice cores. Earth Planet Sci Lett 224:45–54
42. Townsend A, Howarth RW (2010) Fixing the global Nitrogen problem. Sci Am 302:50–57
43. Aki K (1983) Strong motion seismology. In: Kanamori H, Boschi E (eds) Earthquakes: observation, theory and interpretation. North-Holland Publishers, Amsterdam, pp 223–250
44. Sorokhtin OG et al (2007) Global warming and global cooling—evolution of climate on earth. Elsevier, Amsterdam
45. Ahrens TJ (1971) The state of mantle minerals. Technophysics XX:189–219
46. Wang L et al (2010) Nanoprobe measurements of materials at megabar pressures. PNAS 107(14):6140–6145
47. Ringwood AE (1962) The chemical composition and the origin of Earth. In: Hurley PM (ed) Advances in earth science. MIT Press, Cambridge, pp 276–356
48. Urey HC, Craig H (1953) The composition of the stone meteorites and the origin of the meteorites. Geochim Cosmochim Acta 4(1–2):36–82
49. Rees M (2005) Universe—the definitive visual guide. Dorling Kindersley Ltd., New York

Chapter 20
Earth's Atmosphere: Correlated Fluctuations in Worldwide Global Seismicity and Carbon Dioxide Increments

Abstract The crucial step-wise changes in the geochemical evolution of Earth's crust, ocean, and atmosphere can be explained assuming low-energy nuclear reactions (LENR) triggered by the tectonic and seismic activities. The phono-fission reactions are related to Fe and Ni as starting elements, to medium-weight elements (Mg, Al, Si) as intermediate resultants, and to C, N, O as final resultants. Geochemical data and experimental evidence support the phono-fission conjecture. A spectral analysis over the time interval 1955–2013 shows common cycles (same periods) between atmospheric CO_2 growth rate and worldwide global seismic-energy release rate, whereas the trending increasing behavior of the atmospheric CO_2 is in response to the anthropogenic activities. Investigating on the correlation between such seismic and atmospheric time-fluctuations, the latter can be explained by cyclic worldwide seismicity triggering phono-fission reactions massively in the Earth's crust. In this framework, LENR from active faults and volcanoes can be considered as a consistent cause for carbon dioxide time-oscillations. Such a partial and natural carbon production is to be taken into a serious account when discussing about climate change.

Keywords Atmospheric chemical evolution · Worldwide global seismicity · Phono-fission nuclear reactions · Carbon dioxide pollution · Time- series analysis

20.1 Preliminary Remarks

Several geochemistry studies have demonstrated that Earth's crust and atmosphere have undergone significant changes in their chemical composition over the last 4.5 billion years [1–10]. Undoubtedly, one of the basic characteristics of the early stages of the Earth's evolution was the presence of a highly toxic primordial atmosphere, very different from the current one, the origin of which is still being investigated. Various hypotheses have been formulated regarding the variation in the atmospheric composition over the Earth's life-time, such as degassing from volcanism [11–15], heavy bombardment by asteroids [16, 17], and production of free oxygen by

A. Carpinteri, *Terahertz Phonons and Nanomechanical Instabilities*,
https://doi.org/10.1007/978-3-032-14692-2_20

cyanobacterial photosynthesis leading to the Great Oxidation Event [18–20], but none of them is fully convincing and conclusive.

On the other hand, the changes in element concentration appear to be intimately correlated to the tectonic activity. Recent data on chemical composition time-variations in the Earth's atmosphere have shown that CO_2 and H_2O concentrations increased dramatically during the Archean period (3.8 to 2.5 Gyr ago), between the tectonic plate formation and the most severe tectonic activity. This primordial carbon "pollution" is strictly correlated to the depletion in Fe and Ni, as well as to the increment in Si and Al that occurred in two subsequent step-wise transitions: 3.8 Gyr ago, at the beginning of tectonic activity, and then 2.5 Gyr ago, during the most severe tectonic activity [21–26], see Chap. 18 (Fig. 18.2).

Subsequently, between 2.5 and 2.0 Gyr ago, N_2 and O_2 concentrations increased sharply (Great Oxidation Event) in concomitance with the most relevant formation of the Earth's continental crust [18–20]. Thus, a significant coupling appears between the periods of intense tectonic activity and the sudden increments in CO_2 and H_2O, occurred 3.5 Gyr ago, and later, 2.5 Gyr ago, in N_2 and O_2 [1, 19, 20] (Fig. 20.1). The chemical mass balances emphasize the condensed matter transformation (and not the migration) in the Earth's crust, ocean, and atmosphere [27–31] (see Table 18.2) [27]:

$$\mathrm{Fe}_{26}, \mathrm{Co}_{27}, \mathrm{Ni}_{28} \rightarrow \mathrm{Mg}_{12}, \mathrm{Al}_{13}, \mathrm{Si}_{14} \rightarrow \mathrm{C}_{6}, \mathrm{N}_{7}, \mathrm{O}_{8} \qquad (20.1)$$

where the atmospheric elements (C, N, O) can be regarded as resultants of the second set of reactions, involving Mg, Al, and Si as starting elements. In particular, the estimated Mg increment (3.2%, according to [22]) is equivalent to the carbon content

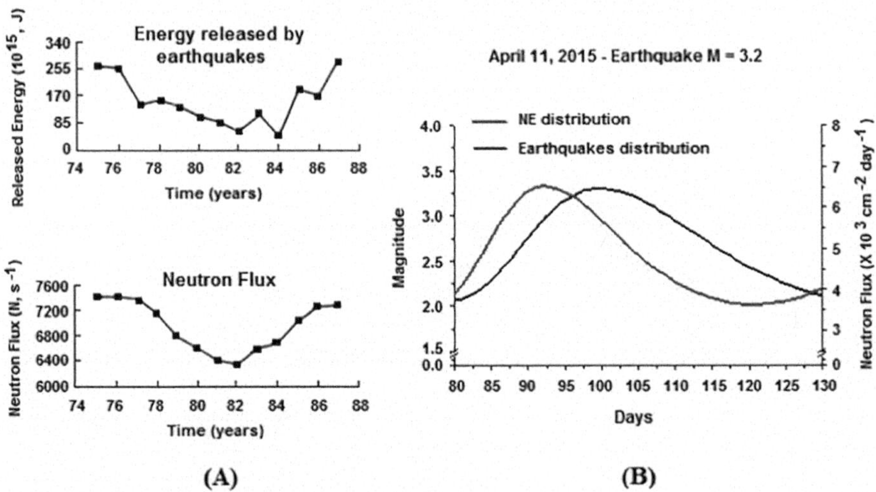

Fig. 20.1 **a** Seismic activity and neutron flux measurements in the period 1975–1987 (Kola Peninsula, Russia [33]); **b** Correlation between neutron flux time-distribution at the Murisengo gypsum mine and a local seismic swarm (main shock magnitude of 3.2) [36]

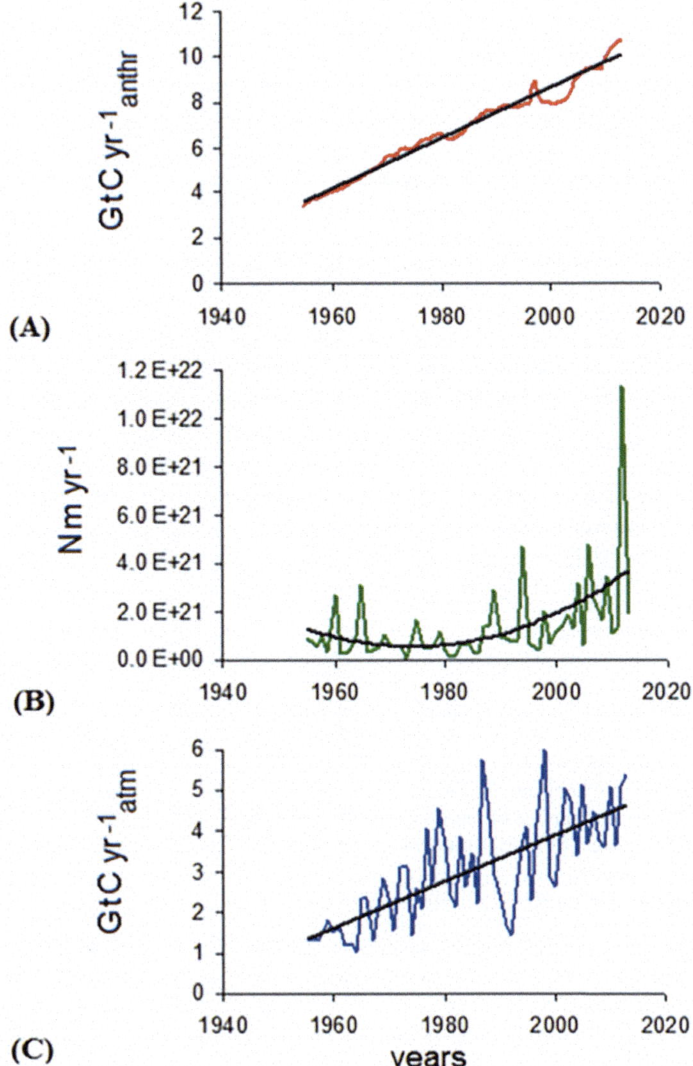

Fig. 20.2 Time-series plots (time period:1955–2013) for: **a** anthropogenic CO_2 emission rate; **b** seismic-energy (-moment) release rate; **c** atmospheric CO_2 growth rate. Trend lines are in black

in the primordial atmosphere (particularly high in the molecules of CO_2 and CH_4). The second set of reactions in Eq. (20.1), involving Mg as the starting element and C as the resultant, can be regarded as an alternative or competing mechanism with volcanic degassing [11–15], by which tectonic forces added CO_2 to the atmosphere at the end of the Hadean period [27, 32]. Interestingly, the comparison between a very rough estimate and the experimental evidence seems to support our hypothesis.

Assuming an average density equal to 3.6 g/cm^3 and a thickness of ~ 60 km for the Earth's protocrust, the mass increment in Mg (~3.5×10^{21} kg, corresponding to 3.2% of the mass of the entire protocrust), and therefore in C, implies a very high atmospheric pressure. Given a terrestrial surface area of 5.1×10^{14} m^2 and conjecturing the same gravitational acceleration for the proto-Earth, a value of ~ 660 atm is obtained for the atmospheric pressure. This very high value, which is plausible and consistent with the conditions prevailing between 3.8 and 2.5 Gyr ago, is corroborated by models presented for a CO_2- and CH_4-rich primordial atmosphere that had ground level pressures of some hundreds atm (~650 atm is the value reported in [1]).

Furthermore, significant neutron emissions were measured, both at the Earth's crust scale, during earthquake preparation stages (Fig. 20.1) [33–36], and at the laboratory scale, during crushing tests on iron-rich natural rock specimens.

In the laboratory experiments, fracto-emissions were measured in correspondence to microcracking and macroscopic fracture [37–41], with a final considerable reduction in the Fe content, consistently balanced by an increment in Mg, Al, and Si [29, 31]. Further evidence supporting the link between seismicity and variations in the atmospheric CO_2 is the spatial organization of the CO_2 release from the ground in Himalayas and Nepal, which was apparently controlled by strong earthquakes [42].

20.2 Atmospheric CO_2 and the Carbon Cycle

Carbon dioxide is an integral part of the carbon cycle, in which carbon is exchanged between Earth's oceans, soils, rocks, and biosphere at two different rates: the slow carbon cycle, which involves geochemical processes between atmosphere, oceans, soils, rocks, and volcanos; and the fast carbon cycle, which refers to the transfer of carbon from the environment to living organisms in the biosphere, including photosynthesis and respiration [43]. The slow-rate geochemical processes, including formation and burial of carbonates, the burial of organic matter (on land or in the ocean), and volcanism have largely determined the flow of CO_2 into and out of the atmosphere on a multi-million-year time scale. As currently accepted, the reduction in atmospheric CO_2 caused by these burial processes, appears to be bound to tectonic processes, which return the carbon from the Earth's mantle and crust to the atmosphere through volcanic degassing (which became less frequent as the Earth's mantle progressively cooled). Later, the flow of atmospheric CO_2 began to be controlled by other natural processes such as the origin and the expansion of forests, which caused an increased burial of organic carbon by photosynthesis, and the respiration of living organisms, which added CO_2 back into the atmosphere [44, 45].

In the pre-industrial era, the flow of atmospheric CO_2 was considered to be largely in balance, with natural sources of CO_2 nearly balanced by natural CO_2 sinks. Since the Industrial Revolution, however, the anthropic activity has perturbed the carbon cycle due to the direct addition of carbon to the atmosphere from the burning of fossil fuels and land-use changes (mainly deforestation). According to carbon cycle models [46], more than 57% of the anthropogenic CO_2 emissions should be removed from the atmosphere by the biosphere (vegetation and land) and the oceans, which behave as carbon sinks.

As current atmospheric CO_2 levels exceed estimates from the last 800,000 years, according to artic ice core measurements [47], and are rising quickly, the anthropogenic perturbation of the global carbon cycle is considered to be the only attributing factor to the dramatic increment in CO_2, as well as that of the other greenhouse gases of the contemporary Earth's atmosphere [48, 49]. According to this paradigm, the global carbon budget, describing the exchanges of CO_2 between the atmosphere and the other major carbon reservoirs, is currently given by the following balance equation:

$$\text{atm growth} = \text{fossil fuel} + \text{land use change} - \text{ocean sink} - \text{land sink} \quad (20.2)$$

where geochemical terms are usually neglected, as their rate is considered too slow to have some relevance on the atmospheric CO_2 concentration over hundred- or thousand-year time-scales, or because they became relatively infrequent, like volcanism. However, some doubts remain regarding the balance of this equation, because the ocean sink is estimated by a combination of global ocean biogeochemistry models and the land sink is often estimated from the residual of the other budget terms [48]. With regards to the latter point, recent studies suggest that the role of tropical forests may have previously been underestimated. They indicate a terrestrial tropical carbon sink larger than the previous estimates [50, 51]. Accounting for this finding while balancing the global carbon cycle implies that the currently considered carbon sources are larger [52], and/or that other carbon sources need to be considered in the budget. It must be said that some extra sources, missing in the balance, can also be caused by the human activity, such as the increasing acidification of the oceans, which makes them carbon sources as well.

20.3 Data Analysis

When observing anthropogenic CO_2 emission rates and atmospheric CO_2 growth rates (compare the diagrams in Fig. 20.2a, c), plotted in Gt C/year, from 1955 to 2013 (the choice of the period is motivated by the need for comprehensive and reliable data, available online at https://earthquake.usgs.gov/earthquakes/search/ and at http://www.globalcarbonproject.org/carbonbudget/archive.htm#CB20147), this evidence casts some doubts on the belief that anthropogenic emissions effectively drive the entire atmospheric CO_2 growth [53–55].

The significant increment in the anthropogenic emission rates, largely ascribed to China's contribution since 2002, is reflected in the analogous plot of atmospheric CO_2 growth rates, although the former presents a steady growth, whereas the latter also evident fluctuations. A sharp decrement in atmospheric CO_2 growth rate is evident in some periods, especially from 1988 to 1993, despite the continuous growth of the anthropogenic emission rate.

However, it can be argued that the atmosphere, as a part of a complex system that includes also oceans and lands, may give a delayed answer to a perturbation such as the change in CO_2 emission. On the other hand, there is scientific evidence about changes in atmospheric composition related to tectonic activity [56], neutron emissions from laboratory crushing tests, and seismic activity at the Earth's crust scale. Consistently, the assumed relationship between phono-fission reactions and active faults may be regarded as the principal cause of magnesium transformation into carbon [27]. Our purpose is investigating any possible correlation between atmospheric CO_2 growth rate and worldwide global seismic-energy (seismic-moment) release rate (Nm/year). Several previous studies focused only onto temporary and localized CO_2 emissions related to limited seismic events [57, 58]. Accordingly, a comparison between the rates of anthropogenic CO_2 emission, worldwide global seismic-moment release (calculated from data provided by the USGS Earthquake Hazards Program [59]), and atmospheric CO_2 growth (refer to xls files in Global Carbon Budget Archive [48]) shows that all the considered rates display an increasing trend, linear for atmospheric and anthropogenic data, quadratic (with a minimum) for seismic data (Fig. 20.2a–c). It is worth noting that the anthropogenic CO_2 emission rate exhibits one-order-of-magnitude lower fluctuations if compared to those of the atmospheric CO_2 growth rate (Fig. 20.2a, c).

A further relevant comment emerges when considering the periodicity hidden behind the fluctuations of the time series. The fast Fourier transform (FFT) analysis of residual data, obtained by subtracting the trend component (identified by a simple curve-fitting approach) from the actual data, shows that the low-frequency components (approximately lying in the 0.00–0.15 year^{-1} interval), in the anthropogenic emission spectrum, prevail over the high-frequency components. A more uniform mixture of components emerges in seismic-moment release and atmospheric CO_2 spectra, where the spectral peaks appear at higher frequencies (compare FFT diagrams in Fig. 20.3).

As shown in Fig. 20.3a, the anthropogenic time series exhibit a dominant spectral peak at 0.05 year^{-1}, equivalent to a 20-year period, which does not correlate to the atmospheric CO_2 fluctuations, since such a periodicity is not relevant in the atmospheric spectrum (compare to the fourth spectral component of atmospheric CO_2 data in Fig. 20.3c). Conversely, the atmospheric fluctuations exhibit a principal periodicity of 3.7 years that is almost absent in the anthropogenic emission spectrum. The 3.7-year periodicity is relevant in both atmospheric CO_2 growth rate and in the seismic-moment release rate spectra, as well as additional periodicities such as the 2.5-year spectral peaks (Fig. 20.3b, c).

Then, residuals are subjected also to coherence analysis [60] to find out more systematically on which frequencies the spectral coherence appears between an input x(t) and an output y(t) time series, i.e., on whose frequencies the output responds to the input. The spectral coherence Cxy(f) (or normalized cross-spectral density) is defined as:

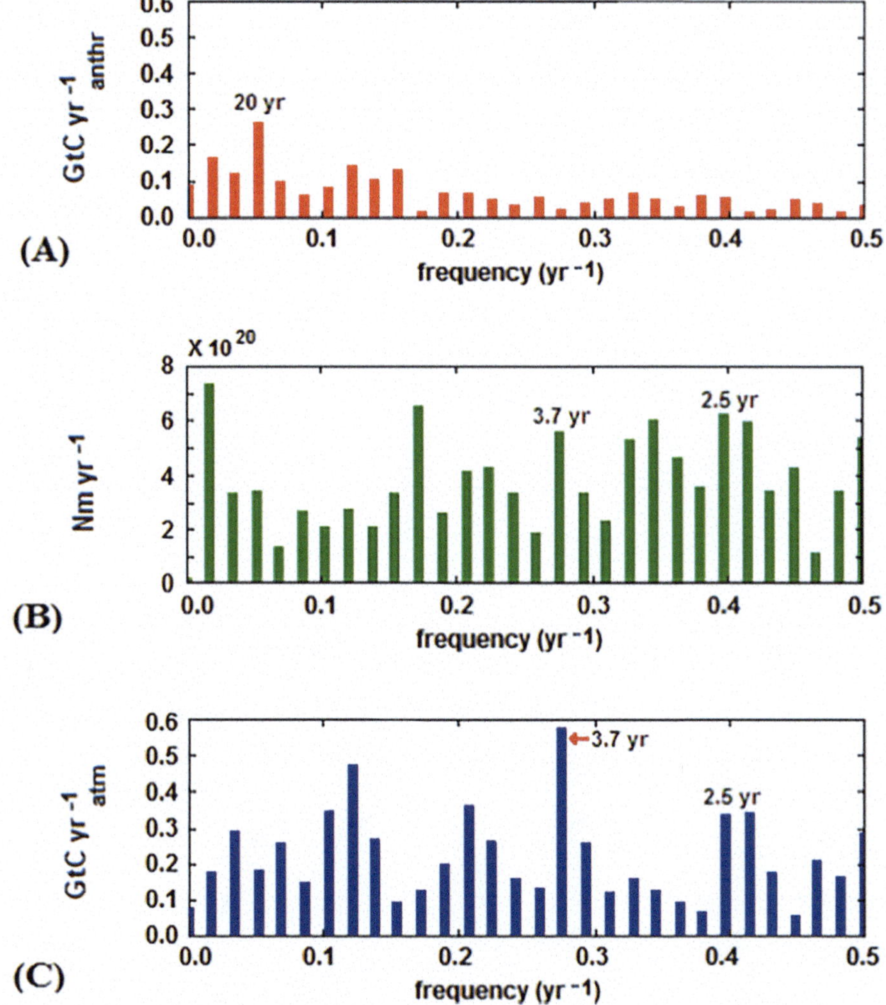

Fig. 20.3 FFT spectra of the de-trended time series for: **a** anthropogenic CO_2 emission rate; **b** worldwide global seismic-energy release rate; **c** atmospheric CO_2 growth rate. Note the discrepancy between the main peaks of anthropogenic (at 20 years) and atmospheric (at 3.7 years) spectra, and the peaks shared by the seismic and atmospheric spectra (at 2.5 and 3.7 years)

$$Cxy(f) = |Gxy(f)|^2 \times Gxx^{-1}(f) \times Gyy^{-1}(f) \tag{20.3}$$

where Gxy(f) is the cross-spectral density between x and y, and Gxx(f) and Gyy(f) are the auto-spectral densities of x and y respectively. Here, Cxy(f) is calculated to estimate the degree of correlation between the perturbation x (the anthropogenic or the seismic time series) and the answer y (the atmospheric time series) in the frequency domain.

Coherence diagrams in Fig. 20.4 differentiate mainly in the interval 0.2–0.4 year^{-1}, where the spectral coherence between seismic and atmospheric data is higher than that between anthropogenic and atmospheric data, consistently with the previous observations. High-coherence values between 0.2 and 0.4 year^{-1} reflect relevant common cycles between 2.5 and 5 years in the spectrum of both seismic and atmospheric fluctuations, almost absent in the anthropogenic time series. Thus, short-term atmospheric CO_2 fluctuations are better correlated to seismic than to anthropogenic activity.

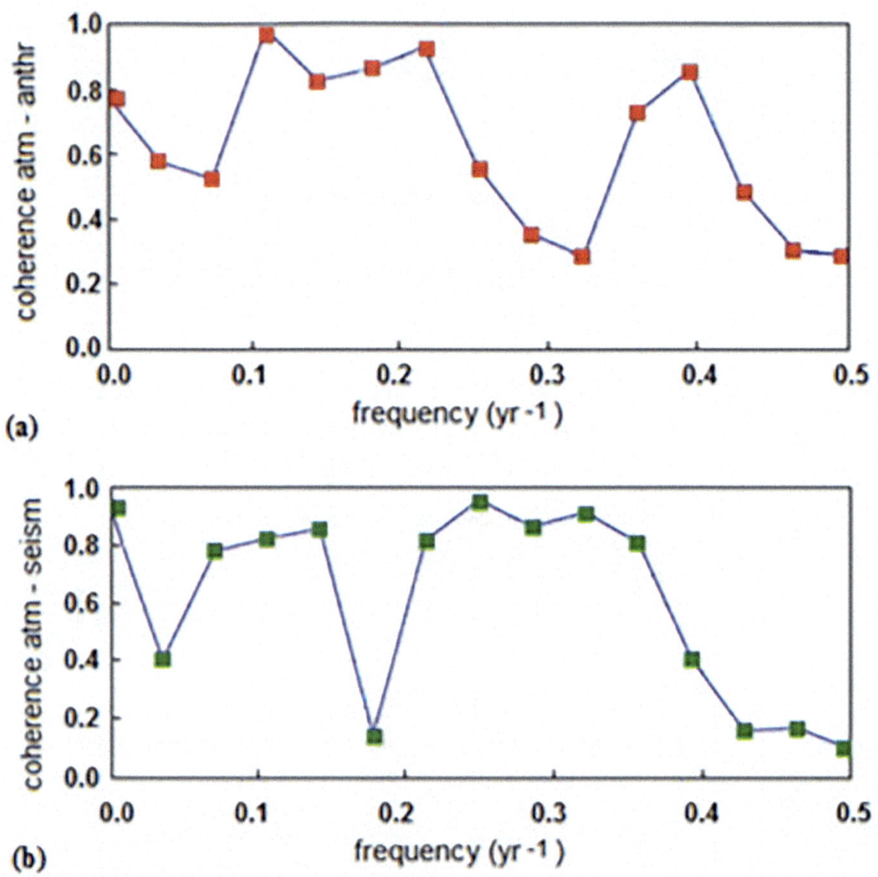

Fig. 20.4 Spectral coherences Cxy(f) between atmospheric CO2 growth rate and: **a** anthropogenic CO2 emission rate; **b** seismic-energy release rate

20.4 Conclusions

The conventional assumptions in relation to the migration of chemical elements in the Earth do not imply the nearly perfect satisfaction of the mass balances before and after the main tectonic events. The atmospheric elements (C, N, O) can be regarded as the resultants of a set of reactions, involving Mg, Al, Si as starting elements, that are massively triggered by the tectonic activity. In particular, this conjecture not only accounts for atmospheric CO_2 formation and growth over the Earth's life-time, but also provides an explanation to some of the observed anomalies in the correlation between atmospheric CO_2 growth rate and anthropogenic emission rate.

By examination of the time series, the fluctuations in atmospheric CO_2 growth rate cannot be explained in terms of low-amplitude and low-frequency anthropogenic emission rate fluctuations.

Although global climate trends are recognized as driven by the atmospheric CO_2 growth rate [61], common cycles are found between atmospheric and seismic variables. The atmospheric CO_2 fluctuations may be ascribed to the cycles of the worldwide global seismicity.

The content of this chapter was preliminarily published in [62].

References

1. Liu L (2004) The inception of the oceans and CO_2-atmosphere in the early history of the Earth. Earth Planet Sci Lett 227:179–184
2. Weast RC (1980) Handbook of chemistry and physics, CRC handbook of chemistry and physics. CRC Press, New York
3. Garrison TS (2005) Oceanography: an invitation to marine science. Thomson Brooks Cole, Belmond
4. Schopf J (1983) Earth's earliest biosphere: its origin and evolution. Princeton University Press, Princeton
5. Kolb E (2000) Blind watchers of the sky: the people and ideas that shaped our view of the universe. Oxford University Press, Oxford
6. Kolb E, Matarrese S, Notari A, Riotto A (2005) Primordial inflation explains why the universe is accelerating today. arXiv
7. Williams RP, Da Silva FJR (2003) Evolution was chemically constrained. J Theor Biol 220:323–343
8. Buesseler KO, Doney SC, Karl DM, Boyd PW, Caldeira K, Chai F, Coale KH, de Baar HJ, Falkowski PG, Johnson KS et al (2008) Ocean iron fertilization moving forward in a sea of uncertainty. Science 319:162
9. Holland HD (2006) The oxygenation of the atmosphere and oceans. Philos Trans R Soc B 361:903–915
10. Abbott D, Burgess L, Longhi J, Smith WH (1994) An empirical thermal history of the Earth's upper mantle. J Geophys Res Solid Earth 99:835–850
11. Kasting JF, Ackerman TP (1986) Climatic consequences of very high carbon dioxide levels in the Earth's early atmosphere. Science 234:1383–1385
12. Yung YL, De More WB (1999) Photochemistry of planetary atmospheres. Oxford University Press, New York
13. Ahrens TJ (1971) The state of mantle minerals. Technophysics 13:189–219

14. Ringwood AE (1962) The chemical composition and the origin of Earth. In: Hurley PM (ed) Advance in earth science. MIT Press, Cambridge, pp 287–356
15. Gaillard F, Scaillet B (2014) A theoretical framework for volcanic degassing chemistry in a comparative planetology perspective and implications for planetary atmospheres. Earth Planet Sci Lett 403:307–316
16. Tera F, Papanastassiou DA, Wasserburg GJ (1974) Isotopic evidence for a terminal lunar cataclysm. Earth Planet Sci Lett 22:1–21
17. Cohen BA, Swindle TD, Kring DA (2000) Support for the lunar cataclysm hypothesis from lunar meteorite impact melt ages. Science 290:1754–1755
18. Saito MA (2009) Less nickel for more oxygen. Nature 458:714–715
19. Kopp RE, Kirschvink JL, Hilburn IA, Nash CZ (2005) The Paleoproterozoic snowball Earth: a climate disaster triggered by the evolution of oxygenic photosynthesis. Proc Natl Acad Sci (USA) 102:11131–11136
20. Konhauser KO, Pecoits E, Lalonde SV, Papineau D, Nisbet EG, Barley ME, Arndt NT, Zahnle K, Kamber BS (2009) Oceanic nickel depletion and a methanogen famine before the great oxidation event. Nature 458:750–753
21. Favero G, Jobstraibizer P (1996) The distribution of aluminum in the Earth: from cosmogenesis to Sial evolution. Coord Chem Rev 149:367–400
22. Taylo SR, McLennan SJ (2009) Planetary crusts: their composition, origin and evolution. Cambridge University Press, Cambridge
23. Hawkesworth CI, Kemp AIS (2006) Evolution of the continental crust. Nature 443:811–817
24. Doglioni C (2007) Interno della Terra. In: Enciclopedia Scienza e Tecnica. Treccani, Milano, pp 595–605
25. Rudnik RL, Fountain DM (1995) Nature and composition of the continental crust: a lower crustal perspective. Rev Geophys 33:267–309
26. Yaroshevsky AA (2006) Abundances of chemical elements in the Earth's crust. Geochem Int 44:54–62
27. Carpinteri A, Manuello A (2015) Evolution and fate of chemical elements in the earth's crust, ocean, and atmosphere. In: Carpinteri A, Lacidogna G, Manuello A (eds) Acoustic, electromagnetic, neutron emissions from fracture and earthquakes. Springer, Basel
28. Carpinteri A, Manuello A (2013) Geomechanical and geochemical evidence of piezonuclear fission reactions in the Earth's Crust. Strain 49:548–551
29. Carpinteri A, Borla O, Lacidogna G, Manuello A (2010) Neutron emissions in brittle rocks during compression tests: monotonic versus cyclic loading. Phys Mesomech 13:264–274
30. Carpinteri A, Manuello A (2011) Geomechanical and geochemical evidence of piezonuclear fission reactions in the Earth's crust. Strain 47:282–292
31. Carpinteri A, Lacidogna G, Manuello A, Borla O (2012) Piezonuclear fission reactions: evidences from microchemical analysis, neutron emission, and geological transformation. Rock Mech Rock Eng 45:445–459
32. Carpinteri A, Manuello A, Negri L (2015) Chemical evolution in the Earth's mantle and its explanation based on piezonuclear fission reactions. In: Carpinteri A, Lacidogna G, Manuello A (eds) Acoustic, electromagnetic, neutron emissions from fracture and earthquakes. Springer, Basel
33. Sobolev GA, Shestopalov IP, Kharin EP (1998) Implications of solar flares for the seismic activity of the Earth. Izvestiya Phys Solid Earth 34:603–607
34. Volodichev NN (1999) Lunar periodicity of the neutron radiation burst and seismic activity on the Earth. In: Proceedings of the 26th International cosmic ray conference, Salt Lake City
35. Sigaeva EA, Nechaev OY, Panasyuk M, Bruns A, Vladimirsky B, Kuzmin Y (2006) Thermal neutrons' observations before the Sumatra earthquake. Geophys Res Abstr 8:00435
36. Carpinteri A, Borla O (2017) Fracto-emissions as seismic precursors. Eng Fract Mech 177:239–250
37. Carpinteri A, Lacidogna G, Manuello A, Niccolini G, Borla O (2012) Time correlation between acoustic, electromagnetic and neutron emissions in rocks under compression. In: Proceedings of the conference & exposition on experimental and applied mechanics (SEM), Costa Mesa, CA, USA, Chapter N. 50, pp 387–393

38. Carpinteri A, Lacidogna G, Borla O, Manuello A, Niccolini G (2012) Electromagnetic and neutron emissions from brittle rocks failure: experimental evidence and geological implications. Sadhana 37:59–78
39. Borla O, Lacidogna G, Di Battista E, Niccolini G, Carpinteri A (2014) Electromagnetic emission as failure precursor phenomenon for seismic activity monitoring. In: Proceedings of the conference & exposition on experimental and applied mechanics (SEM), vol 5. Greenville, SC, USA, pp 221–229
40. Lacidogna G, Borla O, Niccolini G, Carpinteri, A (2015) Correlation between acoustic and other forms of energy emissions from fracture phenomena. In: Carpinteri A, Lacidogna G, Manuello A (eds) Acoustic, electromagnetic, neutron emissions from fracture and earthquakes. Springer, Basel
41. Manuello A, Sandrone R, Guastella S, Borla O, Lacidogna G, Carpinteri A (2015) Neutron emissions and compositional changes at the compression failure of Iron-rich natural rocks. In: Carpinteri A, Lacidogna G, Manuello A (eds) Acoustic, electromagnetic, neutron emissions from fracture and earthquakes. Springer, Basel
42. Girault F, Bollinger L, Bhattarai M, Koirala BP, France-Lanord C, Rajaure S, Gaillardet J, Fort M, Sapkota SN, Perrier F (2014) Large-scale organization of carbon dioxide discharge in the Nepal Himalayas. Geophys Res Lett 41:6358–6366
43. Rothman DH (2001) Atmospheric carbon dioxide levels for the last 500 million years. Proc Natl Acad Sci (USA) 99:4167–4171
44. Royer DL (2014) Atmospheric CO_2 and O_2 during the phanerozoic: tools, patterns, and impacts. In: Turekian K, Holland H (eds) Treatise on geochemistry. Elsevier Science, New York, pp 251–267
45. Bergman NM, Lenton TM, Watson AJ (2004) Copse: a new model of biogeochemical cycling over Phanerozoic time. Am J Sci 301:182–204
46. CDIAC (2016) Global fossil-fuel CO_2 emissions. http://cdiac.ornl.gov/trends/emis/tre_glob_2013.html. Accessed 14 Oct 2017
47. Luthi D, Le Floch M, Bereiter B, Blunier T, Barnola J-M, Siegenthaler U, Raynaud D, Jouzel J, Fischer H, Kawamura K, Stocker TF (2008) High-resolution carbon dioxide concentration record 650,000–800,000 years before present. Nature 453:379–382
48. Global Carbon Budget. http://www.globalcarbonproject.org/carbonbudget/. Accessed 16 Oct 2017
49. Intergovernmental Panel on Climate Change (IPPC) (2007) Intergovernmental panel on climate change, AR4-WG1. IPPC, Geneva
50. Lewis SL, Lopez-Gonzalez G, Sonké B, Affum-Baffoe K, Baker TR, Ojo LO, Phillips OL, Reitsma JM, White L, Comiskey JA et al (2009) Increasing carbon storage in intact African tropical forests. Nature 457:1003–1006
51. Tollefson J (2009) Counting carbon in the Amazon. Nature 461:1048–1052
52. Schuur EA, Vogel JG, Crummer KG, Lee H, Sickman JO, Osterkamp TE (2009) The effect of permafrost thaw on old carbon release and net carbon exchange from tundra. Nature 459:556–559
53. Salby ML (2012) Physics of the atmosphere and climate. Cambridge University Press, Cambridge
54. Oliver JGJ (2015) Trends in global CO_2 emissions: 2015 report. PBL Netherlands Environmental Assessment Agency, The Hague
55. Courtney RS (2008) Limits to existing quantitative understanding of past, present and future changes to atmospheric carbon dioxide concentration. In: International conference on climate change, New York
56. King CH (1986) Gas geochemistry applied to earthquake prediction: an overview. J Geophys Res 91:12269–12281
57. Pierotti L, Botti F, D'Intinosante V, Facc G (2015) Anomalous CO_2 content in the Gallicano thermo-mineral spring (Serchio Valley, Italy) before the 21 June 2013, Alpi Apuane earthquake ($M = 5.2$). Phys Chem Earth Parts A/B/C 85–86:131–140

58. Gheradi F, Pierotti L (2018) The suitability of the Pieve Fosciana hydrothermal system (Italy) as a detection site for geochemical seismic precursors. Appl Geochem 92:166–179
59. Search Earthquake Catalog. https://earthquake.usgs.gov/earthquakes/search/ Accessed 16 Oct 2017
60. Padron E, Melian G, Marrero R, Nolasco D, Barrancos J, Padilla G, Hernandez PA, Perez NM (2008) Changes in the diffuse CO_2 emission and relation to seismic activity in and around El Hierro, Canary Islands. Pure Appl Geophys 165:95–114
61. Keeling CD, Whorf TP, Wahlen M, Van der Plichtt J (1995) Interannual extremes in the rate of atmospheric carbon dioxide since 1980. Nature 375:666–670
62. Carpinteri A, Niccolini G (2019) Correlation between the fluctuations in worldwide seismicity and atmospheric carbon pollution. Science 1:17

Chapter 21
Solar System: Effects of Seismicity (Fracture) and Atmospheric Storms (Turbulence) on the Chemical Evolution of Rocky and Gaseous Planets

Abstract Chemical evolution evidence from the planets of Solar System is presented and interpreted in the light of phono-fission reactions. In particular, data coming from different space missions and investigations are reported and compared for the crust of planet Mars. They were made available by NASA during the years 1995–2010. The concentration increments in certain elements (Fe, Cl, and Ar) and the corresponding decrements in others (Ni and K), together with neutron emissions from the Mars largest faults, can be considered as highly correlated phenomena. The presented findings provide a clear evidence of how seismic activity contributed to the Red Planet's chemical evolution. Analogous evidence regards Mercury, Jupiter, and the Sun itself. The major compositional changes are interpreted according to phono-fission reactions triggered by earthquakes in rocky planets, as well as by atmospheric storms in gaseous planets and in our star. Analogies and differences with respect to the geochemical and geophysical evolution of our planet are emphasized.

Keywords Fracture · Earthquakes · Turbulence · Atmospheric storms · Rocky planets · Gaseous planets · Solar corona · Solar system · Chemical evolution

21.1 Preliminary Remarks

Evidence from the literature is presented on the cosmology of the Solar System [1–11], with particular emphasis on the chemical composition evolution of planet Mars [1–3, 6–8]. The discussion will focus onto the Red Planet's crust studies made possible by NASA space missions during the years 1995–2010, including Mars Odyssey (2001) and Mars Pathfinder (1996) missions [5]. Data about the chemical composition analysis of the planet's crust are correlated to maps of the major faults provided by altimetric surveys, and of the neutron emission data provided by the High Energy Neutron Detector (HEND) carried on Mars Odyssey 2001, which used He^3 proportional counters to measure epithermal neutrons (energy range: 0.4 eV – 100 keV) [3, 4, 11]. The concentration increments in certain elements (Fe, Cl, Ar),

A. Carpinteri, *Terahertz Phonons and Nanomechanical Instabilities*,
https://doi.org/10.1007/978-3-032-14692-2_21

the corresponding decrements in others (Ni, K), and the emission of neutrons from Mars largest faults can be considered as correlated phenomena [1–3, 6–8, 11].

The chemical composition evolution in other planets of Solar System will also be investigated on the basis of the data accumulated during a large number of space missions: Mariner 10, Messenger (2004) for the Sun, Mercury, and Venus; Mars Odyssey (2001), Pioneer 11 (1973), Voyager 1 (1977), and the Galileo probe (1989) for the outer planets Jupiter and Saturn. The assumption of phono-fission reactions will be presented as a new key for the interpretation of certain astrophysical phenomena, which are associated to the chemical composition of each heavenly body and still lacking in a clear explanation [12–22]. The planets can be classified into two categories: the terrestrial, earth-like, or rocky planets (Mercury, Venus, Earth, and Mars), and the gaseous planets (Jupiter, Saturn, Uranus, and Neptune) [1]. There are several differences between the two groups of planets. First, the rocky planets all have a small mass, no or few moons, and a low rotational speed, whereas the gaseous planets have a large mass, several moons, and a high rotational speed. For these reasons, the gaseous planets are flatter at the poles than the rocky planets. In addition, the latter's density is five times that of water on average, whereas the density of the gaseous planets is only 1.2 times that of water [1]. In examining their composition, one realizes that the terrestrial planets essentially consist of rocky and metallic materials, whereas the gaseous planets are mostly made up of hydrogen, helium, and small quantities of ice [1]. Furthermore, the terrestrial planets either have no atmosphere or one that is in any case rarefied, whereas the gaseous planets' dense atmospheres consist of hydrogen, helium, ammonia, phosphine, and methane [1]. In particular, the earth-like planets differ from the gaseous planets in their internal structure. The former have a solid outer crust, a mantle, and a liquid inner core, whereas the latter are called "gas giants" precisely because they consist almost entirely of volatile elements.

As already done for the Earth, we apply the hypothesis of phono-fission reactions to reconstruct the chemical composition evolution of the other planets, as well as of the Sun itself. The planets that will be investigated belong to the rocky category (Mars and Mercury), with the phono-fissions being triggered by seismic activity, as well as to the gaseous category (Jupiter and Saturn), where the phono-fissions are triggered by very turbulent atmospheric storms. The findings presented herein provide further evidence of how fracture in solids and flow instability in fluids have contributed to the planets' chemical composition evolution, as they did for Earth [10, 17, 22]. The major chemical composition changes were generated by high-frequency pressure waves (TeraHertz phonons) in both cases [12–22].

Theoretical interpretations have been proposed of these phenomena. Cardone and coworkers [23, 24] provided a theoretical explanation based on deformed spacetime. Widom et al. explained neutron emissions as a consequence of nuclear fission reactions taking place in iron-rich rocks during brittle micro-cracking or fracture [25, 26]. Iron nuclear disintegration would take place when iron-rich rocks are crushed. The resulting nuclear transmutations are particularly relevant in the case of magnetite and iron-rich rocks [25, 26]. Hagelstein and Chaudhary reported that the anomalous reactions may be considered to occur with large amplitude excitation of vibrational modes considering phonon-electron coupling and resonance [27]. The anisotropic and asynchronous neutron pulses observed at the time of fracture are accompanied by objective signs of chemical composition changes [16]. The changes observed in the laboratory are also confirmed by data of the chemical composition evolution in the Earth's crust [17, 22]. This interpretation is corroborated also by the locations of the Al and Fe mineral reservoirs on the Earth's surface.

21.2 Internal Structure and Chemical Composition of Mars

Mars formed when rocky fragments, which had remained in the nebula immediately after the formation of the outer gaseous giants, clustered together [1]. As a result of their formation, the rocky planets are nearly devoid of any volatile elements [1]. In comparison to the other rocky planets, Mars has almost twice the quantity of volatile elements, a higher level of oxidation, a smaller core, and a primordial core composition with around twice as much iron as that of the Earth. With all these characteristics, Mars has always differed from the Earth substantially, since the time of its formation 4.57 billion years ago [1].

Mars' today stratigraphy and the actual thicknesses of its crust, mantle, and core are still relatively unclear, as there are no data from seismic activity and they have thus been deduced on the basis of mathematical models that are affected by uncertainties [1]. Thanks to the findings of the 1997 Pathfinder mission, however, the data now available on the planet's stratigraphy and density provide a picture that is sufficiently detailed to permit realistic hypotheses [5].

Mars has an average radius of 3,390 km and a volume of 1.632×10^{11} km^3. Its mass is 6.419×10^{23} kg, giving it an average density of 3.934 g/cm^3 (Fig. 21.1).

Concentrations are shown in mass percentage or in ppm.

The thickness of Mars' crust is highly variable, and ranges from a minimum of 20 km to a maximum of 70 km, with a predominantly basaltic composition [1]. An accurate evaluation of the current average composition of Mars' surface crust is shown in Table 21.1. The differences between the composition of the Red Planet's crust and that of the Earth will be discussed below. As indicated in Fig. 21.1, the average density of Mars' crust alone is around 3.0 g/cm^3, approximately 16% more than the density of Earth's present continental crust (2.5 g/cm^3). The density of Mars' crust is close to the estimated density of the terrestrial protocrust during the Archean period, between 3.8 and 2.5 billion years ago (3.2 g/cm^3) [17, 22]. Considering the Solar System's three types of primary, secondary, and tertiary crust described by Taylor and McLennan [1, 17, 22], Mars' crust can be classified as a secondary crust, resembling our planet's oceanic crust, or the Lunar Seas, and thus is at a stage of evolution midway between primary and tertiary crust [1, 17, 22]. On the basis of the planetary-scale chemical evolution [17, 22], we can understand how Mars, which like the Earth saw an extremely intense tectonic activity during the early stages of its formation, is at an intermediate step of the evolutionary time scale, considering the surface composition of the Solar System's planets, or, in other words, between the

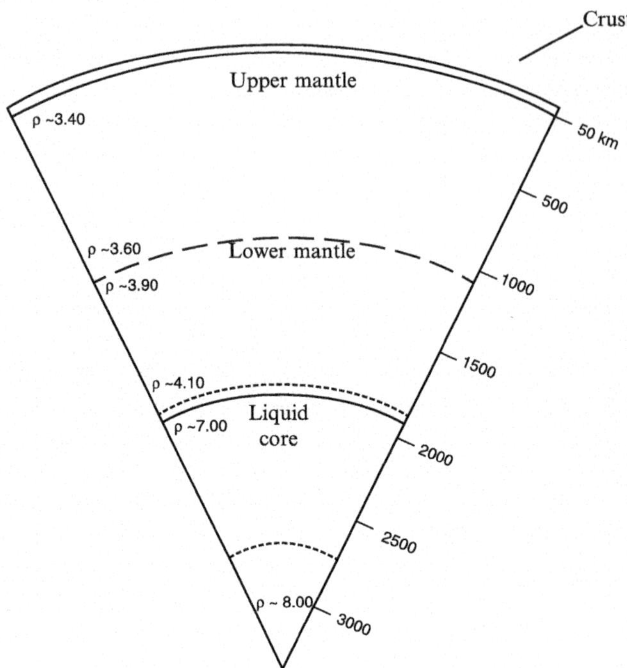

Fig. 21.1 Scheme of Mars' internal structure: crust, mantle, and core [1]

original bodies (protoplanets) and the present-day chiefly sialic composition of the tertiary crusts, whose only known example is the Earth continental crust (Fig. 21.2).

Figure 21.3 shows the concentrations expressed as mass percentages of the elements making up the core, the mantle, and the surface crust of Earth and Mars, the same data being given in Tables 21.1 and 21.2 for the Red Planet. All the indicated data from the specialistic literature [1, 28–33] emphasize several basic differences between the two planets. First, the current quantity of Fe present on the surface of Mars, which accounts for 15% of the planet's crust, is around 1.8 times the amount of Fe present in the Earth's oceanic crust (~7.8% by mass) and around 3.5 times that

Table 21.1 Average abundances of the major elements constituting Mars' crust

Element	Concentration (mass %)	Element	Concentration (ppm)
Si	23.0	Ti	5,880
Fe	15.0	P	3,930
Al	5.5	K	3,740
Mg	5.4	Mn	2,790
Ca	4.9	Cr	2,600
Na	2.2	Ni	337

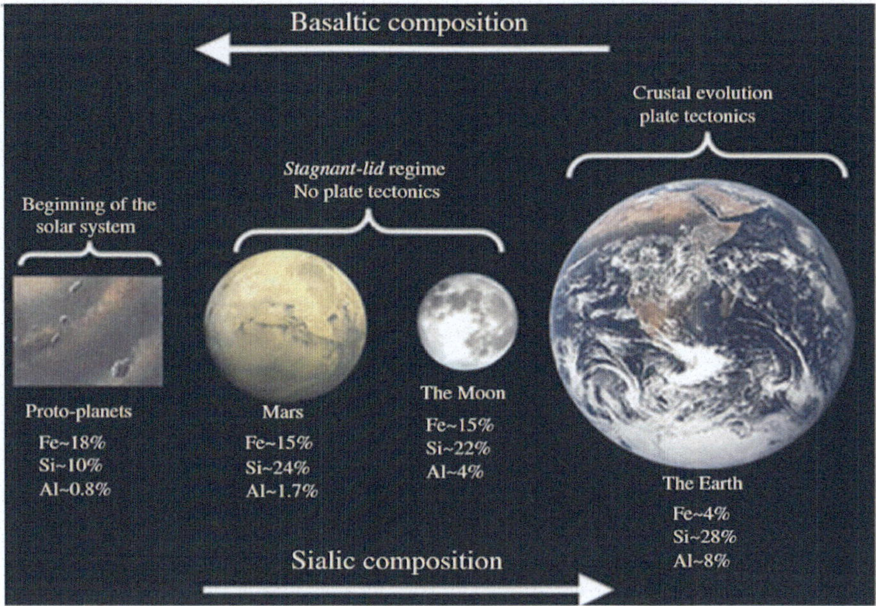

Fig. 21.2 Concentrations of Fe, Si, and Al on the protoplanets, on the surface of the Moon, of Mars, and in the Earth's crust. Note that the concentration of Fe drops, whereas the concentrations of Si and Al jump, passing from the formation of protoplanets to their current status

in the continental crust (~4.0%) [1, 9, 10, 22] (Fig. 21.3c and Table 21.1). About silicon, the concentration on

Mars (23%) is approximately 17% lower than that found in Earth's continental crust (28.8%). The situation for Al is similar, as the 5.5% concentration on Mars is some 30% less than the 8% concentration in the continental crust. The values of Mg, Ca, and Na are also significant. Mars' mantle is approximately 1,050 km thick, and is divided into an upper mantle and a lower mantle with different densities. That of the upper mantle ranges from a minimum of 3.4 g/cm^3 at a depth of around 50 km up to a maximum of 3.6 g/cm^3 at 1,000 km, whereas the lower mantle starts from a density of 3.9 g/cm^3 at a depth of 1,050 km and arrives at the core-mantle boundary layer with an average density of 4.1 g/cm^3. As mentioned earlier, these densities are determined from studies carried out with mathematical models rather than measured experimentally. Hence, for instance, the uncertainty regarding the position of the boundary between lower mantle and core varies over a range of approximately 500 km. The average chemical composition of the mantle presented in the literature, on the other hand, is determined by spectrographic investigations carried out on meteorites that landed on Earth from Mars in the primordial era. Consequently, the chemical composition indicated for the mantle refers to a period differing from that available for the crust (Table 21.2) [1].

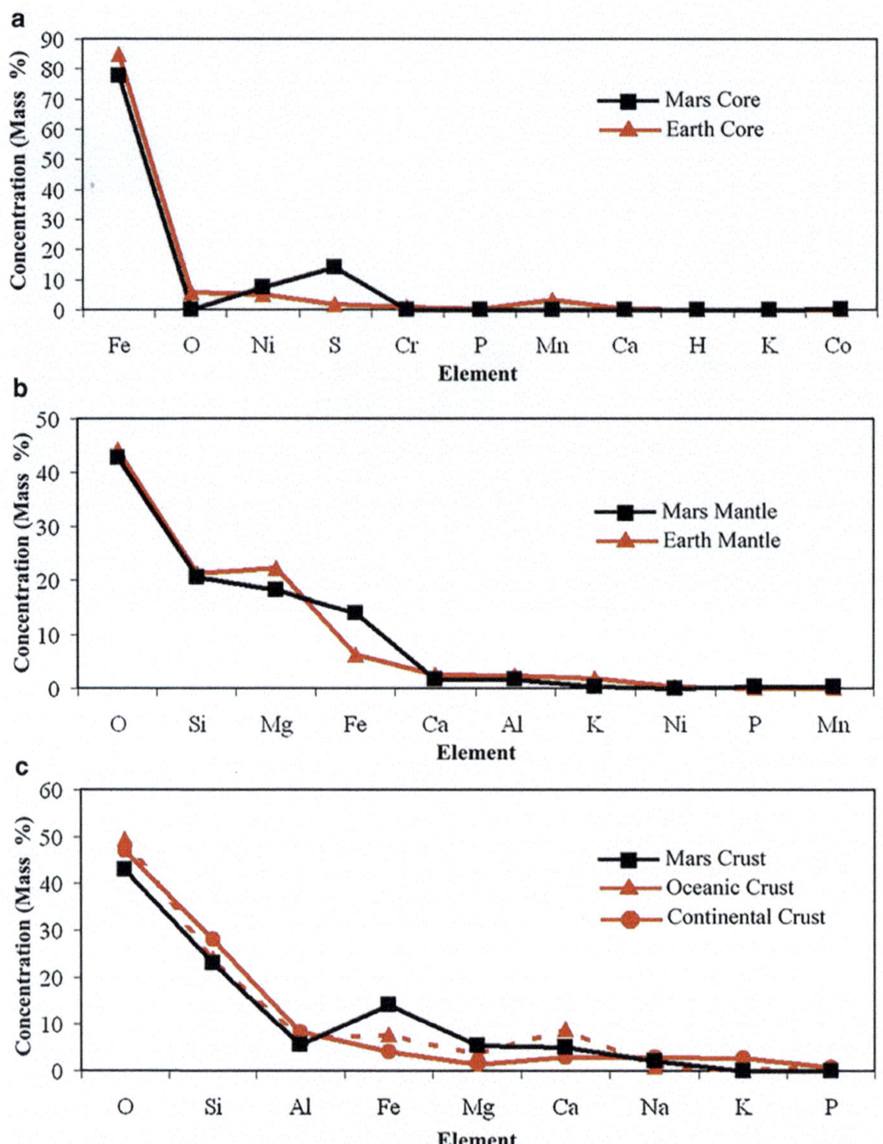

Fig. 21.3 Chemical composition of the layers making up Earth and Mars. Average chemical composition of the core for Earth and Mars **a**. Average composition of the mantle for the two planets **b**. Average composition of Earth's continental and oceanic crusts, and average composition of Mars' surface **c**

Table 21.2 Average chemical composition of Mars' primordial mantle

Element	Concentration (mass %)	Element	Concentration (ppm)
Si	20.6	P	700
Fe	13.9	K	305
Al	1.6	Ti	840
Mg	18.2	Cr	5,200
Ca	1.75	Ni	400
Na	0.37	Zn	62
Mn	0.36	Co	68

Size, chemical composition, and state (i.e., whether it is liquid or solid) of Mars' core are still hotly debated. Mars' core has a radius of about 1,600 km and a density of 7–8 g/cm^3, whereas its inferred chemical composition is shown in Table 21.2 [1].

21.3 Geological History of Planet Mars

The origin of the external Martian surface is still a controversial subject, given that, by contrast with our own planet, it is not possible to establish a time scale that can be considered accurate, since the only way of dating the planet is by counting the craters left behind by meteorite impacts [23, 28]. The geological history of Mars is roughly divided into three main periods: Noachian, Hesperian, and Amazonian [4]. As has been repeatedly confirmed by a number of scientists, Mars is probably no longer an intensely active planet from the tectonic standpoint, but there can be no doubt that it was such in the past and that it continues to show a sporadic tectonic activity today [1, 4, 5, 11, 29, 30].

The fact that it does not show abundant signs of ongoing tectonic activity does not necessarily mean that Mars has always been seismically stable or lacking in internal dynamics. A look at the different epochs, in which the Martian geological history is divided, in fact will show that the Noachian Period was one of dramatic tectonic and volcanic activity [1]. Over 80% of the planet's crust was formed during this process [4]. Figure 21.4 shows a comparison between formation percentages for Earth's continental crust and for Mars' crust. As can be seen, the formation of the terrestrial crust was spread over the last 3.8 billion years, whereas the formation of Mars' crust can be reasonably dated between 4.57 and 3.50 billion years ago. This evidence indicates that the Red Planet's tectonic activity came substantially to an end after the Noachian Period, "crystallizing" the crust's evolution with no further macroscopic changes.

Proof that Mars' crust was the scene of major tectonic events in the past is provided by the presence of a large number of faults on the planet's surface (Fig. 21.5) [4, 11]. Created during the Noachian Period, these faults are the result of both the planet's internal dynamics [1, 4] and of the global cooling that was accompanied by an overall

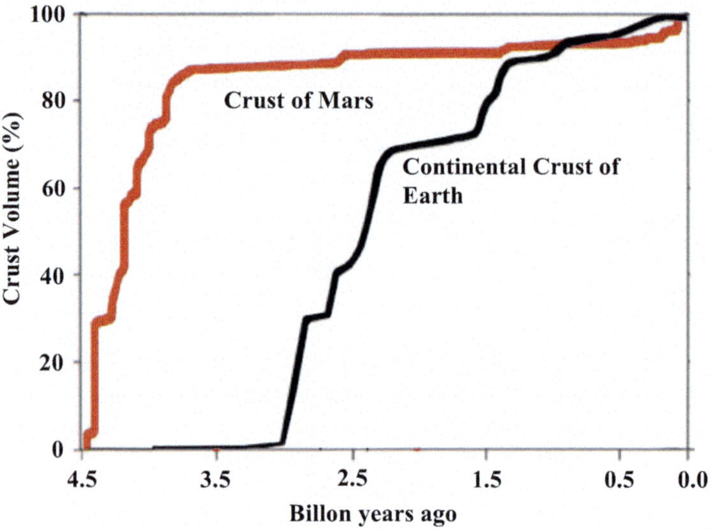

Fig. 21.4 Formation percentage versus time of Mars' crust and Earth's continental crust

contraction of the lithosphere [11]. Figure 21.5 is a geological planisphere of Mars showing the areas with the highest concentration of thrust faults [4].

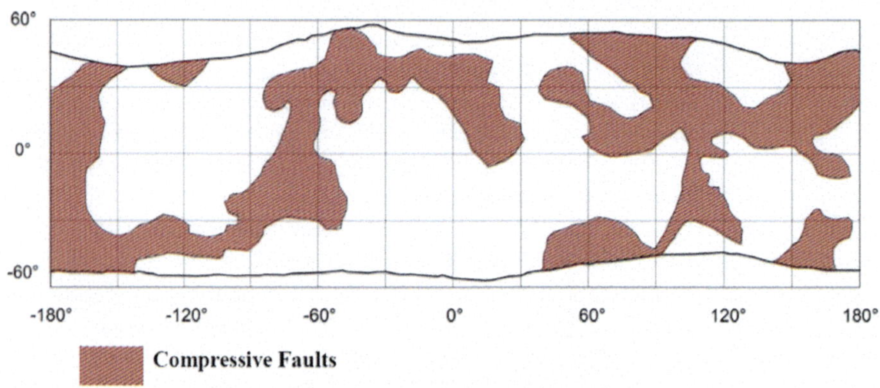

Fig. 21.5 Planisphere map showing the distribution of the major Mars' faults

21.4 Chemical Composition Evolution in the Martian Crust

As already treated in the present part of the book, the major changes in the chemical composition evolution of Earth's continental crust [17, 22] are interpreted in the light of phono-fission reactions and reproduced in the laboratory with crushing failure experiments on natural rocks. Similar plausible assumptions can be formulated for Mars, since the violent tectonic and volcanic activity that it experienced in the Noachian Period, and the sporadic tectonic events that still today take place, have resulted in a surface marked by faults and signs of earthquakes [28–30]. In this context, mention should be made to the fact that Mars is commonly known as the "red planet" because of the characteristic red color of its surface. This color results from the iron oxide, FeO, that covers much of the planet's crust and is the second most abundant element in it [1]. This large quantity of Fe—around 15% by mass, to be compared to ~ 7.8% in the Earth's oceanic crust [1], and to the continental crust (~4.0%)—raises interesting questions about the planet's evolution and the role that unexpected nuclear reactions may have had in this planet.

Evidence of the possible role of these reactions can be provided by the data regarding the distribution of Fe on Mars' surface [1, 4, 8], which we have superimposed onto the map of the compressive faults published by Knapmeyer et al. [4] (Fig. 21.6a). As can be seen, there is a very high correlation between the areas showing extensive signs of tectonic activity and those with a high Fe concentration.

In recent years, numerous studies have addressed the chemical composition of Mars' surface. In particular, some researchers were able to reconstruct maps of the surface concentrations of elements such as Ni, Fe, K, Cl, Ar, Th, Si, and H by analyzing the data provided by the Gamma Ray Spectrometer (GRS) on board of the 2001 Mars Odyssey spacecraft [8].

The exploratory Mars Odyssey mission also analyzed the neutron emission data provided by the HEND ^{3}He High Energy Neutron Detector carried on the spacecraft to measure epithermal neutrons in an energy range of 0.4 eV to 100 keV [2–5, 8].

Figure 21.7a shows a map of epithermal neutron emissions from the planet's surface. In Fig. 21.7b and c, such emissions are superimposed, firstly with the locations of the major faults, and then with the map of prevalent Fe concentrations on the Red Planet's surface. In both cases, the correlations between neutron emissions and fault locations, as well as between iron abundances and fault locations are surprisingly high. In the present case, Fe is not the starting element, as it is on Earth, but it is the resultant of reactions that involve starting elements with a higher atomic number (Ni, Sn, Cu). As a matter of fact, in the scientific literature, several studies report an increment in the concentration of Fe on the Red Planet's surface over the last 4.57 billion years [1, 2, 8, 31–33]. At the same time, a number of Authors report a significant decrement in heavier elements such as Ni, Co, Mo, In, Cu, Zn, and Sn [1, 7, 31, 32].

In particular, the average increment in Fe can be evaluated, with a rise from the approximate 14% in the Noachian crust to the today's ~15%. It results an increment of 1%, corresponding to 2.15×10^{20} kg of Mars' crust (Fig. 21.8b).

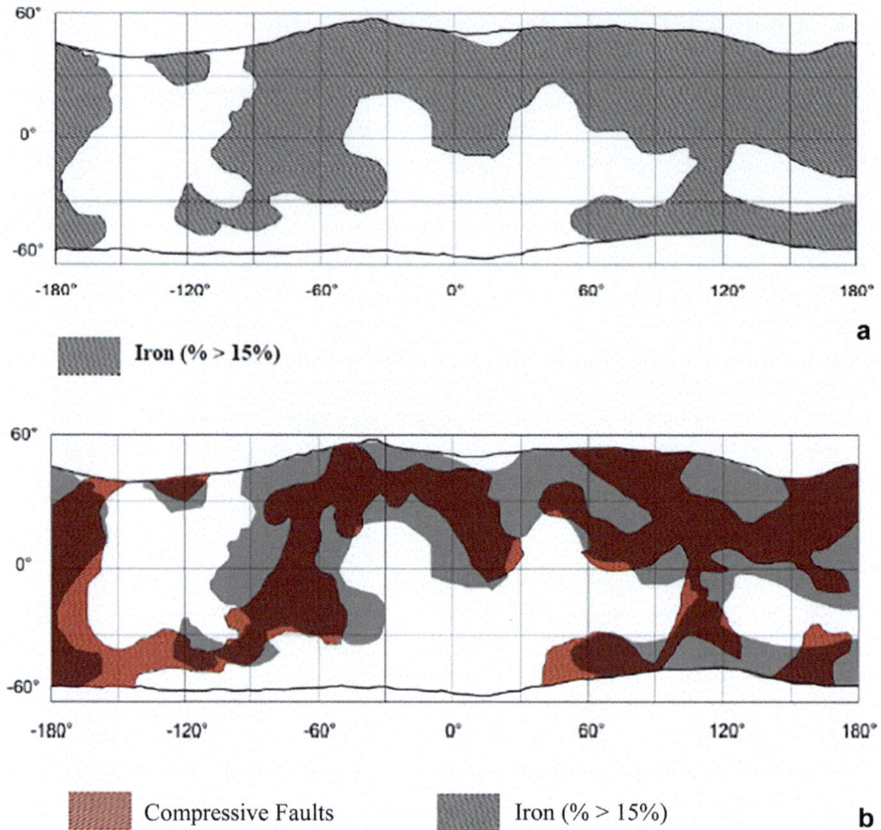

Fig. 21.6 Iron-rich (>15%) areas on Mars's surface **a**. The previous map superimposed onto the map of planet's faults **b**

Likewise, the data for Ni concentrations, as shown in Fig. 21.8a, indicate a roughly equivalent decrement in this element, passing from a concentration of around 1% in the Noachian Period to a negligible or near-zero concentration in the present-day crust. This is thus a relative decrement of about 100% and an absolute decrement of 1% [1].

As was assumed for other elements in the case of Earth's crust, this evidence suggests the following phono-fission reaction:

$$\mathrm{Ni}_{28}^{59} \rightarrow \mathrm{Fe}_{26}^{56} + 2\mathrm{H}_1^1 + 1 \text{ neutron} \tag{21.1}$$

This reaction can explain the anomalous increment in the Fe concentration and the corresponding decrement in Ni by the same amount in terms of mass (Fig. 21.8a and b).

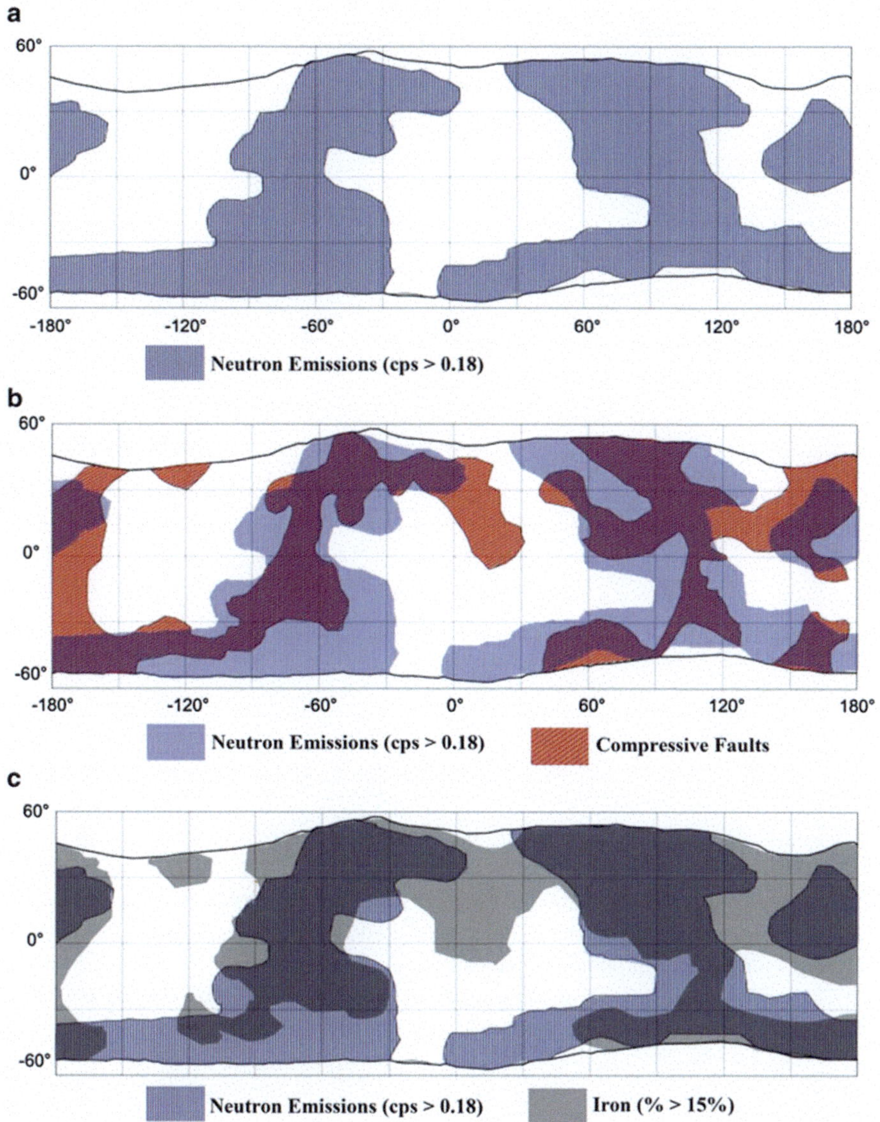

Fig. 21.7 Areas with neutron emissions > 0.18 cps **a**. Superposition of neutron emissions and seismic areas **b**. Neutron map superimposed onto the iron concentration map **c**

In addition, the data presented by Hahn and McLennan (2006) and by Boynton et al. (2007) [2, 8] emphasize anomalous changes also in K and Cl. A small portion of the decrement in K, moreover, can be correlated to the high concentration in Mars' atmosphere of an isotope of argon, Ar^{36}. This evidence makes it possible to assume two further fission reactions that may have affected the planet's crust [1, 2,

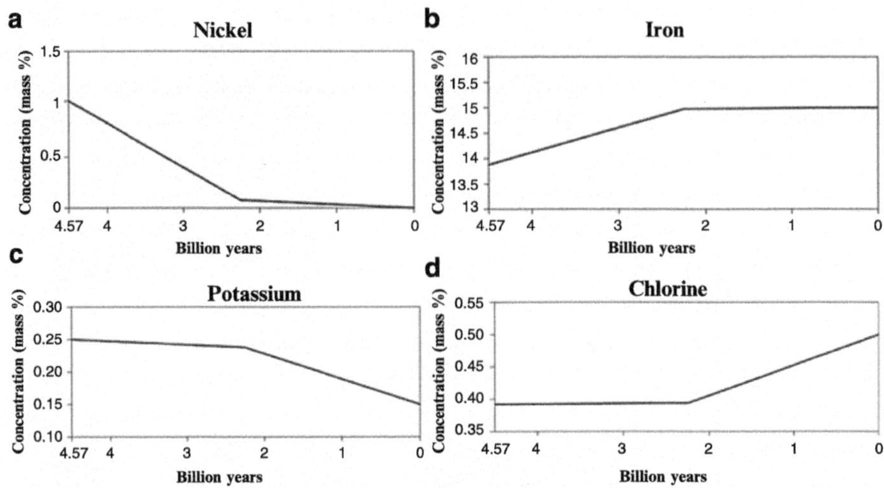

Fig. 21.8 Highly correlated evolution of the concentrations of Ni **a**, Fe **b**, K **c**, and Cl **d** on the Martian surface

8] (Fig. 21.8c and d):

$$K_{19}^{39} \rightarrow Cl_{17}^{35} + 2H_1^1 + 2 \text{ neutrons} \tag{21.2}$$

$$K_{19}^{39} \rightarrow Ar_{18}^{36} + H_1^1 + 2 \text{ neutrons} \tag{21.3}$$

By combining the data from different investigations, we can see an overall decrement in K (~0.1%) at the planet's faults and a practically identical increment in Cl (~0.1%). This evidence suggests that reaction (21.2) is responsible for the decrement in potassium and the increment in chlorine. In addition, several authors mentioned that Mars' atmosphere appears to be laden of an anomalously high concentration of Ar^{36}, approximately 2.5 times higher than that in the atmosphere of Earth. The evident increment in the concentration of Ar^{36}, during the planet's geological evolution, is correlated to tectonic activity and may be linked to the decrement in K, see reaction (21.3) (Fig. 21.9).

21.5 Others Planets of Solar System and the Sun

Analogous evidence can be seen for other bodies in the Solar System. The Earth-like heavenly bodies differ from those referred to as gas giants in their internal structure. The former consist of a solid outer crust, a mantle, and a liquid outer core, whereas the gaseous planets are composed almost entirely of volatile elements.

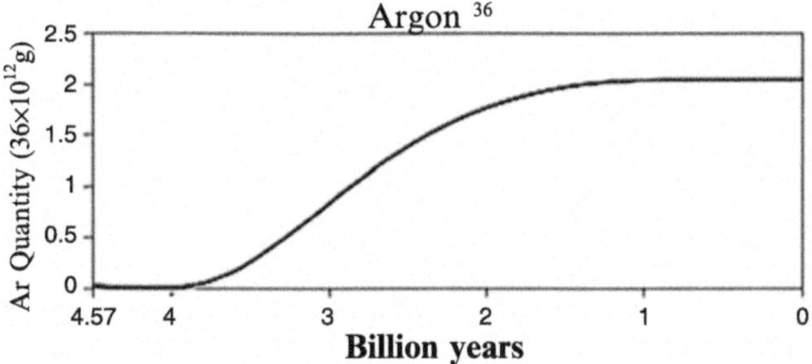

Fig. 21.9 Abundance evolution of Ar^{36} in Mars' atmosphere [31]

As done for Earth and Mars, we attempt to reconstruct the chemical composition evolution of other planets and the Sun on the basis of the data found in the literature and to shed some light onto still inadequately understood changes.

For the planets that have no solid crust (Jupiter and Saturn) and thus cannot be affected by tectonic activity, a different phenomenon of mechanical instability is considered: turbulence, i.e., the storms in their gaseous atmospheric masses.

21.5.1 Mercury's Crust and Atmosphere

Mercury is the planet closest to the Sun, orbiting at a distance of 57.9 million km, in comparison to the Earth's 149.6 million km [1, 34]. Because of the high temperatures of the planet's surface and the major influence that the solar wind has on it, it was not possible in the past to perform in-depth accurate analyses of the chemical composition of Mercury's crust and atmosphere. The only certain data were provided by two space missions: Mariner 10 (1974–1975) and Messenger (2004–2011). The latter, launched by NASA to analyze the planet's exosphere, magnetosphere, surface, and interior, was equipped with an X-ray spectrometer capable of measuring the quantities of elements such as Mg, Al, Ca, Si, Fe, and S on Mercury's surface [35]. In the past, Mercury was always thought of as the Moon's twin, as its basaltic surfaces were believed to be similar in structure and composition. The first results of an analysis of the composition of Mercury's surface and lunar basalt showed that Mercury's crust is richer than the crust of the Moon in elements such as Mg and S, but it presents lower concentrations of other elements such as Al and Ca [1].

Unlike Earth, Mercury has no plate tectonics. This does not mean that the planet does not have, or did not have in the past, some form of tectonic activity. In fact, the observations made with the Mercury Dual Imaging System (MDIS) module on the Messenger probe, which took high-definition photographs of the planet's surface, have provided abundant evidence of Mercury's past tectonic activity. In particular, it has been noted that the planet's surface is marked by numerous cliffs caused by contraction of the planet's crust and accompanying the cooling of its core [36]. This contraction, which doubtless took place in the past and is almost certainly still under way today, results in enormous compression stresses in the surface, forcing the latter upwards to form lobate scarps.

Mercury has a very weak atmosphere consisting of traces of H, He, O [1], and unexpected concentrations of Na ions [37]. There is still no consensus regarding the phenomenon behind the presence of sodium ions in Mercury's exosphere, as there are several contrasting theories for its cause [37]. An alternative explanation, associated to the planet's tectonic activity, is provided by phono-fission reactions. By observing the data from Messenger on the chemical composition of Mercury's crust [35], which is poor in elements such as Al and Ca, and on the Na ion-rich atmosphere [37], it can be assumed that the following reactions took place:

$$Al_{13}^{27} \rightarrow Mg_{12}^{24} + H_1^1 + 2 \text{ neutrons} \tag{21.4}$$

$$Al_{13}^{27} \rightarrow Na_{11}^{23} + He_2^4 \tag{21.5}$$

$$Ca_{20}^{40} \rightarrow S_{16}^{32} + 2He_2^4 \tag{21.6}$$

Through these reactions, it is evident how the decrement in Al and Ca is accompanied by the increments in Mg and S in the crust, and in Na, which Mercury's atmosphere contains in anomalous quantities.

21.5.2 Kelvin–Helmholtz Instability (Turbulence) and Its Role in the Chemical Composition Evolution of Jupiter and Saturn

Kelvin–Helmholtz instability is a type of turbulent flow instability that arises when different layers of fluid are in motion relative to each other. It was discovered and investigated by Lord Kelvin [38] and by Helmholtz [39], and later by Rayleigh. The simplest example that can be imagined in two dimensions is that of a perfect fluid present in two adjacent regions of space: in the first region, the fluid is at rest, whereas in the second it moves at a constant velocity. If a small disturbance is introduced at the boundary separating the two regions, fluid particles that were at rest (i.e., with zero velocity) will be moved to the region at finite velocity (and vice versa). This creates an instability: the amplitude of the disturbance continues to increase, and the particles in the two regions will mix together, forming vortices and causing the original configuration to be lost [38–42]. In this simple case, a configuration such as that just described is always unstable, however small the initial perturbation may be [38, 39]. Less rare than might be thought, this phenomenon is often found in nature and can be readily seen with the naked eyes. An example is represented by the wave-like formations that clouds sometimes show if subjected to particular air currents traveling at different speeds. Kelvin–Helmholtz instability phenomena are very frequent on our planet, and studies on their effects have led to the development of models that can predict transitions from laminar to turbulent flows [38, 39]. A consequence of the turbulence instability between two fluids is the hydrodynamic cavitation, which, by causing the gas bubbles that form at the boundary to implode, increases the level of disorder in the fluid [39].

During the last century, several research groups investigated on cavitation and the effects it has on the surrounding matter. The research team led by Cardone at the Italian National Research Council (CNR) conducted experiments with liquid solutions of iron salts cavitated by ultrasounds. They observed anomalous neutron emissions associated to cavitation [40]. These neutron emissions produced by stable elements such as iron can be explained by assuming that phono-fission reactions take place. In solids, brittle fracture is governed by snap-back instability and scale effects, as well as in fluid-dynamics cavitation is by Kelvin–Helmholtz instability and scale effects. Both represent characteristic dissipative and multi-fractal phenomena in condensed matter. Nuclear effects produced by the cavitation of liquid solutions are reported in Chap. 14.

The Solar System consists of rocky planets—Mercury, Venus, Earth, and Mars—with a solid crust, a mantle, and a core, and of two planets called "gas giants", Jupiter and Saturn. The latter two planets, composed almost entirely of light elements such as H and He, have a mass that is 317.8 times that of the Earth in the case of Jupiter, and 95.1 times in the case of Saturn. Jupiter orbits the Sun at a distance of 778.33 million km, and Saturn at a distance of 1,429.4 million km [35]. These two planets, at the outer edge of the Solar System, for a long time have been the subject of chemical studies in order to achieve a full understanding of their behavior, chemical composition, and origin. Almost all the probes that have visited Jupiter and Saturn made fly-bys (Pioneer 11, Voyager 1, etc.), or, in other terms, observed the planets without entering their orbits. The only exceptions were the Galileo probe, which remained in orbit around Jupiter for over 7 years, and the Cassini-Huygens probe, which was orbiting Saturn for more than 13 years [43]. These several missions revealed a number of aspects of Jupiter and Saturn that are of considerable scientific interest. In particular, it was found that both planets emit around twice as much energy as they receive from the Sun [44, 45], and that they are surrounded by a "belt" of radiation 10 times higher than that around the Earth [46, 47].

The "gas giants", Jupiter and Saturn, take their name from the fact that their mass is composed almost entirely of gas. In both planets, the enormous masses of gaseous fluid that make up their atmospheres cause gigantic storms that in some cases can last for several centuries. An outstanding example is Jupiter's Great Red Spot, an elliptically-shaped storm consisting of gas and dust, seen for the first time more than 300 years ago by Giovanni Cassini [44]. The fluids involved in these huge storms are moving at high velocity, giving rise to currents of gas that flow in relative motion. In this way, instability phenomena are created at the boundary between two currents of gas moving at different velocities, as has been observed several times by the many probes that have visited the two planets [43, 44, 48–50]. These fluid-dynamics instabilities are similar to those observed on the Earth (Fig. 21.10).

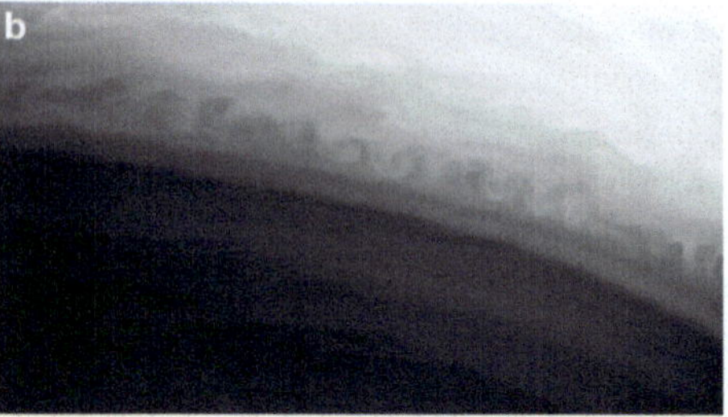

Fig. 21.10 Kelvin–Helmholtz instability around Jupiter's GRS **a**; Kelvin–Helmholtz instability between gaseous currents on Saturn **b** [43]

It was initially thought that both Jupiter and Saturn consisted entirely of light elements such as H and He [48]. However, chemical composition analyses of the atmospheres of the two planets showed that, differently from previous belief, they are also composed of small amounts of heavier elements and molecules: NH_3 (0.08%) and PH_3 (0.0001%) [48].

Subsequently, a group of scientists focused their attention onto quantitative changes in these heavier elements on Jupiter [50]. Observing Jupiter's Great Red Spot, they noted that a decrement in the amount of NH_3 took place in it. They attributed this decrement to convective phenomena that should have transported this heavier element downwards [50]. The Great Red Spot (GRS) is an enormous storm in Jupiter's southern hemisphere [44]. The center of the storm is stagnant and no high-speed gas movements have been detected, whereas current speeds of gas and dust at its edges exceed 350 km/h [44]. Using computational models of the GRS, different scientists have found that this atmospheric phenomenon can produce a Kelvin–Helmholtz instability [51]. In this context, the elimination of heavier elements such as N (deriving from NH_3) in areas affected by Kelvin–Helmholtz instabilities and cavitation may be caused by phono-fission reactions transforming N into lighter elements, e.g., H and He:

$$N_7^{14} \rightarrow H_1^1 + 3He_2^4 + 1\ \text{neutron} \tag{21.7}$$

This reaction can identify the source of neutron radiation that affects Jupiter and has never been considered [46, 47].

21.5.3 Lithium and Beryllium Depletion on the Sun's Surface

The Sun is the Solar System's mother star, around which the eight major planets (including the Earth) orbit along with the dwarf planets, the satellites, innumerable smaller bodies, and the space dust that makes up the interplanetary medium. The

Sun's mass, which amounts to around 2×10^{30} kg [52], accounts on its own for 99.8% of the Solar System's total mass [53, 54].

The energy emitted by the Sun is produced through nuclear fusion processes, which compress the nuclei of two or more atoms sufficiently to enable this strong attracting force to overcome electromagnetic repulsion. Consequently, these atoms are fused together to form a single atom, thus generating a nucleus of a mass greater than that of the reacting nuclei as well as one or more free neutrons. These nuclear fusion reactions take place deeply in the Sun's core at temperatures around 13.6×10^6 °K and pressures of 500 billion atmospheres [49], releasing energy in the form of γ radiation. Once emitted by the core, this radiation is absorbed and re-emitted by the upper layers, contributing to maintain high temperatures. As it travels through the star's layers, the electromagnetic radiation loses energy, assuming longer and longer wavelengths as it passes from the γ to the X and the ultraviolet band, reaching the surface at a temperature of around 5,507 °C, before escaping into the space as visible light [55].

The Sun is a medium-small star consisting essentially of hydrogen and helium [56], as well as traces of heavier elements such as C, O, Li, and Be [56]. Recent studies on the chemical composition of the Sun's surface (Solar Corona) have indicated that, differently from what might be expected from considering these elements as the result of nuclear fusion and from the standard models of a star's evolution, elements such as Li [57–59] and Be [60, 61] are much less abundant than what is predicted. In evaluating the differences between the present composition of the Sun and that of the proto-Sun, it can be seen that the concentrations of Li are 160 times lower than they were 4.57 billion years ago [59], which is not what the canonical models predict [62]. Likewise, the concentrations of Be are also lower than what is expected. The depletion of these elements is usually associated to convective phenomena that transport them from the surface to hotter zones near the core (migration), where they are destroyed by high temperatures and pressures. However, these convective phenomena are not sufficient to explain such low concentrations of Li and Be on the Sun's surface [57–60]. The logic based on matter migration should turn to a logic based on matter transformation.

It has recently been observed that Kelvin–Helmholtz instability phenomena usually take place on the Sun's surface and in the so-called "Solar Corona" [63]. As was indicated previously, these phenomena may be associated to hydrodynamic cavitation and hence to phono-fission reactions (Fig. 21.11):

$$\mathrm{Be}_4^9 \rightarrow 2\mathrm{He}_2^4 + 1 \text{ neutron} \tag{21.8}$$

$$\mathrm{Be}_4^9 \rightarrow \mathrm{Li}_3^6 + \mathrm{H}_1^1 + 2 \text{ neutrons} \tag{21.9}$$

$$\mathrm{Li}_3^6 \rightarrow \mathrm{He}_2^4 + \mathrm{H}_1^1 + 1 \text{ neutron} \tag{21.10}$$

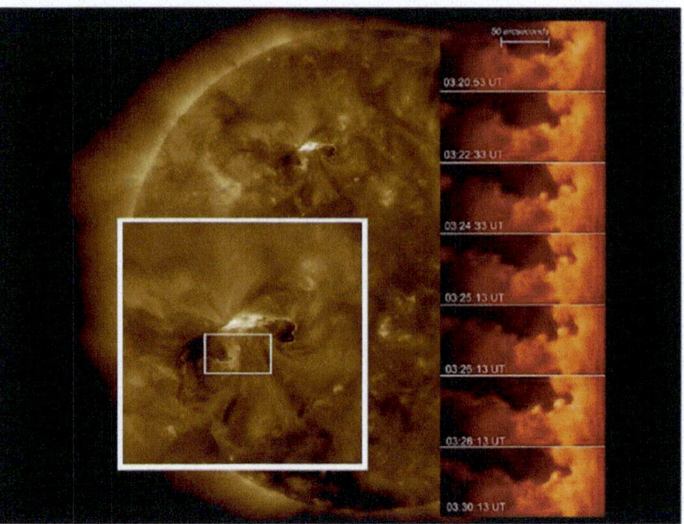

Fig. 21.11 Kelvin–Helmholtz instability in the Solar Corona [52]

21.6 Conclusions

The most important evidence regarding the chemical composition changes in Mars' crust and atmosphere, Mercury crust, Jupiter's Great Red Spot, and Sun's surface may be interpreted in the light of phono-fission reactions.

In particular for the planet Mars, the increment in Fe, corresponding to about 2.15 $\times$ 10^{20} kg of Mars' crust, may be balanced by the Ni concentration decrement that is roughly equivalent. These evidence implies Ni as a starting element and Fe as a resultant with the production of H and neutrons (reaction 21.1). In addition, a small portion of the decrement in K can be correlated, through reactions (21.2) and (21.3), to the increment in Cl and to an high concentration of the argon isotope Ar^{36} in the Mars' atmosphere. From this point of view, the Mars' atmosphere would appear to be laden with an anomalously high concentration of Ar^{36}. The correlations between neutron emissions, Fe concentrations, and fault locations on the planetary crust are surprisingly evident as reported in Fig. 21.7.

A similar evidence may be observed for planet Mercury. It has a very weak atmosphere consisting of traces of H, He, O [1], and unexpected concentrations of Na [37]. There is no consensus yet regarding the phenomenon behind the presence of sodium ions in Mercury's exosphere [37]. An alternative explanation, associated to the planet's tectonic activity, is provided by the phono-fission reactions (21.4–21.6).

Observing Jupiter's Great Red Spot, a decrement in the amount of NH_3 took place around it. In this context, the elimination of heavier elements such as N (deriving from NH_3) in areas affected by Kelvin–Helmholtz instabilities, and thus by turbulence and hydrodynamic cavitation, may be produced by reaction (21.7), transforming N into lighter elements like He and H.

Lastly, it has been recently observed that Kelvin–Helmholtz instabilities habitually take place on the Sun's surface and in the so-called "Solar Corona" [63]. These phenomena may be associated to turbulence and cavitation. It can be thus assumed that the low Li and Be contents in the Sun's surface are associated to phono-fission reactions (21.8–21.10) transforming Li and Be into lighter elements, i.e., He and H.

References

1. Taylor SR, McLennan SM (2009) Planetary crusts: their composition, origin and evolution. Cambridge University Press, Cambridge
2. Hahn BC, McLennan SM (2006) Gamma-ray spectometer elemental abundance correlation with Martian surface age: Implication for Martian crustal evolution. Lunar Planet Sci XXXVII:1–2
3. Mitrofanov I et al (2002) Maps of subsurface hydrogen from the high energy neutron detector, Mars Odyssey. Science 297:78–81
4. Knapmeyer M et al (2006) Working models for spatial distribution and level of Mars' seismicity. J Geophys Res 111:E11006
5. Mars Nasa Exploration Website (2011) http://www.nasa.gov/mission_pages/mars/news/mgs-092005.html
6. McSween HY (2007) Mars (Chapter 1.22). In: Treatise on geochemistry update Holland HD, Turekian KK (eds), vol 1. Elsevier Science, Oxford, pp 601–621
7. Righter K, Drake MJ (2000) Metal/silicate equilibrium in the early Earth – New constraints from the volatile moderately siderophile elements Ga, Cu, P and Sn. Geochim Cosmochim Acta 64(20):3581–3597
8. Boynton WV et al (2007) Concentration of H, Si, Cl, K, Fe and Th in the low—and mid—latitude regions of Mars. J Geophys Res 112:E12S99
9. Taylor SR, McLennan SM (1995) The geochemical evolution of the continental crust. Rev Geophys 33:241–265
10. Favero G, Jobstraibizer P (1996) The distribution of aluminum in the Earth: from cosmogenesis to Sial evolution. Coord Chem Rev 149:367–400
11. Jeffrey C et al (2008) Strike-slip faults on Mars: observation and implication for global tectonics and geodynamics. J Geophys Res 113:E08002
12. Carpinteri A, Cardone F, Lacidogna G (2009) Energy emissions from failure phenomena: mechanical, electromagnetic, nuclear. Exp Mech 50:1235–1243
13. Carpinteri A, Cardone F, Lacidogna G (2009) Piezonuclear neutrons from brittle fracture: early results of mechanical compression tests. Strain 45:332–339
14. Cardone F, Carpinteri A, Lacidogna G (2009) Piezonuclear neutrons from fracturing of inert solids. Phys Lett A 373:4158–4163
15. Carpinteri A, Borla O, Lacidogna G, Manuello A (2010) Neutron emissions in brittle rocks during compression tests: monotic versus cyclic loading. Phys Mesomech 13:264–274
16. Carpinteri A, Chiodoni A, Manuello A, Sandrone R (2011) Compositional and microchemical evidence of piezonuclear fission reactions in rock specimens subjected to compression tests. Strain 47(2):267–281
17. Carpinteri A, Manuello A (2011) Geomechanical and geochemical evidence of piezonuclear fission reactions in the Earth's Crust. Strain 47(2):282–292
18. Carpinteri A, Lacidogna G, Manuello A, Borla O (2011) Energy emissions from brittle fracture: neutron measurements and geological evidences of piezonuclear reactions. Strength Fract Compl 7:13–31
19. Carpinteri A, Lacidogna G, Borla O, Manuello A, Niccolini G (2012) Electromagnetic and neutron emissions from brittle rocks failure: experimental evidence and geological implications. Sadhana 37(1):59–78

20. Carpinteri A, Lacidogna G, Manuello A, Borla O (2012) Piezonuclear fission reactions in rocks: evidences from microchemical analysis, neutron emission, and geological transformation. Rock Mech Rock Eng 45(4):445–459
21. Carpinteri A, Lacidogna G, Manuello A, Borla O (2013) Piezonuclear fission reactions from earthquakes and brittle rocks failure: evidence of neutron emission and nonradioactive product elements. Exp Mech 53(3):345–365
22. Carpinteri A, Manuello A (2012) An indirect evidence of piezonuclear fission reactions: geomechanical and geochemical evolution in the Earth's Crust. Phys Mesomech 15:14–23
23. Cardone F, Calbucci V, Albertini G (2014) Deformed space-time of the piezonuclear emissions. Mod Phys Lett B 28:1450012
24. Cardone F, Cherubini G, Petrucci A (2009) Piezonuclear neutrons. Phys Lett A 373(8–9):862–866
25. Widom A, Swain J, Srivastava YN (2015) Photo-disintegration of the iron nucleus in fractured magnetite rocks with magnetostriction. Meccanica 50:1205–1216
26. Widom A, Swain J, Srivastava YN (2013) Neutron production from the fracture of piezoelectric rocks. J Phys G Nucl Part Phys 40(15006):1–8
27. Hagelstein PL, Chaudhary IU (2015) Anomalies in fracture experiments and energy exchange between vibrations and nuclei. Meccanica 50:1189–1203
28. Hartmann WK, Neukum G (2001) Cratering chronology and the evolution of Mars. Space Sci Rev 96:165–194
29. Morgan WJ (1971) Convection plumes in the lower mantle. Nature 230: 442–443
30. Van Thienen P, Rivoldini A, Van Hoolst T, Lognonné PH (2006) A top-down origin for mantle plumes. Icarus 185:197–210
31. Wanke H (1991) Chemistry, accretion, and evolution of Mars. Space Sci Rev 56:1–8
32. Carr M (2006) The surface of mars. Cambridge University Press, New York
33. Faure G, Mensing TM (2007) Introduction to planetary science: the geological perspective. Springer, Dordrecht
34. Rees M (2005) Universe—the definitive visual guide. Dorling Kindersley Ltd., New York
35. Nittler LR et al (2011) The major-element composition of mercury's surface from MESSENGER X-ray spectrometry. Science 333:1847–1850
36. Watter TR et al (2009) The tectonics of mercury: the view after MESSENGER's first flyby. Earth Planet Sci Lett 285:283–296
37. Paral J et al (2010) Sodium ion exosphere of Mercury during MESSENGER flybys. Geophys Res Lett 37:L19102
38. Thomson W (1871) Hydrokinetic solutions and observations. Philos Mag 42:362–377
39. von Helmholtz H (1868) Über discontinuierliche Flüssigkeits-Bewegungen [On the discontinuous movements of fluids]. Monatsberichte der Königlichen Preussische Akademie der Wissenschaften zu Berlin [Monthly Rep R Prussian Acad Philos Berlin] 23:215–228
40. Cardone F, Mignani R (2003) Possible observation of transformation of chemical elements in cavitated water. Int J Mod Phys B 17:307–317
41. Frank J-P (2006) Physics and control of cavitation. In: Design and analysis of high speed pumps. Educational Notes RTO-EN-AVT-143, Paper 2, Neuilly-sur-Seine
42. Aeschlimann V et al (2011) Velocity field analysis in an experimental cavitating mixing layer. Phys Fluids 23:055105
43. NASA Exploration Site (2011) http://www.nasa.gov/mission_pages/cassini/main/index.html
44. Rogers JH (1995) The giant planet jupiter. Cambridge University Press, Cambridge
45. Rieke GH (1975) The thermal radiation of Saturn and its rings. Icarus 1:37–44
46. Brice N et al (1973) Jupiter radiation belt. Icarus 18:206–219
47. Roussos E (2011) Long and short term variability of Saturn's ionic radiation belts. J Geophys Res 116:A02217
48. Lewis JS et al (2004) Physics and chemistry of the solar system. Elsevier Academic Press, Amsterdam
49. Irwin GJP (2009) Giant planets of our solar system. Springer, Berlin

50. Fletcher LN et al (2010) Thermal structure and composition of Jupiter's Great Red Spot from high-resolution thermal imaging. Icarus 208:306–328
51. Nezlin MV et al (1982) Kelvin—Helmholtz Instability and the Jovian Great Red Spot. JETP Lett 36(6):234–238
52. NASA (2012) http://solarsystem.nasa.gov/planets/profile.cfm?Object=Sun&Display=Facts
53. Woolfson M (2000) The origin and evolution of the solar system. Astron Geophys 41(1):12–19
54. Cohen H (2012) From core to corona: layers of the sun. Princeton Plasma Physics Laboratory (PPPL). http://fusedweb.llnl.gov/cpep/chart_pages/5.plasmas/sunlayers.html
55. Chansen CJ et al (2004) Stellar interiors. Springer, New York
56. Basu S, Antia HM (2008) Helioseismology and solar abundances. Phys Rep 457:217–283
57. Blöcker T et al (1998) Lithium depletion in the Sun: a study of mixing based on hydrodynamical simulation. Space Sci Rev 85:105–112
58. Israelian G et al (2009) Enhanced lithium depletion in Sun-like stars with orbiting planets. Nature 462:189–191
59. Baumann P et al. (2010) Lithium depletion in solar-like stars: no planet connection. Astron Astrophys
60. Boesgaard AM et al. (2002) The case of missing beryllium. http://www.ifa.hawaii.edu/info/press-releases/boes-MissingBeryllium.html
61. Xiong DR, Deng LC (2003) Beryllium depletion in the solar atmosphere. Chin Astron Astrophy 27:1–3
62. D'Antona F, Mazzitelli I (1984) Lithium depletion in stars. Pre-main sequence burning and extra-mixing. Astron Astrophys 138:431–442
63. NASA Website (2012) http://www.nasa.gov/mission_pages/sunearth/news/gallery/sun-surf.html

Part VII
Theoretical Nuclear Models

Chapter 22
TeraHertz Phonons and Plasmons Triggering Low-Energy Phono-Fission Nuclear Reactions in Condensed Matter

Abstract Brittle crushing of iron-rich natural rocks is demonstrated to produce neutron emission, sometimes orders of magnitude larger than the environmental background. Chemical composition changes and a global ponderal equivalence, respecting atomic weight and atomic number balances, are observed. Phonons and/or plasmons induced by brittle crushing, if sufficiently energetic and resonant with the atomic lattice, may split the nucleus into different fragments and neutrons (disintegration). It is relevant to emphasize how the Earth's crust evolution from basaltic to sialic (or granitic) can be consistently explained by these experimental data and theoretical assumptions when applied to tectonics.

Keywords Natural rock crushing · THz phonons · THz plasmons · Resonance · Spalling neutrons · Phono-fission nuclear reactions · Nuclear disintegrations

22.1 Preliminary Remarks

A new explanation to the neutron emissions from stone breaking has recently been proposed [1], i.e., the phono-fission reactions repeatedly revealed in a great number of experiments [2–5]. The approach is based on the spalling effect that is generated by the plasmon-lattice interaction [1]. On the other hand, different detectors have demonstrated the presence of neutron emissions, in some cases by several orders of magnitude higher than the usual environmental background [2–5]. Any solid body is subjected to rapid fluctuations for a few moments when it breaks in a brittle way, even if only partially, or when the formation and propagation of microcracks occurs. Macroscopically, this appears under the form of longitudinal waves of expansion–contraction (tension–compression), in addition to transverse (shear) waves [2–5]. Microscopically, these waves are phonons in relation to the frequency under consideration [6–9]. Even if phonons and plasmons originate from different mechanisms, they are both bosons in the many-body quasi-particle perspective and share several

A. Carpinteri, *Terahertz Phonons and Nanomechanical Instabilities*,
https://doi.org/10.1007/978-3-032-14692-2_22

similarities. In particular, a plasma oscillation can be considered as a sort of longitudinal phonon, for which the elastic modulus, as well as the frequency of transverse oscillation, is equal to zero [8].

At the nanoscale, the frequency measured is that at the order of THz [2–5, 10, 11]. During the experimental activities, significant neutron emissions (phono-fission neutrons) have been repeatedly measured [2–5, 10]. The neutron flux was found to depend, besides on the iron content, on the size of the specimen through the well-known brittleness size effect [11]. Larger sizes imply a higher brittleness, i.e., a more relevant strain energy emission, and consequently more neutrons.

The resulting nuclear transmutations are particularly evident for loadstones as a consequence of photo-disintegrations [12–15]. The electro-strong coupling between electromagnetic fields and nuclear giant dipole resonances is considered as the theoretical core for the explanation of the observed nuclear reactions. The electro-weak interactions can produce neutrons from energetic protons and electrons, thus inducing nuclear transmutations. The electro-strong condensed matter coupling can represent many-body collective nuclear photo-disintegration effects [12, 16–18]. On the other hand, the spalling neutrons approach results to be very useful to explain the experimental results related to neutron emissions from stone breaking. Moreover, it allows us also to link the experimental results to the energy-mass conversion balance inside solid lattices. Indeed, this recalls some new discussions on mass [19] in the Universe, based on the fundamental laws of physics, with particular regards to the principle of least action [19–22]. In this context, the discrete character of nature was proven to be reflected at all levels of its hierarchical organization [19–22]. The basic geometry implies how baryons could pack together in an atomic nucleus. In this way, it has been highlighted how a coordinate of space embodies a closed circulation of energy, and a moment of time will elapse when the circulation opens up either to acquire or discard quantized flux of energy [23]. Consequently, the mass of a body depends on how much its energy density on the least-action path perturbs, i.e., the elastic energy of the solid lattice. Therefore, the curvature and chirality of particular paths invariably relate gravitational and electromagnetic interactions, as well as weak and strong interactions, with each other [19].

These last considerations allow us to introduce further ones on the links among solid lattice and the fundamental interactions inside the solids.

It is well-known one of the key mechanisms, through which low-energy nuclear reactions can occur: for neutron production (and subsequent nuclear transmutations) via weak interactions, special conditions in condensed matter systems must be found, which accelerate an electron to MeV ranges of energy. If certain particular conditions are verified, the electromagnetic energy stored in many relatively slow-moving electrons can be collectively transferred into fewer, much faster electrons with energies sufficient for the latter to combine with protons (or deuterons, if present) to produce neutrons via weak interactions. In this way, electromagnetic and weak interactions can induce low-energy nuclear reactions to occur with observable rates for a variety of processes. The produced neutrons can then initiate low-energy nuclear reactions through further nuclear transmutations.

In this chapter, a novel explanation to some experimental results will be given by means of a theory based on collective oscillations of the electronic density inside the solid lattice, i.e., the so-called *plasmons.*

22.2 Theoretical Considerations on Plasmons

A plasmon is a collective oscillation of the electronic density [7–9] inside the atomic lattice, whereas a phonon is an oscillation of the solid matter. At the equilibrium state, the electric charge of the electronic gas is balanced by the electric charge of the nuclei of the lattice knots [7–9], whereas, when the electron gas moves through a lattice dislocation, a charge asymmetry occurs on the opposite sides of the dislocation itself with the consequent generation of a reactive force, which causes a simple harmonic motion at the plasmonic frequency [6]:

$$\omega_{\mathrm{ph}}^2 = \frac{ne^2}{\varepsilon_0 m_{\mathrm{eff}}} \tag{22.1}$$

where n is the electron density, e is the electric charge of the electron, ε_0 is the electric permittivity in vacuum, and m_{eff} is the effective electron mass. For the crushing of iron-rich natural rocks, the plasmonic frequency was obtained in [1] as (see Appendix):

$$\omega_{\mathrm{ph}}^2 = \frac{n_{\mathrm{eff}} e^2}{\varepsilon_0 m_{\mathrm{eff}}} = \frac{2\pi c^2}{a^2 \ln\left(\frac{a}{r}\right)} \tag{22.2}$$

with:

$$n_{\mathrm{eff}} = \frac{A}{a^2} n \tag{22.3}$$

where a is the lattice constant, c is the speed of light, and n_{eff} is the number of electrons that concur to the plasmonic frequency. On the other hand, this number is not equal to the total number of all electrons, but only to the ones really in motion inside the dislocation, in a direction perpendicular to the dislocation surface itself, of area $A = \pi r^2$. The electronic motion generates a magnetic field, which produces a self-induction effect with a consequent superposition of different forces acting on the electrons. The resulting effect is represented by the apparent effective mass of the electrons [6]:

$$m_{\mathrm{eff}} = \frac{\mu_0}{2\pi} n e^2 A \ln\left(\frac{a}{r}\right) \tag{22.4}$$

where μ_0 is the magnetic permeability in vacuum, introduced in relation to the self-inductance of the current generated by the electron motion in accordance to [6], and r is the radius of the virtual cylinder generated by the electron trajectory. Plasmons produce important effects on the physical properties of matter interacting with electromagnetic fields, because they generate a relative electric permittivity:

$$\varepsilon(\omega) = 1 - \frac{\omega_{\mathrm{ph}}^2}{\omega\,(\omega + \mathrm{i}\gamma)} \tag{22.5}$$

where γ is the damping factor and ω is the electromagnetic frequency. The value of relation (22.5) is negative for frequencies lower than the plasmonic one. The physical meaning of this negative electric permittivity consists in the generation of electromagnetic structures at the border of dislocations.

It is also highlighted how the scattering of the electronic motion inside the dislocation presents the same inertial effect as that of a nucleus of the same velocity [1].

In addition, the usual values of the damping factor and of the plasma frequency for iron (as well as for other metals) are in the THz band [23, 24]. Thus, during the crushing of the iron-rich rocks, a resonant scattering of plasmons occurs with a characteristic frequency of the latter around 1.0 THz. Therefore, the plasmons are absorbed by the lattice nuclei with the following nuclear reaction:

$${}^{\mathrm{A}}_{\mathrm{Z}}\mathrm{X} + \gamma_{\mathrm{ph}} \rightarrow {}^{\mathrm{A}}_{\mathrm{Z}}\mathrm{X}^{*} \rightarrow {}^{\mathrm{A}_1}_{\mathrm{Z}_1}\mathrm{X} + {}^{\mathrm{A}_2}_{\mathrm{Z}_2}\mathrm{X} + N\mathrm{n} \tag{22.6}$$

where X is the lattice nucleus, A is its mass number, Z is its atomic number, n represents the spalling neutrons, N their number, γ_{ph} is the incoming plasmon [7–9, 25, 26], and X* is the excited nucleus. Reaction (22.6) is not anything more than a resonant effect related to the absorption of the plasmon. Consequently, the extraction energy for the neutrons, related to reaction (22.6), can be evaluated as follows [25]:

$$E_{em,N\mathrm{n}} = 931\left[\begin{array}{l} M(Z_1, A_1) + M(Z_2, A_2) + m_{\mathrm{ph}} \\ -Nm_{\mathrm{n}} - M(Z, A) + (Z - Z_1 - Z_2)m_e \end{array}\right] \tag{22.7}$$

and the energy of any neutron can be evaluated as [25]:

$$E_{k,\mathrm{n}} = \frac{1}{N}\frac{A - N}{A}\left(E_{\mathrm{ph}} - E_{em,N\mathrm{n}}\right) \tag{22.8}$$

22.3 Experimental Results

In order to discuss the experimental results, we can consider some possible fission reactions [2–5] for iron:

$$Fe_{26}^{56} \rightarrow 2Al_{13}^{27} + 2 \text{ neutrons} \tag{22.9}$$

$$Fe_{26}^{56} \rightarrow Mg_{12}^{24} + Si_{14}^{28} + 4 \text{ neutrons} \tag{22.10}$$

where the spalled neutrons (with a resultant energy in the GeV order of magnitude) are expelled together with two main fragments of the nucleus. In addition, the plasmons in the lattice under crushing have an effective mass (m_{eff}) between 5 and 7 times the proton mass (m_p) and a frequency in the THz band. In the following, the frequency and the effective mass of the plasmons for the different minerals characterizing the tested rocks are reported: Phengite, $\omega_{ph} = 16.1$ THz and $m_{eff} = 6.9\, m_p$; Biotite, $\omega_{ph} = 15.4$ THz and $m_{eff} = 7.1\, m_p$; Olivine, $\omega_{ph} = 24.6$ THz and $m_{eff} = 5.2\, m_p$; Magnetite, $\omega_{ph} = 17.8$ THz and $m_{eff} = 6.5\, m_p$.

On the other hand, the neutron emissions, experimentally monitored by means of a ^3He proportional counter, refer to the low-energy component. As a matter of fact, the elastic and inelastic interactions of neutrons with matter contribute to energy slowing down, reaching values typical of the thermal component (0.025 eV).

We must consider that reactions (22.9) and (22.10) are the consequence of the absorbed plasmon, and this absorption can be linked to a resonance phenomenon. Consequently, the nucleus jumps to an excited state and after a time of 10^{-14}–10^{-13} s the reaction occurs.

The theoretical results obtained in the previous section are experimentally confirmed. Neutron emissions were measured during laboratory experiments conducted on iron-rich natural rocks. In particular, magnetite specimens were loaded up to the final failure under monotonic displacement control. Also basalt rocks were tested under cyclic loading conditions (2 Hz) up to the final failure. In order to detect neutron emissions, the tests were monitored by two different neutron measurement devices: a ^3He proportional counter and thermodynamic (bubble) detectors. After the experiments, energy dispersive X-ray spectroscopy (EDS) analyses were carried out to possibly detect a direct evidence of low-energy nuclear reactions on the fracture surfaces. In particular, a quantitative evidence of nuclear reactions, involving iron decrement and the corresponding increment in lighter elements, were observed in olivine, the crystalline mineral phase widely diffused in the basalt matrix, and in magnetite. These results reinforce the evidence previously observed for Luserna stone (granitic orthogneiss) and confirm that phono-fissions take place in natural iron-bearing materials subjected to damage accumulation and cracking.

Granite

Preliminary tests on prismatic specimens were presented in early contributions [4, 27, 28] and related to phono-fission reactions occurring in solids containing iron. Four specimens were tested, two made of Carrara marble and two made of Luserna stone [4, 27, 28]. All of them were of the same size and shape, measuring $6 \times 6 \times 10\ \text{cm}^3$. The neutron measurements obtained on the two Luserna stone specimens exceeded the background level by approximately one order of magnitude, when a catastrophic failure occurred. The first specimen reached at time T = 32 min a peak load of ca 400 kN, corresponding to an average pressure on the bases of 111.1 MPa. When failure occurred, the count rate was found to be $(28.3 \pm 0.2)\ 10^{-2}$ cps, corresponding to an equivalent flux of thermal neutrons of $(43.6 \pm 0.3)\ 10^{-4}\ n_{\text{thermal}}\ \text{cm}^{-2}\ \text{s}^{-1}$. The second specimen reached at time T = 29 min a peak load of ca 340 kN, corresponding to an average pressure on the bases of 94.4 MPa. When failure occurred, the count rate was found to be $(27.2 \pm 0.2)\ 10^{-2}$ cps, corresponding to an equivalent flux of thermal neutrons of $(41.9 \pm 0.3)\ 10^{-4}\ n_{\text{thermal}}\ \text{cm}^{-2}\ \text{s}^{-1}$.

In more recent experiments, cylindrical specimens, with different size and slenderness, were selected. For the specimens of larger dimensions, neutron emissions, detected by ^{3}He, were found to be of about one order of magnitude higher than the natural background level at the time of the catastrophic failure. These emissions fully confirmed the preliminary tests [2–5, 27–35]. For specimens with sufficiently large size and/or slenderness, a relatively high mechanical energy emission is expected, and hence a higher probability of neutron emission at the time of failure. Completely different neutron emission levels were found for different rocks [2–5, 27–35]. In particular, for granitic rocks, presenting an iron content of ~1.5%, a neutron emission up to one order of magnitude greater than the natural background level was detected. In the case of basalt and magnetite, where the iron concentration is respectively ~15% and ~72.5%, neutron emissions up to 10^2 times (for basalt) and 10^3 times (for magnetite) the environmental level were measured, whereas for marble, no neutron emissions greater than the background were observed. On the other hand, the results obtained from steel specimens subjected to compression or tension up to the final failure were characterized by neutron emissions not higher than 2.5 times the background level [2].

Furthermore, during compression tests with cyclic loading, an equivalent neutron dose was found at the end of the test by neutron bubble detectors, about twice higher than the background level [4, 5, 32]. Using an ultrasonic horn suitably joined to the specimen, ultrasonic tests were carried out on Luserna stone producing a continuous vibration at 20 kHz. At the end of the test, an equivalent neutron dose about twice higher than the background level was detected [4, 29].

After these tests, energy dispersive X-ray spectroscopy (EDS) analysis was performed on different sample spots of external or fracture surfaces belonging to the granite specimens used in the crushing experiments, in order to get information about possible changes in their chemical composition. The results for phengite and biotite (minerals with a higher Fe concentration in granite) show that the reduction in iron abundance is balanced by the increment in lighter elements such as Al, Si, and Mg.

In the case of phengite (Fig. 22.1), the distribution of Fe concentrations shows two different values equal to 6.2% and 4.0% for external and fracture surface, respectively (Fig. 22.1a). The measurement precision or percent resolution is equal to 0.1%.

Similarly, the Al concentration (Fig. 22.1b) shows an average value changing from 12.5% (external surface) to 14.5% (fracture surface). The evidence that the values of the iron decrement (–2.2%, Fig. 22.1a) and of the Al increment (+2.0%, Fig. 22.1b) are approximately the same is really intriguing.

In Fig. 22.2a–d, the results for Fe, Al, Si, and Mg concentrations of biotite crystalline phase are shown. In this case, the iron decrement is about 3.0%, from 21.2% (external surface) to 18.2% (fracture surface) (Fig. 22.2a). At the same time, the Al content variations show an average increment of about 1.5% (Fig. 22.2b). In Fig. 22.2c, d, it is shown that, in the case of biotite, also Si and Mg contents present considerable variations. The mass percentage of Si changes from a mean value of 18.4% (external surface) to a mean value of 19.6% (fracture surface), with an increment of 1.2% (Fig. 22.2c). Similarly, the mean value of Mg concentration changes from 1.5% (external surface) to 2.2% (fracture surface), with an increment of 0.7% (Fig. 22.2d). Therefore, the iron decrement (–3.0%) in biotite is balanced by increments in aluminum (+1.5%), silicon (+1.2%), and magnesium (+0.7%).

Basalt

After the mechanical loading experiments, energy dispersive X-ray spectroscopy (EDS) was performed on different samples of external and fracture surfaces, belonging to the same specimen used during the cyclic loading test on basalt. The analysis was conducted in order to correlate the neutron emission from the specimen to the variations in rock composition and to detect possible anomalous transformations from iron to lighter elements. The quantitative elemental analyses were performed by a ZEISS Supra 40 field emission scanning electron microscope (FESEM) equipped with an Oxford X-rays microanalysis [5, 32–34].

The first analysis was performed on fracture surfaces of a basalt specimen after the fatigue test and the consequent failure. In this case, similarly to the case of Luserna stone [34], taking into account the heterogeneity of the material, the sample spots were carefully chosen to investigate and compare the same mineral before and after the failure. In particular, olivine was considered due to its high iron content (~24%) and because it is rather abundant within this type of rock [34].

In Fig. 22.3a–c, the distributions of Fe, Si, and Mg concentrations in olivine are reported, for external and fracture surfaces. It can be observed that the distribution of Fe content for the external surface shows an average value of 18.4% (Fig. 22.3a). In the same graph, the distribution of Fe concentrations on the fracture sample shows a significant variation. It can be seen that the mean value of the distribution of measurements performed on the fracture surface is equal to 14.4%, considerably lower than the mean value of external surface measurements (18.4%). Similarly to Fig. 22.3a, in Fig. 22.3b the Si mass percentage concentrations are considered. For Si contents, the observed variations show a mass percentage increment approximately equal to 2.2%. The average value of Si concentration changes from 18.3% on the external surface to 20.5% on the fracture surface. In Fig. 22.3c it is shown that, in the case of olivine, also the Mg content presents considerable variations. Figure 22.3c shows that the mass percentage concentration of Mg changes from a mean value of 21.2% (external surface) to a mean value of 22.8% (fracture surface) with an increment of 1.6%.

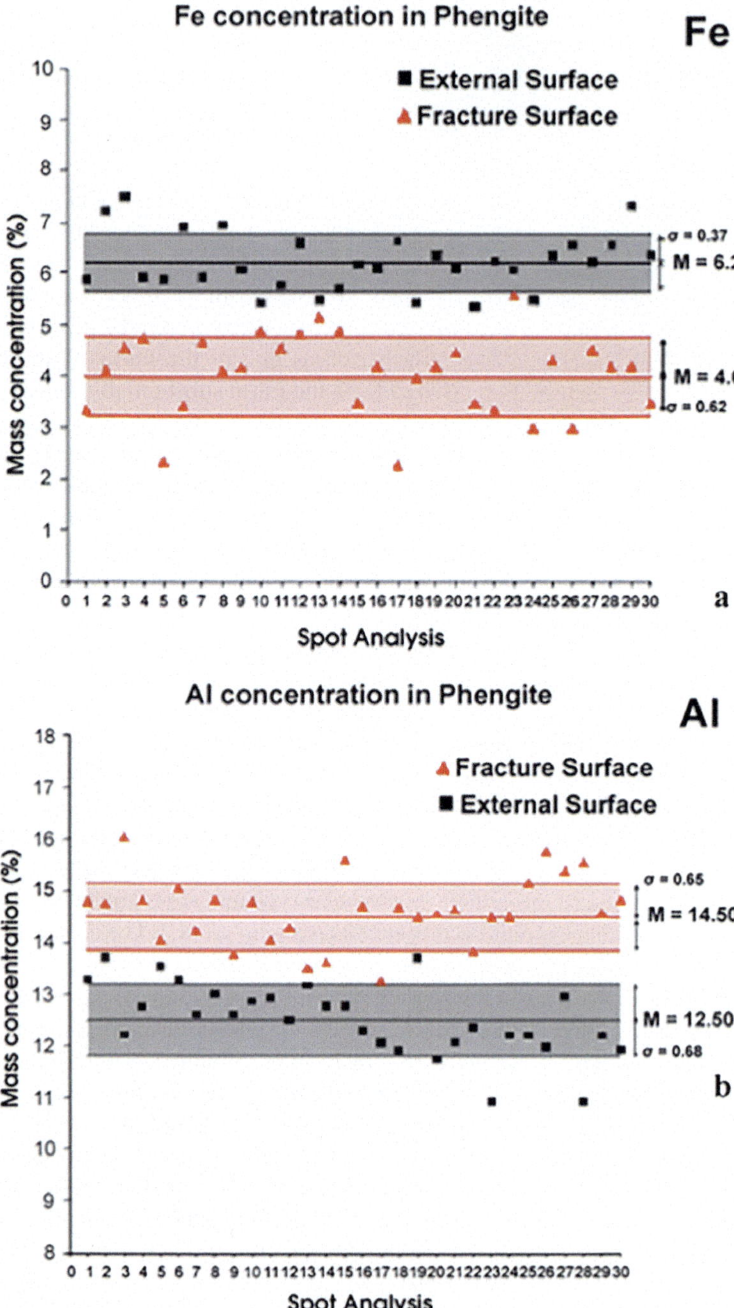

Fig. 22.1 EDS results of Fe and Al concentrations in phengite analyzed on samples coming from crushed specimens of granite: **a** Fe concentration on external surfaces (squares) and on fracture surfaces (triangles). The Fe decrement considering the two mean values of the distributions is equal to 2.2%. **b** Al concentration on external surfaces (squares) and on fracture surfaces (triangles). The Al increment, considering the two mean values of the distributions, is equal to 2.0%

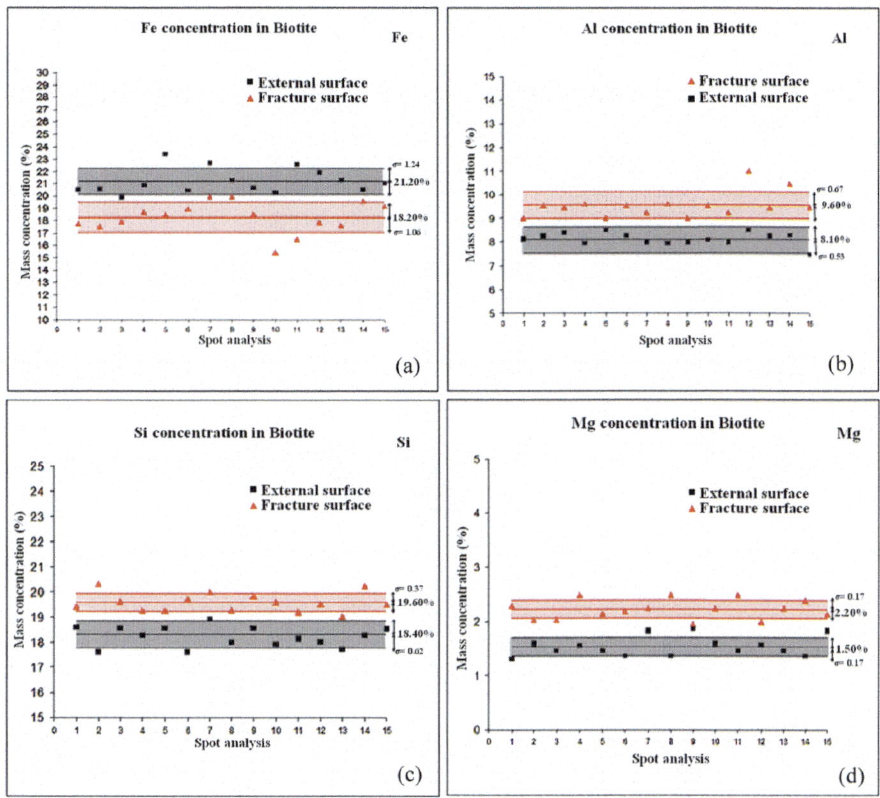

Fig. 22.2 EDS results of Fe (**a**), Al (**b**), Si (**c**), and Mg (**d**) concentrations in biotite analyzed on samples coming from crushed specimens of granite. Considering the analysis done on external and fracture surfaces, the iron decrement (−3.0%) in biotite is balanced by increments in aluminum (+1.5%), silicon (+1.2%), and magnesium (+0.7%)

Therefore, the iron decrement (–4.0%) in olivine is nearly balanced by increments in silicon (+2.2%) and magnesium (+1.6%).

The results reported in Fig. 22.3 are also important from a geophysical point of view. In fact, at the scale of the Earth's crust, the non-homogeneous composition of oceanic and continental crusts could be explained by the transition from basaltic to sialic compositions. Comparing the data presented in the literature concerning the composition of the two different types of terrestrial crust, it can be noted the today iron concentration difference from ~8%, in the oceanic crust, to ~4%, in the continental one [36–43].

Magnetite

Similar quantitative results have been obtained also in the case of magnetite. The fracture surfaces of a magnetite specimen were analyzed in order to recognize a possible evidence of phono-fission reactions. Taking into account the homogeneity

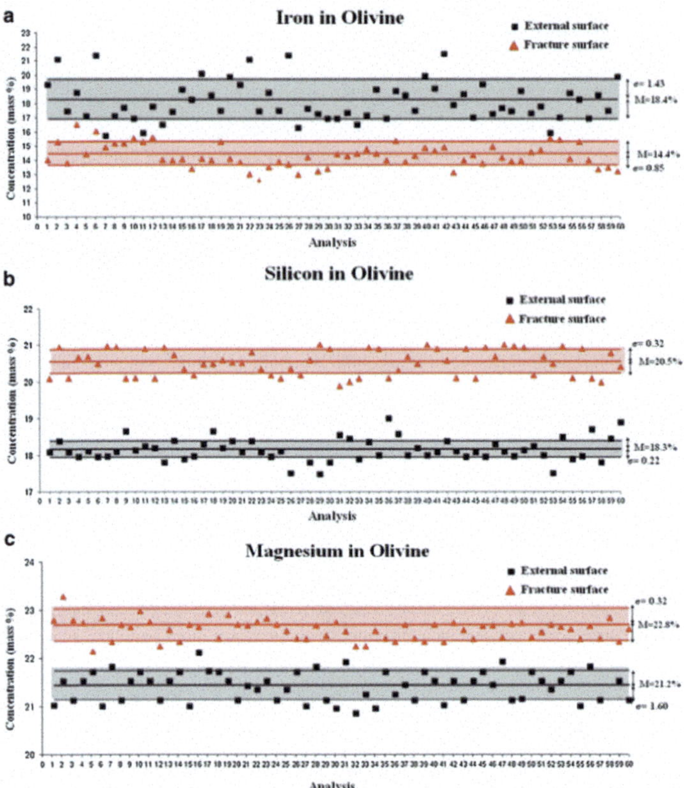

Fig. 22.3 Olivine chemical changes after mechanical loading and crushing: **a** the Fe decrement (−4.0%) is almost perfectly balanced by increments in Si (**b**) (+2.2%) and Mg (**c**) (+1.6%)

of the chemical composition of this mineral (Fe-oxides content ~95%), the analysis on the external surface of this rock was conducted in a different manner with respect to granite and basalt. In this case, in fact, taking into account the homogeneity of the rock, both map and spot analyses were performed to evaluate the composition changes. As mentioned before, the typical composition of magnetite is given by Fe and O, these elements representing about the 95% of this rock. The remaining part is represented by traces of other elements such as Na, Si, Cl, and K. Maps of dimension 250 μm × 200 μm were analyzed to detect the presence of significant changes in the chemical composition.

In Fig. 22.4, the chemical element concentrations are reported indicating the mass percentages for both external and fracture surfaces. At a first glance, it can be noted that, for the fracture surface, consistent contents of Al and Mn appear after the specimen failure (Fig. 22.4), whereas they were absent in the analysis performed on the external surface before the failure test (Fig. 22.4b, e).

At the same time, on the fracture surface, we observe a significant Fe decrement (Fig. 22.4a). In order to evaluate these changes from a statistical point of view, 15

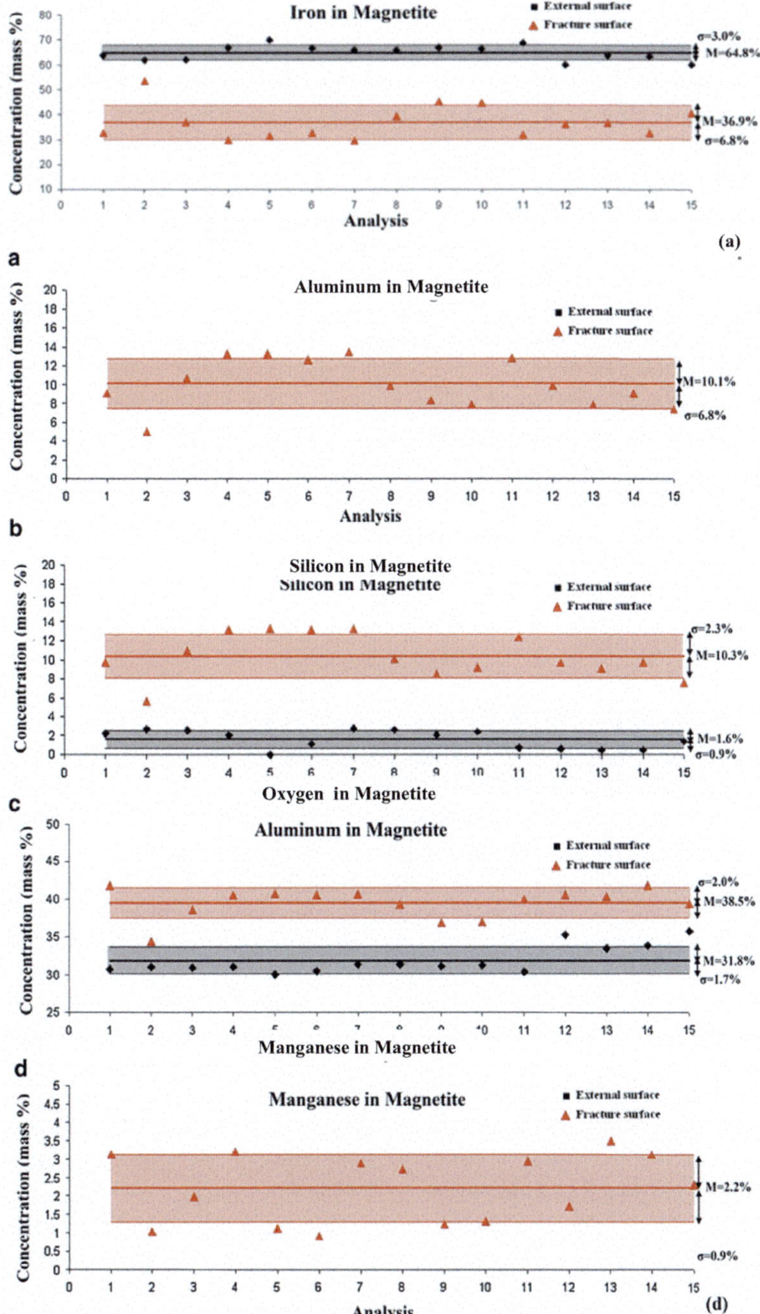

Fig. 22.4 **a** Iron chemical concentration, with mass percentages for both external and fracture surfaces. Fe decreases by 27.9% in magnetite after the brittle fracture of the specimen. The Fe concentration on the external surface is 64.8% and the concentration on the fracture surface, after the test, is 36.9%. **b** Aluminum, silicon, oxygen, and manganese chemical concentrations, with mass percentages for both external and fracture surfaces

analyses on the external surface and 15 on the fracture surface were carried out and reported in Fig. 22.4a–e. In these diagrams, similarly to the results obtained for granite and basalt and reported previously, the mass percentage concentrations for Fe, Al, Si, O, and Mn (Fig. 22.4) are reported making a comparison between external and fracture surfaces. It is interesting to observe a decrement in Fe concentration of 27.9% (from a value of 64.8% down to 36.9%). At the same time, Al concentration increases from zero to a mean value of 10.1%. It is important to consider that approximately one third of the Fe decrement may be balanced by the Al increment. The remaining part of iron decrement is almost perfectly balanced by the increments in the other elements.

The appearance of Al and Mn, both absent before the test, and the almost perfect balance between the Fe decrement (–27.9%) and the increment in the other elements (+27.7%), can be considered as a direct evidence that nuclear reactions of a new type take place. Al, Si, O, and Mn are increasing on the magnetite fracture surface after the experiment. In particular Al and Mn, with concentrations of about 10.1% and 2.2% after the experiment, were both absent on the external surface before the experiment.

22.4 Conclusions

Theoretical explanations to LENR phenomena were given by Widom et al. [12, 16], in order to justify neutron emissions as a consequence of nuclear reactions taking place in iron-rich rocks during brittle fracture. The same authors argued that neutron emissions may be related to piezoelectric effects and that the iron nucleus splitting may be a consequence of a nuclear photodisintegration.

As a conclusion and according to the experimental results, the hypothesis of iron depletion finds a very important evidence and confirmation at the Earth's crust scale, when particular pressure waves originate from fracture phenomena before or in correspondence to earthquakes [3–5, 27–35, 38, 44, 45]. The phono-disintegration reactions are the final forthcomings of the present alternative theoretical approach based on collective oscillations of the electronic density inside the atomic lattice (plasmons).

Phono-disintegration reactions can be therefore considered in order to interpret the most significant geophysical and geological transformations, today still unexplained. The laboratory results, herein reported for granite, basalt, and magnetite, give a new and original interpretation to the evolution of the lithosphere, contribute to explain the difference between oceanic and continental crust compositions due to subductive and tectonic phenomena, and, more generally, to the basaltic-to-sialic transformation of our planet.

The content of this chapter was preliminarily published in [46].

Appendix

Let us consider an electron gas that moves through an atomic lattice dislocation. As a consequence of this electron flow, a charge asymmetry occurs on the opposite sides of the dislocation. As a consequence of the generated opposite sign, a reactive force appears generating a simple harmonic motion at the plasmonic frequency, see Eq. (22.1) as evaluated in [6]. This relation is difficult to be evaluated, so that we introduce relations (22.3) and (22.4) into Eq. (22.1), obtaining:

$$\omega_{\mathrm{ph}}^2 = \frac{\mathrm{n}_{\mathrm{eff}}e^2}{\varepsilon_0 m_{\mathrm{eff}}} = \frac{\frac{A}{a^2}ne^2}{\frac{\varepsilon_0\mu_0}{2\pi}ne^2A\ln\left(\frac{a}{r}\right)} = \frac{2\pi}{\varepsilon_0\mu_0 a^2\ln\left(\frac{a}{r}\right)} = \frac{2\pi c^2}{a^2\ln\left(\frac{a}{r}\right)} \tag{22.11}$$

where the plasmonic frequency can be evaluated only in relation to the characteristics of the atomic lattice [1]. Here, the problem is to evaluate the quantity r. Following the results obtained in [6], at a first approximation, r can be assumed to be of the order of 10^{-6} m. In this way, we are able to evaluate the plasmonic frequency by introducing only one approximation, i.e., the value of parameter r.

References

1. Lucia U, Carpinteri A (2015) GeV plasmons and spalling neutrons from crushing of iron rich natural rocks. Chem Phys Lett 640:112–114
2. Carpinteri A (2015) TeraHertz phonons and piezonuclear reactions from nano-scale mechanical instabilities. In: Carpinteri A, Lacidogna G, Manuello A (eds) Acoustic, electromagnetic, neutron emissions from fracture and earthquakes. Springer International Publishing, Switzerland, pp 1–10
3. Cardone F, Carpinteri A, Lacidogna G (2009) Piezonuclear neutrons from fracturing of inert solids. Phys Lett A 373:4158–4163
4. Carpinteri A, Borla O, Lacidogna G, Manuello A (2010) Neutron emissions in brittle rocks during compression tests: monotonic vs. cyclic loading. Phys Mesomech 13:268–274
5. Carpinteri A, Lacidogna G, Manuello A, Borla O (2013) Piezonuclear fission reactions from earthquakes and brittle rocks failure: evidence of neutron emission and nonradioactive product elements. Exp Mech 53:345–365
6. Pendry JB, Holden AJ, Stewart WJ, Youngs I (1996) Extremely low frequency plasmons in metallic mesostructures. Phys Rev Lett 76(25):4773–4776
7. Ashcroft NW, Mermin ND (1976) Solid state physics. Harcourt, San Diego
8. Kittel C (1996) Introduction to solid state physics, 7th edn. John Wiley & Sons, New York
9. Davydov AS (1980) Theory of solids. Nauka, Moscow
10. Manuello A, Sandrone R, Guastella S, Borla O, Lacidogna G, Carpinteri A (2015) Neutron emissions and compositional changes at the compression failure of iron-rich natural rocks. In: Acoustic, electromagnetic, neutron emissions from fracture and earthquakes. Springer International Publishing, Switzerland, pp 23–37
11. Carpinteri A, Pugno N (2005) Are scaling laws on strength of solids related to mechanics or to geometry? Nat Mater 4:421–423
12. Widom A, Swain J, Srivastava YN (2013) Neutron production from the fracture of piezoelectric rocks. J Phys G: Nucl Part Phys 40(015006):1–8

13. Lucia U (2011) Some considerations on the photofission excitation function. Int J Nucl Energy Sci Technol 6(2):146–152
14. Lucia U (2009) A pn-pair mass evaluation in nuclear photofission. Int J Nucl Energy Sci Technol 4(3):196–200
15. Kaniadakis G, Lucia U, Quarati P (1993) Probability and time of photofission in the quasi deuteron energy region. Int J Mod Phys E 2(4):827–834
16. Widom A, Swain J, Srivastava YN (2015) Photo-disintegration of the iron nucleus in fractured magnetite rocks with magnetostriction. Meccanica 50:1205–1216
17. VanGordon JA, Gall BB, Kovaleski SD, Baxter EA, Kim BH, Kwon JW, Dale GE (2011) Electrical analysis of piezoelectric transformers and associated high-voltage output circuits. In: Pulsed power conference (PPC). IEEE, Chicago, pp 475–479
18. Cirillo D, Germano R, Tontodonato V, Widom A, Srivastava YN, Del Giudice E, Vitello G (2012) Experimental evidence of a neutron flux generation in a plasma discharge electrolytic cell. Key Eng Mater 495:104–107
19. Annila A (2012) The meaning of mass. Int J Theor Math Phys 2(4):67–78
20. Annila A (2010) All in action. Entropy 12:2333–2358
21. Annila A, Salthe S (2010) Cultural naturalism. Entropy 12:1325–1343
22. Tuisku P, Pernu TK, Annila A (2009) In the light of time. Proc R Soc A, Math, Phys Eng Sci 465:1173–1198
23. Alabastri A, Tuccio S, Giugni A, Toma A, Liberale C, Das G, De Angelis F, Di Fabrizio E, Proietti Zaccaria R (2013) Molding of plasmonic resonances in metallic nanostructures: dependence of the non-linear electric permittivity on system size and temperature. Materials 6:4879–4910
24. Ordal MA, Long LL, Bell RJ, Bell SE, Bell RR, Alexander RW, Ward CA (1983) Optical properties of the metals Al, Co, Cu, Au, Fe, Pb, Ni, Pd, Pt, Ag, Ti, and W in the infrared and far infrared. Appl Opt 22:1099–1119
25. Levin VE (translated by Bell LN) (1981) Nuclear physics and nuclear reactors. MIR Publisher, Moscow
26. Segre E (1977) Nuclei and particles, 2nd edn. W.A. Benjamin Inc., San Francisco
27. Carpinteri A, Cardone F, Lacidogna G (2009) Piezonuclear neutrons from brittle fracture: early results of mechanical compression tests. Strain 45:332–339
28. Carpinteri A, Cardone F, Lacidogna G (2010) Energy emissions from failure phenomena: mechanical, electromagnetic, nuclear. Exp Mech 50:1235–1243
29. Carpinteri A, Lacidogna G, Manuello A, Borla O (2011) Energy emissions from brittle fracture: neutron measurements and geological evidences of piezonuclear reactions. Strength, Fract Complex 7:13–31
30. Carpinteri A, Lacidogna G, Manuello A, Borla O (2010) Piezonuclear transmutations in brittle rocks under mechanical loading: microchemical analysis and geological confirmations. In: Kounadis AN, Gdoutos EE (eds) Recent advances in mechanics. Springer, Dordrecht, pp 361–382
31. Carpinteri A, Lacidogna G, Borla O, Manuello A, Niccolini G (2012) Electromagnetic and neutron emissions from brittle rocks failure: experimental evidence and geological implications. Sadhana 37:59–78
32. Carpinteri A, Lacidogna G, Manuello A, Borla O (2012) Piezonuclear fission reactions: evidences from microchemical analysis, neutron emission, and geological transformation. Rock Mech Rock Eng 45:445–459
33. Carpinteri A, Borla O, Lacidogna G, Manuello A (2012) Piezonuclear reactions produced by brittle fracture: from laboratory to planetary scale. In: Proceedings of the 19th European conference of fracture, Kazan, Russia
34. Carpinteri A, Chiodoni A, Manuello A, Sandrone R (2011) Compositional and microchemical evidence of piezonuclear fission reactions in rock specimens subjected to compression tests. Strain 47(2):267–281
35. Carpinteri A, Borla O (2018) Nano-scale fracture phenomena and TeraHertz pressure waves as the fundamental reasons for geochemical evolution. Strength, Fract Complex 11:149–168

36. Verkaeren J, Bartholomè P (1979) Petrology of the San Leone magnetite skarn deposit (S. W. Sardinia). Econ Geol 74:53–66
37. Tanguy JC (1978) Tholeiitic basalt magmatism of Mount Etna and its relations with the alkaline series. Contrib Miner Petrol 66:51–67
38. Carpinteri A, Manuello A (2011) Geomechanical and geochemical evidence of piezonuclear fission reactions in the earth's crust. Strain 47:282–292
39. Catling CD, Zahnle KJ (2009) The planetary air leak. Sci Am 300:24–31
40. Taylor SR, McLennan SM (1995) The geochemical evolution of the continental crust. Rev Geophys 33:241–265
41. Taylor SR, McLennan SM (2009) Planetary crusts: their composition, origin and evolution. Cambridge University Press, Cambridge
42. Fowler CMR (2005) The solid earth: an introduction to global geophysics. Cambridge University Press, Cambridge
43. Rudnick RL, Fountain DM (2005) Nature and composition of the continental crust: a lower crustal perspective. Rev Geophys 33:267–309
44. Carpinteri A, Borla O (2017) Fracto-emissions as seismic precursors. Eng Fract Mech 177:239–250
45. Carpinteri A, Borla O (2019) Acoustic, electromagnetic, and neutron emissions as seismic precursors: the lunar periodicity of low-magnitude seismic swarms. Eng Fract Mech 210:29–41
46. Carpinteri A, Borla O, Lucia U, Zucchetti M (2023) TeraHertz vibrations and phono-fission reactions from crushing of iron-rich natural rocks. J Condens Matter Nucl Sci 37:84–97

Chapter 23
Solid Lattice Model of the Atomic Nucleus: Comparison to Liquid Drop and Gaseous Shell Models

Abstract An enhanced version is presented of the face-centered-cubic (fcc) lattice model—originally developed by Norman Cook—for the analysis of the structure of atomic nuclei. A survey of the main theoretical models of the nuclear structure, such as the independent particle model (IPM) and the liquid drop model (LDM), is provided. A brief description is given for each of these models, highlighting how they are able or fail in explaining the available evidence on nuclear phenomena, such as the mean free path of nucleons within the nucleus, the nuclear size, shape, and skin, the nature of the nuclear force, the occurrence of asymmetric nuclear fission, and the so-called low-energy nuclear reactions (LENR). Then, we explore in more detail the Cook's solid lattice model. Starting from the basic assumptions and features of this model, which are associated to the shell levels of the IPM, we see how the nucleus can be geometrically defined by filling the sites of an fcc lattice with alternating isospin layers of protons and neutrons. The predictions arising from this lattice model are compared to the available experimental evidence, as shown in the original Cook's work. Eventually, we report the development of an enhanced version of this model, developed within an in-house Matlab code (https://github.com/domenicoscaramozzino/FCC_nuclear_structure) with the main purpose of studying nuclear fission. We describe how the randomization of the nuclear structure, coupled with an energy-based statistical analysis of the fission fragments, leads to a simple evaluation of the fission results. It is also discussed how to take into account decaying phenomena in order to consider the stability of the fission fragments.

Keywords Nuclear structure · Independent particle model (IPM) · Liquid drop model (LDM) · Face-centered-cubic lattice model · Phono-fission nuclear reactions · Low-energy nuclear reactions (LENR) · Nuclear decay of fragments · Randomization of nuclear skin

A. Carpinteri, *Terahertz Phonons and Nanomechanical Instabilities*,
https://doi.org/10.1007/978-3-032-14692-2_23

23.1 Theoretical Models of the Nuclear Structure

Since the first half of the twentieth century, a plethora of theoretical models has been proposed to study and explain the structure of the atomic nucleus. These include models where the nucleus is described as a liquid drop, as a dynamic gas, or even as a solid lattice of protons and neutrons. Figure 23.1 shows an outline of the main nuclear models adopted in nuclear structure theory, and their evolution since the 1930s [1]. These models can be grouped into two main families, depending on their different assumptions on the nuclear force. Weak interaction models assume the nuclear force between nucleons to be relatively weak. Conversely, strong interaction models require the nuclear force to be strong in nature.

The weak interaction models correspond to the family of the independent particle models (IPM). In the IPM, the nucleus is described as a set of nucleons orbiting around the center due to a net attractive force acting in a potential-well. Like the atomic structure, which is made up of shells of electrons at different energy levels, the IPM describes the nucleus as a quantum–mechanical gas of point nucleons orbiting on well-defined shells, whose position is associated to the specific energy level of each nucleon [2]. For the nucleons to be able to orbit on these shells rather freely, these need to experience a weak interaction force. Therefore, the IPM, which later became known as the shell model, is part of the weak interaction models.

On the other hand, the strong interaction models assume that the nuclear force is strong. The nucleons are hold in place in a condensed environment, where the motion of each nucleon is hindered by the motion of the others. The family of the strong

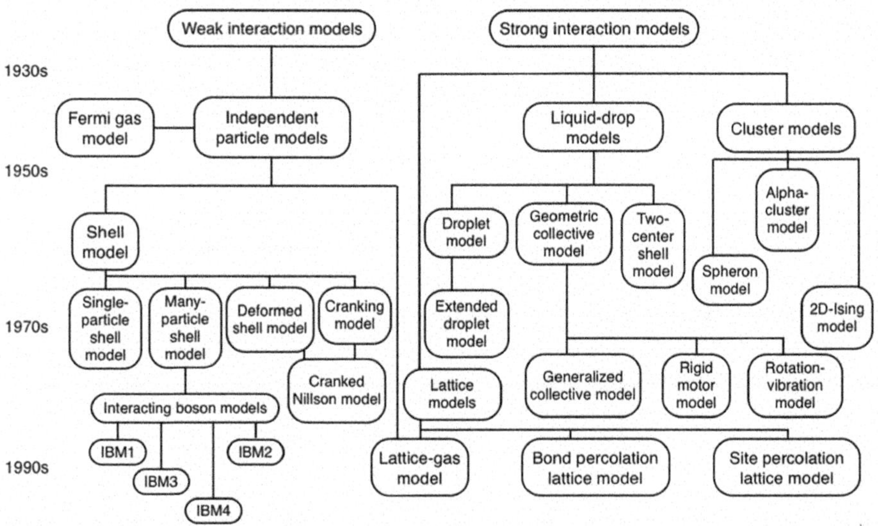

Fig. 23.1 Outline of the main models in nuclear structure theory. The models are grouped into two major categories, depending on the different assumptions on the nature of the nuclear force, i.e., weak or strong. Adapted from [1]

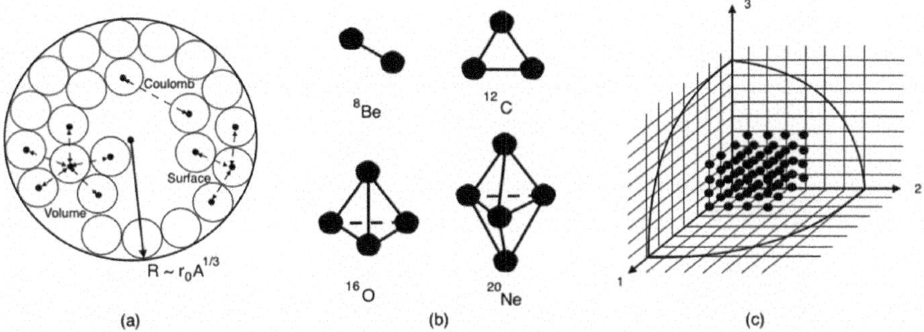

Fig. 23.2 **a** The basic components of the LDM in the atomic nucleus: volume, surface, and electrostatic forces. **b** Possible geometrical arrangements of nucleons in the lighter 4n-nuclei based on alpha-particle clusters. Each black dot represents an alpha particle. **c** Arrangement of nucleons in the basic simple-cubic lattice nuclear structure

interaction models can be subdivided into different groups, depending on how the interaction between nucleons is thought to affect the nuclear structure (Fig. 23.1). The liquid drop model (LDM) assumes that the nucleus can be described as a liquid, where all the particles are held together by volume, surface, and electrostatic forces (Fig. 23.2a) [3]. The cluster models assume that the nuclei can be described not only as an assembly of individual protons and neutrons, but also as clusters of densely packed particles. Among these, the alpha-particle model (Fig. 23.2b) describes the nucleus as an aggregate of interacting alpha particles [4]. Another example of strong interaction model is the lattice model, where the nucleus is represented as a union of protons and neutrons occupying well-defined positions within a solid lattice (Fig. 23.2c) [5]. In these models, it is the strong nuclear force between nucleons that is expected to hold them in the proper place to give the lattice its geometrical properties.

As can be observed from the brief summary reported above, there is a large difference—and perhaps contradiction—between these nuclear models. The nucleus is arbitrarily thought of as a gas of freely orbiting particles, as a liquid drop of interacting neighbors, or even as a solid lattice of nucleons. Nevertheless, all these models have a reason for being proposed, since each of them can explain certain nuclear properties known experimentally. On the other hand, they have shortcomings as well, insofar they are unable to address all the available evidence on the nuclear structure. Now, we will look at these models in more detail, since all of them provide important information for the understanding of the Cook's lattice model.

Starting from the quantum–mechanical nature of the shell levels and based on the Schrödinger wave-equation with corrections related to the spin–orbit coupling, the IPM (and the various shell models) were rather successful in predicting the shell and sub-shell closures observed for the so-called "nuclear magic numbers" [6]. These magic numbers refer to the most common numbers of protons and neutrons that are observed in stable isotopes and isotones, such as Z, $N = 2, 8, 20$, etc. These magic numbers resemble the shell closures observed for electrons in the atomic realm. Since the IPM was developed as a transfer of the atomistic theory to the nuclear realm, the assumptions behind this model seemed a rational choice to predict nuclear shell closures. However, one of the main problems of the IPM is that it assumes that nucleon–nucleon interactions are weak, which results in predicting a rather long mean free path (MFP) of nucleons due to their free orbiting [5]. On the other hand, the experiments show that MFP of nucleons within the nucleus is short, suggesting that the nucleon–nucleon interaction is strong enough to prevent a free orbiting. Another problem of the IPM is related to the description of the potential-well [7], which is supposed to hold all the nucleons in orbit. If in the much larger atomic environment the potential-well is simply given by the Coulomb attraction between the positively charged atomic center and the negatively charged orbiting electrons, what is the potential-well of a set of positively charged protons and uncharged neutrons orbiting around a fictitious nuclear center? All these issues pose theoretical challenges to the basic assumptions of the IPM. Nevertheless, due to the elegant formalism of this model, which is grounded on quantum mechanics, energy shell levels, and Schrödinger equation, the IPM has been one of the most accepted and used models in nuclear structure theory.

Conversely, the strong interaction models abandon the quantum–mechanical formalism of the nuclear interaction and describe the nucleus as a collection of corporeal particles interacting through a strong nuclear force (Fig. 23.2) [3]. Among these, the LDM has been the dominant. As shown in Fig. 23.2a, the LDM describes the nucleus as a tight quasi-spherical collection of particles. These particles interact with each other based on various forces contributing to the overall binding energy of the nucleus. All these forces play a role in the well-known Weizsäcker's mass formula [8], which is used to calculate the binding energy (*BE*) of a nucleus for a given number of protons Z, neutrons N, and mass number A ($A = Z + N$):

$$BE(Z, N) = k_1 A + k_2 A^{2/3} + k_3 \frac{Z(Z-1)}{A^{1/3}} + k_4 \frac{(A-2Z)^2}{A} + \frac{k_5}{A^{1/2}} \quad (23.1)$$

where k_1, k_2, k_3, k_4, and k_5 are constants that are set up to obtain the best fit with experimental data. One set of these parameters is: $k_1 = 15.8$, $k_2 = -18.3$, $k_3 = -0.72$, $k_4 = -23.2$, and $k_5 = -11.2$ for Z and N odd, $k_5 = +11.2$ for Z and N even, $k_5 = 0$ otherwise [9].

The first term of the mass formula, which is known as the volume term, is proportional to the total number of nucleons within the nucleus, i.e., the mass number A. The second term is referred to as the surface term, which considers the decrement in binding energy due to the external surface of the drop, where the nucleons feel a weaker interaction due to a lower number of neighbors. This term is proportional to the surface of the drop and, since the volume is proportional to the mass number, this term turns out to be proportional to $A^{2/3}$. The third term is related to the Coulomb repulsion between protons, which are tightly packed within the nucleus and whose pairwise repulsion is expected to decrease the stability of the nucleus. Counting all the pairwise proton–proton interactions, the Coulomb term is proportional to Z $(Z-1)$, whereas it is inversely proportional to $A^{1/3}$, which is an estimate of the radius of the drop. The first three terms of the mass formula are graphically represented in

Fig. 23.2a. The fourth term is known as the symmetry term and is proportional to $(A-2Z)^2/A$. Since $(A-2Z)$ represents the excess of neutrons in a nucleus with more neutrons than protons, this term is meant to bias the stability toward a symmetric nucleus where $N = Z$, i.e.,$A = 2Z$. However, being the mass number at the denominator, this term becomes more important for lighter nuclei than for heavier ones. This comes from the experimental observation that, for lighter nuclei, the stable ones are symmetric, i.e., $N = Z$, whereas for heavier ones the stable ones have more neutrons, i.e.,$N > Z$. The last is called the pairing term and is based on the experimental evidence that nuclei with even-numbers of protons or neutrons have binding energies higher than those with odd-numbers. Correction terms to the mass formula have been proposed, which arise from the shell and sub-shell texture of the nucleus as described by the IPM.

The Weizsäcker's mass formula is more commonly expressed in terms of binding energy per nucleon, *BE*/*A*, leading to:

$$\frac{BE(Z,N)}{A} = k_1 + \frac{k_2}{A^{1/3}} + k_3\frac{Z(Z-1)}{A^{4/3}} + k_4\left(1 - 2\frac{Z}{A}\right)^2 + \frac{k_5}{A^{3/2}} \tag{23.2}$$

Figure 23.3a shows the average binding energy per nucleon as a function of the nuclear mass number, as computed from Eq. (23.2), highlighting the contribution of the five terms. This LDM-based formula is found to be in general agreement with experimental data, except for some nuclei lying in correspondence to the so-called "magic numbers" that manifest a higher stability [10].

Another nuclear property, of which the LDM was found to provide good estimates, is the value of nuclear radius [10]. Since the nuclear structure is assumed to be a quasi-spherical liquid drop, its radius should be proportional to the cubic root of the

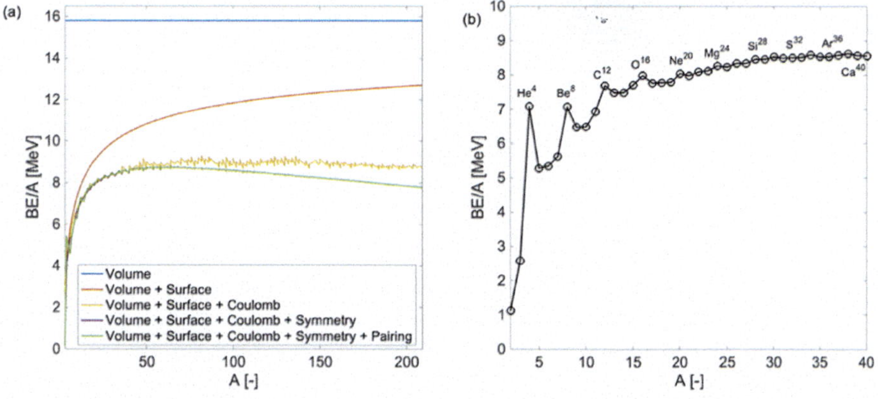

Fig. 23.3 **a** Average binding energy per nucleon (*BE*/*A*) evaluated from the LDM-formula: contribution of volume, surface, Coulomb, symmetry, and pairing terms. **b** Average binding energy per nucleon for nuclei with mass *A* up to 40. The peaks in the curve are related to nuclei with 4n-multiplicity, indicating the higher stability of nuclei made up of clusters of alpha particles

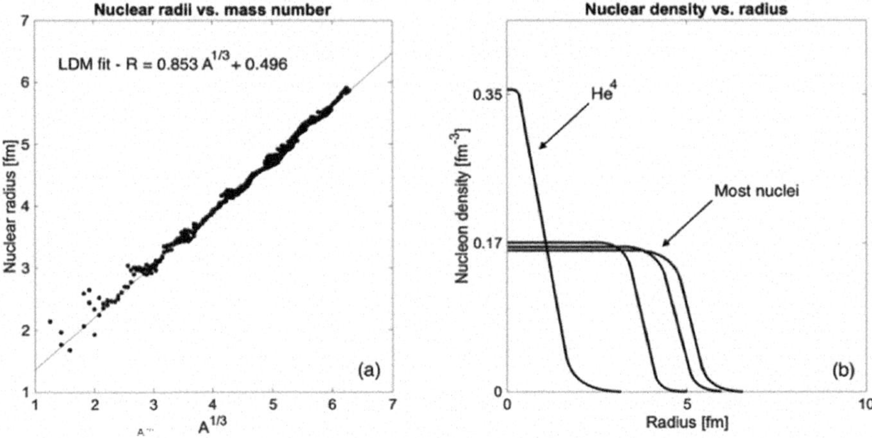

Fig. 23.4 **a** Linear dependence between the nuclear radius and the cubic root of the mass number A: the LDM-based formula is compared to experimental data. **b** Schematic representation of nuclear density, showing the constant value in the core region (which is double for He^4) and the decay associated to the surface skin

volume, hence proportional to the cubic root of the mass number, i.e., $R = r_0 A^{1/3}$, r_0 being a constant that gives the best fit to experimental data. Figure 23.4a shows the agreement between experimental nuclear radii and the theoretical estimation from the LDM formula, with $r_0 = 0.853$ fm. As can be seen, apart from small discrepancies especially for light nuclei, the LDM-based formula provides an accurate estimate of nuclear radii.

The cluster models were also developed based on the assumption of a strong interaction force between nucleons. However, differently from the LDM, the cluster models imagine the nucleus to be made up of a texture of clusters of particles. The interaction forces are very strong within each cluster. The most famous example of cluster model is the alpha-particle model, where the nucleus is described as a union of alpha particles (Fig. 23.2b), plus eventual exceeding neutrons and protons. This model came into play because of the experimental evidence showing that peaks in the average binding energy per nucleon are observed for small 4n-nuclei, such as He^4, Be^8, C^{12}, and O^{16} (Fig. 23.3b), indicating an unusual stability for nuclei made up of multiples of alpha particles [11]. Moreover, the experimental evidence that some heavier nuclei often emit alpha particles upon nuclear decay pointed in the direction that alpha particles might be clustered within the nuclear structure.

Finally, the lattice models postulate that the nuclear structure can be represented as a solid lattice of nucleons in well-defined positions [5]. These positions are the sites of the lattice. The lattice models lie within the family of the strong interaction models. The first lattice models were developed to explain data from high-energy multi-fragmentation experiments [12]. These experiments were simulated by modeling the nucleus as a three-dimensional lattice of nucleons, and then simulating the fragmentation phenomenon either by assigning a breaking probability to each nucleon–nucleon bond (bond percolation lattice model), or by assigning a de-population probability to each nucleon in the lattice sites (site percolation lattice model) [13]. A lattice-gas model was also developed, in which the lattice sites are occupied by nucleons in a dynamically random way, therefore simulating the gas-like behavior advocated by the IPM, whereas adopting the solid-like framework of the lattice model [14]. These lattice models were developed by following different lattice symmetries, such as simple-cubic-packing (scp), body-centered-cubic (bcc), and face-centered-cubic (fcc). In the remainder of this chapter, the fcc lattice model developed by Norman Cook will be thoroughly reviewed [5]. First, let us survey the main issues that may pose serious challenges to some of the classical models of nuclear structure theory described so far.

As mentioned above, one of the shortcomings of the weak interaction models regards the explanation to the MFP of nucleons within the atomic nucleus [5]. The MFP is a measure of the distance over which a nucleon can travel within the nuclear environment before colliding with another nucleon. Since the shell model assumes the free orbiting of nucleons, the MFP predicted by this model is rather long. For large nuclei, e.g., with radii of 5–6 fm, the IPM-predicted MFP is in the order of tens of fm. However, experimental data indicate that MFPs of nucleons are generally shorter, in the order of 3–4 fm. This experimental evidence seems then to suggest that the nuclear structure is a crowded environment, where nucleons interact over short distances. Thus, the assumption of freely orbiting nucleons around a potential-well, which is the basis of the IPM, seems against this experimental evidence. A possible solution for this paradox was provided by Weisskopf, who invoked the application of Pauli exclusion principle [15]. In this view, the nucleus can be assimilated to a dense and crowded aggregation of nucleons (thus explaining the short MFP), whereas the exclusion principle prevents the collisions between close nucleons and allows them to orbit onto the IPM shell levels. Since the IPM was still able to explain a variety of additional nuclear properties, such as nuclear spins and parities, it was considered as an acceptable model of the nuclear structure. On the other hand, the paradox of the short vs. long MFP of nucleons still remains.

Another issue that needs to be addressed by any model of the nuclear structure is the prediction of nuclear size, nuclear density, and skin thickness. As shown in Fig. 23.4a, the size of most nuclei is well-predicted by the LDM, with the average radius proportional to $A^{1/3}$. Another interesting experimental evidence came from studies on the nuclear density values [16]. Figure 23.4b reports the charge density of some nuclei, showing that they exhibit a constant value for most of their radial extension, after which a density decay is observed. This decay characterizes the so-called nuclear skin, where the nuclear density smoothly decreases from its core value to zero. Since the LDM describes the nucleus as a closely packed quasi-spherical ensemble of nucleons (Fig. 23.2a), it is shown to provide a correct estimate of the constant nuclear density if nucleons are given a radial value of about 0.862 fm, leading to the core nucleon density of 0.17 nucleons/fm^3 observed experimentally. However, two aspects of the curves reported in Fig. 23.4b cannot be explained straightforwardly by the LDM. First, the surprisingly high-density value of He^4, which exhibits a density of around 0.35 nucleons/fm^3 [5]. Second, since the nucleus resembles a sphere in the LDM, the skin region is expected to experience an abrupt density decay with infinite slope. Thus, in the LDM framework, the smooth decay observed experimentally can only be explained by inducing an artificial polarization of the drop, resulting into a prolate or oblate deformation. On the other hand, the dimensions of nuclear skin are experimentally observed to be rather constant (about 2.3–2.4 fm), regardless of the size of the nucleus, which might be hard to explain theoretically [5].

Another matter that needs to be accounted for is related to the very nature of the nuclear force. As we have already seen, the nature of the nuclear force is what makes the two main families of nuclear models different. The first question becomes: is the nucleus dominated by strong short-range interaction forces that have significant effects up to about 2 fm, thus leading the nuclear structure to behave as a liquid drop or as a solid lattice? Or is the nucleon–nucleon interaction a force that has effects up to 7–8 fm, such that all nucleons interact with each other weakly in a diffuse gas-like environment [5]? Despite we have some information about the nuclear force between isolated nucleons (proton-proton, neutron-neutron, proton-neutron), the nuclear potential in the complex nuclear environment is still an open question. Some experimental evidence (short MFP of nucleons within the nucleus, high stability of alpha particles, etc.) seem to point in the direction of a short-range strong nuclear force acting over short distances in the order of few fm. However, much more research is needed to provide a conclusive answer to this open problem.

The last two subjects that we need to address in relation to the predictions of these models are related to the phenomenon of nuclear fission and the occurrence of the so-called low-energy nuclear reactions (LENR).

Nuclear fission occurs when a nucleus splits up into fragments, resulting sometimes in a release of energy and radioactivity. This can either occur spontaneously for very large and unstable nuclei, or it can be induced by external perturbations. In either case, experimental evidence suggests that the fragments resulting from nuclear fission are different in mass, i.e., nuclear fission is generally asymmetric. Figure 23.5a shows the fragment yield in terms of nuclear mass for the neutron-induced fission of U^{233}, U^{235}, and Pu^{239} [17], highlighting the asymmetry of fission. This asymmetry poses problems to both strong interaction models, such as the LDM, and weak interaction models, such as the shell model. The LDM, which can account for the energy released in the fission process [18], predicts that the dominant mode of fission is symmetrical. Differently from the shell model, in the LDM the nucleus does not contain any sub-structure: the nucleus is a structureless, amorphous, and collective distribution of interacting particles. As a result, upon fission this structure should then be fragmented in a quasi-symmetrical way. In order to explain the asymmetry observed experimentally, various attempts have been made to tweak the LDM to predict the experimental results, e.g., it has been suggested that, while undergoing fission, the drop might become elongated due to the internal Coulomb repulsion between protons, inducing one lobe of the drop to become larger than the other [5]. On the other hand, such manipulation attempts were not completely successful in explaining all the details of asymmetric fission. Differently from the LDM, the shell model implies a nuclear sub-structure, which was the reason why various researchers

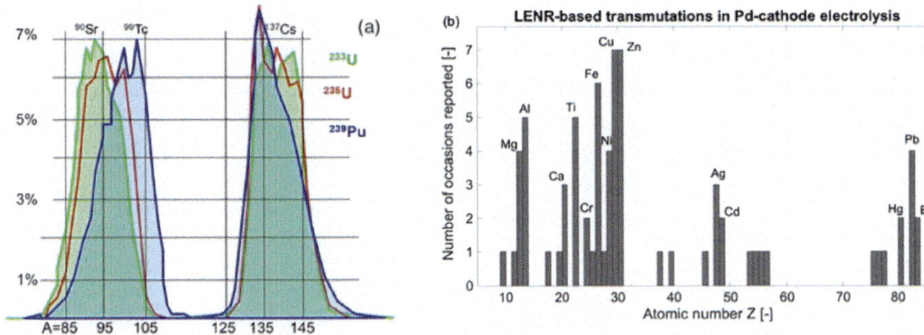

Fig. 23.5 **a** Mass yield vs. fragment mass for fission of U^{233}, U^{235}, and Pu^{239}. Reproduced with permission from [17]. **b** Meta-analysis of the transmutation results from LENR. Most studies used Pd cathodes and Pt anodes, so that most of the transmutations could be explained according to Cook as Pd-plus-several-deuterons, Pt-plus-several-deuterons, or Pd fission fragments

advocated the plausibility that the shell model could explain nuclear fission and its asymmetry. This was carried out by adjusting the parameters of the potential-well associated to the IPM [19]. Although some insights on the asymmetric distribution of fission fragments could be obtained in this way, this procedure showed that the asymmetry could not be explained entirely based on the first principles, on which the shell model relies. In conclusion, the asymmetry of nuclear fission cannot be easily explained by the main models of nuclear structure theory, unless some of their fundamental parameters are properly tuned [5].

The final subject we mention here is related to the so-called low-energy nuclear reactions (LENR), i.e., the subject of the present volume. The first observation of these anomalous reactions was in 1989, when Fleischmann and Pons reported the occurrence of nuclear reactions from the loading of deuterium into Pd electrodes [20]. These experiments provided an "anomalous excess heat", which was observed to be orders of magnitude too large to arise from conventional chemical reactions. However, traditional models of nuclear structure theory were all unable to provide a theoretical explanation to such excess heat phenomena. This, together with the fact that LENR have not been found to be always replicable, is also the reason why much debate is still going on about the scientific reliability of LENR. Interestingly, besides excess heat LENR were also generally found to involve nuclear transmutations, i.e., elemental products, absent at the beginning of the experiments, were retrieved after the test. Figure 23.5b shows the results of a meta-analysis reported by Storms [21], where transmutations were observed from Pd–Pt electrolysis experiments. As can be seen, a variety of different elements was found after the experiment, such as Mg, Al, Ti, Fe, Cu, Zn, etc., which were absent before the test. The presence of these elements, which was thought to be a fingerprint of LENR taking place, was explained by Cook as the possible result of fission of Pd or as the "cold fusion" of Pd and Pt nuclei with one or several deuterons. As in the case of the excess heat, none of the traditional models reported above was able to provide a satisfactory explanation to the occurrence of such anomalous reactions.

In the remainder of this chapter, we will examine in more detail the face-centered-cubic (fcc) lattice model developed by Norman Cook for the visualization and modeling of the nuclear structure [5]. As we will see, despite its simplicity, this model is found to provide reasonable explanations to most of the issues mentioned above (MFP of nucleons within the nucleus, nuclear size, shape, and skin, nuclear force, nuclear fission, and LENR).

23.2 The Cook's Face-Centered-Cubic (Fcc) Lattice Model

The Cook's model belongs to that family of models that describe the nucleus as a set of nucleons occupying well-defined sites in a three-dimensional lattice. This model was proposed with the purpose of overcoming most of the paradoxes highlighted above and as a possible unification of the existing nuclear models [5]. As a matter of fact, it makes use of the quantization of nucleon energies in the shell levels of the IPM, while retaining several characteristics typical of the LDM and cluster models, e.g., short-range nuclear forces and closely packed interacting nucleons. While developing this lattice model, Cook explicitly refers to the symmetries associated to the shells and sub-shells that constitute the nuclear texture [5, 22]. Quoting from [5]: "The conventional application of quantum mechanics in the nuclear realm results in a pattern of symmetries (energy shells and sub-shells with specific particle-occupancies), and the entire pattern is reproduced in one particular, closed-packed lattice of nucleons". Hence, the model starts from the quantum–mechanical shell-based IPM approach, but the features of the resulting lattice of nucleons will lead to certain nuclear properties that are typical of liquid/solid models.

The IPM is based on Schrödinger wave equation and, as we have seen above, it relies on the quantum–mechanical framework, which is universally acknowledged to be the correct theory of reality at the sub-atomic level. This framework provides each nucleon with a quantification of its energy levels through five main quantum numbers, i.e., n, j, m, s, and i [5]. The first quantum number,$n = 0, 1, 2, \ldots$, is related to the shell level occupied by the nucleon and comes directly from the energy values arising from the Schrödinger wave equation. The second quantum number j represents the nucleon's total angular momentum and, due to the spin–orbit coupling assumption, it is equal to $|l + s|$, where l is the orbital angular momentum ($l = 0, 1, \ldots, n$) and s is the fourth quantum number, which represents the spin of the nucleon ($s = +½$ or $-½$). As a result, j expresses the total content of angular momentum for each nucleon within a defined n-shell ($j = 1/2, 3/2, 5/2, \ldots$). The third quantum number, m, is related to the orientation state of the nucleon, with $m = \pm 1/2, \pm 3/2, \pm 5/2, \ldots$, whereas the fifth quantum number, i, represents the isospin of the nucleon, with $i = +1$ or -1. Each nucleon state can be uniquely identified by a set of the five quantum numbers. These in turn affect the total occupancies of the nuclear shells. Figure 23.6 shows the energylevels, degenaracy and parity values calculated considering a harmonic potential. The degeneracy, which reflects the number of distinct wave-functions at each level, depends on the permutations between the n and l quantum numbers. The

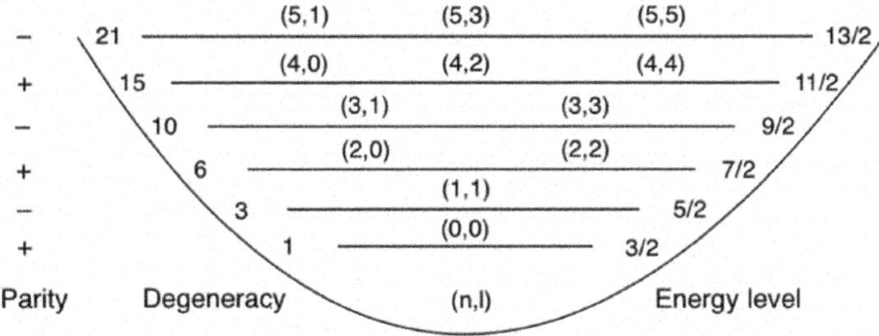

Fig. 23.6 Energy levels, degeneracy, and parity values for the nucleon states considering the harmonic oscillator. The energy levels are expressed as multiples of the fundamental energy level, i.e., $h\omega_0$

occupancy of each shell reflects these degeneracy values, also considering the spin and isospin mutual occupation rules. Taking these into account leads to the total occupancy values shown in Table 23.1 [5].

Given the discrete (quantized) shell and sub-shell texture of the nucleus, the question is: can this mathematical representation of the nucleon states be reflected by any geometrical description of the arrangement of nucleons? The answer provided by Cook was based on the observation that the x, y, z components of the principal quantum number n, i.e., n_x, n_y, n_z, can be directly associated to the Cartesian coordinates of the sites of an anti-ferromagnetic face-centered-cubic (fcc) lattice with alternating isospin layers [5, 22]. This isomorphism between the quantum nucleon states and the geometrical positions in an fcc lattice has already been proposed previously. Wigner was the first one suggesting in 1937 that the IPM quantum numbers relate strictly to the close-packed geometry of spheres in a 2D array, also noting that these two-dimensional layers could be piled up in an fcc fashion [23]. Everling showed that fcc geometries might explain the first three doubly magic nuclei, pushing further the idea that nucleons could be arranged in fcc dynamic lattices [24]. In 1974, Lezuo [25] developed a consistent lattice model, where he showed that the close-packing of Gaussian nucleon-probability spheres arranged in an fcc or hexagonal lattice could reproduce several nuclear features. Cook's fcc lattice model fits itself within this theoretical framework [22].

Based on the isomorphism between the IPM quantum numbers and the fcc lattice geometry, the three-dimensional Cartesian coordinates x, y, z of each nucleon can be associated to its five quantum numbers n, j, m, s, i. Taking (0,0,0) as the origin of the reference system and assuming that the x, y, z coordinates of the nucleons are odd integers (1, –1, 3, –3, etc.) corresponding to the sites of the lattice, we have:

$$n = (|x| + |y| + |z| - 3)/2 \tag{23.3a}$$

$$j = (|x| + |y| - 1)/2 \tag{23.3b}$$

$$m = (-1)^{(x-1)/2}|x|/2 \quad (23.3c)$$

$$s = (-1)^{(x-1)/2}/2 \quad (23.3d)$$

$$i = (-1)^{(z-1)/2} \quad (23.3e)$$

As a direct consequence of Eq. (23.3a), the occupancy of the shell levels reported in Table 23.1 can be straightforwardly derived based on the quantum numbers analogy with fcc lattice sites. The first shell level ($n = 0$) can be occupied by nucleons with parity (which is defined as the product of the signs of the nucleon coordinates, see below) of + 1 (Fig. 23.6). Thus, considering only the sites whose coordinates satisfy Eq. (23.3a), the nucleons of the first shell can occupy the positions (1,1,1), (1, –1, –1), (–1,1, –1), and (–1, –1,1), confirming that the occupancy of the first shell is equal to 4. The second shell level ($n = 1$) has a parity of –1. Again, in order to satisfy Eq. (23.3a), the nucleons in this level can occupy only the positions (1,1,–3), (1,3,–1), (3,1,–1), (1,–1,3), (1,–3,1), (3,–1,1), (–1,1,3), (–1,3,1), (–3,1,1), (–1,–1,–3), (–1,–3,–1), and (–3,–1,–1). This confirms that the second shell level counts a total of 12 nucleons. Cumulatively, the first two shell levels have an occupancy of 16 nucleons. Proceeding in the same way for the higher shell levels ($n = 2, 3, \ldots$), the whole occupancy set of the IPM levels can be translated into occupation of these lattice sites.

Equation (23.3a) also shows that the principal quantum number n can be thought of as the overall distance of the nucleon from the origin of the reference system. Figure 23.7a shows some of the first nucleons in the build-up procedure color-coded according to their shell levels (from $n = 0$ to $n = 3$). In this way, n identifies shell levels that resemble concentric spheres: the higher n, the larger the sphere, and the more distant the nucleon from the center of the nucleus (Fig. 23.7b). By filling all these sites of the lattice for increasing n values, these shells turn out to have an approximate octahedral shape of increasing size (Fig. 23.7c). Notably, some of

Table 23.1 Energy and angular momentum quantum numbers, and their influence on degeneracy and total occupancy of the nucleon states in the IPM nucleus build-up

n	0	1	2	3	4	5	6
l	0	0,1	0,1,2	0,1,2,3	0,1,2,3,4	0,1,2,3,4,5	0,1,2,3,4,5,6
Degeneracy	1	3	6	10	15	21	28
Degeneracy with spin mutual exclusion	2	6	12	20	30	42	56
Degeneracy with spin and isospin mutual exclusion	4	12	24	40	60	84	112
Total occupancy	4	16	40	80	140	224	336

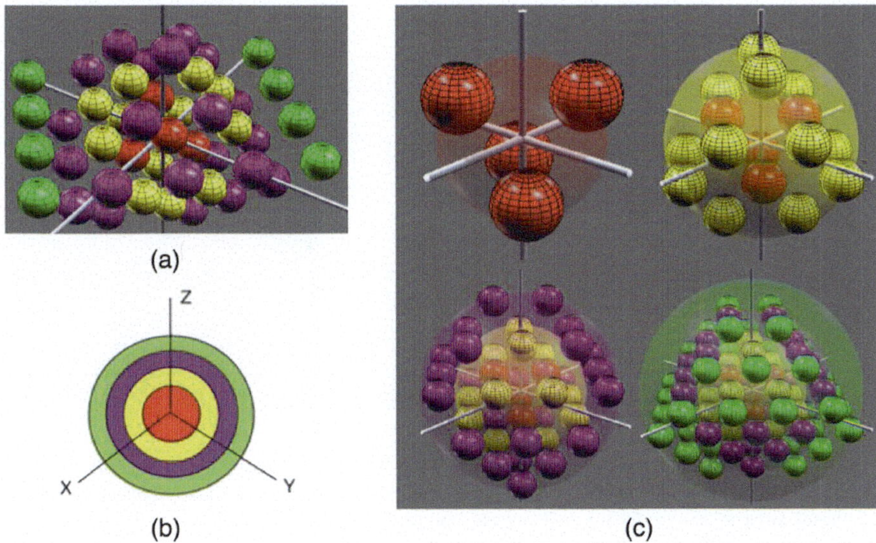

Fig. 23.7 The principal quantum number n associated to the occupation of certain sites of the fcc lattice. **a** Some of the nucleons contained in the first four shells, colored according to the different shell levels (red for $n = 0$, yellow for $n = 1$, purple for $n = 2$, green for $n = 3$). **b** Schematic representation of the shell levels, resembling a quasi-spherical closely packed structure. **c** Actual geometrical arrangement of nucleons for increasing n values, highlighting the approximate octahedral shape of the higher shells when completely filled

the values associated to the closures of shells and sub-shells are related to some experimental evidence related to magic numbers [5].

Equation (23.3b) shows that the total angular momentum j is related to the distance of the nucleon from the z-axis, which is assumed to represent the spin direction. As a result, nucleons with the same value of j are arranged along concentric quasi-cylindrical forms of increasing radius (Fig. 23.8a). The angular momentum quantum number j leads to sub-shell arrangements of nucleons within each n shell. On the other hand, the azimuthal quantum number m reported in Eq. (23.3b) depends on the distance of the nucleon along one of the axes perpendicular to the spin axis, which is conventionally taken as the x-axis. Nucleons with the same m values are arranged along quasi-conical sub-shells with vertices in the center of the nucleus (Fig. 23.8b). As the nucleon becomes more distant from the spin axis, the absolute value of m increases and its sign changes with alternating layers perpendicular to the x-axis.

Equation (23.3d) shows that the alternating layering of the m sign reflects the layering of the spin values s of the nucleons (Fig. 23.9a). As a result, the nucleons are layered according to their spin in an alternating fashion along the x-axis. Finally, Eq. (23.3e) refers to the isospin layering along the z-axis of the nucleus (Fig. 23.9b). Nucleons with $i = +1$ are protons, whereas nucleons with $i = -1$ are neutrons. As a consequence, this alternating isospin layering in the nuclear lattice leads to an alternating layering of protons and neutrons along the z-axis (Fig. 23.9b).

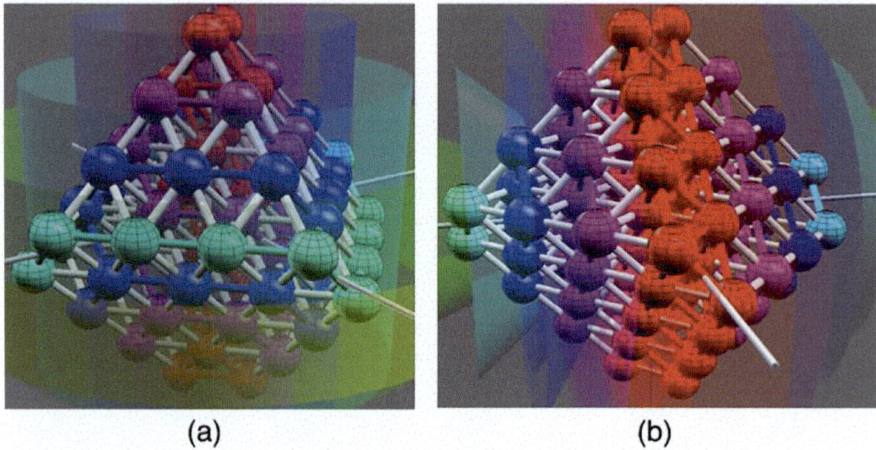

Fig. 23.8 Nuclear lattice structure of Zr^{80}, with color-coding according to: **a** j values (red for $j = 1/2$, purple for $j = 3/2$, blue for $j = 5/2$, light-blue for $j = 7/2$); **b** m values (red for $m = \pm 1/2$, purple for $m = \pm 3/2$, blue for $m = \pm 5/2$, light blue for $m = \pm 7/2$)

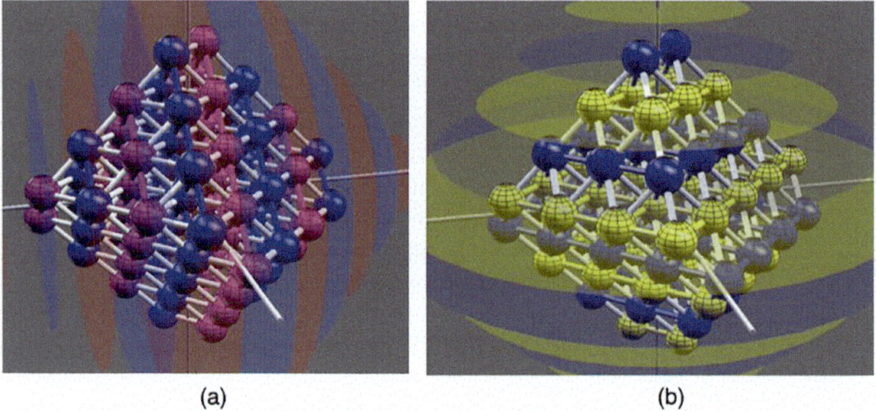

Fig. 23.9 Nuclear lattice structure of Zr^{80}, with color-coding according to: **a** s values (purple for $s = +1$, blue for $s = -1$); **b** i values (yellow for $i = +1$, blue for $i = -1$)

As briefly mentioned above, each nucleon and assembly of nucleons is characterized by a certain parity value. The parity p of each nucleon can be simply defined as the product of the signs of its Cartesian coordinates, whereas the overall parity of the nucleus with mass number A, P_A, is the product of all the individual parities [5], i.e.:

$$p = \text{sign}(x) \times \text{sign}(y) \times \text{sign}(z) \quad (23.4a)$$

$$P_A = \prod_{k=1}^{A} p_k = \prod_{k=1}^{A} sign(x_k) \times sign(y_k) \times sign(z_k) \qquad (23.4b)$$

Equations (23.3) provide a one-to-one relationship between the nucleon coordinates in the lattice and the corresponding quantum numbers. The inverse relationships can also be obtained, i.e., those providing the Cartesian coordinates of the nucleon, x, y, z, for a given set of quantum numbers n, j, m, s, and i:

$$x = 2|m|(-1)^{m-1/2} \qquad (23.5a)$$

$$y = |2j + 1 - 2|m||(-1)^{n-m+1/2+i/2-|n-j+1|} \qquad (23.5b)$$

$$z = 2|n - j + 1|(-1)^{|n-j+1|-i/2} \qquad (23.5c)$$

These equations provide a straightforward way to build the three-dimensional arrangement of nucleons for a given set of quantum numbers.

In the previous section, we explored a variety of experimental nuclear phenomena, that some of the traditional models fail to explain. How does the Cook's lattice model perform in relation to these experimental issues? In the following, we will describe the predictions arising from the lattice model in relation to nuclear density, shape, and size, internal textures of the nucleus, nuclear binding energy, and we will conclude by reporting how the fcc model can address the phono-fission nuclear reactions [5].

As far as nuclear density is concerned, we have seen above that the nuclei have a rather constant value of the mass density (around 0.17 nucleons/fm^3) across most of their core, almost regardless of their size [16]. In order to illustrate the basic features of the fcc lattice model, we have previously used odd integer numbers for the nucleon coordinates. Now, in order for the core nuclear density to match the experimental value of 0.17 nucleons/fm^3, the inter-nucleon distance between close neighbors in the lattice needs to be set at 2.026 fm. Using this value, one finally obtains the nuclear density value detected experimentally. Additionally, due to a simple geometrical property of the fcc lattice, the "anomalous" double density of He^4 (Fig. 23.4b) finds here a simple justification. As shown in Fig. 23.10a, the lattice of a typical nucleus larger than He^4 results into a mixed arrangement of octahedra and tetrahedra, whereas the smaller He^4 is a simple tetrahedron with the two protons and two neutrons filling the first shell ($n = 0$). Octahedra and tetrahedra exhibit different densities due to their spatial configuration, the former being less dense than the latter. Since He^4 is the only nucleus acquiring the tetrahedral configuration, whereas all the other larger nuclei correspond to an arrangement of high-density tetrahedra and low-density octahedra, this simply explains why He^4 is expected to exhibit a larger value of the core nuclear density.

We have also discussed above how the traditional LDM generally fails in predicting the decay of the nuclear density value across the nuclear skin, being the liquid drop a quasi-spherical object [5]. We have mentioned that adjustments of

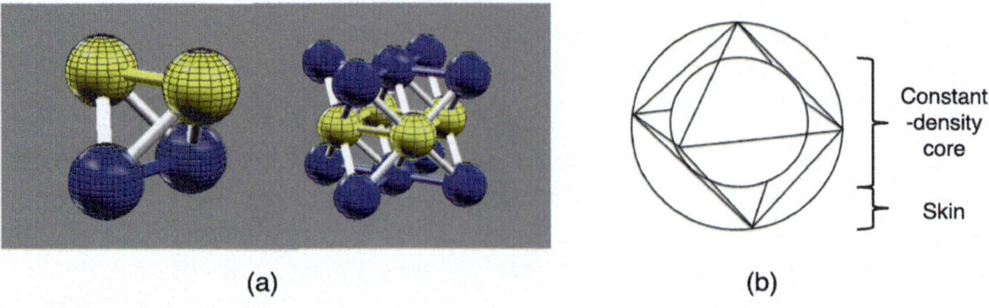

Fig. 23.10 **a** Geometrical arrangement within the fcc lattice (tetrahedra and mixed tetrahedra-octahedra) giving rise to different nucleon densities. **b** The geometrical explanation of the smooth density decay in the nuclear skin due to the pseudo-octahedral arrangement of nucleons in the fcc lattice model

the LDM, including oblate or prolate deformations of the drop, were proposed for the purpose of explaining the smooth decrement of nuclear density across the skin. In the framework of the fcc lattice model, this experimental evidence can be again obtained only based on geometric considerations. Figure 23.10b illustrates that, since the three-dimensional arrangement of nucleons in the fcc lattice model resembles an octahedron, there is a region on the surface where the nuclear density smoothly decreases from the value pertaining to the nuclear core to zero. Since the fcc lattice is by-default a non-spherical object (Figs. 23.7, 23.8 and 23.9), such density decrement is not as abrupt as in the LDM.

The lattice model requires the progressive placement of nucleons at specific lattice sites depending on the values of the nucleon quantum numbers. Adding sequentially more nucleons to an existing nucleus makes it larger, as the nucleons are usually placed within progressively larger shells and sub-shells (Figs. 23.7, 23.8 and 23.9). The overall size of each nucleus can be estimated simply via geometrical considerations, and it can be compared to the experimentally known values of nuclear size (Fig. 23.4a). This comparison shows that the nuclear size predicted by the fcc lattice model is in great agreement with experimental data, with errors generally below 0.2 fm [5].

In Fig. 23.3b, we observe that 4n-nuclei generally exhibit higher values of binding energy, suggesting the possible existence of clusters of alpha particles within the nucleus [11]. Remarkably, the fcc lattice model is also able to identify such internal clusters, simply due to the internal octahedral-tetrahedral arrangements of protons and neutrons within the lattice (Fig. 23.10). Figure 23.11 shows the lattice model of Ca^{40} and how the nucleons placed in the fcc lattice positions can be clustered together to form an overall aggregate of ten alpha particles, eight of them arranged in a larger octahedral configuration (the light blue spheres in Fig. 23.11b) and four of them into a smaller tetrahedron (the green spheres).

As shown in Eqs. (23.1) and (23.2), the LDM provides a simple formula for evaluating the binding energy of the nucleus, depending on volume, surface, Coulomb repulsion, symmetry, and pairing terms. Due to the analogies between LDM and lattice model, the latter turns out to reproduce all the terms in the binding energy

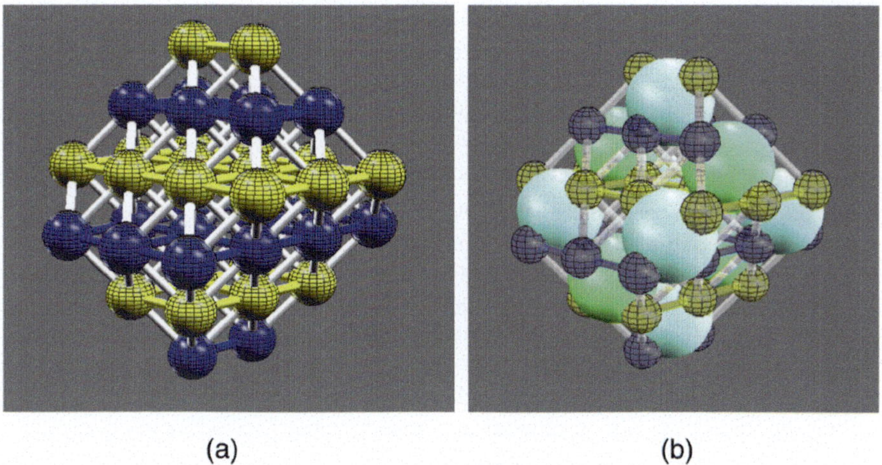

Fig. 23.11 Representation of the Ca^{40} nucleus as an fcc lattice (**a**) and highlighting the internal clustering of alpha particles resulting from the spatial arrangement of protons and neutrons within the nucleus (**b**). Nucleons are depicted as small yellow and blue particles, whereas the resulting clusters of alpha particles are drawn as big green and light-blue spheres

formula with simple geometrical considerations. The volume and surface terms are merely related to the bonds connecting nucleons within the nuclear core and on the surface, respectively. Due to the lattice properties, nucleons embedded within the core have a maximum of 12 neighboring nucleons, whereas surface nucleons can be bonded with a maximum of 9 neighbors. The energy contained in each of these bonds, either within the core or on the surface, represents the two first terms in Eqs. (23.1) and (23.2). As a first approximation, nucleon–nucleon bonds can be assigned an average binding energy per bond of about 2.8 MeV, irrespective of the spin/isospin properties of the nucleons [5]. This value can then be adjusted to provide a more accurate estimation of the total binding energy of the nucleus, to be compared to the values known experimentally. The third term of the binding energy formula, the Coulomb term, can be straightforwardly associated to the proton-proton repulsion forces, that are calculated based on the positions of protons in the lattice. By subtracting the Coulomb repulsion from the total bond-dependent binding energy, one finally obtains an estimate of the net binding energy of the nucleus. Due to the spin-isospin layering of the fcc lattice model, symmetry and pairing effects can also be traced directly from the model. The symmetry term biases the preference of small nuclei to have the same number of protons and neutrons ($Z = N$) and heavier nuclei to have an excess of neutrons ($N > Z$). This is explained based on the isospin layering of protons and neutrons in the lattice model: the addition of neutrons or protons in heavier nuclei leads to a different increment in the number of generated connections. In turn, this biases toward an excess of neutrons in heavy nuclei, whereas no bias is traced for the lighter nuclei that are less crowded. Finally, pairing effects are explained by adding unpaired nucleons to previously closed shells and sub-shells and by observing the difference in the number of bonds generated, which reflects the difference in the binding energy increment, when an even or odd number of nucleons is added [5].

The lattice model also provides a rather simple way to address the problem of nuclear fission. Following the suggestion by Winans, who proposed that nuclear fission might be the result of the splitting of the atomic nucleus along cleavage planes [26], the fission of the nucleus is simulated by following three basic steps: (i) build the three-dimensional lattice model of the mother nucleus following the build-up procedure based on the IPM quantum numbers; (ii) split the nucleus with planes associated to the fundamental lattice directions; (iii) analyze the obtained fragments and carry out the statistics of fragment abundances. The planes of fission are associated to the seven main crystallographic directions passing through the origin of the reference system. We can consider a total of 21 fission planes with the seven main planes plus the two planes running parallel to each of the main ones. Figure 23.12 shows an example of fission for the O^{16} nucleus, that is undergoing fission along four of these cleavage planes. The atomic number and the mass of the fragments are simply evaluated by counting the number of protons and neutrons on each side of the fission plane.

Two additional features are also employed for the simulation of nuclear fission in the Cook's lattice model. The first is related to the randomization of the position of the surface nucleons. This strategy assumes that the nuclear surface behaves like a liquid, whereas the core of the nucleus resembles an immutable solid [5]. The core nucleons are placed in the lattice sites according to the IPM build-up, whereas the nucleons in the surface-skin region occupy the lattice positions randomly. This generates a variety of slightly different structures for the same nucleus with given Z and N. By repeating this randomization many times and fracturing each structure, one can collect thousands/millions fragments and carry out a large statistical analysis of the fragment abundances.

The second additional feature is associated to assigning different probabilities to different fission events depending on the required fission energy [5]. Not all fission

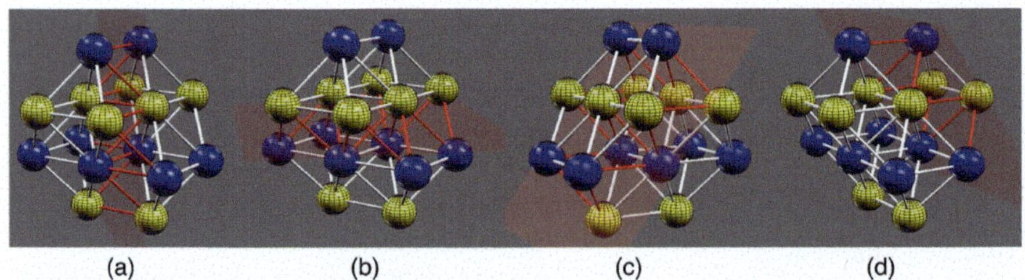

Fig. 23.12 Fission of O^{16} nucleus along four of the 21 main crystallographic planes. The fission fragments are evaluated by counting the number of protons and neutrons on each side of the plane. In this case, we obtain the following fragments: **a** Be^8 and Be^8; **b** He^8 and C^8; **c** Li^7 and B^9; **d** He^3 and C^{13}. Some of the obtained fragments are unstable and experience further decay. Broken bonds due to the fission phenomenon are drawn in red. Protons are in yellow, whereas neutrons are in blue

events have the same probability to occur. In the framework of the lattice model, this probability is estimated based on the fission energy values. These are computed based on two contributions: (i) the total Coulomb repulsion forces between protons belonging to the two fragments, and (ii) the energy required to break the bonds crossed by the fission plane. The former contribution, i.e., the Coulomb repulsion between the fragments favors fission and therefore increases the fission probability. Conversely, the latter, i.e., the bonds to be broken along a specific fission plane, hinders fission and thus it lowers its probability along that plane. Following this criterion, the probability of fission P_{fission} along a certain plane by definition is the inverse of the fission energy E_{f} [5]:

$$P_{\text{fission}} = \frac{1}{E_{\text{f}}} = \frac{1}{\beta n_{\text{bonds}} - C} \tag{23.6}$$

where n_{bonds} is the total number of broken bonds (calculated via geometrical considerations, see Fig. 23.12), C is the total Coulomb energy resulting from the repulsion between the positively charged fragments (calculated from all the proton-proton repulsion energies between the two fragments), and β is the average binding energy per bond (whose typical value is around 2.8 MeV). As can be seen from Eq. (23.6), the most probable fission events are those involving the lowest number of broken bonds and the largest Coulomb energy. This competition between bond breakage and Coulomb repulsion can also explain the observed asymmetry in fission fragments. Symmetric fission events usually exhibit high Coulomb repulsion between the symmetric fragments, but a very large number of bonds to be broken. Conversely, asymmetric fission involves much fewer bonds to be broken, but it exhibits a lower repulsion between the asymmetric fragments. Depending on the specific size and shape of the nucleus under investigation, asymmetric fission can become favored as a result of this energy competition.

We conclude this section by reporting how the Cook's lattice model can address those anomalous phenomena that are generally referred to as low-energy nuclear reactions (LENR). These reactions usually involve a few MeV of energy, giving rise to the occurrence of nuclear transmutations [21]. Based on this observation, Cook postulated that the lattice model can explain LENR insofar these can be seen as the result of fission events [5]. In this way, nuclear transmutations accompanying these LENR could in principle be verified by fracturing the lattice of the mother nucleus and by comparing the obtained fission fragments with the elements arising from the transmutations observed in the experiments. A typical experimental investigation concerning LENR is associated to the loading of Pd with deuterium. By inducing fission on all the isotopes of Pd lattices and carrying out the statistics of the obtained fission fragments, Cook found that the lattice model is indeed able to capture the astonishingly large abundance of Cr [5], which was observed in the experimental tests by Mizuno's group [27]. This was meant to provide both an additional validation of the lattice model and a confirmation of the fact that LENR might be associated to fission reactions of atomic nuclei.

23.3 The In-House Enhanced Version of the Cook's Lattice Model

As we have seen above, the Cook's fcc lattice model provides a simple way to visualize the three-dimensional structure of each atomic nucleus, with the nucleons placed at specific sites of the lattice depending on their quantum states. A dedicated software, the nuclear visualization software (*NVS*), was developed by Norman Cook in order to visualize atomic nuclei, calculate some of their fundamental nuclear properties (quantum states, radius, stability, etc.), and carry out fission analyses [5]. Unfortunately, the available version of the *NVS* was not provided with two vital features for the analysis of nuclear fission: (i) the randomization of the positions of external nucleons, and (ii) the stability analysis of post-fission fragments, which can undergo nuclear decay. To overcome these shortcomings and to be successively able to carry out comprehensive fission analyses of a variety of nuclei, we developed an in-house enhanced version of the lattice model. This code, entirely developed within the Matlab environment, allows to: (i) build the three-dimensional structure of an atomic nucleus with given Z and N, following the framework of the Cook's fcc lattice model; (ii) calculate the binding energy of the nucleus; (iii) randomize the positions of external nucleons, based on permutation calculations; and (iv) analyze the fission of the nucleus, with the subsequent probability-weighted statistics of the collected fragments. An additional code was developed to carry out the post-processing of the fission fragments, considering their stability and possible nuclear decay (see below).

Points (i) and (ii) above have been implemented by following the original Cook's formulation thoroughly described in [5]. The lattice sites are firstly evaluated based on Eqs. (23.5) relating the five quantum numbers (n, j, m, s, i) to the three Cartesian coordinates (x, y, z). After these available positions have been assessed, the default IPM build-up is implemented. This allows to fill the shells and sub-shells by adding progressively the nucleons in the proper positions in order to reach the total number of protons Z and neutrons$N$ for the given nucleus. Following Cook's implementation, this is done by filling progressively lattice positions for increasing n-values, decreasing j-values, and increasing m-values. This leads to obtain the pseudo-octahedral shapes with a slightly oblate form (Figs. 23.7, 23.8 and 23.9) [5]. Once all protons and neutrons are placed into the proper positions, the total Coulomb repulsion and binding energies are calculated. The former is simply evaluated by considering all the pairwise interactions between protons and summing up the individual proton-proton Coulomb repulsion energy. The latter is evaluated by counting how many nucleon-nucleon connections are included in the lattice. It is worth noting that the binding energy was evaluated by Cook in two different ways [5]. In the simplest form, it considers the number of bonds between first neighbor nucleons and assigns an average value of binding energy per bond (β) to each connection. In the second form, the binding energy is evaluated by counting how many neighbors each nucleon has (between 1 and 12) and assigning a different energy term for each of them. The procedure implemented into the *NVS* was to evaluate the binding energy via the second approach, by using energy coefficients for each number of neighbors

(c_1, c_2, ...,c_{12}). These coefficients have been set up by comparison with experimental data and are available in the *NVS* code [5]. This neighbor-based total binding energy (second approach) is then converted into a bond-based binding energy (first approach), thus allowing to estimate the value of β that best reproduces the binding energy of the overall nucleus. The advantage of this approach is that it does not require a user-based definition of the β value. This value is then simply obtained a posteriori, so that the binding energies evaluated with the two approaches (neighbor-based and bond-based) match. Finally, the net binding energy can be simply obtained as the difference between the total binding energy and the Coulomb energy. This net binding energy value gives an estimate of the stability of the nucleus.

By following the computational procedure explained above, the fcc lattice model can be built for every atomic nucleus, whose lattice structure can be visualized in the three-dimensional space. As an example, Fig. 23.13a shows the lattice model of Fe^{56} ($Z = 26$, $N = 30$) obtained via default IPM build-up. Protons are shown in yellow, whereas neutrons are in blue. Black lines represent nucleon-nucleon connections between first neighbors. The energy calculations provide a total proton-proton Coulomb repulsion energy of about 116 MeV, and a net total binding energy of about 460 MeV. The average binding energy per nucleon (*BE*/*A*) thus corresponds to about 8.21 MeV, which is only 6.5% lower than the value observed experimentally (8.79 MeV) [28].

As remarked above, this in-house enhanced version of the Cook's lattice model also includes the possibility to randomize the positions of external nucleons. This has been developed in a straightforward way, following the shell texture of the lattice model. A core of nucleons is kept fixed, whereas nucleons in the external surface (skin) are placed in different random sites of the fcc lattice. The fixed core can be

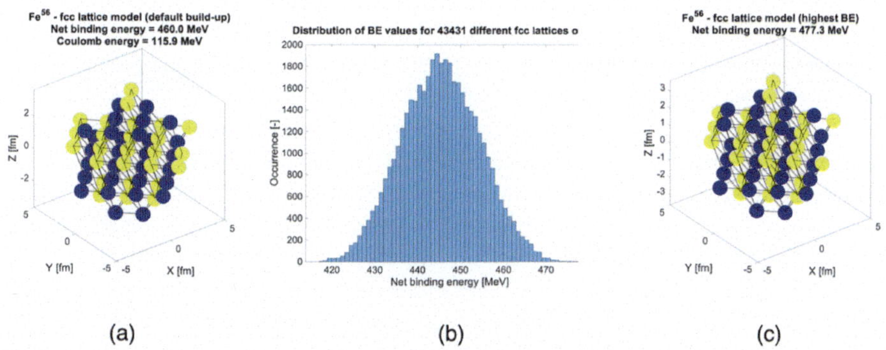

Fig. 23.13 **a** Lattice model of Fe^{56} ($Z = 26$, $N = 30$) built with the in-house code following the default IPM build-up. Yellow spheres represent protons, the blue ones are neutrons. Note the isospin layering and overall pseudo-octahedral shape of the nucleus. The size of the spheres is adjusted for visualization purposes and does not correspond to the actual nucleon dimension. **b–c** Randomization of external nucleons of Fe^{56}, with $n_{core} = 1$ and $n_{max} = 4$, for both protons and neutrons. **b** Distribution of the computed binding energy values over 43,431 randomized lattices. **c** The lattice structure corresponding to the maximum value of net binding energy ($BE = 477$ MeV)

simply identified by selecting a value of the principal quantum number *n,* which is assumed to be filled with nucleons (Fig. 23.7c). A different value of n_{core} can be chosen in principle for protons ($n_{\mathrm{core,P}}$) and neutrons ($n_{\mathrm{core,N}}$). As a result, all nucleons filling lattice sites with principal quantum numbers $n \le n_{\mathrm{core}}$ are retained in fixed positions. On the other hand, the positions of all the exceeding protons and neutrons are modified in order to occupy randomly all the lattice sites lying between the fully occupied core (n_{core}) and the largest shell that can be occupied by the external nucleons (n_{max}). The random occupation of the sites with principal quantum numbers $n_{\mathrm{core}} \le n \le n_{\mathrm{max}}$ is carried out via exclusive permutation techniques. Given the high number of available lattice sites, this approach leads to a high number of slightly different lattices for each nucleus. This number generally increases for larger nuclei, as well as for increasing values of n_{max} and decreasing values of n_{core}. Among these hundreds/thousands different lattice structures, only those with the same angular momentum of the default IPM build-up and with no unbonded nucleons are retained for further analyses.

Each of these hundreds/thousands generated lattice structures is associated to a certain value of total net binding energy, and therefore to a slightly different value of the average binding energy per nucleon (*BE*/*A*). Figure 23.13b shows the distribution of net binding energy values for the population of the generated Fe^{56} structures, with randomization parameters n_{core} ($n_{\mathrm{core,P}} = n_{\mathrm{core,N}}$) and n_{max} ($n_{\mathrm{max,P}} = n_{\mathrm{max,N}}$) equal to 1 and 4, respectively. These values of binding energies follow a quasi-Gaussian distribution, ranging from a minimum value of about 419 MeV up to a maximum value of about 477 MeV, with a peak around 445 MeV. The structures associated to the highest values of binding energy are supposed to represent stabler lattice configurations. As remarked above and shown in Fig. 23.13a, the default IPM build-up provides a value of Fe^{56} binding energy of about 460 MeV (about 8.21 MeV per nucleon). Figure 23.13c reports a slightly different arrangement of the nucleons in the Fe^{56} lattice, which corresponds to the maximum value of binding energy (477 MeV). This configuration leads to a value of 8.52 MeV per nucleon, which is only 3% lower than the 8.79 MeV found experimentally.

Once we have built all these thousands slightly different lattices for a certain atomic nucleus, the code allows to investigate the fission phenomenon simply by fracturing each structure along its 21 crystallographic planes (Fig. 23.12) and collecting the resulting fragments. For each structure, we then obtain $21 \times 2 = 42$ fragments, thus leading to a large variety of fragments for each atomic nucleus. As an example, fracturing the 43,431 lattices of Fe^{56} structures provides a total number of fission fragments equal to 1,824,102. In turn, each of these fragments is associated to a certain value of fission energy E_{f}, which is calculated depending on the number of fractured bonds and the value of Coulomb repulsion energy between the fragments, as from Eq. (23.6). Figure 23.14 shows the distribution of fission energies obtained by fracturing the 43,431 lattice structures of Fe^{56}. As can be seen, in this case these energies all exhibit positive values, implying that the fission of Fe^{56} requires an input energy to be provided. This graph also suggests that most of the fission events require an amount of energy between 20 and 80 MeV, with a peak value around 40 MeV. A small number of events is also found to lie within the 100–140 MeV energy range.

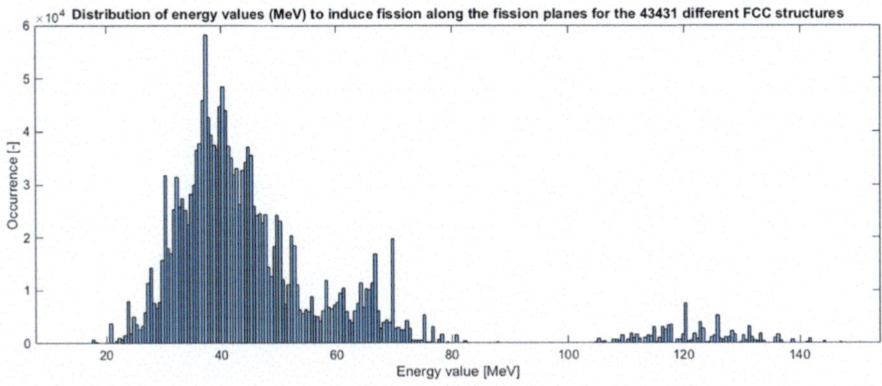

Fig. 23.14 Distribution of fission energies (in MeV) for Fe^{56}. Fission events with higher energies are associated to a lower probability of occurrence, according to Eq. (23.6)

Based on the obtained fission energies associated to each fission event, the fragments can then be collected, and their abundance can be weighted by considering the fission probability. As a result, we obtain the estimates of fragment abundances in terms of fragment atomic number Z_f and mass number A_f. From these abundance plots, we can finally get insights about the nature of the atomic elements generated from the fission of the mother nucleus and about the masses of the fission fragments. Note that, as described in the previous section, this approach implies a two-body fission phenomenon, with two fragments generated from each mother nucleus [5]. Therefore, multi-body fission phenomena cannot be directly retrieved from the present approach, unless further particle emissions arise from the nuclear decay of unstable fragments.

After fission of the mother nucleus, it is observed that not all the resulting fragments are stable. Depending on the number of protons Z_f and neutrons N_f of the fragment, one generally experiences three alternative situations: (i) the couple of values (Z_f, N_f) corresponds to a stable nucleus; (ii) the couple of values (Z_f, N_f) corresponds to an unstable nucleus, which can reach stability via a certain decay mode; (iii) the couple of values (Z_f, N_f) does not correspond to any known stable or unstable nucleus. In case (i), the stable nucleus corresponding to the couple of values (Z_f, N_f) is simply retained for further analyses and collected for the final abundance statistics. In case (ii), the unstable nucleus is allowed to undergo its specific decay mode. In this way, a new couple of values (Z_f', N_f') is found, which is again associated to one of the above cases. In case (iii), the unstable fragment is simply discarded from the subsequent statistical analysis. Due to the complex geometrical arrangement of the nucleons within the lattice, case (ii) is usually the most common, the fragment obtained directly from the fracture of the lattice often exhibiting a couple of values (Z_f, N_f) associated to an unstable nucleus. Therefore, several nuclear decay modes have been included into this in-house version of the lattice model, specifically: alpha decay ($Z_f' = Z_f - 2$, $N_f' = N_f - 2$), proton emission ($Z_f' = Z_f - 1$, $N_f' = N_f$), double proton emission ($Z_f' = Z_f - 2$, $N_f' = N_f$), neutron emission ($Z_f' = Z_f$, $N_f' = N_f - 1$), double neutron emission ($Z_f' = Z_f$, $N_f' = N_f - 2$), β^+ decay ($Z_f' = Z_f - 1$, $N_f' = N_f + 1$), β^- decay ($Z_f' = Z_f + 1$, $N_f' = N_f - 1$), and ε decay ($Z_f' = Z_f - 1$, $N_f' = N_f + 1$). For each couple of values (Z_f, N_f) of the unstable fragment, the corresponding decay mode is evaluated, and the numbers of protons and neutrons are adjusted accordingly for the decay mode to take place. The proper decay modes are evaluated from the NUBASE2020 database [28] and inserted into the code for the post-processing of the unstable fragments.

Besides containing the information about the stability or possible decay modes of each nucleus, the NUBASE2020 database also provides the half-life ($T_{1/2}$) of unstable nuclei [28]. This information is also exploited within this enhanced version of the lattice model: the decay of an unstable nucleus is not seen as a simplistic binary condition (decay or not decay), but it is assumed to be dependent on the characteristic time of the observation. In general terms, the process of nuclear decay can be described via an analytical equation of the form:

$$N(t) = N_0 \mathrm{e}^{-\lambda t} \tag{23.7}$$

where N_0 is the total number of nuclei at time $t = 0$, $N(t)$ is the number of (non-decayed) nuclei at time t, and λ is the decay constant, which is a measure of the decay speed and is measured in $\sec^{-1}$. The decay constant λ is directly related to the half-life ($T_{1/2}$), through the relationship $\lambda = \ln(2)/T_{1/2}$. The half-life ($T_{1/2}$) is defined as that value of t at which half the population of the initial nuclei has decayed, i.e., $N(T_{1/2}) = N_0/2$. If we are interested in the number of decayed nuclei at time t, we can simply take the complementary part of N(t) with respect to N_0, i.e., $N_0 - N(t)$, thus obtaining:

$$N_{\mathrm{dec}}(t) = N_0\left(1 - \mathrm{e}^{-\lambda t}\right) = N_0\left(1 - \mathrm{e}^{-\ln 2 \frac{t}{T_{1/2}}}\right) \tag{23.8}$$

If we are not focusing on a set of N_0 potentially decaying nuclei, but we are looking at a single unstable fragment, Eq. (23.8) can also be interpreted as the probability that a single nucleus undergoes decay as a function of time and its half-life. The decay probability $P_{\mathrm{dec}}(t)$ for the unstable nucleus can then be written as:

$$P_{\mathrm{dec}}(t) = \left(1 - \mathrm{e}^{-\ln 2 \frac{t}{T_{1/2}}}\right) \tag{23.9}$$

Figure 23.15 shows a graphical representation of Eq. (23.9). The probability of decay for an unstable nucleus is 0 at time $t = 0$, increases over time, and shows an asymptote toward $P_{\mathrm{dec}} = 1$ for $t \to \infty$. The half-life $T_{1/2}$ rules the rate of this decay. Shorter half-lives imply a faster decay and therefore a faster increment in P_{dec}. On the other hand, longer half-lives imply a higher stability of the nucleus and a slower increment in P_{dec}. Depending on the half-life $T_{1/2}$ of the unstable fragment, we then calculate the probability P^* that the unstable nucleus will undergo a specific decay mode after a certain amount of time T_{char}, according to Eq. (23.9). T_{char} represents the characteristic instant of time, at which we make the observation, hence it can

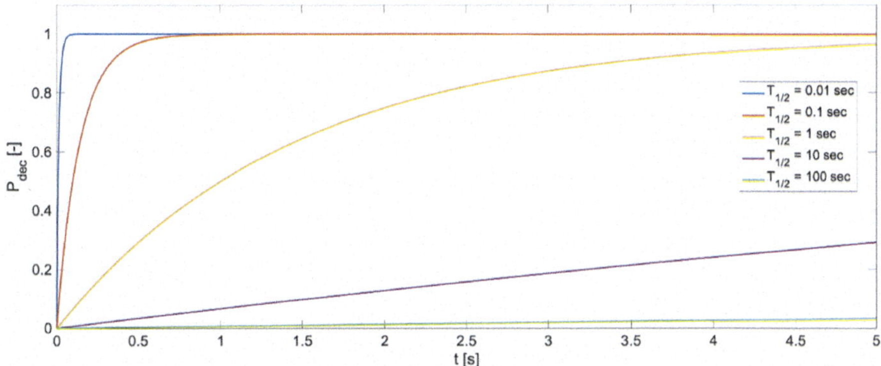

Fig. 23.15 P_{dec} (dimensionless) as a function of time t (in sec). The five lines refer to different values of the half-life $T_{1/2}$ (0.01, 0.1, 1, 10, and 100 s)

be thought of as the time interval, after which the fission takes place. The decay probability P^* is then computed at each of these times of observation T_{char} as:

$$P^* = \left(1 - \mathrm{e}^{-\ln 2 \frac{T_{\mathrm{char}}}{T_{1/2}}}\right) \tag{23.10}$$

The $P*$ value is eventually compared to a random number r extracted from a uniform distribution in the range (0,1). In this way, if P^* is found to be lower than the random number r, the probability of decay can be considered low enough, so that the nucleus does not undergo decay at T_{char}. Conversely, if P^* is found to be higher than r, the probability of decay is assumed to be high enough to trigger the nuclear decay, and the fragment undergoes nuclear transformation. As can be appreciated, this procedure has a certain degree of randomness in it, because of the random nature of number r, which needs to be lower than P^* for the decay to take place. On the other hand, this randomness is a consistent feature in the description of the phenomenon, being nuclear decay an intrinsically stochastic process.

23.4 Conclusions

In the present chapter, we have addressed the problem of modeling and analyzing the structure of the atomic nucleus via an enhanced version of the fcc lattice model originally developed by Norman Cook. After a brief survey of the main models of nuclear structure theory (IPM, LDM, alpha-particle model) and their interpretation of various nuclear phenomena, we have thoroughly reviewed the Cook's lattice model, its fundamental assumptions and properties, as well as its description of the main nuclear characteristics, such as nuclear shape, size, and binding energy. We have also seen how the phenomenon of nuclear fission can be straightforwardly addressed by this model simulating the breakage of the nucleus into two fragments along specific crystallographic planes. The details of a recently developed version of this model, entirely written within the Matlab environment, have then been reported, where the atomic nucleus can again be visualized as an fcc lattice in the three-dimensional space.

In this enhanced version, the model also supports the automatic randomization of the positions of external nucleons, which are based on combinatorial calculus and gives rise to hundreds/thousands slightly different lattices for each nucleus. The phenomenon of nuclear fission can be simulated by considering the energy-weighted statistics of the collected fragments, as well as the nuclear decay of unstable fragments.

In the next two chapters, the model presented here will be applied to the fission of various nuclei, which is the underlying cause for the chemical evolution in our planet with the ocean formation, as well as for the element transmutations observed during electrolysis experiments [29].

References

1. Cook ND, Hayashi T, Yoshida N (1999) Visualizing the atomic nucleus. IEEE Comput Graphics Appl 19(5):54–60
2. Mayer MG, Jensen JHD (1955) Elementary theory of the nuclear shell structure. Wiley, London
3. Das A, Ferbel T (2005) Introduction to nuclear and particle physics, 2nd edn. World Scientific Publishing, Singapore
4. Brink DM, Friedrich H, Weiguny A, Wong CW (1970) Investigation of the alpha-particle model for light nuclei. Phys Lett B 33(2):143–146
5. Cook ND (2006) Models of the atomic nucleus: unification through a lattice of nucleons, 2nd edn. Springer, Berlin/Heidelberg
6. Blin-Stoyle RJ (1959) The structure of the atomic nucleus. Contemp Phys 1(1):17–34
7. Nilsson SG, Tsang CF, Sobiczewski A, Szymanski Z, Wycech S, Gustaffson C, Lamm IL, Möller P, Nilsson B (1969) On the nuclear structure and stability of heavy and superheavy elements. Nucl Phys A 131:1–66
8. Weizsäcker CF (1935) Zur Theorie der Kernmassen. Z Phys 96(7–8):431–458
9. Yang F, Hamilton JH (1996) Modern atomic and nuclear physics. McGraw-Hill, New York
10. Myers WD (1977) Droplet model of atomic nuclei. Plenum, New York
11. Goldhammer P (1968) The structure of light nuclei. Rev Mod Phys 35(1):40–107
12. Bauer W, Kellogg WK (1988) Extraction of signals of a phase transition from nuclear multifragmentation. Phys Rev C 38(3):1297–1303
13. Campi X (1986) The percolation approach to nucleus break-up. J Phys 47:419–422
14. Pan J, Gupta DS (1995) Unified description for the nuclear equation of state and fragmentation in heavy-ion collisions. Phys Rev C 51:1384–1392
15. Weisskopf V (1951) Nuclear models. Science 113:101–102
16. Hofstadter R (1956) Electron scattering and nuclear structure. Rev Mod Phys 28(3):214–254
17. Kienzler B, Geckeis H (2018) Radioactive wastes and disposal options. EPJ Web Conf 189(October):00014. https://doi.org/10.1051/epjconf/201818900014
18. Bohr N, Wheeler JA (1939) The mechanism of nuclear fission. Phys Rev 56:426–450
19. Ragnarsson I, Nilsson SG (1995) Shapes and shells in nuclear structure. Cambridge University Press, Cambridge

20. Fleischmann M, Pons S (1989) Electrochemically induced nuclear fusion of deuterium. J Electroanal Chem Interfacial Electrochem 261(2):301–308
21. Storms EK (2007) The science of low-energy nuclear reactions. World Scientific Publishing, Singapore
22. Cook ND, Dallacasa V (1987) Face-centered-cubic solid-phase theory of the nucleus. Phys Rev C 35(5):1883–1890
23. Wigner E (1937) On the consequences of the symmetry of the nuclear Hamiltonian on the spectroscopy of nuclei. Phys Rev 51:106–119
24. Everling F (2008) Clusters in the nuclear dynamic model. In: First workshop on state of the art in nuclear cluster physics, Strasbourg, France
25. Lezuo K (1974) On the structure of the atomic nuclei. Atomkernenergie 23:285–290
26. Winans JG (1947) The nucleus as a crystalline solid. Phys Rev 72:435–436
27. Mizuno T, Kurokawa K, Akimoto T, Kitaichi M, Inoda K, Azumi K, Shimokawa S, Ohmori T, Enyo M (1996) Anomalous isotopic distribution of elements deposited on palladium induced by cathodic electrolysis. Denki Kagaku oyobi Kogyo Butsuri Kagaku 64:1160–1165
28. Kondev FG, Wang M, Huang WJ, Naimi S, Audi G (2021) The NUBASE2020 evaluation of nuclear physics properties. Chin Phys C 45(3):030001
29. Carpinteri A, Lacidogna G, Manuello A (eds) (2015) Acoustic, electromagnetic, neutron emissions from fracture and earthquakes, Springer, Heidelberg

Chapter 24
Fission Nuclear Reactions of Medium-Weight Chemical Elements: Implications to Geochemistry

Abstract The geochemical evolution of Earth's crust, ocean, and atmosphere is closely correlated to the occurrence of major tectonic events and due to low-energy nuclear reactions (LENR). By looking at the chemical evolution of the Earth's crust over the past 4.5 billion years, one can observe sharp decrements in certain heavier elements, such as Fe, Ni, Mg, Ca, and increments in other lighter elements, such as Al, Si, Na, K, C, and O. Notably, these chemical discontinuities in the Earth's composition have been found to occur during the periods of formation and of most intense activity of the tectonic plates. This suggests that LENR, in the form of phono-fission reactions, triggered by tectonic and seismic events, might explain the elemental changes at the planetary scale. Similar suggestions involving the fission of medium-weight atoms are proposed to explain the transmutations detected on the fracture surfaces of rock specimens after crushing failure in the laboratory. In this chapter, we make use of the enhanced version of Cook's lattice model, presented in Chap. 23, in order to investigate on the potential fission of the atomic nuclei involved in the geochemical evolution. The outcomes of these computational simulations provide new insights on the energy aspects behind the assumed fission reactions and on the expected fission fragments.

Keywords Earth's geochemical evolution · Rock compression failure · Low-energy nuclear reactions (LENR) · Phono-fission nuclear reactions · Lattice model · Medium-weight chemical elements

24.1 Preliminary Remarks

In the last decade, a certain number of experimental investigations at the laboratory scale have revealed that, when rock specimens fail catastrophically under compression loading, subatomic particles such as neutrons and alpha particles are emitted [1–6]. This particle emission is found to be correlated to transmutations in the chemical element composition on the fracture surfaces [3–7]. This experimental evidence suggests the occurrence of low-energy nuclear reactions (LENR) taking place at the

A. Carpinteri, *Terahertz Phonons and Nanomechanical Instabilities*,
https://doi.org/10.1007/978-3-032-14692-2_24

fracture surface, as a result of nano-scale fracture instabilities and consequent THz vibrations [8]. In addition, the observed chemical shifts are generally found to involve a decrement in the content of the chemical elements with higher atomic numbers and a corresponding increment in lighter ones [1, 7]. This supports the hypothesis that low-energy fission reactions could be the underlying phenomena explaining both particle emissions and chemical changes upon brittle rock failure.

Similar assumptions involving fission reactions are also put forward at the much larger scale of our planet. By analyzing the chemical evolution of the Earth's crust [9–12] and the abrupt chemical shifts in concomitance with the Great Oxidation Event [13, 14], it is a logical deduction that similar fission-based LENR could be the cause of these geochemical transformations [3, 4, 15–17]. In this case, since the most important chemical shifts in the Earth's crust are observed during the periods of formation and of most intense tectonic activity, strong seismic events are expected to trigger the fission of atomic nuclei, thus inducing the modification of the chemical composition of the Earth's crust [8]. Additional support to this hypothesis also comes from the detection of strong neutron emissions during, or right before, intense seismic activities, corroborating the idea of the occurrence of nuclear reactions triggered by earthquakes [1, 8, 18]. This LENR assumption was thus applied to explain the iron and calcium depletions in the Earth's crust [8], with the consequent formation of oceans [16] and primordial carbon "pollution" in the atmosphere [16, 19].

In the previous chapter, we described an enhanced version of the Cook's lattice model to address the problem of nuclear fission in general. Here, we present the outcomes arising from the application of this model to study the fission of the nuclei that are most important for the geochemical evolution. The most stable isotopes will be investigated for each of the following nuclei: Mg^{24}, Al^{27}, Si^{28}, Ca^{40}, and Fe^{56}. The nuclei are modeled as fcc lattices by using the default build-up proposed by Cook [20] and described in the previous chapter. The randomization of external nucleons is also carried out here, by setting up the values of n_{core} and n_{max} for both protons and neutrons. Based on the selected values of n_{core} and n_{max}, thousands of slightly different lattice structures are obtained for each nucleus. Table 24.1 reports the values of n_{core} and n_{max} adopted for the randomization of the external nucleons for all these nuclei, with the corresponding number of obtained randomized fcc structures.

After all these structures have been generated, each of them is fractured along its 21 main crystallographic planes in order to generate the fission fragments. The energy required to induce these fission events, and the consequent fission probability, are also computed as described previously, and these probability values are taken into account to carry out the statistics of the resulting fragment abundances. Finally, each of the obtained fragments is analyzed in terms of its potential nuclear decay. The decay is assessed at eight specific times of observation T_{char}, namely: 1 min, 1 h, 1 day, 1 month, 1 year, 1000 years, 1 million years, and 1 billion years, after the fission event. Note that the Cook's model has already been applied to study the fission of these nuclei in [21, 22]. However, these studies present two main limitations: (i) no randomization of the external nucleons was carried out, therefore the statistics of the fission fragments is limited to only one static structure and 42 fission fragments for each nucleus; and (ii) the decay of the unstable fragments was not considered at all.

Table 24.1 Nuclei analyzed via the enhanced version of the Cook's fcc lattice model, also considering the randomization of external nucleons. For each nucleus, the values of n_{core} and n_{max} with the randomization of external nucleons are reported. Note that the same values of n_{core} and n_{max} have been applied to both protons and neutrons, i.e., $n_{core,P} = n_{core,N}$ and $n_{max,P} = n_{max,N}$. The number of obtained randomized structures is reported for each nucleus

Nucleus	Z	N	n_{core}	n_{max}	Number of randomized fcc structures
Mg^{24}	12	12	0	3	5,261
Al^{27}	13	14	0	3	11,507
Si^{28}	14	14	0	3	17,385
Ca^{40}	20	20	1	4	20,625
Fe^{56}	26	30	1	4	43,431

These two points are both overcome here with the enhanced version of Cook's lattice model.

In the following sections, the results of fission of the investigated nuclei will be presented individually, and comparisons will be made against the available geochemical evidence.

24.2 Fission of Mg^{24}

Mg^{24} is the most abundant isotope of Mg in nature, with a relative abundance of 79.0% [23]. It comprises 12 protons and 12 neutrons, which, according to the shell model, fill the two first shells completely ($n = 0$, 1), with the remaining 4 protons and 4 neutrons partially filling the third shell ($n = 2$). The default build-up, whose three-dimensional representation is shown in Fig. 24.1a, provides a value of the net binding energy of the nucleus equal to 182 MeV, corresponding to an average value of binding energy per nucleon (*BE*/*A*) equal to 7.57 MeV. This value is 8.4% lower than the experimental value of 8.26 MeV [23]. Figure 24.1b shows the distribution of fission energy values, calculated over the 220,962 fission fragments (21 fission events for each of the 5,261 randomized Mg^{24} lattice structures). In this case, most of the fission events are found to require 20–40 MeV, which is in a general agreement with experiments and calculations carried out by previous authors [24, 25]. As can be appreciated, fission of Mg^{24} requires external energy to occur, with a most probable value of about 30 MeV. Notably, some fragments are also generated with a higher value of fission energy, e.g., in the range between 60 and 90 MeV. It is also interesting to note that just a few fission events are found to be exo-energetic, i.e., the required energy to induce fission along these crystallographic planes is negative (only a few MeV). However, these cases are associated to the generation of highly asymmetric fragments, where the number of fractured bonds is very low and the total repulsion energy between fragments is rather high (see the definition of fission energy E_f in the previous chapter).

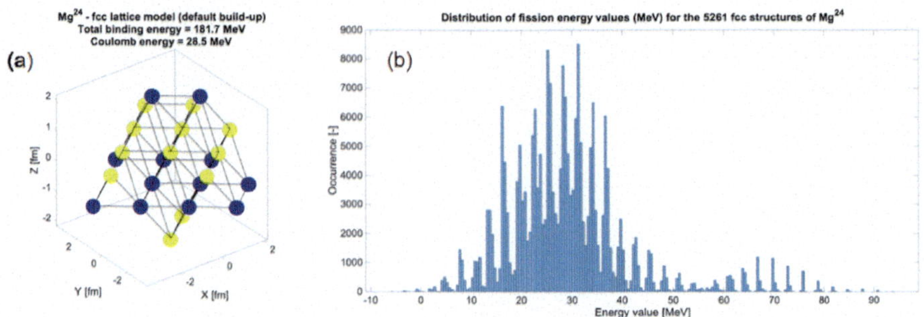

Fig. 24.1 **a** Fcc lattice model of Mg^{24} ($Z = 12, N = 12$) built with the in-house MatLab code following the default build-up. Yellow spheres represent protons, whereas blue ones are neutrons. The size of the spheres is adjusted for visualization purposes and does not correspond to the actual nucleon dimensions. **b** Distribution of fission energies (in MeV) for the 5,261 randomized structures of Mg^{24}

Figure 24.2 shows the distribution of the fission fragments obtained from the fission of the 5,261 randomized structures of Mg^{24}, in terms of relative occurrence of the atomic number Z_f of the fragment (left panel) and mass number A_f (right panel). The lines with different colors are associated to the different characteristic times at which the observation is carried out, T_{char}, spanning from 1 min to 1 billion years. Table 24.2 reports the six most abundant elements found from the statistics of the fission fragments, for each of the eight characteristic times T_{char}. Note that the distribution of the fragment atomic number Z_f is somewhat dependent on the time at which the observation takes place, whereas A_f is almost independent of this parameter. As a matter of fact, since most of the radioactive decay modes are found to involve mainly β^{+} and β^{-} decays, which imply the "transformation" of a proton into a neutron and vice versa, and thus do not induce a change in the total mass of the fragment, the distribution of fragment masses A_f is mostly independent of T_{char}. Obviously, this is not the case for the distribution of atomic numbers Z_f, where the decay modes play a major role in "transforming" the element at different times of observation.

From these distributions, we observe that the largest peak in the mass of the fragments is associated to $A_f = 4$. This information, together with the presence of a large peak at $Z_f = 2$ in the left panel, and a second large peak at $A_f = 20$ and $Z_f = 10$, suggests a high occurrence probability of the reaction $Mg^{24} \rightarrow Ne^{20} + He^{4}$ (Table 24.2). This means an asymmetric splitting of the Mg^{24} nucleus into a Ne^{20} nucleus and an alpha particle. This fission reaction was also reported in [24] to be the one requiring the least amount of energy among all possible fission modalities of Mg^{24}. The large peak at $Z_f = 6$ (Fig. 24.2, left panel) points in the direction of the symmetric fission reaction $Mg^{24} \rightarrow 2C^{12}$, which was also proposed by previous authors in the literature [1, 24, 25]. These findings suggest that the adopted fcc model is highly effective in assessing the most asymmetric ($Mg^{24} \rightarrow Ne^{20} + He^{4}$) and the symmetric ($Mg^{24} \rightarrow 2C^{12}$) fission modes of Mg^{24}. Note that the symmetric fission

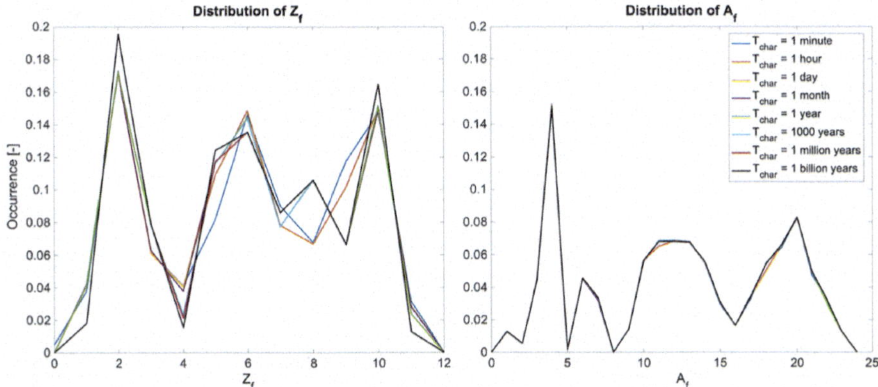

Fig. 24.2 Distribution of the collected fragments from the fission of Mg^{24} in terms of atomic number Z_f (left panel) and mass number A_f (right panel). The eight different curves refer to the different characteristic times, T_{char}

Table 24.2 Most abundant elements arising from the fission of Mg^{24} for different times of observation T_{char}. Values in parentheses represent the relative abundances expressed in percentage

T_{char}	1st	2nd	3rd	4th	5th	6th
1 min	He (17.3%)	Ne (14.7%)	C (14.6%)	F (11.8%)	N (9.0%)	B (8.2%)
1 h	He (17.3%)	Ne (15.0%)	C (14.9%)	B (10.9%)	F (10.2%)	N (7.9%)
1 day	He (17.1%)	Ne (14.8%)	C (14.4%)	B (11.6%)	O (10.6%)	N (7.7%)
1 month	He (17.2%)	Ne (14.9%)	C (14.4%)	B (11.6%)	O (10.6%)	N (7.7%)
1 year	He (17.2%)	Ne (15.2%)	C (14.4%)	B (11.6%)	O (10.6%)	N (7.7%)
1,000 years	He (19.5%)	Ne (16.4%)	C (14.4%)	B (11.6%)	O (10.6%)	Li (7.8%)
1 million years	He (19.5%)	Ne (16.4%)	C (13.5%)	B (11.7%)	O (10.6%)	N (8.6%)
1 billion years	He (19.5%)	Ne (16.4%)	C (13.5%)	B (12.4%)	O (10.6%)	N (8.6%)

reaction $Mg^{24} \rightarrow 2C^{12}$, which has frequently been found from the present calculations, was also proposed in [8, 15] to explain the evolution of the Earth's atmosphere and the high level of C in the primordial atmosphere. This fission reaction was even put into correlation to the increment in seismic activity that has occurred over the last centuries and the corresponding increment in atmospheric CO_2. In this context, the increment in CO_2 registered in correspondence to intense seismic activities was suggested to underlie the fission reaction $Mg^{24} \rightarrow 2C^{12}$, where the fracturing of Mg-rich calcareous rocks is expected to be the origin of the release of CO_2 into the atmosphere [15, 19]. Furthermore, the $Mg^{24} \rightarrow 2C^{12}$ reaction was suggested to occur at the laboratory scale, when crushing Carrara marble specimens under compression loading [6]. In this case, the fractured rock exhibited a decrement in Mg content, with a notable increment in C on the fracture surface. All this experimental evidence agrees well with the results from the present calculations, hence corroborating the large occurrence probability of the symmetric fission reaction $Mg^{24} \rightarrow 2C^{12}$. The experimental detection of alpha particles from the brittle failure of Carrara marble specimens [6] can find a justification in the asymmetric fission reaction $Mg^{24} \rightarrow Ne^{20} + He^4$ frequently found in the present computations.

Table 24.2 shows that also other elements, such as B ($Z_f = 5$), N ($Z_f = 7$), and O ($Z_f = 8$), can be retrieved from the fission simulation of Mg^{24}, albeit with lower occurrence probability values. The presence of these elements suggests both direct reactions, such as the $Mg^{24} \rightarrow B^{10} + N^{14}$ reaction previously reported in [24], or just the result of radioactive decays involving unstable forms of C and Ne. It must also be noted that the large occurrence of O from the fission of Mg was also postulated in [8, 15], where the authors proposed the fission reaction $Mg^{24} \rightarrow O^{16} + 4p + 4n$, as a reason for the Mg depletion at the Earth's crust level, with the consequent generation of oceanic water. The large occurrence of O from the fission of Mg^{24} lattice supports the conjecture that fission of alkaline-earth elements might in fact be behind the Great Oxidation Event (GOE), as suggested in [8].

24.3 Fission of Al^{27}

Al^{27} represents the only stable isotope of Al in nature, Al^{26} being present only in traces [23]. Its nucleus is made up of 13 protons and 14 neutrons. These fill the two first nuclear shells completely ($n = 0$, 1), with the remaining 5 protons and 6 neutrons partially filling the third shell ($n = 2$). The three-dimensional representation of the fcc lattice model of Al^{27}, obtained by following the default build-up, is shown in Fig. 24.3a. This provides a value of the net binding energy of about 201 MeV, corresponding to an average value of binding energy per nucleon equal to 7.44 MeV, which is 10.6% lower than the experimental value of 8.33 MeV [23]. Figure 24.3b shows the distribution of fission energy values from the fission of Al^{27}, calculated over the 483,294 fission fragments (21 fission events for each of the 11,507 randomized Al^{27} lattice structures). The fission energies required to split the nucleus of Al^{27} in two fragments are very similar to those of Mg^{24} (Fig. 24.1b), with most of the fission events occurring in the 20–40 MeV energy range. Also in this case, some fragments can be retrieved at much higher energies, between 60 and 90 MeV. However, differently from Mg^{24}, the fission of Al^{27} does not show any fission event with negative energy, suggesting that the fracture of Al^{27} always requires energy to be provided from the external environment. Hence, fission of Al^{27} is always an endo-energetic phenomenon. As we will see in the remainder of this chapter, this is the most common case for all nuclei investigated here.

Figure 24.4 shows the abundance of the fragments obtained from the fission of Al^{27} in terms of atomic and mass numbers, whereas Table 24.3 reports the six most abundant elements found at the eight different times of observation. As already for Mg^{24}, the distribution of fragment mass numbers A_f is almost independent of the time of observation, again suggesting that most of the nuclear decay modes imply "transformations" of protons into neutrons or vice versa, without any change in the mass number, i.e., without any subatomic particle emission. On the other hand, the

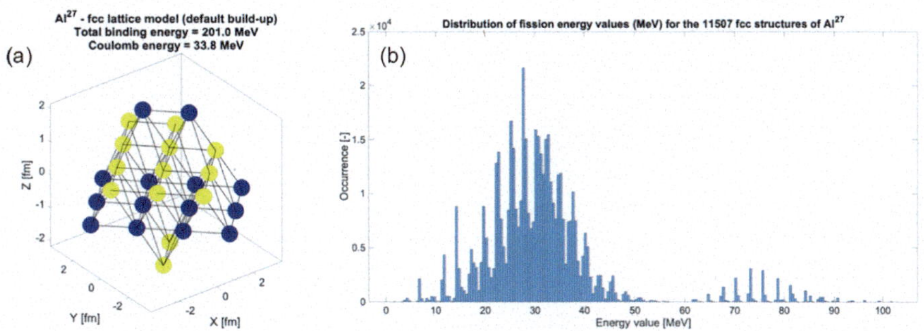

Fig. 24.3 **a** Fcc lattice model of Al^{27} ($Z = 13$, $N = 14$) built with the in-house MatLab code following the default build-up. Yellow spheres represent protons, whereas blue ones are neutrons. The size of the spheres is adjusted for visualization purposes and does not correspond to the actual nucleon dimensions. **b** Distribution of fission energies (in MeV) for the 11,507 randomized structures of Al^{27}

distribution of the fragment atomic numbers has a strong dependence on T_{char}. As we will see in the remainder of the chapter, this behavior is found for all investigated nuclei. By looking at Fig. 24.4 and Table 24.3, we see that the most common elements obtained from the fission of Al are C, N, Ne, Li, and He.

From the outcomes of the present calculations, C and N are mainly found to appear from the direct fission reactions $Al^{27} \rightarrow C^{13} + N^{14}$ and $Al^{27} \rightarrow C^{12} + N^{15}$, or as by-products of the radioactive decay of lighter and heavier elements, such as B and O. It has to be noted that the fission of Al involving the generation of C and N fragments was proposed in [15] as a possible explanation of the chemical evolution in the composition of the Earth's atmosphere. According to this hypothesis, LENR involving the fission of Al nuclei at the level of the Earth's crust, probably induced

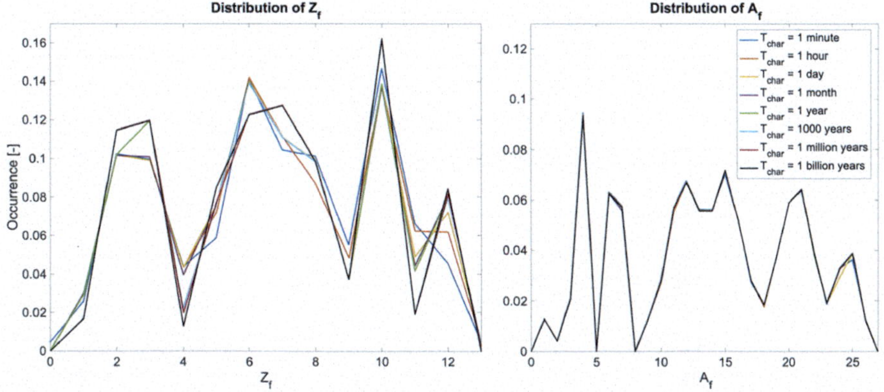

Fig. 24.4 Distribution of the collected fragments from the fission of Al^{27} in terms of atomic number Z_f (left panel) and mass number A_f (right panel). The eight different curves refer to the different characteristic times, T_{char}

Table 24.3 Most abundant elements arising from the fission of Al^{27} for different times of observation T_{char}. The relative abundances are expressed in percentage

T_{char}	1st	2nd	3rd	4th	5th	6th
1 min	Ne (14.7%)	C (14.1%)	N (10.4%)	He (10.3%)	O (10.1%)	Li (10.0%)
1 h	C (14.2%)	Ne (13.7%)	N (11.1%)	He (10.2%)	Li (9.9%)	O (8.7%)
1 day	C (14.0%)	Ne (13.7%)	N (11.1%)	He (10.2%)	Li (9.9%)	O (9.9%)
1 month	C (14.0%)	Ne (13.7%)	N (11.1%)	He (10.2%)	Li (10.1%)	O (9.9%)
1 year	C (14.0%)	Ne (13.9%)	Li (11.9%)	N (11.1%)	He (10.2%)	O (9.8%)
1,000 years	Ne (16.2%)	C (13.9%)	Li (12.0%)	He (11.5%)	N (11.1%)	O (9.8%)
1 million years	Ne (16.2%)	N (12.7%)	C (12.3%)	Li (12.0%)	He (11.5%)	O (9.8%)
1 billion years	Ne (16.2%)	N (12.7%)	C (12.3%)	Li (12.0%)	He (11.5%)	O (9.8%)

by intense tectonic activity, might be one of the main causes of the release of large amounts of C and N in gaseous form into the atmosphere [15].

On the other hand, the other abundant elements found from the results of the calculations based on the Cook's lattice model, i.e., Ne and Li, are observed from the direct fission reactions $Al^{27} \rightarrow Li^7 + Ne^{20}$ and $Al^{27} \rightarrow Li^6 + Ne^{21}$, or, again, as by-products of the decay of other unstable nuclei. It is also worthy of note to observe the relatively high abundance of He, which is mostly found to come from the splitting of Al into unstable He and Na nuclei, which then undergo nuclear decay. An example of one of these frequently observed reactions is $Al^{27} \rightarrow He^5 + Na^{22}$, with He^5 then decaying into an alpha particle by emitting one neutron and Na^{22} experiencing a β^+ decay into the more stable Ne^{22}.

We can then summarize the results of fission of the Al^{27} nuclei observing that the main fission mode involves a quasi-symmetric splitting of Al into C and N fragments, and a secondary asymmetric splitting into Li and Ne. Alpha particles are also found from the splitting of Al into He and Na. As can be seen from Table 24.3, the relative abundance of the fission fragments varies greatly depending on the time of observation, suggesting that most of the generating fragments are unstable and undergo a subsequent nuclear decay.

24.4 Fission of Si^{28}

Si^{28} is the most stable isotope of silicon in nature, with an abundance of 92.2%. The other isotopes, Si^{29} and Si^{30}, have an abundance of 4.7% and 3.1%, respectively [23]. The nucleus of Si^{28} is made up of 14 protons and 14 neutrons, thus filling the two first shells completely ($n = 0, 1$), with the remaining 6 protons and 6 neutrons partially filling the third shell ($n = 2$). The default build-up, whose three-dimensional representation is visualized in Fig. 24.5a, provides a value of the net binding energy equal to 203 MeV approximately, corresponding to an average value of BE/A of 7.25 MeV, which is 14.2% lower than the experimental value of 8.45 MeV [23].

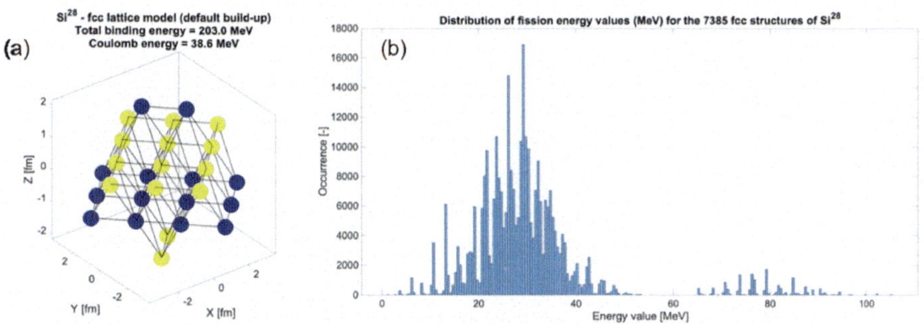

Fig. 24.5 **a** Fcc lattice model of Si^{28} ($Z = 14$, $N = 14$) built with the in-house MatLab code following the default build-up. Yellow spheres represent protons, whereas blue ones are neutrons. The size of the spheres is adjusted for visualization purposes and does not correspond to the actual nucleon dimensions. **b** Distribution of fission energies (in MeV) for the 17,385 randomized structures of Si^{28}

Figure 24.5b reports the distribution of fission energy values, calculated over the 730,170 fragments (21 fission events for each of the 17,385 randomized Si^{28} lattice structures). As in the previous cases, fission energies have a high probability in the range between 20 and 40 MeV, where most of the fission events are clustered. Again, some fission events are found in the range 60–90 MeV. Finally, like the case of Al^{27}, all fission events are found to be endo-energetic, presenting the calculated fission energy always as a positive value.

Figure 24.6 reports the abundances of the fission fragments from Si^{28} in terms of atomic number (left panel) and mass number (right panel). Like Tables 24.2 and 24.3, so table 24.4 provides the six most abundant elemental species found from the fragment statistics. Again, the mass number distribution is mostly independent of T_{char} and shows the largest peak in correspondence to $A_f = 4$ (alpha particles, or very light nuclei such as Li). Figure 24.6 and Table 24.4 suggest that the most abundant elements arising from the fission of Si are C, N, O, Ne, and He. C and O arise from the quasi-symmetric fission reaction $Si^{28} \rightarrow C^{12} + O^{16}$. This reaction is frequently found as one of the fissions of the fcc lattices of Si^{28}. This fission reaction was also suggested in [15] as a possible explanation to the composition of the primordial atmosphere due to Si depletion. Being C and O among the most abundant fragments for all times of observation (Table 24.4), the present calculations based on the Cook's lattice model provide a computational confirmation of the fission reaction previously proposed. It is also worthy to note that other direct fission reactions involving C and O are: $Si^{28} \rightarrow C^{11} + O^{17}$ and $Si^{28} \rightarrow C^{13} + O^{15}$. Both produce unstable fragments (C^{11} and O^{15}), which undergo β^{+} radioactive decay. The total occurrence of C and O for different times of observation is due to the nuclear (β^{+} and β^{-}) decay of other lighter or heavier unstable fragments.

The appearance of N is the easiest to explain, since it simply arises from the symmetric fission reaction $Si^{28} \rightarrow 2N^{14}$. This reaction is found to occur many times upon fracturing the lattice structure of Si^{28}, together with the slightly less-symmetric reaction $Si^{28} \rightarrow N^{13} + N^{15}$. However, in the latter case, the N^{13} nucleus is unstable

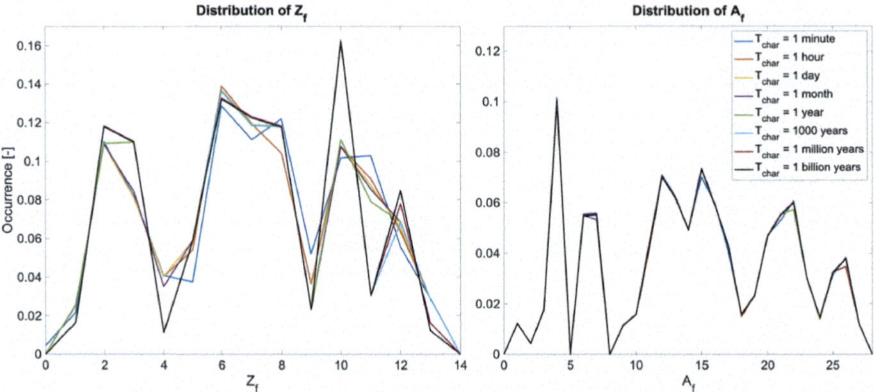

Fig. 24.6 Distribution of the collected fragments from the fission of Si^{28} in terms of atomic number Z_f (left panel) and mass number A_f (right panel). The eight different curves refer to the different characteristic times, T_{char}

Table 24.4 Most abundant elements arising from the fission of Si^{28} for different times of observation T_{char}. The relative abundances are expressed in percentage

T_{char}	1st	2nd	3rd	4th	5th	6th
1 min	C (12.9%)	O (12.2%)	N (11.1%)	He (11.0%)	Na (10.3%)	Ne (10.2%)
1 h	C (13.9%)	N (11.9%)	He (10.9%)	Ne (10.8%)	O (10.4%)	Na (9.1%)
1 day	C (13.6%)	N (11.9%)	O (11.8%)	He (10.9%)	Ne (10.7%)	Na (8.7%)
1 month	C (13.7%)	N (11.9%)	O (11.8%)	He (10.9%)	Ne (10.8%)	Na (8.5%)
1 year	C (13.7%)	N (11.9%)	O (11.8%)	Ne (11.1%)	Li (11.0%)	He (10.9%)
1,000 years	Ne (16.2%)	C (13.6%)	N (11.9%)	He (11.8%)	O (11.8%)	Li (11.0%)
1 million years	Ne (16.3%)	C (13.3%)	N (12.3%)	He (11.9%)	O (11.8%)	Li (11.0%)
1 billion years	Ne (16.2%)	C (13.2%)	N (12.3%)	He (11.8%)	O (11.8%)	Li (11.0%)

($T_{1/2} = 10$ min), and it is prone to undergo a β^+ nuclear decay into the more stable C^{13}. It is important to note that the symmetric fission reaction $Si^{28} \rightarrow 2N^{14}$ was suggested also in [15], again as an additional explanation to the formation of the primordial atmosphere upon Si reduction in the Earth's crust. This reaction, together with the above mentioned $Si^{28} \rightarrow C^{12} + O^{16}$, was proposed in order to explain the increment in gaseous elements (C, N, and O) in the terrestrial atmosphere, in concomitance with the decrement in heavier elements like Si at the level of Earth's crust. These reactions find confirmation by the results of the present calculations based on the fcc lattice model.

Finally, Ne and He are found to appear rather frequently among the population of the obtained fission fragments. They mainly arise from the highly asymmetric reaction $Si^{28} \rightarrow Ne^{20} + Be^8$. Be^8 is a highly unstable nucleus, having a very short half-life of about 82 attoseconds (82×10^{-18} s) and exhibiting a nuclear decay mode that involves alpha particle emission and Be^8 splitting into two alpha particles, i.e., $Be^8 \rightarrow He^4 + He^4$, in a very short time. As a consequence, the above fission reaction $Si^{28} \rightarrow Ne^{20} + Be^8$ practically becomes equivalent to $Si^{28} \rightarrow Ne^{20} + 2He^4$, which is observed to occur with high probability for all considered times of observation. This explains the large abundance of Ne and He for all times of observation, as well as the large peak at $A_f = 4$ observed in Fig. 24.6 (right panel), which suggests a large number of alpha particles.

24.5 Fission of Ca^{40}

Ca^{40} is the most stable isotope of calcium, with an abundance of 96.9%. The other isotopes Ca^{41}, Ca^{42}, Ca^{43}, Ca^{44}, Ca^{46}, and Ca^{48} are either present in traces or with abundances lower than 1%, except for Ca^{44}, whose abundance is 2.1% [23]. The nucleus of Ca^{40} is particularly stable, since it is made up of 20 protons and 20 neutrons, which fill completely the three first shells ($n = 0, 1, 2$), with no additional exceeding nucleons. This confers a particular stability to this nucleus. The three-dimensional fcc representation of Ca^{40} is reported in Fig. 24.7a, which provides a value of the net binding energy equal to 348 MeV. This corresponds to an average value of binding energy per nucleon equal to 8.50 MeV, which is only 0.6% lower than the experimental value of 8.55 MeV [23]. Note the high degree of compactness of the lattice structure of Ca^{40} (Fig. 24.7a), which is due to the full closure of the three first nuclear shells [20]. Figure 24.7b reports the distribution of fission energy values, calculated over the 866,250 fragments (21 fission events for each of the 20,625 randomized Ca^{40} lattice structures). Differently from the previous cases, the fission energy distribution does not manifest a double quasi-Gaussian distribution (see the large clustering at 20–40 MeV, and the smaller one at 60–90 MeV in the above figures for Mg^{24}, Al^{27}, and Si^{28}). In this case, fission events associated to Ca^{40} show a main clustering in the region between 35 and 45 MeV, with the distribution exhibiting a broad continuous tail for energies up to 85 MeV. Again, all fission events are found to be endo-energetic, being the values of the fission energies all positive. This implies that Ca^{40}, similarly to all nuclei investigated above (with some exceptions for a few fission events of Mg^{24}), does not suffer of spontaneous fission, and therefore nuclear fission can occur only if energy is provided in some way to the system.

Figure 24.8 shows the distribution of the fragments obtained from the fission of Ca^{40}, in terms of atomic number (left panel) and mass number (right panel). As usual, Table 24.5 reports the six most abundant elements found at different times of observation, after the occurrence of nuclear decay. Differently from the cases reported above (Mg^{24}, Al^{27}, and Si^{28}), the distribution of mass numbers A_f for the fission fragments of Ca^{40} is now getting rather symmetrical. As can be appreciated from the right panel of Fig. 24.8, despite a large peak centered at $A_f = 4$ (which suggests a large occurrence of alpha particles either due to the direct production of He nuclei or to the nuclear decay of other elements), the distribution of mass numbers A_f is almost symmetrical with respect to half the mass of Ca^{40}, i.e., with respect to

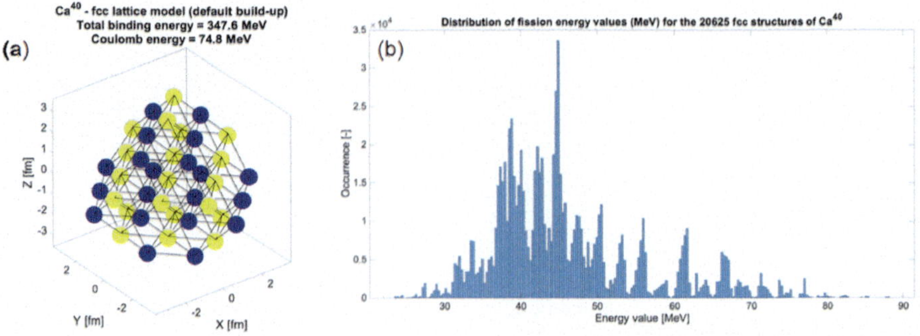

Fig. 24.7 **a** Fcc lattice model of Ca^{40} ($Z = 20$, $N = 20$) built with the in-house MatLab code following the default build-up. Yellow spheres represent protons, whereas blue ones are neutrons. The size of the spheres is adjusted for visualization purposes and does not correspond to the actual nucleon dimensions. **b** Distribution of fission energies (in MeV) for the 20,625 randomized structures of Ca^{40}

$A_f = 20$. Conversely, the left panel of Fig. 24.8 does not show any sign of symmetry, reflecting the asymmetry in Z_f generated by the decay of unstable fragments. These data show that the main elements arising from the fission of Ca are O, Mg, Na, S, and He.

As can be observed from Fig. 24.8 and Table 24.5, O and Mg are always found to be the most abundant fission product elements of Ca^{40} at all times of observation after the fission event. These mainly come from the occurrence of the fission reaction $Ca^{40} \rightarrow O^{16} + Mg^{24}$, which is observed several times as a result of the fracturing of the Ca^{40} fcc lattice. This reaction might provide an additional explanation to the Ca depletion at the level of the Earth's crust found in concomitance with the formation of the oceans. As explained in [8], the release of O nuclei from the fission of Ca

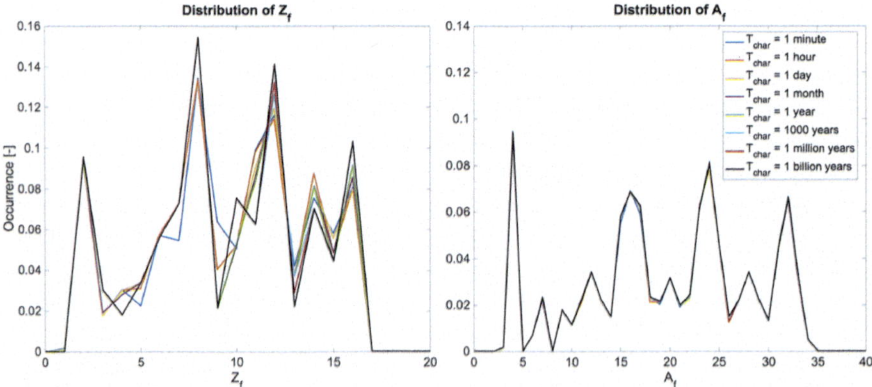

Fig. 24.8 Distribution of the collected fragments from the fission of Ca^{40} in terms of atomic number Z_f (left panel) and mass number A_f (right panel). The eight different curves refer to the different characteristic times, T_{char}

Table 24.5 Most abundant elements arising from the fission of Ca^{40} for different times of observation T_{char}. The relative abundances are expressed in percentage

T_{char}	1st	2nd	3rd	4th	5th	6th
1 min	O (13.4%)	Mg (11.6%)	Na (9.9%)	He (9.5%)	S (8.1%)	Si (7.6%)
1 h	O (13.4%)	Mg (11.5%)	Na (9.8%)	He (9.4%)	Si (8.8%)	S (8.0%)
1 day	O (15.4%)	Mg (12.0%)	He (9.4%)	Na (9.1%)	Si (8.1%)	S (8.0%)
1 month	O (15.5%)	Mg (12.7%)	He (9.4%)	S (8.6%)	Na (8.6%)	Si (8.1%)
1 year	O (15.4%)	Mg (12.6%)	He (9.4%)	S (9.2%)	Na (8.3%)	Si (8.1%)
1,000 years	O (15.4%)	Mg (12.6%)	S (10.3%)	He (9.5%)	Ne (7.5%)	N (7.3%)
1 million years	O (15.4%)	Mg (13.2%)	S (10.3%)	He (9.6%)	Ne (7.5%)	N (7.3%)
1 billion years	O (15.4%)	Mg (14.1%)	S (10.3%)	He (9.5%)	Ne (7.5%)	N (7.3%)

induced by strong tectonic activity might be one of the main causes behind the Great Oxidation Event and the formation of the oceans on our planet. The abundance of O and Mg as a result of the fission of the Ca^{40} nucleus is also found to occur as a consequence of two additional fission reactions, namely, $Ca^{40} \rightarrow O^{15} + Mg^{25}$ and $Ca^{40} \rightarrow O^{17} + Mg^{23}$. However, these reactions give rise to the fragments O^{15} and Mg^{23}, which are unstable (with half-lives in the order of seconds) and thus decay into N^{15} and Na^{23}, respectively. Moreover, the outcomes of the present computations reveal that O and Mg fragments can also result from fission reactions involving F, such as $Ca^{40} \rightarrow F^{17} + Na^{23}$, and Al, such as $Ca^{40} \rightarrow N^{14} + Al^{26}$ and $Ca^{40} \rightarrow N^{15} + Al^{25}$. These generate F and Al fragments that are unstable and undergo β^{+} decay into O and Mg, respectively, thus enriching the population of O and Mg fragments.

Another abundant element from the fission of Ca^{40}, especially for early times of observation, is Na. As we have seen above, this might arise from the reaction $Ca^{40} \rightarrow F^{17} + Na^{23}$, which generates the only stable isotope of Na in nature, i.e., Na^{23}. Other reactions involving Na are also frequently found to occur from the outcomes of the calculations, such as $Ca^{40} \rightarrow F^{18} + Na^{22}$, which generate unstable isotopes of Na, such as Na^{22}. This explains why Na is found to be among the most abundant elements for early times of observation, whereas it becomes less frequent afterwards.

Note that the large occurrence of oxygen fragments as a result of the fission of Ca nuclei can provide support for the hypothesis according to which Ca depletion at the level of the Earth's crust is strictly correlated to the formation of the oceans, as suggested in [8, 15].

Table 24.5 also shows that He is always present as one of the most abundant elements, almost independently of the time at which the observation takes place. The presence of He mainly arises as a result of the frequent highly asymmetric fission reaction $Ca^{40} \rightarrow He^{8} + Ar^{32}$, which generates the highly unstable He^{8} and Ar^{32} fragments. The former undergoes a series of nuclear decays, which can be summarized by the reaction chain $He^{8} \rightarrow Li^{8} + e^{-} \rightarrow Be^{8} + 2e^{-} \rightarrow 2He^{4} + 2e^{-}$ (the electron antineutrino has not been reported in the reactions for sake of simplicity). This chain of reactions starts from the highly unstable nucleus of He^{8}, which has a short half-life of about 120 ms and experiences a β^{-} decay into Li^{8}. Li^{8} is also highly unstable, with a half-life of about 850 ms, which then undergoes a subsequent β^- decay into Be^8. Finally, as we have already remarked above, Be^8 is susceptible of a very quick alpha particle emission, splitting almost immediately into two alpha particles. Given the high rapidity of these three subsequent decays, the abundance of stable He^4 nuclei from early times of observation is then straightforwardly justified (Table 24.5 and the peaks at $Z_f = 2$ and $A_f = 4$ in Fig. 24.8). As for Ar^{32}, this nucleus is also highly unstable, with a half-life of about 100 ms, and it undergoes a series of two β^+ decays in the form $Ar^{32} \rightarrow Cl^{32} + e^+ \rightarrow S^{32} + 2e^+$ (also in this case the electron neutrino has not been reported for sake of simplicity). This conversion between the unstable nucleus Ar^{32} into the stable S^{32} also provides an explanation to the large abundance of S, especially at longer times of observation (Table 24.5).

Finally, regarding the large occurrence of He fragments, it must be noted that alpha particle emission was experimentally detected as a result of the brittle failure of Ca-rich Carrara marble [6]. Based on this experimental evidence and given the large abundance of Ca in this rock, the authors assumed that fission reactions of Ca nuclei, triggered by nanomechanical instabilities and consequent THz phonons/plasmons, could be the reason for this observation of alpha particles. The results of the present calculations based on the simplified lattice model of Ca^{40} provide a confirmation that alpha particles are emitted from the fission of Ca.

24.6 Fission of Fe^{56}

The last element that we analyze in this chapter is iron. Fe^{56} is the most abundant isotope of iron in nature, with an abundance of 91.8%. The other isotopes Fe^{54}, Fe^{57}, and Fe^{58} are present with abundances lower than 3%, except for Fe^{54} whose abundance is 5.9% [23]. The nucleus of Fe^{56} is made up of 26 protons and 30 neutrons that fill completely the three first shells ($n = 0, 1, 2$), with the remaining 6 protons and 10 neutrons filling the upper shell level ($n = 3$). The spatial representation of the Fe^{56} fcc lattice is shown in Fig. 24.9a. As already reported in the previous chapter, this configuration leads to a total binding energy of the nucleus of about 460 MeV, with an average value of BE/A equal to 8.21 MeV, which is 6.5% lower than the experimental value [23]. Figure 24.9b was already commented in the previous chapter and shows the distribution of the fission energy values for Fe^{56}, calculated over the 1,824,102 fission fragments (21 fission events for each of the 43,431 randomized lattice structures). As already reported above, this energy distribution shows a major clustering in the region between 30 and 50 MeV, and a minor clustering at lower probabilities in the range between 100 and 140 MeV. Again, all fission events are endo-energetic, i.e., Fe^{56} requires energy from the external environment to undergo fission.

Figure 24.10 displays the abundance values of the fragments obtained from the fission of the Fe^{56} lattice, whereas Table 24.6 shows the six most frequent fragments. From these data, we can observe that the most abundant elements arising from fission of Fe are Mg, S, Ca, Si, O, P, and Na. The abundance of Mg and Si is mainly found to

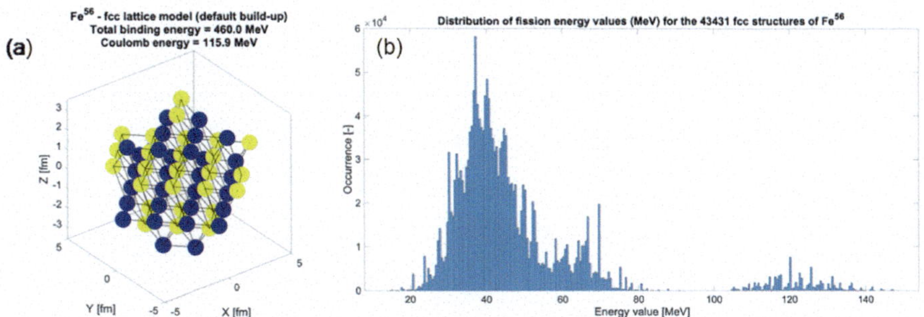

Fig. 24.9 **a** Fcc lattice model of Fe^{56} ($Z = 26$, $N = 30$) built with the in-house MatLab code following the default build-up. Yellow spheres represent protons, whereas blue ones are neutrons. The size of the spheres is adjusted for visualization purposes and does not correspond to the actual nucleon dimensions. **b** Distribution of fission energies (in MeV) for the 43,431 randomized structures of Fe^{56}

occur due to the fission reactions $Fe^{56} \rightarrow Mg^{24} + Si^{32}$, $Fe^{56} \rightarrow Mg^{25} + Si^{31}$, and $Fe^{56} \rightarrow Mg^{26} + Si^{30}$. All these three reactions produce three stable isotopes of Mg (Mg^{24}, Mg^{25}, and Mg^{26}), whereas they trigger the formation of nuclei of Si with different stability. The first one gives rise to Si^{32}, which has a rather long half-life (around 157 years), but it is prone to β^- decay. When this β^- decay takes place, Si^{32} becomes P^{32}, which is also unstable (having a half-life of about two weeks) and decays into the stable S^{32}. Therefore, for early times of observation we can consider the fission reaction $Fe^{56} \rightarrow Mg^{24} + Si^{32}$ to generate two stable fragments, whereas for longer times of observation this reaction can be rewritten as $Fe^{56} \rightarrow Mg^{24} + S^{32} + 2e^-$. The second fission reaction involving Mg and Si, i.e., $Fe^{56} \rightarrow Mg^{25} + Si^{31}$, gives rise to Si^{31}, which is more unstable than Si^{32}, having a half-life of about two hours and a half, and decays into the stable P^{31}. Therefore, for times of observation longer than a few hours, the fission reaction $Fe^{56} \rightarrow Mg^{25} + Si^{31}$ is equivalent to $Fe^{56} \rightarrow Mg^{24} + P^{31} + e^-$. Finally, the third fission reaction considered above, i.e., $Fe^{56} \rightarrow Mg^{26} + Si^{30}$, gives rise to stable isotopes for both Mg and Si, hence no radioactive decay takes place in this case. It has to be noted that the additional fission reactions $Fe^{56} \rightarrow Mg^{27} + Si^{29}$ and $Fe^{56} \rightarrow Mg^{28} + Si^{28}$ are also observed a few times from the results of the present simulation. Although both Si nuclei in these reactions are stable, Mg^{27} has a half-life of about 10 min and is prone to experience a β^- decay into the stable Al^{27}, whereas Mg^{28} has a half-life of about 21 h and experiences a double β^- decay into the unstable Al^{28} first, and then into the stable Si^{28}. As a result, for times of observation of a few hours or a few days, these fission reactions might be written as $Fe^{56} \rightarrow Al^{27} + Si^{29} + e^-$ and $Fe^{56} \rightarrow 2Si^{28} + 2e^-$, where the former is very important for its geochemical implications.

Finally, it should be noted that Mg and Si are also found to occur as by-products of the nuclear decay of other unstable fragments, such as unstable isotopes of Na and Al. Mg^{24} and Mg^{25} are often found from the decay of the unstable Na^{24} and Na^{25}, whereas Si^{28} and Si^{29} are also obtained from the unstable Al^{28} and Al^{29} nuclei. It is important to observe that the occurrence of Mg and Si as a result of the potential

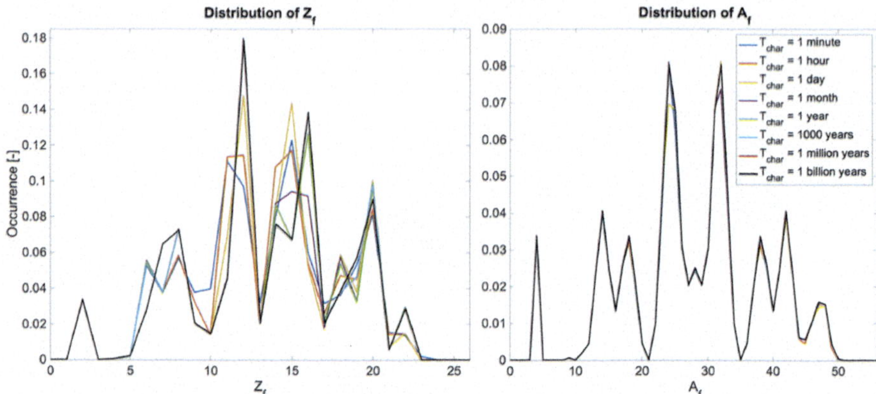

Fig. 24.10 Distribution of the collected fragments from the fission of Fe^{56} in terms of atomic number Z_f (left panel) and mass number A_f (right panel). The eight different curves refer to the different characteristic times, T_{char}

Table 24.6 Most abundant elements arising from the fission of Fe^{56} for different times of observation T_{char}. The relative abundances are expressed in percentage

T_{char}	1st	2nd	3rd	4th	5th	6th
1 min	P (12.3%)	Na (11.1%)	Mg (9.7%)	Ca (8.1%)	Si (8.0%)	S (6.1%)
1 h	P (11.7%)	Mg (11.5%)	Na (11.4%)	Si (10.8%)	Ca (8.5%)	O (5.9%)
1 day	Mg (14.8%)	P (14.4%)	Ca (10.1%)	Si (8.5%)	O (7.3%)	Na (6.9%)
1 month	Mg (18.0%)	Ca (9.7%)	P (9.4%)	S (9.2%)	Si (8.8%)	O (7.3%)
1 year	Mg (17.8%)	S (12.7%)	Ca (9.6%)	Si (8.7%)	O (7.2%)	P (6.7%)
1,000 years	Mg (17.8%)	S (13.8%)	Ca (9.9%)	Si (7.6%)	O (7.2%)	P (6.7%)
1 million years	Mg (17.8%)	S (13.8%)	Ca (9.0%)	Si (7.6%)	O (7.2%)	P (6.7%)
1 billion years	Mg (17.8%)	S (13.8%)	Ca (9.0%)	Si (7.6%)	O (7.2%)	P (6.7%)

fission of Fe was frequently proposed in the preceding parts of the book, in relation to a variety of geochemical processes. For example, it was proposed to explain the observed chemical changes after the brittle compression failures of Luserna stones [1, 4, 5], as well as the iron depletion at the level of the Earth's crust with the formation of the lighter Si-Al elements and sialic rocks [8, 15–18]. As confirmed by Fig. 24.10 and Table 24.6, Mg and Si elements are found to appear very frequently in the simulation from the fission of the Fe^{56} lattice, both at temporal scales associated to laboratory experiments and subsequent analyses (days, weeks, and months), as well as at longer temporal scales associated to the chemical evolution at the planetary scale (million and billion years). The present results can provide a theoretical explanation to the possibility that the Fe nucleus splits into Mg and Si fragments, if enough energy is provided (Fig. 24.9b).

As briefly mentioned above, Fig. 24.10 and Table 24.6 show a relatively high abundance of S. The presence of this element can usually be explained in two ways: either with the formation of a stable isotope from the direct fission of Fe^{56}, or as a by-product of nuclear decays of unstable fragments. In the former case, eight fission reactions directly generating S have been observed from the outcomes of the simulation, namely: $Fe^{56} \rightarrow S^{31} + Ne^{25}$, $Fe^{56} \rightarrow S^{32} + Ne^{24}$, $Fe^{56} \rightarrow S^{33} + Ne^{23}$, $Fe^{56} \rightarrow S^{34} + Ne^{22}$, $Fe^{56} \rightarrow S^{36} + Ne^{20}$, $Fe^{56} \rightarrow S^{37} + Ne^{19}$, and $Fe^{56} \rightarrow S^{38} + Ne^{18}$. Among these reactions, four are found to generate stable isotopes of S, i.e., S^{32}, S^{33}, S^{34}, and S^{36}. On the other hand, S^{31} has a half-life of 2.5 s and is prone to experience a β^+ decay into P^{31}, whereas S^{37} decays into Cl^{37}, and S^{38} experiences a double β^- decay into the stable Ar^{38}. Note that, in the above reactions, a variety of Ne isotopes is also generated. Two of them, i.e., Ne^{20} and Ne^{22}, are stable, whereas all the others are prone to undergo β decay into elements with lower (O and F) or higher (Na and Mg) atomic number.

Another element that is always found among the most abundant fragments from the fission of Fe^{56} is Ca. The direct fission reactions giving rise to Ca are found to involve a splitting of the Fe nucleus into Ca and C. Most of the generated Ca fragments, such as Ca^{40}, Ca^{41}, Ca^{42}, Ca^{43}, Ca^{44}, and Ca^{48}, are stable, whereas others, such as Ca^{39} and Ca^{47}, are prone to undergo a β^+ decay into K^{39} or a double β^- decay into Ti^{47}. Note that the corresponding C nuclei, which are generated as a result of the Fe $\rightarrow$ Ca + C reactions, correspond to both stable isotopes, e.g., in the reactions $Fe^{56} \rightarrow Ca^{44} + C^{12}$ and $Fe^{56} \rightarrow Ca^{43} + C^{13}$, and unstable ones, e.g., $Fe^{56} \rightarrow Ca^{40} + C^{16}$ and $Fe^{56} \rightarrow Ca^{48} + C^{8}$. In [8, 16, 17], the authors suggested the fission reactions of Fe into Ca and C as a possible explanation to the geochemical evolution of the Earth's crust, which was characterized by intense iron depletion correlated to the observed increment in atmospheric carbon. Although C is not found to be among the most abundant fragments from the results of the simulations (Table 24.6), the large occurrence of Ca seems to provide a theoretical explanation to the Fe splitting into Ca and C fragments, if enough energy is provided. Moreover, although not exceptionally high, the peak corresponding to C ($Z_f = 6$) is in fact present in the left panel of Fig. 24.10a, thus suggesting that C can be generated, albeit not commonly, as a result of the fission of the Fe^{56} lattice.

Figure 24.10 and Table 24.6 show that there is a relatively high abundance of P and O. As we have already mentioned, P can be obtained as a by-product of the β^+ decay of unstable S^{31} fragments. On the other hand, P is also found to occur as a result of four direct fission reactions, namely: $Fe^{56} \rightarrow P^{30} + Na^{26}$, $Fe^{56} \rightarrow P^{31} + Na^{25}$, $Fe^{56} \rightarrow P^{32} + Na^{24}$, and $Fe^{56} \rightarrow P^{33} + Na^{23}$. The only stable isotope of P in these reactions is P^{31}, whereas P^{30} ($T_{1/2} = 2.5$ min) generally undergoes a β^+ decay into Si^{30}, P^{32} ($T_{1/2} = 2$ weeks) a β^- decay into S^{32}, and P^{33} ($T_{1/2} = 25$ days) a β^- decay into S^{33}. P is also found to occur as a by-product of the β^+ decay of the unstable S^{31}.

As regards O, the results of the simulation suggest the occurrence of the direct fission reactions: $Fe^{56} \rightarrow O^{14} + Ar^{42}$, $Fe^{56} \rightarrow O^{15} + Ar^{41}$, $Fe^{56} \rightarrow O^{16} + Ar^{40}$, $Fe^{56} \rightarrow O^{17} + Ar^{39}$, $Fe^{56} \rightarrow O^{18} + Ar^{38}$, and $Fe^{56} \rightarrow O^{19} + Ar^{37}$. The stable isotopes of O in these reactions are O^{16}, O^{17}, and O^{18}, whereas the unstable O^{14} ($T_{1/2} =$

70 s), O^{15} ($T_{1/2} = 122$ s), and O^{19} ($T_{1/2} = 26$ s) undergo β decays into N^{14}, N^{15}, and F^{19}, respectively. Note that some of the corresponding isotopes of Ar generated from these Fe $\rightarrow$ O + Ar fission reactions are stable, such as Ar^{38} and Ar^{40}, whereas the others, i.e., Ar^{37}, Ar^{39}, Ar^{41}, and Ar^{42}, undergo nuclear decay into more stable nuclei.

24.7 Conclusions

In the present chapter, we have described the outcomes of fission of medium-weight elements, such as Mg^{24}, Al^{27}, Si^{28}, Ca^{40}, and Fe^{56}, according to the enhanced version of the Cook's lattice model. The results of the present computations are described in terms of fission energies and fragments, also considering the potential nuclear decay of unstable fragments. These outcomes reveal that the lighter nuclei, i.e., Mg^{24}, Al^{27}, and Si^{28}, give rise to large occurrences of fission products such as C, N, and O, which are very abundant in the Earth's atmosphere, and other particularly stable elements such as Ne and He. Fission of these light nuclei is always found to be endo-energetic (with just a few exo-energetic events for the fission of Mg^{24}), thus implying that energy must be provided from the external environment to the nucleus in order for the fission to take place. This information agrees well with some of the experimental conclusions of previous studies [8, 16, 19]. According to the latter, the intense tectonic activity in the Earth's crust provided the required energy—through the activation of nanomechanical instabilities and consequent THz phonons and/or plasmons—to induce the fission of these Mg, Al, and Si nuclei. The generation of C, N, and O fission products was then put forward to explain the chemical evolution of the terrestrial atmosphere [8, 16].

On the other hand, the computational simulation of fission of the heavier nuclei reveals a large abundance of Mg, O, and He from the splitting of Ca^{40}, and a large occurrence of Mg, Si, and Ca fragments from fission of Fe^{56}. Both these outcomes provide ample justification to the geochemical evidence reported and summarized previously [1]. In particular, the predominant occurrence of O from fission of Ca nuclei agrees well with the Ca depletion in the Earth's crust and the correlated Great Oxidation Event and ocean formation [8], whereas the abundance of He nuclei provides a theoretical explanation to the detection of alpha particles during compression failure of Ca-rich Carrara marble [6]. As regards fission of iron, the observation of Mg and Si within the population of the most frequent fragments agrees well with previous experimental observations of the decrement in iron in Luserna stone and other iron-rich rocks upon brittle failure and consequent increment in Mg and Si elements on the fracture surface [2, 7], as well as with the suggestion that Fe depletion in the Earth's crust is triggered by intense seismic events [2, 7, 8]. Again, looking at the distribution of fission energies involved in fracture of the Ca^{40} and Fe^{56} lattices, we find only endo-energetic fissions. In these cases, the fission-required energy is provided by the phonons and/or plasmons originated from nanomechanics instabilities. These fracture phenomena occur at the laboratory scale for rock specimens, such as Carrara marble and Luserna stone, or they can take place at the much larger tectonic scale in the case of seismic events.

Regarding the last analysis involving fission of iron, it is important to note the low abundance of Al fragments for all times of observation (Fig. 24.10), Al being always absent from the population of the most frequent fission fragments (Table 24.6). In the context of the geochemical evidence discussed above, Al has been observed to play a rather important role both at the scale of laboratory and at the planetary scale. As a matter of fact, high Al contents have been often detected on the fracture surface of iron-rich rock specimens [2, 4, 5, 7, 8]. On the other hand, Al increment in the Earth's crust was suggested to result from the Fe depletion induced by an intense seismic activity [8, 15, 16, 18]. In both these very different contexts, the fission reaction $Fe^{56} \rightarrow 2Al^{27} + 2n$ emerged as a logical proposal [1]. Unfortunately, as can be seen from the absence of high peaks for $Z_f = 13$ in the left panel of Fig. 24.10, Al is only rarely found to be a result of the direct fission of Fe^{56}. This has a simple explanation in the context of the present computational model. The only stable isotope of Al is Al^{27}. This stable fragment can be obtained from the present simulation either as a result of the direct fission reaction $Fe^{56} \rightarrow Al^{27} + Al^{29}$, in which the unstable Al^{29} ($T_{1/2} = 6.5$ min) decays into Si^{29}, or as a by-product of the radioactive decay of Mg^{27} (as reported above). Both these reaction pathways, allowing for a large presence of Al^{27} within the population of the fission fragments, are detected only a few times. On the other hand, another fission reaction is found to be very frequent from the outcomes of the simulation, namely, $Fe^{56} \rightarrow 2Al^{28}$. Although this reaction does involve the splitting of the Fe nucleus into two symmetric Al fragments, Al^{28} is unstable, with a short half-life of only 135 s, and hence it is prone to a β^- decay into the stable Si^{28}. This is the reason why, although found to some extent within the population, Al is not one of the most important outcomes of the Fe fission (Table 24.6). This cannot provide a sufficient theoretical corroboration to experimental evidence of Al increment, both in the laboratory and in the Earth's crust [1].

The Cook's view of nuclear fission relies on fracture of lattice of the mother nucleus along one of its crystallographic planes. As we have thoroughly described in the previous chapter, this implies a two-body fission phenomenon, where each mother nucleus splits into two fragments. The additional emitted particles that are normally observed in fission events, such as neutrons and alpha particles, can only arise here as a result of the decay of primary unstable fragments. However, for the medium-weight nuclei investigated in this chapter, the generated fragments lie within a portion of the nuclear table where the most common nuclear decays are the β^- andβ^+. It is now easy to understand how certain multi-body fission reactions, which might involve neutron and alpha particle emissions, such as the $Fe^{56} \rightarrow 2Al^{27} + 2n$ reaction, are simply not retrievable by the present model.

References

1. Carpinteri A, Lacidogna G, Manuello A (eds) (2015) Acoustic, electromagnetic, neutron emissions from fracture and earthquakes. Springer
2. Carpinteri A, Lacidogna G, Manuello A, Borla O (2012) Piezonuclear fission reactions in rocks: evidences from microchemical analysis, neutron emission, and geological transformation. Rock Mech Rock Eng 45(4):445–459
3. Carpinteri A, Lacidogna G, Borla O, Manuello A, Niccolini G (2012) Electromagnetic and neutron emissions from brittle rocks failure: experimental evidence and geological implications. Sadhana 7:59–78
4. Carpinteri A, Lacidogna G, Manuello A, Borla O (2013) Piezonuclear fission reactions from earthquakes and brittle rocks failure: evidence of neutron emission and non-radioactive product elements. Exp Mech 53(3):345–365
5. Manuello A, Sandrone R, Guastella S, Borla O, Lacidogna G, Carpinteri A (2015) Neutron emissions and compositional changes at the compression failure of iron-rich natural rocks. In: Carpinteri A, Lacidogna G, Manuello A (eds) Acoustic, electromagnetic, neutron emissions from fracture and earthquakes. Springer International Publishing, Switzerland, pp 23–37
6. Carpinteri A, Lacidogna G, Borla O (2015) Alpha particle emissions from Carrara marble specimens crushed in compression and X-ray photoelectron spectroscopy of correlated nuclear transmutations. In: Carpinteri A, Lacidogna G, Manuello A (eds) Acoustic, electromagnetic, neutron emissions from fracture and earthquakes. Springer International Publishing, Switzerland, pp 57–71
7. Carpinteri A, Chiodoni A, Manuello A, Sandrone R (2011) Compositional and microchemical evidence of piezonuclear fission reactions in rock specimens subjected to compression tests. Strain 47:282–292
8. Carpinteri A, Borla O (2018) Nano-scale fracture phenomena and TeraHertz pressure waves as the fundamental reasons for geochemical evolution. Strength Fracture Complexity 11(2–3):149–168
9. Taylor SR, Mclennan SM (1981) The composition and evolution of the continental crust: rare Earth element evidence from sedimentary rocks. Philos Trans R Soc London Ser A Math Phys Sci 301:381–399
10. Taylor SR, McLennan SM (1995) The geochemical evolution of the continental crust. Rev Geophys 33(2):241–265
11. Yaroshevsky AA (2006) Abundances of chemical elements in the Earth's crust. Geochem Int 44(1):48–55
12. Hawkesworth CJ, Kemp AIS (2006) Evolution of the continental crust. Nat Rev 443:811–817
13. Konhauser KO, Pecoits E, Lalonde SI, Papineau D, Nisbet EG, Barley ME, Arndt NT, Zahnle K, Kamber BS (2009) Oceanic nickel depletion and a methanogen famine before the great oxidation event. Nature 458(7239):750–753
14. Saito MA (2009) Less nickel for more oxygen. Nature 458:714–715
15. Carpinteri A, Manuello A (2011) Geomechanical and geochemical evidence of piezonuclear fission reactions in the Earth's crust. Strain 47:67–81
16. Carpinteri A, Manuello A (2015) Evolution and fate of chemical elements in the Earth's crust, ocean, and atmosphere. In: Carpinteri A, Lacidogna G, Manuello A (eds) Acoustic, electromagnetic, neutron emissions from fracture and earthquakes. Springer International Publishing, Switzerland, pp 163–181
17. Carpinteri A, Manuello A, Negri L (2015) Chemical evolution in the Earth's mantle and its explanation based on piezonuclear fission reactions. In: Carpinteri A, Lacidogna G, Manuello A (eds) Acoustic, electromagnetic, neutron emissions from fracture and earthquakes. Springer International Publishing, Switzerland, pp 183–196
18. Borla O, Lacidogna G, Carpinteri A (2015) Piezonuclear neutron emissions from earthquakes and volcanic eruptions. In: Carpinteri A, Lacidogna G, Manuello A (eds) Acoustic, electromagnetic, neutron emissions from fracture and earthquakes. Springer International Publishing, Switzerland, pp 135–151

19. Carpinteri A, Niccolini G (2019) Correlation between the fluctuations in worldwide seismicity and in atmospheric carbon pollution. Sci 1(1):17
20. Cook ND (2006) Models of the atomic nucleus: unification through a lattice of nucleons. Springer, Berlin, Heidelberg
21. Cook ND, Manuello A, Veneziano D, Carpinteri A (2015) Piezonuclear fission reactions simulated by the lattice model of the atomic nucleus. In: Carpinteri A, Lacidogna G, Manuello A (eds) Acoustic, electromagnetic, neutron emissions from fracture and earthquakes. Springer International Publishing, Switzerland, pp 219–235
22. Carpinteri A, Manuello A, Veneziano D, Cook ND (2015) Piezonuclear fission reactions simulated by the lattice model. J Condens Matter Nucl Sci 15:149–161
23. Kondev FG, Wang M, Huang WJ, Naimi S, Audi G (2021) The NUBASE2020 evaluation of nuclear physics properties. Chin Phys C 45(3):030001
24. Litherland AE (1972) The electrofission of magnesium. Revista Brasileira de Física 2:101–118
25. Sandorfi AM, Calarco JR, Rand RE, Schwettman HA (1980) Fission modes of 24Mg. Phys Rev Lett 45(20):1615–1618

Chapter 25
Fission Nuclear Reactions of Nickel and Palladium: Implications to Electrochemistry

Abstract Experimental investigations performed in the last four decades have shown that low-energy nuclear reactions (LENR) take place during electrolysis experiments. During these experiments, anomalous excess heat, emission of subatomic particles, such as neutrons and alpha particles, as well as appreciable variations in the chemical and isotopic compositions at the electrode surfaces were repeatably detected. Based on this experimental evidence and observing that micro- to macrocracks usually form at the electrode surfaces, it is assumed that LENR occur as fission reactions of the atomic nuclei lying on the electrode surfaces. Similarly to what done in the previous chapter, here we make use of the enhanced version of the Cook's lattice model to study the fission of two chemical elements often employed as electrodes in electrochemical experiments, Ni and Pd. The results of the computations provide useful information on the expected fission-based transmutations of these elements and are commented in the light of experimental electrolysis data.

Keywords Electrolysis · Low-energy nuclear reactions (LENR) · Phono-fission nuclear reactions · Lattice model · Nickel · Palladium

25.1 Preliminary Remarks

In 1989, Fleischmann and Pons reported for the first time the occurrence of anomalous nuclear reactions in electrolysis experiments during the loading of deuterium into Pd electrodes [1]. The anomalous excess heat, the magnitude of which was too high to be explained solely in terms of chemical reactions, made the two scientists assume that low-energy nuclear reactions (LENR) were taking place [1, 2]. In the following decades, several research groups tried to reproduce these electrochemical experiments looking for a solid scientific explanation behind these puzzling phenomena. At the end of the twentieth century, the Mizuno's group was very active in this field of investigation [3–7]. Although the outcomes of the experiments were not always showing large excess heat and subatomic particle emissions, nevertheless

A. Carpinteri, *Terahertz Phonons and Nanomechanical Instabilities*,
https://doi.org/10.1007/978-3-032-14692-2_25

these events were detected several times [2, 5, 8]. Element transmutations on the electrode surfaces were frequently observed at the end of the experiments with notable deviations from the natural chemical composition and isotopic distribution [3, 4, 6, 7]. This copious experimental evidence suggested that LENR take place during electrolysis in the form of phono-fission reactions [8, 9]. Additional experiments in the following years corroborated this hypothesis, where it was argued that hydrogen embrittlement and nanomechanics instabilities are the reasons behind the activation of these fission reactions [10–13]. Electrolysis tests were carried out by using a variety of electrodes, such as Ni–Fe and Co–Cr, as well as Ni and Pd. These experiments gave rise to a repeatable detection of subatomic particle emissions (neutrons and alpha particles), largely exceeding the natural background level. Significant transmutations were repeatedly detected on the electrode surfaces. The manifestation of lighter elements at the end of the test, which were totally absent on the electrode surface before the experiment, demonstrate the occurrence of phono-fission nuclear reactions [12].

As we have already seen in the previous chapter, the Cook's face-centered-cubic (fcc) lattice model offers a straightforward way to investigate nuclear fission on the basis of the structural properties of the nucleus [8]. Our goal is here to study the fission of Ni and Pd nuclei via the enhanced version of the Cook's lattice model, in order to get more insights into the expected fission fragments and the energies required for the activation of the fission events. The outcomes of these simulations will be directly compared to the available experimental data.

We will investigate on the stable isotope Ni^{58}, which shows the highest natural abundance. On the other hand, Pd does not have a prevailing isotope in the natural composition. For this reason, all six stable isotopes, i.e., Pd^{102}, Pd^{104}, Pd^{105}, Pd^{106}, Pd^{108}, and Pd^{110}, will be considered, and the results will be weighted considering the percentages associated to the natural abundances. Similarly to what done in the previous chapter, these nuclei are modeled as fcc lattices by using the default build-up of the independent particle model (IPM), as suggested by Cook [8]. The randomization of external nucleons is also carried out, by setting up the values of n_{core} and n_{max} (see previous chapters). Table 25.1 reports the values of n_{core} and n_{max} adopted for the randomization of the external nucleons, in the case of Ni and Pd isotopes, with the corresponding number of obtained randomized fcc structures. In the following sections, the results of the computational simulations for fission of Ni and Pd are presented and compared to experimental electrochemical data.

25.2 Fission of Ni^{58}

Ni^{58} is the most abundant isotope of nickel, with a prevailing natural abundance of 68.1% [14]. The nucleus of Ni^{58} is made up of 28 protons and 30 neutrons, which totally fill the three first nuclear shells ($n = 0, 1, 2$), with the remaining 8 protons and 10 neutrons filling the upper shell level ($n = 3$). The default built-up, whose three-dimensional representation is shown in Fig. 25.1a, provides a value of net

Table 25.1 Randomization of external nucleons in the lattice models of Ni^{58} and Pd isotopes. For each nucleus, the value of n_{core} and n_{max} for the randomization of external nucleons is reported. Note that the same values of n_{core} and n_{max} are applied to both protons and neutrons, i.e., $n_{core,P} = n_{core,N}$ and $n_{max,P} = n_{max,N}$. The number of obtained randomized structures is reported for each nucleus

Nucleus	Z	N	n_{core}	n_{max}	Number of randomized fcc structures
Ni^{58}	28	30	1	4	58,014
Pd^{102}	46	56	2	5	183,629
Pd^{104}	46	58	2	5	187,860
Pd^{105}	46	59	2	5	188,611
Pd^{106}	46	60	2	5	188,911
Pd^{108}	46	62	2	5	190,279
Pd^{110}	46	64	2	5	191,030

binding energy approximately equal to 469 MeV, corresponding to an average value of binding energy per nucleon (BE/A) equal to 8.09 MeV. This value is 7.4% lower than the experimental value of 8.73 MeV [14]. Figure 25.1b reports the distribution of the fission energy values, calculated over 2,436,588 fission fragments (21 fission events for each of the 58,014 randomized Ni^{58} lattice structures). This distribution is very similar to that of Fe^{56} (see the previous chapter), with a major clustering of fission energies in the 30–50 MeV energy spectrum, and a minor clustering in the high-energy range between 100 and 140 MeV. Again, all fission events are found to be endo-energetic.

This suggests that, similarly to most of the nuclei investigated in the previous chapter, energy must be provided from the external environment in order to activate fission of Ni. Nanomechanical instabilities and resonance phenomena induced by THz phonons are suggested to be the cause of phono-fissions [13, 15].

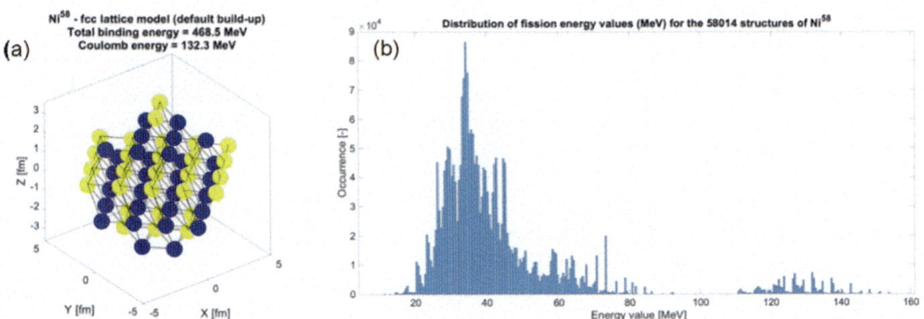

Fig. 25.1 **a** Fcc lattice model of Ni^{58} ($Z = 28$, $N = 30$) built with the in-house MatLab code following the default build-up. Yellow spheres represent protons, whereas the blue ones are neutrons. The size of the spheres is adjusted for visualization purposes and does not correspond to the actual nucleon dimensions. **b** Distribution of fission energies (in MeV) for the 58,014 randomized structures of Ni^{58}

Figure 25.2 reports the abundance values of the fission fragments obtained from Ni^{58}, whereas Table 25.2 shows the six most frequent fragments depending on the values of T_{char}. The most abundant chemical elements obtained upon fracturing the fcc lattice of Ni are S, Mg, Ca, O, P, Si, and Al. S and Mg, which represent the most abundant elements for all times of observation, are found to arise from the four fission reactions $Ni^{58} \rightarrow Mg^{24} + S^{34}$, $Ni^{58} \rightarrow Mg^{25} + S^{33}$, $Ni^{58} \rightarrow Mg^{26} + S^{32}$, and $Ni^{58} \rightarrow Mg^{27} + S^{31}$. The first three reactions involve three stable isotopes of magnesium (Mg^{24}, Mg^{25}, and Mg^{26}) and of sulfur (S^{34}, S^{33}, and S^{32}). The fourth one, i.e., $Ni^{58} \rightarrow Mg^{27} + S^{31}$, gives rise to unstable fragments of both elements. The former experiences β^- decay (Mg^{27} has a half-life $T_{1/2}$ of approximately 10 min, then decaying into Al^{27}), whereas the latter undergoes a β^+ decay (S^{31} has a half-life $T_{1/2}$ of 2.5 s, then decaying into P^{31}). Therefore, considering times of observation greater than a few minutes, this fourth fission reaction is equivalent to $Ni^{58} \rightarrow Al^{27} + P^{31}$. Note that the observed abundance of S and Mg is not only due to the direct reactions $Ni \rightarrow Mg + S$, but also to the β decays of unstable isotopes of other elements, such as Na, Al, P, and Cl, obtained from additional fission events.

Similarly to the occurrence of S and Mg, that of Ca and O is due to the asymmetric fission reactions $Ni \rightarrow Ca + O$. From the results of the simulations, this is found out to occur due to the six reactions $Ni^{58} \rightarrow Ca^{39} + O^{19}$, $Ni^{58} \rightarrow Ca^{40} + O^{18}$, $Ni^{58} \rightarrow Ca^{41} + O^{17}$, $Ni^{58} \rightarrow Ca^{42} + O^{16}$, $Ni^{58} \rightarrow Ca^{43} + O^{15}$, and $Ni^{58} \rightarrow Ca^{44} + O^{14}$. Among these reactions, $Ni^{58} \rightarrow Ca^{39} + O^{19}$ gives rise to unstable fragments that undergo β^+ (Ca^{39}—$T_{1/2} = 860$ ms—decays into K^{39}) and β^- (O^{19}—$T_{1/2} = 26.5$ s—decays into F^{19}) decays, respectively. $Ni^{58} \rightarrow Ca^{40} + O^{18}$ produces stable forms of Ca and O. $Ni^{58} \rightarrow Ca^{41} + O^{17}$ produces a stable isotope of O and a slightly unstable isotope of Ca, which undergoes an electron-capture decay mode with $T_{1/2} = 99{,}400$ years. $Ni^{58} \rightarrow Ca^{42} + O^{16}$ produces stable forms of both nuclei. $Ni^{58} \rightarrow Ca^{43} + O^{15}$ produces a stable form of Ca, but an unstable isotope of O, which undergoes β^+ decay into N^{15}

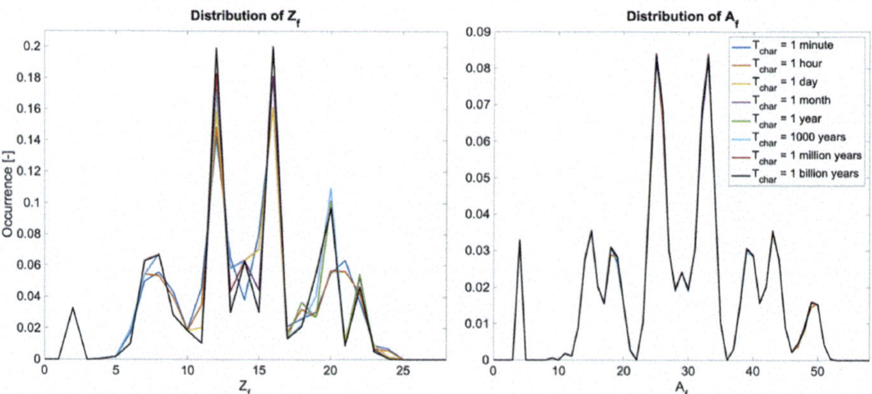

Fig. 25.2 Distribution of the collected fragments from the fission of Ni^{58} in terms of atomic number Z_f (left panel) and mass number A_f (right panel). The eight different curves refer to the different characteristic times, T_{char}

Table 25.2 Most abundant elements arising from the fission of Ni^{58} for different times of observation T_{char}. Values in parentheses represent the relative abundances expressed in percentage

T_{char}	1st	2nd	3rd	4th	5th	6th
1 min	S (16.1%)	Mg (14.2%)	P (8.0%)	Al (6.5%)	Sc (6.3%)	O (5.6%)
1 h	S (16.1%)	Mg (14.8%)	P (7.0%)	Si (6.3%)	Al (5.8%)	Ca (5.7%)
1 day	S (16.1%)	Mg (15.9%)	Ca (10.0%)	P (7.0%)	O (6.7%)	Si (6.3%)
1 month	S (18.1%)	Mg (17.3%)	Ca (10.1%)	O (6.7%)	Si (6.3%)	Al (5.8%)
1 year	S (19.9%)	Mg (17.2%)	Ca (10.1%)	O (6.7%)	Si (6.3%)	Al (5.8%)
1,000 years	S (20.0%)	Mg (17.2%)	Ca (11.0%)	O (6.7%)	Si (6.3%)	Al (5.8%)
1 million years	S (20.0%)	Mg (18.3%)	Ca (9.7%)	O (6.7%)	N (6.3%)	Si (6.3%)
1 billion years	S (19.9%)	Mg (19.9%)	Ca (9.7%)	O (6.7%)	N (6.3%)	Si (6.3%)

with $T_{1/2} = 122$ s. Finally, $Ni^{58} \rightarrow Ca^{44} + O^{14}$ produces a stable isotope of Ca and an unstable nucleus of O, which undergoes β^{+} decay into N^{14} with $T_{1/2} = 70$ s. Again, the total abundance of Ca and O is not only retrieved by the direct fission reaction Ni $\rightarrow$ Ca + O, but is also associated to the decay of other unstable isotopes of fission fragments such as N, F, K, and Sc. Note that the relatively high occurrence of O as a resultant of fission from the Ni nucleus can be associated to the LENR hypothesis behind the Great Oxidation Event (GOE) [15]. As a matter of fact, the decrement in Ni observed around 3.8 billion years ago in the Earth's crust might underly a fission of the Ni nucleus into lighter elements. The simultaneous increment in O associated to the GOE could therefore be explained also on the basis of the fission of Ni^{58}, whereas the large generation of O nuclei should be due to the fission of iron.

Finally, the relatively high abundance of P, Si, and Al can be explained via the direct fission reactions Ni $\rightarrow$ P + Al and Ni $\rightarrow$ Si + Si, or as the result of nuclear decay of unstable isotopes close in the periodic chart. We have already seen above how the reaction $Ni^{58} \rightarrow Mg^{27} + S^{31}$ can turn out to be $Ni^{58} \rightarrow Al^{27} + P^{31}$ if decay modes are taken into account. As for the direct fission reactions generating P and Al, five of them are observed from the results of the simulations, namely: $Ni^{58} \rightarrow P^{29} + Al^{29}$, $Ni^{58} \rightarrow P^{30} + Al^{28}$, $Ni^{58} \rightarrow P^{31} + Al^{27}$, $Ni^{58} \rightarrow P^{32} + Al^{26}$, and $Ni^{58} \rightarrow P^{33} + Al^{25}$. Note that only the reaction $Ni^{58} \rightarrow P^{31} + Al^{27}$ generates stable isotopes of P and Al, whereas all the others give rise to unstable fragments of P, which decay into Si or S, and of Al, which decay into Mg or Si. After the decays are properly taken into account, the reactions Ni $\rightarrow$ P + Al contribute to increase the abundance of the two most occurring elements, i.e., S and Mg. On the other hand, as for the fission reactions giving rise to Si, these are either $Ni^{58} \rightarrow Si^{28} + Si^{30}$ or $Ni^{58} \rightarrow 2Si^{29}$. Both give rise to stable isotopes of Si, so that no nuclear decay is expected to take place.

The fission of Ni into the lighter Si, Mg, and O nuclei was assumed by Carpinteri et al. based on their experimental observations during electrolysis experiments with Ni and Ni–Fe electrodes [10–13]. In those experiments, high levels of Si, Mg, and O were recorded on the surface of the Ni electrode. As we have seen above, Si, Mg, and O fragments are observed from the results of the computations (Fig. 25.2 and Table 25.2). During the experimental tests, the authors observed large amounts of alpha-particle emissions [10–13]. Also this aspect is captured by the present calculations, as can be confirmed by the presence of a high peak in the abundance graphs of Fig. 25.2 in correspondence to $Z_f = 2$ and $A_f = 4$. On the other hand, strong neutron emissions were detected during the electrolysis experiments [10, 11], which cannot be predicted by the present model.

25.3 Fission of Pd (Stable Isotopes and Natural Mix)

Pd does not have a prevailing stable isotope in nature. Six isotopes of Pd, i.e., Pd^{102}, Pd^{104}, Pd^{105}, Pd^{106}, Pd^{108}, and Pd^{110}, are all stable, with abundances of 1.0%, 11.1%, 22.3%, 27.3%, 26.5%, and 11.7%, respectively [14]. The nuclei of these isotopes are made of 46 protons and 56, 58, 59, 60, 62, and 64 neutrons, respectively. These nucleons totally fill the four first shells ($n = 0, 1, 2, 3$), with the remaining 6 protons and 16–24 neutrons filling the upper shell level ($n = 4$). The three-dimensional representations of the fcc lattices of these six Pd isotopes are shown in Figs. 25.3a, 25.4a, 25.5a, 25.6a, 25.7a and 25.8a. These structural configurations provide values of total net binding energies of 854 MeV (Pd^{102}), 882 MeV (Pd^{104}), 892 MeV (Pd^{105}), 904 MeV (Pd^{106}), 920 MeV (Pd^{108}), and 938 MeV (Pd^{110}). These numbers lead to values of the computed BE/A of 8.38 MeV (Pd^{102}), 8.48 MeV (Pd^{104}), 8.50 MeV (Pd^{105}), 8.53 MeV (Pd^{106}), 8.52 MeV (Pd^{108}), and 8.53 MeV (Pd^{110}). The experimental values of BE/A for these six isotopes are 8.58 MeV (Pd^{102}, Pd^{104}, Pd^{106}), 8.57 MeV (Pd^{105}, Pd^{108}), and 8.55 MeV (Pd^{110}) [14]. The differences between experimental and computed values of BE/A are only a few percentage points.

Figures 25.3b, 25.4b, 25.5b, 25.6b, 25.7b and 25.8b report the distribution of fission energy values, calculated over about 8 million fission fragments per isotope (21 fission events for each of the ~190,000 randomized Pd lattices). These distributions are all very similar to each other, presenting three main clusters of energy values: a major clustering in the 5–20 MeV energy spectrum, and two minor ones in the ranges between 30 and 80 MeV, and between 100 and 150 MeV. Note that in this case, although most of the fission events are associated to positive energies (suggesting the prevailing occurrence of endo-energetic fission reactions), some events are associated to negative values (only few MeV), potentially indicating fission with release of energy.

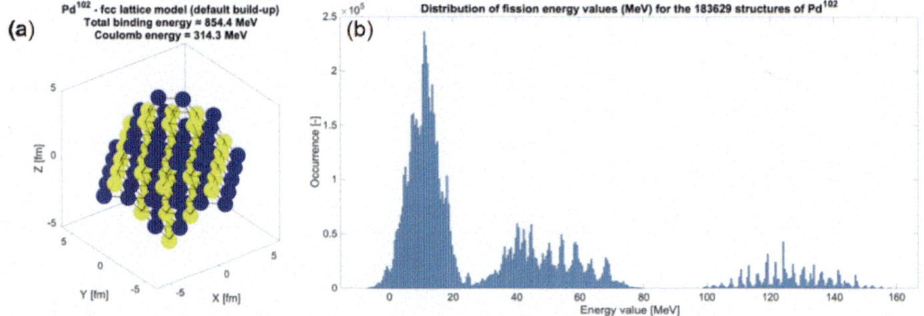

Fig. 25.3 **a** Fcc lattice model of Pd^{102} ($Z = 46$, $N = 56$) built with the in-house MatLab code following the default build-up. Yellow spheres represent protons, whereas the blue ones are neutrons. The size of the spheres is adjusted for visualization purposes and does not correspond to the actual nucleon dimensions. **b** Distribution of fission energies (in MeV) for the 183,629 randomized structures of Pd^{102}

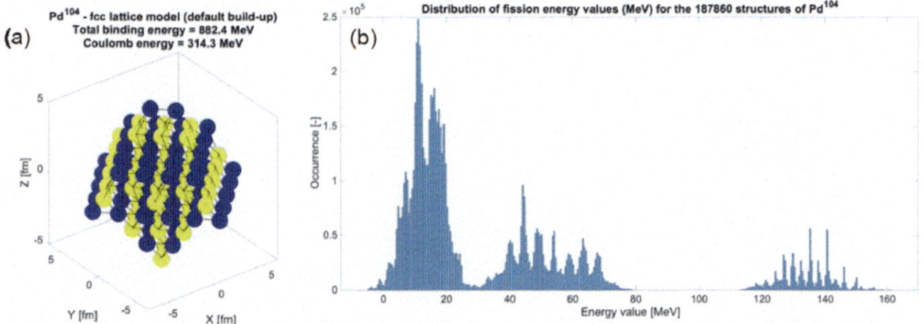

Fig. 25.4 **a** Fcc lattice model of Pd^{104} ($Z = 46$, $N = 58$) built with the in-house MatLab code following the default build-sup. Yellow spheres represent protons, whereas the blue ones are neutrons. The size of the spheres is adjusted for visualization purposes and does not correspond to the actual nucleon dimensions. **b** Distribution of fission energies (in MeV) for the 187,860 randomized structures of Pd^{104}

Figures 25.9, 25.10, 25.11, 25.12, 25.13 and 25.14 show the distribution of collected fragments from each of the six isotopes of Pd, for different times of observation T_{char}, whereas Fig. 25.15 reports the abundance of fragments from the natural mix of Pd. This is computed on the basis of the data from the single isotopes, and considering the relative natural abundances reported above. Tables 25.3, 25.4, 25.5, 25.6, 25.7 and 25.8 show the six most abundant fragments detected at the different times of observation for the six individual Pd isotopes, whereas Table 25.9 reports the most abundant fission fragments from the Pd natural mix.

In the following, detailed comments on the observed fission reactions and on the nature of the obtained fission fragments will be given only for the three most abundant isotopes, i.e., Pd^{105} (Fig. 25.11 and Table 25.5), Pd^{106} (Fig. 25.12 and Table 25.6), and Pd^{108} (Fig. 25.13 and Table 25.7), which together cover more than 75% of the

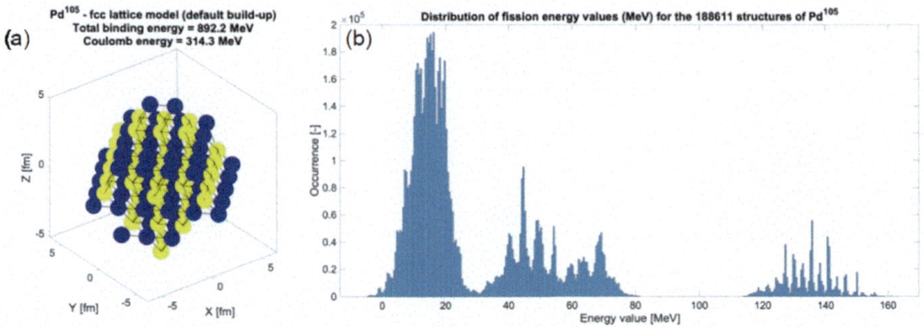

Fig. 25.5 **a** Fcc lattice model of Pd^{105} ($Z = 46$, $N = 59$) built with the in-house MatLab code following the default build-up. Yellow spheres represent protons, whereas the blue ones are neutrons. The size of the spheres is adjusted for visualization purposes and does not correspond to the actual nucleon dimensions. **b** Distribution of fission energies (in MeV) for the 188,611 randomized structures of Pd^{105}

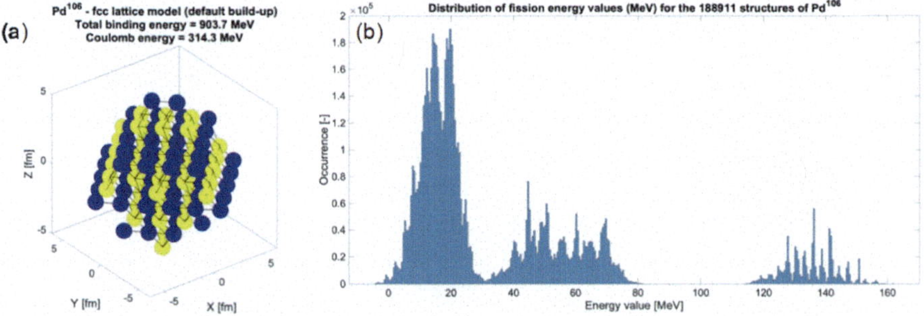

Fig. 25.6 **a** Fcc lattice model of Pd^{106} ($Z = 46$, $N = 60$) built with the in-house MatLab code following the default build-up. Yellow spheres represent protons, whereas the blue ones are neutrons. The size of the spheres is adjusted for visualization purposes and does not correspond to the actual nucleon dimensions. **b** Distribution of fission energies (in MeV) for the 188,911 randomized structures of Pd^{106}

natural abundance of Pd. Final comments will be given on the outcomes arising from the fission of the Pd natural mix (Fig. 25.15 and Table 25.9).

Figure 25.11 and Table 25.5 report the results from the fission of the fcc lattice model of Pd^{105}. The calculations reveal that Ti and Fe are the most abundant elements among the collected fragments, together with Sc and Zn.

The large abundance of Ti comes mostly from fission reactions where Pd splits into Ti and Cr. In this case, seven fission reactions are observed, i.e., $Pd^{105} \rightarrow Ti^{47} + Cr^{58}$, $Pd^{105} \rightarrow Ti^{48} + Cr^{57}$, $Pd^{105} \rightarrow Ti^{49} + Cr^{56}$, $Pd^{105} \rightarrow Ti^{50} + Cr^{55}$, $Pd^{105} \rightarrow Ti^{51} + Cr^{54}$, $Pd^{105} \rightarrow Ti^{52} + Cr^{53}$, and $Pd^{105} \rightarrow Ti^{53} + Cr^{52}$. Note that, in these reactions, Ti^{47}, Ti^{48}, Ti^{49}, and Ti^{50} represent stable isotopes, whereas Ti^{51}, Ti^{52}, and Ti^{53} are unstable, with half-lives of about five minutes, two minutes, and thirty seconds, respectively, and are prone to undergo β^- decays into V^{51} (stable), V^{52}

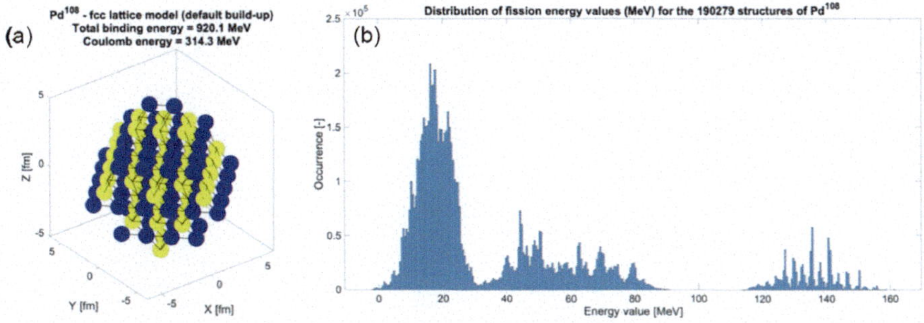

Fig. 25.7 **a** Fcc lattice model of Pd^{108} ($Z = 46$, $N = 62$) built with the in-house MatLab code following the default build-up. Yellow spheres represent protons, whereas the blue ones are neutrons. The size of the spheres is adjusted for visualization purposes and does not correspond to the actual nucleon dimensions. **b** Distribution of fission energies (in MeV) for the 190,279 randomized structures of Pd^{108}

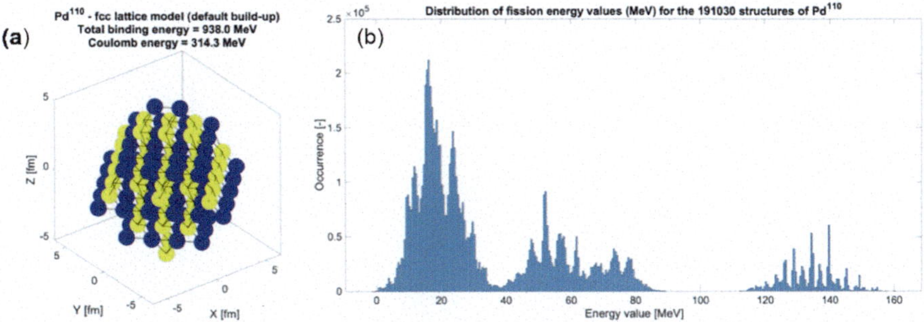

Fig. 25.8 **a** Fcc lattice model of Pd^{110} ($Z = 46$, $N = 64$) built with the in-house MatLab code following the default build-up. Yellow spheres represent protons, whereas the blue ones are neutrons. The size of the spheres is adjusted for visualization purposes and does not correspond to the actual nucleon dimensions. **b** Distribution of fission energies (in MeV) for the 190,279 randomized structures of Pd^{110}

($T_{1/2} = 3.7$ min), and V^{53} ($T_{1/2} = 1.5$ min). V^{52} and V^{53} being unstable, they are susceptible to undergo a subsequent β^- decay into Cr^{52} and Cr^{53}, which are both stable. On the other hand, out of the seven Cr fragments obtained from the above fission reactions, only Cr^{52}, Cr^{53}, and Cr^{54} are stable, whereas the unstable Cr^{55}, Cr^{56}, Cr^{57}, and Cr^{58} undergo β^- decay. In particular, Cr^{55} ($T_{1/2} = 210$ s) undergoes a β^- decay into the stable Mn^{55}. Cr^{56} ($T_{1/2} = 6$ min) undergoes a β^- decay into Mn^{56} ($T_{1/2} = 2.5$ h), which subsequently experiences another β^- decay into Fe^{56}. Cr^{57} ($T_{1/2} = 21$ s) undergoes a β^- decay into Mn^{57} ($T_{1/2} = 1.5$ min), which subsequently decays into Fe^{57}. Finally, Cr^{58} ($T_{1/2} = 7$ s) decays into Mn^{58} ($T_{1/2} = 3$ s), which subsequently decays into Fe^{58}. As usual, the abundance of Ti is also due to the nuclear decay of other unstable fragments, such as Sc^{46}, Sc^{47}, Ca^{47}, Sc^{48}, Sc^{49}, V^{48}, V^{49}, and V^{50}.

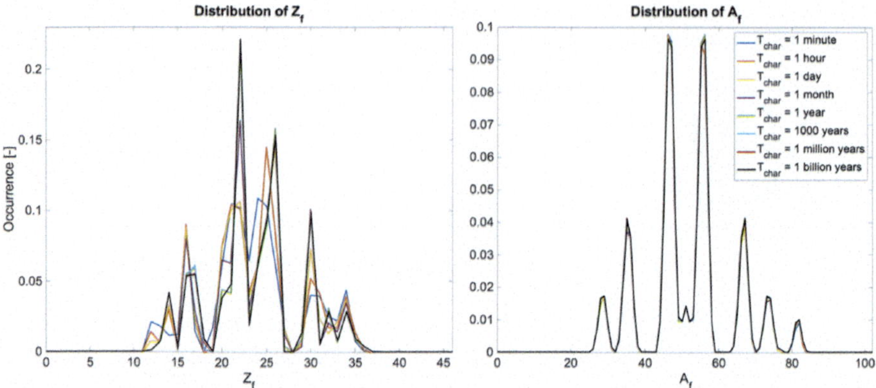

Fig. 25.9 Distribution of the collected fragments from the fission of Pd^{102} in terms of atomic number Z_f (left panel) and mass number A_f (right panel). The eight different curves refer to the different characteristic times, T_{char}

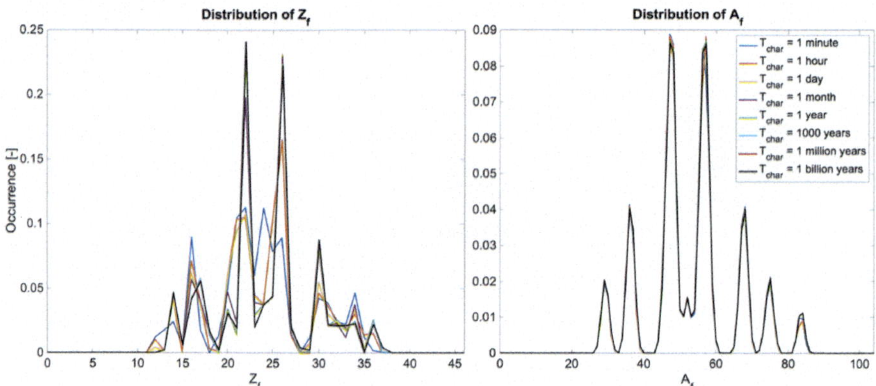

Fig. 25.10 Distribution of the collected fragments from the fission of Pd^{104} in terms of atomic number Z_f (left panel) and mass number A_f (right panel). The eight different curves refer to the different characteristic times, T_{char}

As regards the abundance of Fe, this element arises from the four fission reactions: $Pd^{105} \rightarrow Fe^{57} + Ca^{48}$, $Pd^{105} \rightarrow Fe^{58} + Ca^{47}$, $Pd^{105} \rightarrow Fe^{59} + Ca^{46}$, and $Pd^{105} \rightarrow Fe^{60} + Ca^{45}$. In these reactions, Fe^{57} and Fe^{58} are stable isotopes, whereas Fe^{59} ($T_{1/2}$ = 45 days) experiences a β^- decay into the much more stable Co^{59}. On the other hand, Fe^{60} is also prone to undergo a β^- decay into Co^{60}, but it has a very long half-time $T_{1/2}$ of 60 million years, so that it can be considered stable for most of the investigated times of observation. Ca^{46} and Ca^{48} are the only stable Ca fragments generated from these fission reactions, whereas Ca^{45} ($T_{1/2}$ = 163 days) and Ca^{47} ($T_{1/2}$ = 1.5 days) are likely to undergo β^- decay into Sc^{45} (stable) and Sc^{47} ($T_{1/2}$ = 3.3 days), respectively. Sc^{47} being unstable, it is then observed to undergo a β^- decay

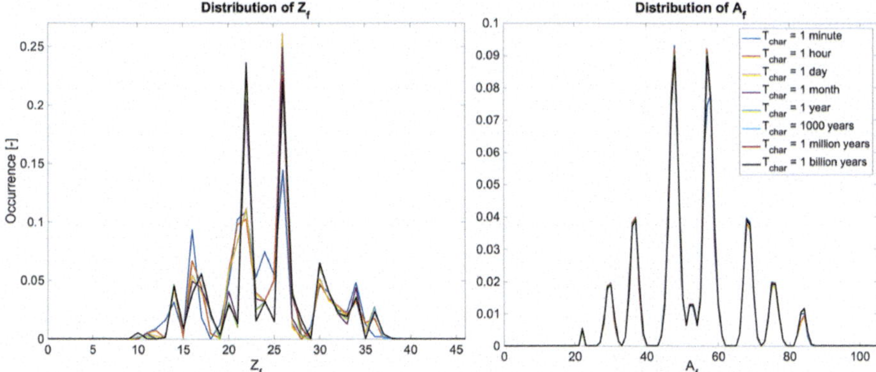

Fig. 25.11 Distribution of the collected fragments from the fission of Pd^{105} in terms of atomic number Z_f (left panel) and mass number A_f (right panel). The eight different curves refer to the different characteristic times, T_{char}

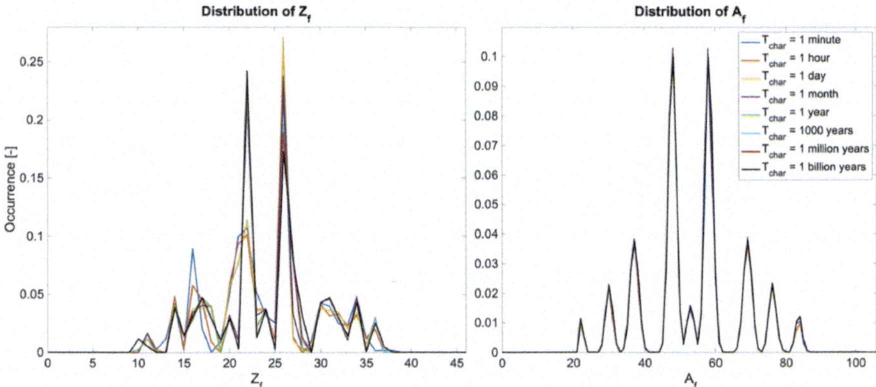

Fig. 25.12 Distribution of the collected fragments from the fission of Pd^{106} in terms of atomic number Z_f (left panel) and mass number A_f (right panel). The eight different curves refer to the different characteristic times, T_{char}

into the stable Ti^{47}. Notice how the Ca fragments are among the six most abundant fragments only for early times of observation, e.g., one hour or one day. This is due to the considerable instability of most of the collected Ca fission fragments, i.e., Ca^{45} and Ca^{47}. Finally, the abundant generation of Fe fragments is also observed to come from the decay of other unstable elements, such as the β^- decay of Mn^{56}, Mn^{57}, Mn^{58}, and Mn^{59}, the double β^- decay of Cr^{56}, Cr^{57}, and Cr^{58}, as well as from the triple β^- decay of V^{56} and V^{57}.

The presence of Sc among the most abundant elements at early times of observation is found to arise from the four fission reactions: $Pd^{105} \rightarrow Sc^{46} + Mn^{59}$, $Pd^{105} \rightarrow Sc^{47} + Mn^{58}$, $Pd^{105} \rightarrow Sc^{48} + Mn^{57}$, and $Pd^{105} \rightarrow Sc^{49} + Mn^{56}$. Notice that none of these Sc fission fragments is stable for long times of observation. As a matter of fact,

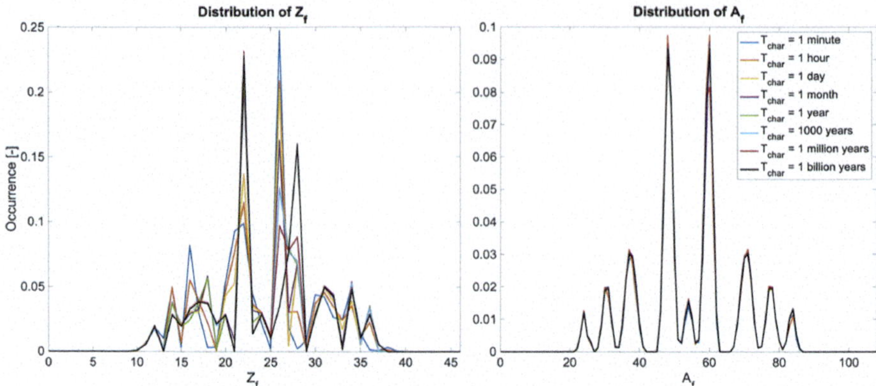

Fig. 25.13 Distribution of the collected fragments from the fission of Pd^{108} in terms of atomic number Z_f (left panel) and mass number A_f (right panel). The eight different curves refer to the different characteristic times, T_{char}

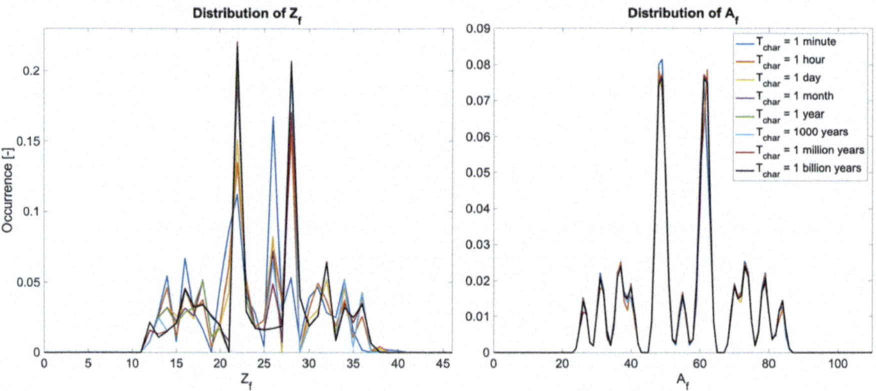

Fig. 25.14 Distribution of the collected fragments from the fission of Pd^{110} in terms of atomic number Z_f (left panel) and mass number A_f (right panel). The eight different curves refer to the different characteristic times, T_{char}

all these Sc fragments decay into the more stable Ti nuclei, with half-lives ranging from about one hour for Sc^{49} up to about three months for Sc^{46}. On the other hand, Mn^{56}, Mn^{57}, and Mn^{58} undergo a β^- decay into the more stable Fe nuclei, whereas Mn^{59} undergoes a double β^- decay into the stable Co^{59}. Given the highly unstable nature of the generated Sc and Mn fragments, it is easy to understand why these are not among the most abundant elements for longer times of observation.

Finally, the presence of Zn is mainly associated to the highly asymmetric fission reactions $Pd^{105} \rightarrow Zn^{66} + S^{39}$, $Pd^{105} \rightarrow Zn^{67} + S^{38}$, $Pd^{105} \rightarrow Zn^{68} + S^{37}$, $Pd^{105} \rightarrow Zn^{69} + S^{36}$, and $Pd^{105} \rightarrow Zn^{70} + S^{35}$. In these reactions, Zn^{66}, Zn^{67}, Zn^{68}, and Zn^{70} are all stable isotopes of zinc, whereas Zn^{69} ($T_{1/2} = 56$ min) is susceptible

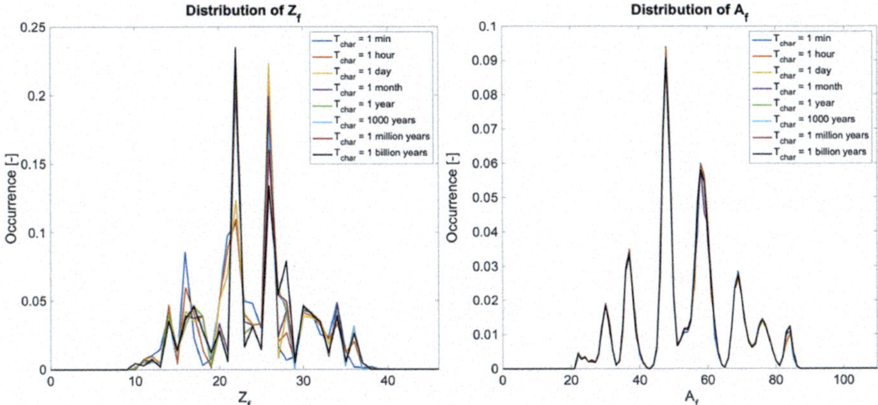

Fig. 25.15 Distribution of the collected fragments from the fission of Pd (abundance-weighted natural mix) in terms of atomic number Z_f (left panel) and mass number A_f (right panel). The eight different curves refer to the different characteristic times T_{char}

Table 25.3 Most abundant elements arising from the fission of Pd^{102} for different times of observation T_{char}. Values in parentheses represent the relative abundances expressed in percentage

T_{char}	1st	2nd	3rd	4th	5th	6th
1 min	Cr (10.9%)	Sc (10.5%)	Mn (10.3%)	Ti (10.1%)	S (8.7%)	V (6.5%)
1 h	Mn (14.6%)	Sc (10.4%)	Ti (10.2%)	S (9.1%)	Fe (8.7%)	Ca (4.3%)
1 day	Fe (14.7%)	Ti (10.7%)	Sc (10.1%)	Mn (9.1%)	S (8.8%)	Ca (7.7%)
1 month	Ti (16.4%)	Fe (15.2%)	Zn (10.1%)	Mn (9.1%)	S (8.0%)	Ca (6.6%)
1 year	Ti (21.4%)	Fe (15.9%)	Zn (9.9%)	Mn (8.9%)	Cr (6.1%)	Cl (5.9%)
1,000 years	Ti (22.1%)	Fe (15.3%)	Zn (9.9%)	Mn (9.4%)	Cr (6.2%)	Cl (6.1%)
1 million years	Ti (22.1%)	Fe (15.3%)	Zn (9.9%)	Mn (9.4%)	Cr (6.5%)	Cl (5.5%)
1 billion years	Ti (22.1%)	Fe (15.3%)	Zn (9.9%)	Mn (9.4%)	Cr (6.5%)	Cl (5.5%)

of undergoing a β^- decay into the stable Ga^{69}. On the other hand, out of the five S fragments obtained above, only S^{36} is stable, whereas S^{35}, S^{37}, S^{38}, and S^{39} all experience β^- decays into Cl. In turn, this explains the occurrence of Cl at higher times of observation (Table 25.12). Again, Zn nuclei are also found to be generated as a by-product of the nuclear decay of other unstable elements, such as Cu^{66}, Cu^{67}, Cu^{68}, and Ge^{68}.

Table 25.4 Most abundant elements arising from the fission of Pd^{104} for different times of observation T_{char}. Values in parentheses represent the relative abundances expressed in percentage

T_{char}	1st	2nd	3rd	4th	5th	6th
1 min	Ti (11.3%)	Cr (11.2%)	Sc (10.4%)	S (9.0%)	Fe (8.9%)	Mn (7.9%)
1 h	Fe (16.5%)	Ti (10.5%)	Sc (10.4%)	Mn (10.3%)	S (7.2%)	Ca (6.0%)
1 day	Fe (23.2%)	Ti (10.8%)	Sc (9.5%)	S (6.3%)	Ca (6.1%)	Zn (5.5%)
1 month	Fe (23.0%)	Ti (19.8%)	Zn (7.9%)	S (5.7%)	Ca (4.8%)	Si (4.7%)
1 year	Ti (22.8%)	Fe (22.4%)	Zn (8.2%)	Cl (5.7%)	Si (4.7%)	Mn (4.4%)
1,000 years	Ti (24.0%)	Fe (22.2%)	Zn (8.7%)	Cl (5.8%)	Si (4.6%)	Mn (4.3%)
1 million years	Ti (24.0%)	Fe (22.2%)	Zn (8.7%)	Cl (5.6%)	Si (4.6%)	Mn (4.3%)
1 billion years	Ti (24.0%)	Fe (22.2%)	Zn (8.7%)	Cl (5.5%)	Si (4.6%)	Mn (4.3%)

Table 25.5 Most abundant elements arising from the fission of Pd^{105} for different times of observation T_{char}. Values in parentheses represent the relative abundances expressed in percentage

T_{char}	1st	2nd	3rd	4th	5th	6th
1 min	Fe (14.5%)	Ti (10.9%)	Sc (10.2%)	S (9.4%)	Cr (7.5%)	Mn (5.5%)
1 h	Fe (22.3%)	Ti (10.3%)	Sc (9.8%)	S (6.7%)	Ca (5.8%)	Mn (5.1%)
1 day	Fe (26.2%)	Ti (11.2%)	Sc (8.3%)	Ca (5.8%)	S (5.5%)	Zn (5.2%)
1 month	Fe (25.0%)	Ti (20.4%)	Zn (6.2%)	S (5.0%)	Si (4.7%)	Se (4.4%)
1 year	Fe (22.9%)	Ti (22.5%)	Zn (6.3%)	Cl (5.6%)	Si (4.7%)	Co (4.3%)
1,000 years	Ti (23.6%)	Fe (22.7%)	Zn (6.5%)	Cl (5.7%)	Si (4.5%)	S (4.0%)
1 million years	Ti (23.6%)	Fe (22.6%)	Zn (6.5%)	Cl (5.5%)	Si (4.5%)	S (4.0%)
1 billion years	Ti (23.6%)	Fe (22.0%)	Zn (6.5%)	Cl (5.5%)	Si (4.5%)	S (4.0%)

Table 25.6 Most abundant elements arising from the fission of Pd^{106} for different times of observation T_{char}. Values in parentheses represent the relative abundances expressed in percentage

T_{char}	1st	2nd	3rd	4th	5th	6th
1 min	Fe (22.0%)	Ti (10.7%)	Sc (10.0%)	S (9.0%)	V (5.2%)	Ca (5.0%)
1 h	Fe (26.6%)	Ti (10.2%)	Sc (9.4%)	Ca (5.8%)	S (5.8%)	Si (4.9%)
1 day	Fe (27.1%)	Ti (11.5%)	Sc (7.7%)	Ca (5.6%)	Si (4.0%)	Ar (3.9%)
1 month	Fe (23.8%)	Ti (21.5%)	Se (4.8%)	Ga (4.6%)	Zn (4.3%)	Si (4.2%)
1 year	Ti (23.0%)	Fe (19.6%)	Co (8.5%)	Ga (4.8%)	Cl (4.7%)	Se (4.4%)
1,000 years	Ti (24.2%)	Fe (19.5%)	Co (7.9%)	Cl (4.7%)	Ga (4.7%)	Se (4.3%)
1 million years	Ti (24.2%)	Fe (19.0%)	Co (7.9%)	Ga (4.7%)	Cl (4.7%)	Se (4.3%)
1 billion years	Ti (24.2%)	Fe (17.3%)	Co (7.9%)	Ga (4.7%)	Cl (4.7%)	Se (4.3%)

Table 25.7 Most abundant elements arising from the fission of Pd^{108} for different times of observation T_{char}. Values in parentheses represent the relative abundances expressed in percentage

T_{char}	1st	2nd	3rd	4th	5th	6th
1 min	Fe (24.7%)	Ti (9.7%)	Sc (9.3%)	S (8.2%)	Ca (5.4%)	Se (5.0%)
1 h	Fe (20.9%)	Ti (11.5%)	Sc (7.7%)	S (5.5%)	Si (5.0%)	Ca (4.7%)
1 day	Fe (20.2%)	Ti (13.7%)	Ni (6.7%)	Ar (5.7%)	Sc (5.3%)	Ga (4.9%)
1 month	Ti (20.5%)	Fe (16.3%)	Ni (6.8%)	Ar (5.8%)	Se (5.4%)	Ga (5.0%)
1 year	Ti (12.7%)	Fe (12.6%)	Co (8.0%)	Ni (6.7%)	Ar (5.7%)	Se (5.3%)
1,000 years	Ti (22.8%)	Fe (12.6%)	Co (7.7%)	Ni (7.0%)	Se (5.3%)	Ga (5.0%)
1 million years	Ti (23.1%)	Fe (9.7%)	Ni (8.8%)	Co (7.8%)	Ga (5.1%)	Se (5.0%)
1 billion years	Ti (22.8%)	Ni (16.0%)	Co (7.7%)	Ga (5.0%)	Se (4.8%)	Ge (4.3%)

Table 25.8 Most abundant elements arising from the fission of Pd^{110} for different times of observation T_{char}. Values in parentheses represent the relative abundances expressed in percentage

T_{char}	1st	2nd	3rd	4th	5th	6th
1 min	Fe (16.8%)	Ti (11.2%)	Sc (8.6%)	S (6.7%)	Si (5.4%)	Ni (5.3%)
1 h	Ni (15.5%)	Ti (13.5%)	Fe (7.9%)	Sc (6.2%)	V (5.3%)	Ga (4.9%)
1 day	Ni (20.0%)	Ti (15.1%)	Fe (8.3%)	Ge (5.2%)	Ar (5.2%)	V (5.0%)
1 month	Ni (20.0%)	Ti (19.4%)	Fe (7.2%)	Ge (6.4%)	Se (5.2%)	Ar (5.2%)
1 year	Ti (20.7%)	Ni (19.9%)	Fe (6.4%)	Ge (6.4%)	Se (5.1%)	Ar (5.1%)
1,000 years	Ti (21.8%)	Ni (19.8%)	Fe (6.4%)	Ge (6.4%)	Se (5.1%)	S (4.5%)
1 million years	Ti (22.1%)	Ni (17.0%)	Ge (6.4%)	Fe (4.9%)	S (4.6%)	Cu (3.9%)
1 billion years	Ti (21.8%)	Ni (20.7%)	Ge (6.4%)	S (4.5%)	Cu (3.9%)	Kr (3.4%)

Table 25.9 Most abundant elements arising from the fission of Pd (abundance-weighted natural mix) for different times of observation T_{char}. Values in parentheses represent the relative abundances expressed in percentage

T_{char}	1st	2nd	3rd	4th	5th	6th
1 min	Fe (18.8%)	Ti (10.7%)	Sc (9.7%)	S (8.6%)	Ca (5.0%)	V (5.0%)
1 h	Fe (20.6%)	Ti (11.0%)	Sc (8.8%)	S (6.0%)	Ca (5.1%)	Si (4.7%)
1 day	Fe (22.3%)	Ti (12.3%)	Sc (7.0%)	Ca (5.0%)	S (4.2%)	Ni (4.3%)
1 month	Ti (20.5%)	Fe (19.9%)	Se (4.8%)	Zn (4.5%)	Ni (4.3%)	Si (4.1%)
1 year	Ti (22.2%)	Fe (17.2%)	Co (5.9%)	Cl (4.6%)	Zn (4.6%)	Se (4.3%)
1,000 years	Ti (23.4%)	Fe (17.1%)	Co (5.4%)	Ni (4.7%)	Cl (4.7%)	Zn (4.6%)
1 million years	Ti (23.5%)	Fe (16.0%)	Co (5.5%)	Ni (4.9%)	Zn (4.7%)	Cl (4.6%)
1 billion years	Ti (23.4%)	Fe (13.4%)	Ni (7.9%)	Co (5.4%)	Zn (4.6%)	Cl (4.6%)

Figure 25.12 and Table 25.6 report the results of fission of Pd^{106}, which represents the most abundant isotope of palladium in nature, with a relative abundance of 27.3%. Again, Ti and Fe are the most abundant elements among the collected fission fragments, with a prevalence of Fe for shorter times of observation (up to few months) and Ti for longer times of observation (from one year onwards). Other relatively abundant elements are Sc for shorter times of observation and Co for longer times.

Like in the previous case, Ti is mostly found to arise from Ti–Cr fission reactions, i.e., $Pd^{106} \rightarrow Ti^{47} + Cr^{59}$, $Pd^{106} \rightarrow Ti^{48} + Cr^{58}$, $Pd^{106} \rightarrow Ti^{49} + Cr^{57}$, $Pd^{106} \rightarrow Ti^{50} + Cr^{56}$, $Pd^{106} \rightarrow Ti^{51} + Cr^{55}$, $Pd^{106} \rightarrow Ti^{52} + Cr^{54}$, and $Pd^{106} \rightarrow Ti^{53} + Cr^{53}$. As reported previously, Ti^{47}, Ti^{48}, Ti^{49}, and Ti^{50} are stable isotopes, whereas Ti^{51}, Ti^{52}, and Ti^{53} have half-lives in the order of few minutes and are susceptible to undergo β^- decay into isotopes of V. On the other hand, out of the seven Cr fragments obtained from the above reactions, only Cr^{53} and Cr^{54} are stable, whereas Cr^{55}, Cr^{56}, Cr^{57}, Cr^{58}, and Cr^{59} undergo β^- decay. Once again this explains why Ti is found to be more abundant compared to Cr as a resultant of the fission of Pd^{106}. Simply, the obtained Ti fragments are statistically more stable than the corresponding Cr fragments. As usual, the abundance of Ti is also due to the nuclear decay of other unstable fragments, such as Sc^{46}, Sc^{47}, Ca^{47}, K^{47}, Sc^{48}, Sc^{49}, V^{48}, V^{49}, and V^{50}.

As regards the large abundance of Fe, this element is found to arise from the fission reactions $Pd^{106} \rightarrow Fe^{58} + Ca^{48}$, $Pd^{106} \rightarrow Fe^{59} + Ca^{47}$, and $Pd^{106} \rightarrow Fe^{60} + Ca^{46}$. Fe^{58} is a stable isotope of iron. Conversely, as already reported above, Fe^{59} ($T_{1/2} = 45$ days) experiences a β^- decay into the stable Co^{59}, whereas Fe^{60} exhibits a preferred β^- decay mode into Co^{60}, but with the very long half-time of 60 million years. Therefore, this fragment can be considered stable for most of the investigated times of observation. If, however, a decay of Fe^{60} into Co^{60} does occur, the latter is unstable, with a rather short half-life of around 5 years, and then it is susceptible to undergo a subsequent β^- decay into the stable Ni^{60}. The lower occurrence of stable fragments of Fe, compared to the ones of Ti reported above, ultimately explains why iron fragments are found for shorter times of observation, whereas Ti is more abundant for longer times. On the other hand, as regards Ca fragments, Ca^{46} and Ca^{48} are the only stable calcium fragments generated from these fission reactions, whereas Ca^{47} undergoes a β^- decay into Sc^{47}, which subsequently decays into the stable Ti^{47}. As usual, the large generation of Fe fragments is also observed to arise from the nuclear decay of other heavier and lighter unstable elements.

The presence of Sc at early times of observation mainly arises from the fission reactions: $Pd^{106} \rightarrow Sc^{46} + Mn^{60}$, $Pd^{106} \rightarrow Sc^{47} + Mn^{59}$, $Pd^{106} \rightarrow Sc^{48} + Mn^{58}$, and $Pd^{106} \rightarrow Sc^{49} + Mn^{57}$. Notice that none of these fragments is stable for longer times of observation. All Sc fragments decay into more stable Ti nuclei, whereas Mn fragments undergo β^- decay into Fe nuclei. Some of these Fe nuclei are stable, such as Fe^{57} and Fe^{58}, whereas others (Fe^{59} and Fe^{60}) undergo subsequent β^- decays, as already pointed out repeatedly above. Finally, the abundance of Co either arises from the direct fission reactions: $Pd^{106} \rightarrow Co^{59} + K^{47}$, $Pd^{106} \rightarrow Co^{60} + K^{46}$, and $Pd^{106} \rightarrow Co^{61} + K^{45}$, or it comes from the nuclear decay of other unstable elements. In the former case, it can be noted that Co^{59} is the only stable fragment, whereas Co^{60} and

Co^{61} can undergo β^- decays into the more stable Ni^{60} and Ni^{61} nuclei. In the latter case, the stable isotope of Co, i.e., Co^{59}, is mainly found to arise from the decay of unstable elements such as Fe^{59}, Mn^{59}, Cr^{59}, and V^{59}, which are frequently found from the results of the present calculations.

Figure 25.13 and Table 25.7 report the results of fission of Pd^{108}, which is the second most abundant isotope in nature after Pd^{106}. As usual, Ti and Fe are the most abundant elements among the collected fragments with a prevalence of Fe for shorter times of observation and Ti for longer times. Other relatively abundant elements are Sc, Co, and Ni.

Ti is found to occur mostly from the eight fission reactions: $Pd^{108} \rightarrow Ti^{47} + Cr^{61}$, $Pd^{108} \rightarrow Ti^{48} + Cr^{60}$, $Pd^{108} \rightarrow Ti^{49} + Cr^{59}$, $Pd^{108} \rightarrow Ti^{50} + Cr^{58}$, $Pd^{108} \rightarrow Ti^{51} + Cr^{57}$, $Pd^{108} \rightarrow Ti^{52} + Cr^{56}$, $Pd^{108} \rightarrow Ti^{53} + Cr^{55}$, and $Pd^{108} \rightarrow Ti^{54} + Cr^{54}$. In these reactions, Ti^{47}, Ti^{48}, Ti^{49}, and Ti^{50} are stable isotopes, whereas Ti^{51}, Ti^{52}, Ti^{53}, and Ti^{54} undergo β^- decays into isotopes of V. On the other hand, out of the eight Cr isotopes obtained from the fissions above, only Cr^{54} is stable, whereas all the others undergo one or multiple β^- decays into Mn, Fe, Co, or Ni. The abundance of Ti is also due to the decay of other unstable fragments that are detected with relatively high abundance among the outcomes of the fission simulations, such as Sc^{46}, Sc^{47}, Ca^{47}, K^{47}, Sc^{48}, Sc^{49}, V^{48}, Ca^{49}, V^{49}, Sc^{50}, and V^{50}.

As regards iron, these fission fragments mainly arise from the three direct reactions: $Pd^{108} \rightarrow Fe^{59} + Ca^{49}$, $Pd^{108} \rightarrow Fe^{60} + Ca^{48}$, and $Pd^{108} \rightarrow Fe^{61} + Ca^{47}$. Note that none of these Fe isotopes is stable, all of them being susceptible to undergo a β^- decay into Co (Fe^{59} with a half-life of 45 days, Fe^{60} with a long half-life of 60 million years, and Fe^{61} with a half-life of about 6 min). This explains why Fe is the most abundant fragment for shorter times of observation, whereas it becomes progressively less frequent for longer times. As regards the obtained Ca fragments, Ca^{48} is the only stable isotope generated from these fission reactions, whereas Ca^{47} and Ca^{49} undergo β^- decay into Sc^{47} and Sc^{49}, and subsequently into Ti^{47} and Ti^{49}. As usual, Fe fragments are also found to arise marginally from the decay of other unstable elements.

As in the previous case of Pd^{106}, the presence of Sc at early times of observation mainly arises from the fission reactions: $Pd^{108} \rightarrow Sc^{46} + Mn^{62}$, $Pd^{108} \rightarrow Sc^{47} + Mn^{61}$, $Pd^{108} \rightarrow Sc^{48} + Mn^{60}$, $Pd^{108} \rightarrow Sc^{49} + Mn^{59}$, and $Pd^{108} \rightarrow Sc^{50} + Mn^{58}$. Again, none of these Sc and Mn fragments is stable for long times of observation. All Sc fragments decay into more stable Ti nuclei, whereas Mn fragments undergo β^- decays into Fe nuclei. Eventually, some of these Fe nuclei undergo subsequent β^- decays into Co and Ni elements. Since the direct fission reaction involving Co, i.e., $Pd^{108} \rightarrow Co^{X} + K^{108-X}$, gives rise only to unstable isotopes of Co, such as Co^{60}, Co^{61}, and Co^{62}, the abundance of Co from the simulation is mostly found to occur as a consequence of the decay of other lighter unstable elements, such as Fe^{59}, Mn^{59}, Cr^{59}, and V^{59}, which explains why Co is abundant at longer times of observation and not at shorter times. Finally, the abundance of Ni, especially for longer times of observation, can also be explained through the decay of lighter unstable elements such as Co^{60}, Fe^{60}, Mn^{60}, Cr^{60}, V^{60}, Co^{61}, Fe^{61}, Mn^{61}, Cr^{61}, Co^{62}, and Mn^{62}.

Based on the outcomes arising from the fission of the six Pd isotopes reported above, Fig. 25.15 and Table 25.9 show the results related to the natural mix of palladium. These have been obtained simply by putting together the relative abundances of chemical elements for each Pd isotope, weighted according to their natural abundances. As expected, Ti and Fe are by a large extent the most abundant elements arising from the fission of the Pd natural mix. These two elements account for about the 30–40% of the total fragments. Table 25.9 shows that Fe is the most abundant expected fragment for shorter times of observation (up to a few days), whereas Ti becomes the most abundant for longer times. Other elements that have been found among the large dataset of the collected fission fragments, with abundance values between 4 and 10%, are Sc, S, Ni, Ca, Se, Co, Zn, V, Si, and Cl.

We have also deeply commented above on the fact that each of these elements can also be obtained as the resultant of nuclear decay of unstable isotopes of other elements, e.g., Ti as the resultant of decay of unstable Sc and V fragments, etc.

Let us now compare the elemental abundance found from the outcomes of the present simulation with the experimental tests performed by Carpinteri et al. [11] and related to the electrolysis experiments with Pd–Ni electrodes. In these experiments, the elemental analysis at the Pd electrode after electrolysis revealed a remarkable decrement in the Pd content on the electrode surface, accompanied by the appearance of lighter elements, which were absent before the test, such as Fe, Ca, O, Mg, K, and Si. The test lasted few hours and the elemental analysis on the electrode was performed a few weeks after the experiment. According to the lattice fission simulation and considering the results related to times of observation between one day and one month, we can observe a strong abundance of Ti, Fe, Sc, Ca, Se, Zn, Ni, S, Si, and other elements with lower abundances. As can be seen, Fe, Ca, and Si are common elements between the experimental investigation and the results of the present simulation. Carpinteri et al. proposed a fundamental primary fission reaction involving the splitting of Pd nuclei into Fe and Ca fragments, plus several neutrons, to explain the supposed large increments in Fe and Ca and the detected neutron emissions [11]. On the other hand, the remaining elements observed experimentally, i.e., O, Mg, and K, are not found among the most abundant elements from the present simulation. This is mainly due to the fact that these elements are much lighter than the mother nucleus of Pd, so that they could be generated only by an asymmetrical fission of the Pd nucleus. Alternatively, as already suggested in [11], these fragments might be the result of secondary fission reactions, where the fragments arising from primary fission reactions undergo a subsequent secondary fission, thus generating these lighter elements. We have already seen in the previous chapter that the potential fissions of Ca and Fe are expected to provide large abundances of both O and Mg, while Si can give rise to O fragments. This might then suggest that it is the sequential splitting of unstable Fe, Ca, and Si nuclei that gives rise to the additional O and Mg fragments. On the other hand, the large presence of K at the electrode surface is not easily justifiable by the results of the present simulation. It could be more simply explained as a result of the chemical deposition of K atoms from the electrolytic solution, which was made of a mixture of water and potassium carbonate (K_2CO_3) [10]. It is also worth noting that the present simulation has given rise to a series of additional fission fragments, such as Ti, Sc, Se, Zn, Ni, and S, which were not observed in the experiments conducted in [11]. Nevertheless, some of these elements were detected in the experimental campaign by Mizuno's group. In particular, large amounts of Ti, Zn, and S were reported in [3, 6], whose abundance values were too large to be explained as a result of sample impurities. All these results suggest that the fission of a lattice structure of the Pd nucleus has the potential to provide important confirmations about the experimental fission fragments.

25.4 Conclusions

After presenting and discussing the nuclear transmutations observed upon the fission of Ni and Pd lattice structures and comparing them with the available observations from electrolytic experiments, a few fundamental problems remain to be addressed.

The first question is related to one of the central experimental observations about electrolysis tests: how to explain the anomalous excess heat produced in these reactions? Although the present computational approach cannot provide a definite answer to this question, our calculations reveal that some of the fission events involving Pd isotopes are exo-energetic (Figs. 25.3b, 25.4b, 25.5b, 25.6b, 25.7b and 25.8b), hence suggesting a release of energy into the external system as a result of Pd fission. In particular, we observed that exo-energetic events account for 1.19%, 0.45%, 0.27%, 0.24%, 0.05%, and 0.01% of the total fission events for Pd^{102}, Pd^{104}, Pd^{105}, Pd^{106}, Pd^{108}, and Pd^{110}, respectively. Although these percentages can seem low at a first sight, if calculated over the eight-million fission events per isotope, these actually represent thousands of fission events. The magnitude of the negative energies associated to these events is usually a few MeV, with maximum (negative) values of about 6.9 MeV, 4.9 MeV, 4.7 MeV, 4.4 MeV, 2.1 MeV, and 1.4 MeV for Pd^{102}, Pd^{104}, Pd^{105}, Pd^{106}, Pd^{108}, and Pd^{110}, respectively. If considered together, these results suggest that there can be thousands of fission events from Pd nuclei, which are able to release a few MeV each. Is this enough to explain the anomalous excess heat detected in electrolysis experiments with Pd electrodes? Unfortunately, we cannot provide an answer to this question, for which more theoretical research is needed.

The second fundamental question is: What does actually trigger these fission events? Where does the energy required to activate the endo-energetic fission events come from? In Chap. 22, it is argued that THz phonons and/or plasmons produced by nanomechanics instabilities might be the cause behind these nuclear reactions [13, 15]. THz phonons and/or plasmons present, in fact, the same energy as that of thermal neutrons, as well as the same frequency as that of resonant atomic lattices. This means that nuclear fission is always a phenomenon of resonance. The energy needed to trigger endo-energetic reactions is cumulated during the resonance time, so that the experimental reaction-delays are due to incubation time intervals.

References

1. Fleischmann M, Pons S (1989) Electrochemically induced nuclear fusion of deuterium. J Electroanal Chem Interfacial Electrochem 261(2):301–308
2. Storms EK (2007) The science of low-energy nuclear reactions. World Scientific Publishing, Singapore
3. Mizuno T, Ohmori T, Enyo M (1996) Isotopic changes of the reaction products induced by cathodic electrolysis in Pd. J New Energy 1(3):1–31
4. Mizuno T, Kurokawa K, Akimoto T, Kitaichi M, Inoda K, Azumi K, Shimokawa S, Ohmori T, Enyo M (1996) Anomalous isotopic distribution of elements deposited on palladium induced by cathodic electrolysis. Denki Kagaku Oyobi Kogyo Butsuri Kagaku 64:1160–1165
5. Mizuno T (1997) Nuclear transmutation: the reality of cold fusion
6. Mizuno T, Akimoto T, Ohmori T, Enyo M (1998) Confirmation of the changes of isotopic distribution for the elements on palladium cathode after strong electrolysis in D_2O solution. Int J Soc Mater Eng Resour 6(1):45–59
7. Ohmori T, Mizuno T, Kurokawa K, Enyo M (1998) Nuclear transmutation reaction occurring during the light water electrolysis on Pd electrode. Int J Soc Mater Eng Resour 6(1):35–44
8. Cook ND (2006) Models of the atomic nucleus: unification through a lattice of nucleons. Springer, Berlin/Heidelberg
9. Carpinteri A, Lacidogna G, Manuello A (eds) (2015) Acoustic, electromagnetic, neutron emissions from fracture and earthquakes. Springer
10. Carpinteri A, Borla O, Goi A, Manuello A, Veneziano D (2015) Cold nuclear fusion explained by hydrogen embrittlement and piezonuclear fissions in metallic electrodes. Part I: Ni-Fe and Co-Cr electrodes. In: Acoustic, electromagnetic, neutron emissions from fracture and earthquakes. Springer International Publishing, Switzerland, pp 99–121
11. Carpinteri A, Borla O, Goi A, Guastella S, Manuello A, Veneziano D (2015) Cold nuclear fusion explained by hydrogen embrittlement and piezonuclear fissions in metallic electrodes. Part II: Pd and Ni electrodes. In: Acoustic, electromagnetic, neutron emissions from fracture and earthquakes. Springer International Publishing, Switzerland, pp 123–134
12. Carpinteri A, Borla O, Manuello A, Veneziano D, Goi A (2015) Hydrogen embrittlement and piezonuclear reactions in electrolysis experiments. J Condens Matter Nucl Sci 15:162–182
13. Carpinteri A, Borla O (2020) Strong correlation between LENR and nano-mechanics instabilities/THz phonons in condensed matter: applications in geophysics, geochemistry, energetics, biology. Infin Energy 153:32–37
14. Kondev FG, Wang M, Huang WJ, Naimi S, Audi G (2021) The NUBASE2020 evaluation of nuclear physics properties. Chin Phys C 45(3):030001
15. Carpinteri A, Borla O (2018) Nano-scale fracture phenomena and TeraHertz pressure waves as the fundamental reasons for geochemical evolution. Strength, Fract Complex 11(2–3):149–168